AF518332

Prospects in Bioscience: Addressing the Issues

Abdulhameed Sabu • Anu Augustine
Editors

Prospects in Bioscience: Addressing the Issues

Editors
Abdulhameed Sabu
Department of Biotechnology
and Microbiology
Kannur University
Kannur, Kerala, India

Anu Augustine
Department of Biotechnology
and Microbiology
Kannur University
Kannur, Kerala, India

ISBN 978-81-322-0809-9 ISBN 978-81-322-0810-5 (eBook)
DOI 10.1007/978-81-322-0810-5
Springer New Delhi Heidelberg New York Dordrecht London

Library of Congress Control Number: 2012956293

Printed on acid-free paper

Springer is part of Springer Science+Business Media (www.springer.com)

Foreword

In late 2011, I was very lucky to be invited to be a speaker at the International Conference on Advances in Biological Sciences (ICABS – 2012), which was organized by the Department of Biotechnology and Microbiology and the Inter University Centre for Bioscience, Kannur University from March 15 to 17, 2012. As for the meeting, I can say that it was a very well-organized and convivial conference and a highly vibrant scientific gathering. I was extremely impressed with both the quality and the very wide diversity of presentations, which were given in both oral and poster form. I was also pleasantly surprised by the quality of the students who were present in large numbers at the meeting. During the conference, I learned a tremendous amount about the diversity of Indian science in the biological area, but I also learned a lot about how Indian scientists themselves regard the Indian biological research arena. There were some interesting historical perspectives given at the beginning of the conference, such as one entitled the "Development of Structural Biology in India: A Personal Perspective" by Professor M. Vijayan from the Indian Institute of Science.

We live in an age of increasing threats to our fragile environment, and we hear a great deal about the future impact of the rapidly increasing populations in India, other parts of SE Asia, and Africa. It is therefore gratifying that many of the talks addressed the questions of how to increase crop and domestic animal productivity, how to ameliorate the environmental effects of human activity, how to provide more food and energy for the future, and how to assess the biodiversity of many of India's important plant and animal species. It is also encouraging that native species are being screened for potentially useful bioactives and other potentially useful compounds.

This book *Prospects in Bioscience: Addressing the Issues* is a compendium of papers presented at the conference, representing a set of 50 papers which really illustrate well the diversity of Indian science and how it is going about addressing the issues which face us at this critical time in human history. The topics range widely and include immunology, molecular biology, structural biology, agricultural biotechnology, and microbial technology. They reflect the results of research conducted in highly specialized areas of biology. Some of the chapters cover topics such as partitioning of proteins from fish effluents using aqueous two-phase systems, Naïve Bayes – a novel method for prediction of short genes from noncoding DNA, RFLP of 16S RNA for detection of porcine meat, an HMG-CoA reductase inhibitor from a plant source, immunoglobulin level studies in vitiligo patients, development of SSR markers in bambara groundnut, novel antimicrobial peptides from *Clitoria ternatea*, a plate assay for β-lactamase inhibitors from natural sources, isolation of a *Pseudomonas* sp. against sheath blight of rice, immunoinformatics prediction of HLA-II binding epitopes of *S. aureus*, and role of lymphocytes in tumor rejection. The scientific content in the various chapters of this book is certainly worth exploring, and this book will serve as a good reference source. This book is further designed to give its readers a unique exposure to the emerging prospects of modern biology.

Dr. Glenn Bryan
Group Leader – Potato Genetics and Breeding
The James Hutton Institute
Invergowrie, Dundee DD2 5DA
Scotland, UK, July 2012

Preface

Throughout the past century, researchers have made considerable efforts to learn the biological intricacies of nature for making the existence of humankind more and more comfortable and sustainable. It started with the traditional fermentation of food to the commercial exploitation of all types of biological cells and molecules. However, research has quite often neglected opportunities to develop technologies for the upliftment of the rural poor, representing approximately 80% of the world population. A change in biotechnology concepts, which are now commercially driven, to a more human-oriented future, combining "old" and "modern" biology for the improvement in the socioeconomic status of our rural population, is very much required these days.

Developments in modern biology including environmental and industrial biotechnology, food and agricultural biotechnology, medical biotechnology, and advances in cell and molecular biology are of high relevance in the current global context. The world is striving hard to manage the adverse impacts due to the discharge of toxic effluents to the environment as a result of urbanization and industrialization. Diseases due to adoption of modern lifestyle and environmental stress are emerging along with reemergence of contagious diseases. Equally important are the socioeconomic problems created by reckless utilization of food grains for fuel production. To feed the growing population, we need to have our warehouses filled with quality food grains, and there is a need for advanced postharvest technology also. Advances in cell and molecular biology armed with the breakthroughs in structural biology are extending services to mankind in combating many grave problems it is facing today.

In this volume, the editors have organized a series of research-based articles which seek to address the hot issues in biosciences the whole world is facing due to urbanization and industrialization. The scientific content in the book is expected to help the researchers in bioscience across the globe in orienting their perspective for developing advancements in the field which will help us in fostering sustainable development of the global society. What we are proposing through this book is that the philosophy of science and scientific research need to orient toward nature.

Dr. Abdulhameed Sabu
Dr. Anu Augustine

About the Editors

Dr. Abdulhameed Sabu is currently teaching Biotechnology at the Department of Biotechnology and Microbiology, School of Life Sciences, Kannur University, Kerala, India and earned his doctorate in Biotechnology from Cochin University of Science and Technology (CUSAT), India. Prior to joining Kannur University, he was a scientist at the National Institute for Interdisciplinary Science and Technology (CSIR). He has published several research papers in reputed Indian and international journals and owns a patent to his credit. He along with his colleagues Dr. S Roussos and Dr. CN Aguilar edited the volume "Chemistry and Biotechnology of Polyphenols" in 2011. Dr. Sabu has also worked at Kerala Biotechnology Commission, India; Technical University of Budapest, Hungary; University of Georgia, USA; University of Mediterranean, Marseilles, France; and University of Paul Cezanne, France. His current research interest includes production, purification, and characterization of industrial enzymes and isolation of bioactive molecules from microorganisms. Dr. Sabu is a member of the panel of experts, IUCN, Switzerland; ecosystem service expert of World Resources Institute, USA; life member of Association of Microbiologists of India; life member of Society for Biotechnologists, India; member of Board Studies in Biotechnology and Faculty of Science of Kannur University; and life member of Biotech Research Society of India. He also served as member of the Academic Council and Senate of Cochin University of Science and Technology (e-mail: drsabu@gmail.com).

Dr. Anu Augustine is teaching biotechnology at the Department of Biotechnology and Microbiology, School of Life Sciences, Kannur University, Kerala, India. She did her Ph.D. at the Indian Institute of Spices Research, Calicut. She has 17 years of research experience in tissue culture of spices and has published several papers in reputed international journals. Dr. Anu Augustine did her postdoctoral studies at the James Hutton Institute, Dundee, UK, as a Boyscast Fellow of Department of Science and Technology, Govt. of India. Her areas of research interest include investigations into the genome of Graminae members, conservation, micropropagation, and characterization of important medicinal plants/RET plants, anti-obesity and anti-cholesterol principles from plant sources, overexpression of glyoxylase I gene in *Hevea brasiliensis*, and alpha-amylase inhibitors from Zingiberaceae members. She is a member of board of studies in Kannur University and external examiner in the universities of Kerala, and Calicut.

Contents

1 **Fish-Processing Industrial Effluent Characterization and Partitioning of Proteins Using Aqueous Two-Phase System** 1
H. Nagaraja Viswanatha and I. Regupathi

2 **Bioremediation of Zinc Using *Bacillus* sp. Isolated from Metal-Contaminated Industrial Zone** 11
M.P. Krishna, Rinoy Varghese, V. Arun Babu, S. Jyothy, and A.A. Mohamed Hatha

3 **Evaluation of Bryophyte Protein-Based Defense Against Selected Phytophagous Insects** 19
Remya Krishnan and K. Murugan

4 **A Novel Algorithm for Prediction of Hub Proteins from Primary Structure in Eukaryotic Proteome Using Dipeptide Compositional Skew Information and Amino Acid Sequence Likeness** 33
B.L. Aswathi, Baharak Goli, Renganayaki Govindarajan, and Achuthsankar S. Nair

5 **Naïve Bayes-Based Classification for Short Microbial Genes Using Chaos Game Representation** 41
Baharak Goli, B.L. Aswathi, and Achuthsankar S. Nair

6 **PCR-RFLP of 16S rRNA Gene for Identification of Speciation of Meat of Porcine Origin in Processed and Admixed Meat** 49
Anjana Sharma, Shiv Kumar, Neha Sharma, and Deepak Sharma

7 **A Novel HMG-CoA Reductase Inhibitor from Methanolic Extract of *Holoptelea integrifolia* (Roxb.) Planch** 55
K.S. Arun and Anu Augustine

8 **Optimisation of Cultural and Nutritional Parameters for the Production of Protease from Newly Isolated Bacterial Strain *Bacillus* SDR 10** 63
Dinta Daniel, B.S. Rakhi, S. Subramaniyan, and G.S. Sandhia

9 **Identification of a PPAR Delta Agonist: An *In Silico* Approach Towards Drug Design for Metabolic Syndrome** 79
Serena D'Souza, Uma Maheshwari, Santhosh Rebello, and Asha Abraham

10 **Preliminary Phytochemical and Antimicrobial Activity of Ethanolic Extract of *Pittosporum viridulum* Stem Bark** 89
C.T. Sadashiva, Satheesh George, Nitha Balan, U. Prakash Kumar, and Indira Balachandran

11 **Biochemical and Nutritional Evaluation of *Monochoria vaginalis* Presl.** 93
S. Pradeesh, G. Nair Archana, M. Devi Chinmayee, C.S. Sarika, I. Mini, and T.S. Swapna

12 **Assessment of Genotypic Variability for Growth, Biophysical Parameters, Yield and Yield-Attributing Characters Under Drought Stress in Cotton** 103
B.C. Patil, A.G. Babu, and K.N. Pawar

13 **Study of Serum Immunoglobulin Levels: Complements C3 and C4 in Vitiligo Patients in Northern India** 111
Usha Singh, Suman Singh, and S.S. Pandey

14 **Evaluation of Antimicrobial Potential of an Invasive Weed *Amaranthus spinosus* L.** 117
A. Mary Sheeba, S.R. Deepthi, and I. Mini

15 **Evolutionary Genomics of Avian MHC BLB2 Gene by Molecular Phylogenetic Analysis** 125
Sudina Kizhakkayil, Arun Raveendran, and E. Jayadevi Variyar

16 **Phytoremediation of Cadmium Using *Pteris* sp.** 131
Sudha Sajeev, P.V. Ramya, B.C. Sunitha, J.S. Melo, and Smitha Hegde

17 **Seasonal Variation of Spore Density and Root Colonisation of Arbuscular Mycorrhizae in Crop Plants in Relation to Soil Edaphic Factors** 141
Deepake Ujwala Sheshrao and G. Gyananath

18 **Computational Studies on Structural Basis of Isoniazid Resistance in Tuberculosis** 151
Vidhyasree Rajasekharan, J. Abhithaj, G. Renganayaki, and Vipin Thomas

19 **Phylogenetic Analysis of Few Actinobacteria with Potential Antimicrobial Properties, Isolated from the Forest Soils of Western Ghats of Kerala** 159
Sugathan Shiburaj and Sreenivasan Preethi

20 Binding of Apolipoprotein A-1 to Jacalin-Sepharose in a Sugar-Dependent Manner 169
P. Priya and V.M. Kannan

21 Morphological Basis for Identification of Cup Quality Characteristics in F_1 Hybrids Derived from *Coffea arabica* L. Crosses 173
Anil Kumar, S. Ganesh, K. Basavraj, M.K. Mishra, and Jayarama

22 Microbial Transformation of Rifamycin by a Bacterial Strain *Chryseobacterium* Species 181
A.H. Jobanputra and B.A. Karode

23 Isolation, Molecular Characterization and Expression of a New *Cry*3a Gene from Native Isolates of *Bacillus thuringiensis* and Its Toxicity Against Asian Grey Weevil *Myllocerus undecimpustulatus undatus* Marshall (Coleoptera: Curculionidae) 191
H.M. Mahadeva Swamy, R. Asokan, Geetha G. Thimmegowda, Riaz Mahmood, and Dilip K. Arora

24 Characteristics of a Newly Isolated Bacterium for Decolorization of Textile Dyes 199
Joshni T. Chacko, S. Kalidass, and Lubna Lukman

25 Analysis of the Pathway of Phenol Biodegradation by *Alcaligenes* sp. *d2* 209
Merlin Antony, Indu C. Nair, and K. Jayachandran

26 Improvement of Vermicompost: Influence of Feeding Materials and Inoculation of Nitrogen-Fixing and Phosphate-Solubilising Bacteria 221
Md. Nasim Ali, S. Chakraborty, Prosanta Saha, and Niladri Lodh

27 Development of SSR Markers and Genetic Variability in Physiological Traits in Bambara Groundnut (*Vigna subterranea* L. Verdc) 229
R. Beena, M.S. Sheshshayee, J.N. Madhura, T.G. Prasad, and M. Udayakumar

28 Biochemical Characterization of Oxidative Burst During Interaction Between Sesame (*Sesamum indicum* L.) in Response to *Alternaria sesami* 243
A.S. Lubaina and K. Murugan

29 Mass Spectrometric Characterization of a Novel Antimicrobial Peptide Isolated from *Clitoria ternatea* 251
Neethu Ajayakumar, Walter Schrenk, Bipin G. Nair, and Sudarslal Sadasivan Nair

30 A Novel Plate Assay System for Screening of Antimicrobial Agents and β-Lactamase Inhibitors from Natural Sources 257
S. Prasanth, Shahana Valsan, Ambika Devi, D.S. Preethidan, M. Haridas, and Abdulhameed Sabu

31 Diurnal Phototactic Migration of Phytoplankton in Cochin Waters and Its Correlation with Limiting Nutrient Flux 263
M. Ratheesh Kumar, V.S. Sudhanandh, A.K. Faisal, S. Raj Vishnu, R. Shibu, M.I. Sreejith, and K. Anoop Krishnan

32 Cryo-preservation of Pollen for Hybrid Seed Production in Hot Pepper 273
Rakesh C. Mathad, S.N. Vasudevan, and S.B. Patil

33 Evaluation of *Pseudomonas* sp. VSMKU-4046 for Suppression of Sheath Blight (ShB) of Rice in Detached Leaf Assay 277
C. Sathish Kumar, H. Harikrishnan, R. Charulatha, and V. Shanmugaiah

34 Antimicrobial and Immunomodulatory Activities of Jellyfish (*Chrysaora quinquecirrha*) Venom 283
K. Suganthi and S. Bragadeeswaran

35 *Diplazium esculentum*: A Wild Nutrient-Rich Leafy Vegetable from Western Ghats 293
G. Nair Archana, S. Pradeesh, M. Devi Chinmayee, I. Mini, and T.S. Swapna

36 *Agrobacterium*-Mediated Genetic Transformation and Somatic Embryogenesis from Leaf Callus of *Hevea brasiliensis*: Effect of Silver Nitrate 303
R.G. Kala, V. Abraham, S. Sobha, P.K. Jayasree, A.M. Suni, and A. Thulaseedharan

37 Immunoinformatics Prediction and Structure-Based Modeling of HLA-II Binding Epitopes of Iron Surface Determinant B (IsdB) Protein of *Staphylococcus aureus* 317
Francis Dileep and Surekha Kuyyalil

38 Damage Assessment of the Gall Mite *Aceria pongamiae* Keifer 1966 (Acari: Eriophyidae) on *Pongamia pinnata* (L.) Pierre 325
P.N.M. Nasareen, Y. Shibu Vardhanan, and N. Ramani

39 *Streptomyces* sp.VSMGT1014-Mediated Antifungal Activity Against Fungal Plant Pathogens 335
H. Harikrishnan and V. Shanmugaiah

40 Taxonomic Significance of Foliar Micromorphology and Their Systematic Relevance in the Genus Solanum (Solanaceae) ... 343
V.S. Anil Kumar and K. Murugan

41 Investigation of Functional Microbial Diversity and Related Abiotic Factors in Coastal and Desert Ecosystem of Gujarat ... 351
D. Biswas, A. Mandal, H. Akbari, and N. Munshi

42 Experimental and Theoretical Investigations on the Anti-oomycetous Activity of Paromomycin Targeted Against *Pythium* Tubulin ... 363
A.K. Resna, C. Geethu, S. Sumna, U.C. Jaleel, and R. Aswati Nair

43 Prenatal Genetic Studies on Intrauterine Growth Retardation ... 373
P. Jisha, R. Manikkumar, D. Dinesh Roy, and K. Ittyavirah Alex

44 Performance of West African Dwarf Sheep Fed with Selected Tropical Roughage Sources: Rumen Microbial Analysis and Nutrient Utilization ... 383
O.A. Isah and R.M. Akinbode

45 Reproductive Toxicity of Anamirta *cocculus* (L.) Wight and Arn and *Sphagneticola trilobata* (L.) Pruski Extracts on *Culex pipiens* Linn. (Diptera: Culicidae) ... 389
E. Pushpalatha, D. Aiswarya, E.K. Sharihan, and B.K.P. Priya

46 Epidemiological Typing of Methicillin-Resistant *Staphylococcus aureus* (MRSA) Isolated from Kerala Using Phage Typing ... 393
J. Jeshina and Kuyyalil Surekha

47 Hypolipidemic Effect of Methanol Fraction of *Acorus calamus* Linn. in Diet-Induced Obese Rats ... 399
K.S. Arun and Anu Augustine

48 Biotyping and Phage Typing of *Salmonella enterica* Serotype Typhi Isolates from Kerala, South India ... 405
N. Ayana and Kuyyalil Surekha

49 Lymphocyte Teamwork in Tumor Rejection ... 411
Anil Shanker

50 Biosorption of Nickel (II) from Aqueous Solution Using *Bacillus laterosporus* ... 415
Rajeswari M. Kulkarni, G. Srinikethan, and K. Vidya Shetty

Subject Index ... 419

1 Fish-Processing Industrial Effluent Characterization and Partitioning of Proteins Using Aqueous Two-Phase System

H. Nagaraja Viswanatha and I. Regupathi

Abstract

Characterization of fish-processing industrial effluent was carried out according to Bureau of Indian Standards (BS.2690:1965). The effluent contained high strength of chemical oxygen demand, biological oxygen demand and proteins. The effect of protein solubility was measured by a function of pH ranging from 2 to 12. Partitioning of proteins from fish industrial effluent in aqueous two-phase systems (ATPS) was investigated. Sodium sulphate was chosen due to its ability of promoting the hydrophobic difference between the phases. Phase compositions including polyethylene glycol (PEG) molecular weight (2000, 4000, 6000 and 8000) and sodium sulphate affecting the protein partitioning were studied. Varying the polyethylene glycol molecular weight and concentrations with sodium sulphate salt affects the partition coefficient. The tie lines for above-said system were developed at 25°C for different compositions. Further, tie-line lengths were calculated and effects of this parameter over partition coefficient were studied. From the experimental results, it was revealed that the PEG molecular weight and concentrations, salt concentrations, pH (5–8) and temperature (20–50°C) of system had strong impacts on partition features.

Keywords

Aqueous two-phase system • Partitioning • Fish proteins • Tie-line length

H.N. Viswanatha (✉) • I. Regupathi
Department of Chemical Engineering, National Institute of Technology – Karnataka, Srinivasnagar Post, Surathkal, Mangalore, Karnataka, 575025, India
e-mail: vishwa.bt@gmail.com

A. Sabu and A. Augustine (eds.), *Prospects in Bioscience: Addressing the Issues*,
DOI 10.1007/978-81-322-0810-5_1, © Springer India 2013

Introduction

The effluent generated from fish-processing industry contains a high organic load, and it should be treated/recovered before discharging the effluent into the environment in order to prevent a negative impact on the environment. On the other hand, this effluent contains a large amount of potentially valuable proteins. The degree of effluent generation from fish-processing industry depends on several parameters, such as type of operation involved and type and amount of seafood processed. Typically, a plant of 100 ton fish-processing per hour capacity generates 10–40 m^3/h effluent with protein loads of 0.5–20 g/l [1–3]. Fish-processing industry wastes are an important environmental contamination source. Research has been carried out in order to develop methods to convert these wastes into useful products. The fish protein hydrolysates have been reported to exhibit various bio-functionalities like antioxidative, antibacterial, anti-hypersensitive and anticancer properties [4]. The recovered molecules also have potential for several pharmaceuticals as well as feed formulations apart from providing value addition to fish-processing waste. Conventional methods were used for recovery and purification of proteins from fish effluents including enzymatic hydrolysis, centrifugation [5], filtration, flocculation, precipitation [6], pH shifting [7], membrane filtration [1, 2] and emerging technologies such as subcritical water hydrolysis, supercritical fluid extraction and ohmic heating [8]. Concerning fish protein recovery, there exist many conventional techniques which give rise to material with poor functionality and nutritive value [9]. Therefore, it is essential to devise reliable separation strategies that would improve yield, selectivity, economy and feasibility of the process.

Aqueous two-phase systems (ATPS) have been successfully used for separation and purification of proteins, metal ions, nanoparticles and dyes from various sources such as wastewater, fermentation broth and animal and plant cell organelles [10]. ATPS are formed by addition of two water-soluble polymers or polymer and salt, with critical concentrations of aqueous solution occurring in two phases. Among those, polymer/salt systems are more preferred over polymer/polymer systems because they are easy to separate and require lesser time to separate, and there is high density difference between phases. The ATPS have advantages over other conventional separation techniques: due to the high water content of both phases, which means high biocompatibility and low interfacial tension, thus minimizing degradation of biomolecules. ATPS also provide good resolution, high yield and a relatively high capacity. In addition, this system is easily scaled up. ATPS could be a good alternative for a first purification step since such systems allow removal of several contaminants by a simple and low-cost process [11–16].

The mechanism governing the partition of biomolecules in ATPS is still not fully understood and largely unknown. Partition is driven by the van der Waals, hydrophobic, hydrogen bond and ionic interactions between the biomolecules and the surrounding phase. Therefore, the partition may also be influenced by the concentrations and molecular weight of phase-forming polymer, concentration of phase-forming salt, type and concentration of added salts, temperature and pH. Several studies have focused on the influences of the above-mentioned parameters on the separation of various biomolecules. Tubio et al. [17] studied the effect of different factors such as polyethylene glycol molecular weight, pH, tie-line length (TLL), temperature and the presence of an inorganic salt on the protein partition coefficient. Chia-Kai and Been [18] studied extraction behaviour of lysozyme from chicken egg white by PEG/sulphate system. Optimized parameters of lysozyme from chicken egg white extraction were found at a temperature of 25.8°C and pH 10. Shahbaz and Omidinia [19] reported influences of PEG molecular weight and its concentrations, pH, $(NH_4)_2SO_4$ concentrations and TLL on recombinant enzyme phenylalanine dehydrogenase partitioning. Silva et al. [20] studied the influence of PEG 1500 with various type of salts (potassium phosphate, sodium citrate, lithium sulphate or sodium sulphate) and temperature (278.15–318.15 K) on the caseinomacropeptide partition coefficient. In addition, thermodynamic parameters (ΔH^o, ΔS^o and ΔG^o) as a function of temperature were calculated for the system PEG 1500–sodium citrate at different PEG concentrations. Nowadays, ATPS have been used to recover valuable biomolecules from various processed industrial wastewaters. Saravanan et al. [21] studied the recovery of proteins from tannery wastewater using a PEG 4000/$MgSO_4$. They analysed effects of PEG molecular weight and concentration, $MgSO_4$ concentration, pH and NaCl concentration on protein partition. The same research group [22] reported the recovery of value-added globular proteins from tannery wastewaters using PEG 6000–sulphate salt ATPS and concluded that sodium sulphate system was the best one compared to ammonium sulphate. Recently, development and evaluation of PEG–lithium citrate salt-based ATPS and its application in partitioning of proteins from fish industry effluent were reported by our research group [23]. In that study, partitioning effect on different PEG molecular weights (4,000, 6,000 and 8,000) and its varying concentrations and lithium citrate salt on fish-processing industrial effluent were studied, and the maximum partition coefficient of 7.82 was obtained at TLL 38%. Sulphate was chosen as the phase-forming salt because of its ability of promoting the hydrophobic difference between the phases [18]. In a phase diagram, sodium sulphate showed a larger biphasic area as compared to potassium phosphate and lithium sulphate salt systems [24].

In the present report, it was proposed to study fish-processing industrial effluent characterization and partition of proteins from fish-processing industrial wastewater employing a PEG–sodium sulphate salt-based ATPS. The effects of polymer molecular weights, its concentrations and salt concentrations, TLL, pH and temperatures of the system on the partitioning behaviour were investigated.

Materials and Methods

Sodium sulphate (>99% purity) of molecular weight obtained from Merck, Mumbai, India. Analytical grade polyethylene glycols of average molecular weight 1,900, 3,850, 5,800 and 7,750 g/mol (PEG 2000, PEG 4000, PEG 6000 and PEG 8000) were purchased from Sigma Chem. Co. (USA). All the reagents were of analytical grade. Double-distilled water was used for this work. Bradford reagent (Sigma-Aldrich Inc., Germany) was used to estimate the protein content of the equilibrium phases and effluent. The fish effluent used in these experiments was obtained from a fishmeal processing industry located at Mangalore, India.

Characterization of Fish-Processing Industrial Effluent and Protein Solubility

Characterization of fish-processing industrial effluent was carried out according to Bureau of Indian Standards (BS.2690:1965). In order to find the effect of different pH on the solubility of proteins from fish protein, a solubility curve was prepared, as described by Thawornchinsombut and Park [25]. Protein-containing fish-processing industry effluent was taken and homogenized (IKA T18@ Basic, Ultra-Turrax, Germany) at a setting of 5× for 1 min. The homogenate (5 mL) was adjusted from pH 2 to 12.0 in 1 interval, using 0.2 M and 1 M HCl or NaOH, with pH metre (Eutech Instruments, Singapore). The homogenate was centrifuged at 10,000 rpm at 4°C for 20 min. The protein concentration of the supernatant was determined by Bradford method using bovine serum albumin (BSA) as the standard protein. The protein solubility was expressed as milligram per gram of effluent. Four replications were carried out for each measurement.

Preparation of ATPS

The system compositions were taken from the literature according to the binodal diagrams for PEG 4000 and 8000 with sodium sulphate [26, 27]. ATPS were prepared in 15-mL centrifuge tubes by adding the desired weight fraction of PEG, salts and fish-processing industry effluent. To study the effect of salts on partitioning of protein from fish-processing effluent, varying molecular weights of PEG (2,000, 4,000, 6,000 and 8,000) and its concentrations with sodium sulphate, fish industry effluent (50% w/w) and distilled water were used to attain 10 g of the aqueous system. The mixtures were dissolved completely using vortex mixer. Phase separation was achieved by centrifugation for 10 min at 3,500 rpm. The samples were incubated to attain equilibrium in a thermostat maintained with an uncertainty of ±0.05°C (Lab Companion, model RW-0525 G, Korea) of constant temperature of 25°C for 24 h. The top phase was carefully separated using a pipette and the interfacial volume was discarded. Volumes of the separated phases were measured, and aliquots from each phase were taken for further studies. The individual phases were separated, and concentrations of polymer and salts were determined by digital refractometer (RX-500, ATAGO Co. Ltd., Japan) and flame photometer (model CL-378, Elico Ltd., India), respectively.

Protein Estimation

Protein estimation was carried out by Bradford method using Coomassie Brilliant Blue G-250 [28]. The samples were read at 595 nm against the blanks with the same compositions as the samples, but without any proteins, to avoid the interference of PEG and salt. Assays were performed in triplicate and the averages were used in the calculations. The results are expressed in terms of bovine serum albumin (BSA) equivalents.

Determination of Partition Parameters

Partition coefficient (K) was calculated by Eq. 1.1.

$$K = \frac{C_{\text{tp}}}{C_{\text{bp}}} \quad (1.1)$$

Partition coefficient is defined as ratio of protein concentration in top phase (C_{tp}) to protein concentration in bottom phase (C_{bp}).

Phase volume ratio (VR) is defined as the ratio of volume in the top phase (V_{tp}) to that in the bottom phase (V_{bp}) Eq. 1.2.

Table 1.1 Composition of fish-processing industrial effluent

Sl no.	Parameters	Units
1	COD	180,775 ppm
2	BOD	54,000 ppm
3	Total solids	22,408 ppm
4	Total dissolved solids	19,275 ppm
5	Density	1,011.7 kg/m^3
6	pH	7.25
7	Sodium	790 ppm
8	Potassium	1,460 ppm
9	Calcium	80 ppm
10	Phosphate	1,517 ppm
11	Total protein	450 ppm
12	Oil and grease	964 ppm

$$\text{VR} = \frac{V_{\text{tp}}}{V_{\text{bp}}} \quad (1.2)$$

Tie-line length (TLL) is proportional to the difference between intensive thermodynamic properties for the upper and lower phases. The tie-line lengths are estimated using the following relationship [29] Eq. 1.3.

$$\text{TLL} = [(W_{P(T)} - W_{P(B)})^2 + (W_{S(T)} - W_{S(B)})^2]^{1/2} \quad (1.3)$$

Results and Discussion

Characterization of Fish-Processing Industrial Effluent

Fish-processing industrial effluent is a critical factor in establishing a corresponding effective effluent treatment process. Effluent characterization was performed according to the Bureau of Indian Standards (BS.2690:1965). Salt concentrations were analysed by flame photometer. The concentrations of Na, Ca and K in the samples were measured in suitable dilutions. Table 1.1 illustrates the average value of various parameters determined for the taken sample of effluent. The BOD to COD ratio is 0.298, indicating that organic matter in the sample is biodegradable and pH around 7.25. All the above parameters are favourable for

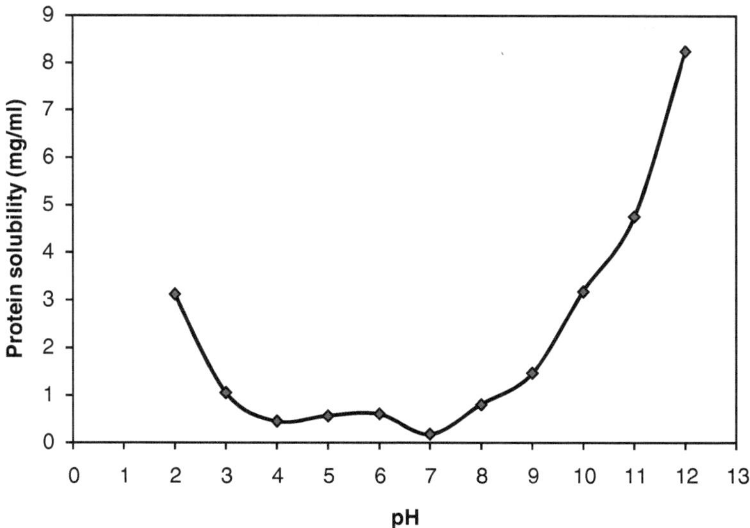

Fig. 1.1 Solubility of proteins from fish-processing industrial effluent at various pHs

biological treatment and suitable for extraction of valuable compounds from the effluent [30].

Effect of pH on Protein Solubility

Protein solubility in an aqueous solution is dependent on pH. In order to obtain a higher recovery of fish proteins by acidic/alkaline processing, protein solubility at various pH conditions was investigated. Solubility of protein was lowest of pH 4 and 7 (Fig. 1.1). The solubility increased gradually as the pH was shifted away from the isoelectric point (pI). The protein solubility profile as a function of pH for water-soluble proteins in fish-processing industry effluent exhibited a characteristic parabolic shape (Fig. 1.1) with a minimum solubility at (or maximum precipitation) pH 7. Isoelectric point hinders protein–water interaction and favours protein–protein bonds, which leads to low protein solubility and their gradual precipitation. Therefore, pH during precipitation is important to achieve maximum protein recovery. Generally, pI of fish muscle proteins ranges between pH 5 and 7; a dramatic increase in solubility was observed from pH 4 to 2 or from pH 8 to 12. Maximum solubility was observed at pH 2 and 12. Similar solubility profile was seen for various protein molecules such as fish sarcoplasmic proteins [31], fish proteins [32] and meat proteins [33].

Effect of Phase Composition and PEG Molecular Weights on Protein Partition Coefficient

PEG of different molecular weight was added in order to form degree of polymerization which affects partition, separation and purification efficiency of proteins. ATPS were prepared with different molecular weight of PEG (2,000, 4,000, 6,000 and 8,000) and varying concentrations from 15 to 25% (w/w), sodium sulphate salt concentration 12.5% (w/w) and 50% (w/w) of fish-processing industrial effluent. Prepared compositions were centrifuged at 3,500 rpm for 10 min and kept in constant temperature water bath at 25°C for attaining equilibrium. Figure 1.2 showed that the partitioning of soluble proteins from fish-processing industrial effluent can be altered by varying the molecular weight of PEG. With increase in the molecular weight of PEG, the partition coefficient of protein was decreased and also observed that partition coefficient of protein decreased with increase in the PEG concentration.

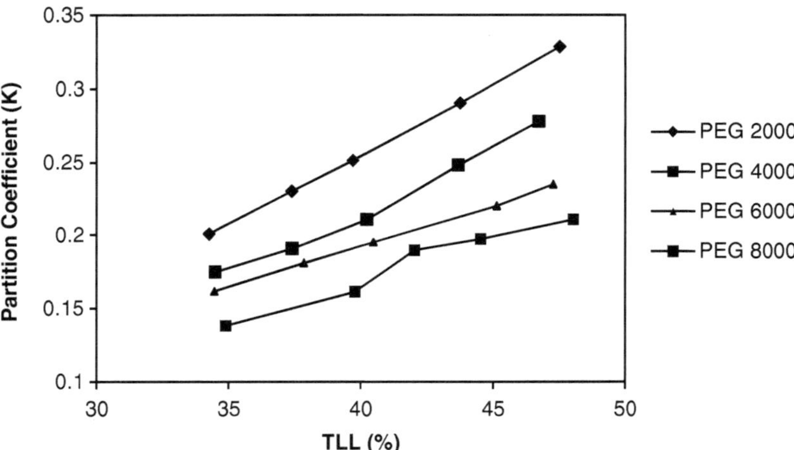

Fig. 1.2 Effect of TLL on protein partition coefficients in sodium citrate system at different molecular weight PEGs

Saravanan et al. [21, 22] have observed similar effect for tannery waste containing proteins, and Regupathi et al. [23] reported for fish-processing industrial effluent proteins. As per these reports, as PEG molecular weight increases, the free volume of top phase of system decreases, causing the migration of proteins towards the bottom phase. On the other hand, the increase in the chain length will also cause the reduction of the free volume, means less space available for the protein interaction. As TLL increases, the free volume at both system phases decreases. A similar behaviour was observed by Madhusudhan et al. [34], Gholamreza et al. [35] and Perumalsamy and Batcha [36] for other ATPS.

Influence of Phase-Forming Salt Concentration on Protein Partitioning

Figure 1.3 shows the effect of phase-forming salt concentration on the partitioning behaviour of fish protein at 25°C. The increase in salt concentration from 12.5 to 22.5% resulted in an increase in partition coefficient from 0.142 to 1.181, respectively. This could be due to "salting-out effect" in bottom phase, as a result of increased partition of protein to top phase. With increases in salt concentration in bottom phase, the solubility of protein in the bottom phase decreases, and the protein shows affinity towards PEG-rich phase or top phase. In addition, an increase in the salt concentration leads to increased bottom phase volume and decreased volume ratios. However, when the top phase volume of proteins is lower, proteins may get precipitated in the interface due to nonsufficient top phase volume for solubility. A similar behaviour was reported by Saravanan [22], Perumalsamy and Batcha [36] and Kohler et al. [37] for other ATPS.

Effect of the System Temperature and pH on Protein Partition Coefficient

The system pH is another important factor, which influences partitioning of biomolecules. The pH affected the charge of target protein depending on its ion composition and surface character of contaminating materials and caused variation in their partitioning into the top and bottom phases. Figure 1.4 reveals that the function of system pH ranging from 5 to 8 and temperature ranging from 20 to 50°C directed the fish protein to bottom phase ($K < 1$), while the protein was concentrated in the top phase ($K > 1$) of PEG 2000/sodium sulphate salt system. The partition coefficient decreases with increase in pH of the system as shown in Fig. 1.4. Therefore, to better understand the relationship between the partition coefficient of a biomolecule and the system pH, the nature of charged amino acid residues at the biomolecule

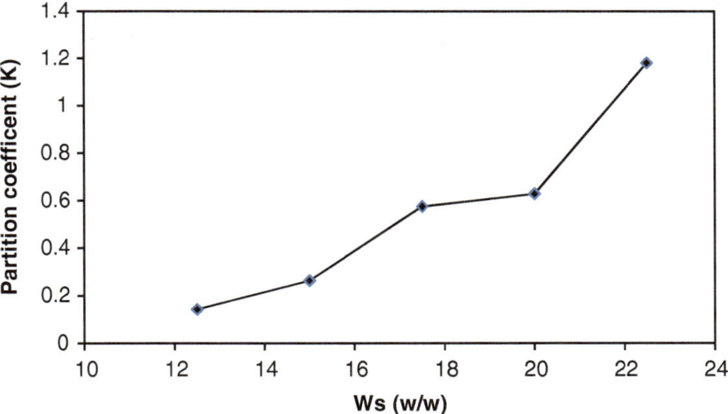

Fig. 1.3 Effect of phase-forming salt concentration on partition coefficient of protein in PEG–sodium sulphate system at 25 °C

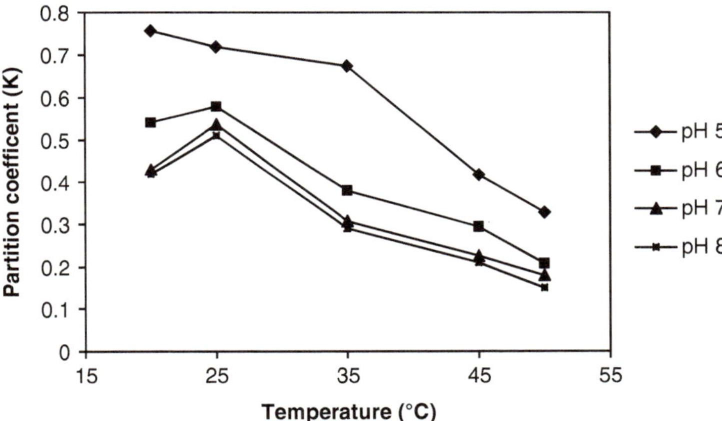

Fig. 1.4 Effect of temperature and pH on the partition coefficient of protein in PEG 2000–sodium sulphate system

surface must be considered. The net charge of the surface-exposed amino acid residues determines the protein surface properties, and it can be altered by changing the system pH. Additionally, electrostatic interactions between the charged biomolecules and the phases, which also carry the electrostatic charges, contribute to the partitioning behaviour. The top phase of an aqueous biphasic system, composed of PEG and sodium sulphate salt, carries a positive charge, while the bottom phase has a negative charge. Consequently, the positively charged proteins will partition to the bottom phase, while negatively charged proteins are directed to the top phase [38]. Yucekan and Onal [39] and Perumalsamy and Batcha [35] observed similar behaviour for other systems. To select the optimal temperature for fish protein partitioning, different temperatures were studied and the partition coefficients of protein with respect to temperature are shown in Fig. 1.4. With increase in temperature, the binodal curve expands to biphasic area resulting in increased differences in the phase compositions. The highest partition coefficient of fish protein was found at 20–25°C. The temperature has an indirect effect on the entropy of the water molecules in their interaction with PEG [40], which drive partition of the protein. It was observed that the partition coefficient of fish protein decreases with an increase in temperature. At low temperatures, the hydrophobic side chains were shielded from water as they were covered inside the proteins. When temperature increases, the hydrophobic side chains were exposed to the solvent, and the water molecules in the solvent formed clusters around these hydrophobic residues. This behaviour is due to the variation in the structure of PEG components with increase in temperature. As the temperature increases, more molecules have enough kinetic energy to undergo the reaction. If the temperature is raised above the optimum point, the kinetic energy of the enzyme or protein and water molecules is so great that the structure of the enzyme molecule starts to disrupt. Therefore, a decrease in partition was determined. Additionally, with further increase in temperature, proteins undergo denaturation and lose their activity. Similar behaviour was observed by Pandey and Banik [41], Ketnawaa et al. [42], Perumalsamy and Batcha [35] and Gautam and Simon [40] on various other types of biomolecules in ATPS.

Conclusions

In the present study, characterization of fish-processing industrial effluent has been done which confirms the presence of proteins favourable for biological processes. Partitioning of fish protein from fish-processing industrial effluent was carried out using PEG–sodium sulphate and water-based ATPS. Various factors affecting the partitioning of the proteins have been studied. The partitioning coefficient of fish protein has shown decrease with an increase in PEG molecular mass, due to increase in hydrophobic nature of PEG with higher molecular weight. It was found that with increase in the PEG concentrations and TLL (%), partition coefficient of fish protein increased. Partition coefficient of fish protein increased with the increase in salt concentration also. The most favourable operating conditions for partition of fish protein were found at pH 5 and temperature 25°C.

Acknowledgement The authors acknowledge a grant (Scheme No. 01(2339)/09/EMR-II) from the Council of Scientific and Industrial Research (CSIR), Government of India, for this research.

References

1. Mateusz K, Daniela S. Application of microfiltration and ceramic membranes for treatment of salted aqueous wastewaters from fish processing. Desalination. 2009;241:227–35.
2. Maria DA, Rodrigo B. Nanofiltration of wastewaters from the fish meal industry. Desalination. 2002;151: 131–8.
3. Maria DA, Javier F, Rodrigo B. An economic assessment of protein recovery from fish meal wastewaters by ultra filtration. Trends Food Sci Technol. 2004; 15:506–12.
4. Kim SK, Mendis E. Bioactive compounds from marine processing by-products-a review. Food Res Int. 2006;39:383–93.
5. Guerrerof L, Omil F, Mendez R, Lema JM. Protein recovery during the overall treatment of waste water from fish–meal factories. Bioresour Technol. 1998;63:221–9.
6. Taskaya L, Jaczynski J. Flocculation-enhanced protein recovery from fish processing by-products by isoelectric solubilization/precipitation. Food Sci Technol. 2009;42:570–5.
7. Helgi N, Ingrid U. The acid and alkaline solubilisation process for the isolation of muscle proteins: state of the art. Food Bioprocess Technol. 2009;2:1–27.
8. Kobsak K, Tipaporn Y, Suvit T, Pisit WN. Protein removal from fish mince wash water using ohmic heating. Songklanakarin J Sci Technol. 2008;30(3): 413–9.
9. Sanmartın E, Carlos J, Arboleya VM, Moreno FJ. Recent Advances in the recovery and improvement of functional proteins from fish processing by-products: use of protein glycation as an alternative method. Compr Rev Food Sci Food Saf. 2009;8:332–44.
10. Selvaraj R, Vytla RM, Varadavenkatesan T, Vinayagam R, Vinayagam R. Aqueous two-phase systems for the recovery of biomolecules – a review. Sci Technol. 2011;1(1):7–16.
11. Albertsson PA. Partition of proteins in liquid polymer/polymer two-phase systems in partition of cell particles and macromolecules. 3rd ed. New York: Wiley Intersciences; 1986. p. 709–71.
12. Raghavarao KSMS, Rastogi GK, Karanth NG. Aqueous two-phase extraction for downstream processing of enzymes/proteins. Adv Appl Microbiol. 1995;41:97–171.
13. Raghavarao KSMS, Martin RG, Paul T. Recent developments in aqueous two-phase extraction in bioprocessing. Separ Purif Method. 1998;27(1):1–49.
14. Rajni HK. Aqueous two-phase systems methods and protocols. Totowa: Humana Press; 2001. p. 32–46.
15. Rosa Piee AJ, Ferreira IF, Azevedo AM, Aires-Barros MR. Aqueous two-phase systems: a viable platform in the manufacturing of biopharmaceuticals. J Chromatogr A. 2010;1217:2296–305.
16. Rosa PAJ, Azevedo AM, Sommerfeld S, Backer W, Aires-Barros MR. Aqueous two-phase extraction as a platform in the biomanufacturing industry: economical and environmental sustainability. J Biotech Adv. 2011. doi:10.1016/j.biotechadv.2011.03.006.
17. Tubío G, Nerli PBG. Partitioning features of bovine trypsin and α-chymotrypsin in polyethylene glycol-sodium citrate aqueous two-phase systems. J Chromatogr B. 2007;852:244–9.

18. Chia-Kai S, Been HC. Partitioning and purification of lysozyme from chicken egg white using aqueous two-phase system. Process Biochem. 2006;41:257–63.
19. Shahbaz MH, Omidinia E. Development and application of aqueous two-phase partition for the recovery and separation of recombinant phenylalanine dehydrogenase. Iran J Chem Chem Eng. 2008;27(2):119–27.
20. Silva CAS, Coimbra JSR, Rojas EEG, Minima LA, Silva LHM. Partitioning of caseinomacropeptide in aqueous two-phase systems. J Chromatogr B. 2007;858:205–10.
21. Saravanan S, Raghava Rao J, Murugesan T, Balachandran UN, Ramasamia T. Partition of tannery wastewater proteins in aqueous two-phase poly (ethylene glycol)-magnesium sulfate systems: effects of molecular weights and pH. Chem Eng Sci. 2007; 62:969–78.
22. Saravanan S, Rao JR, Murugesan T, Nair BU, Ramasami T. Recovery of value-added globular proteins from tannery wastewaters using PEG – salt aqueous two-phase systems. J Chem Technol Biotechnol. 2006;81:1814–9.
23. Regupathi I, Girish B, Prasanna B, Vishwanatha HN. Development and evaluation of PEG-lithium citrate salt based aqueous two phase system and its application in partitioning of proteins from fish industry effluent. Separ Sci Technol. 2012;47(4):591–8.
24. Carvalho CP, Coimbra JSR, Costa IAF, Minim LA, Silva LHM, Maffia MC. Equilibrium data for PEG 4000+salt+water systems from (278.15 to 318.15) K. J Chem Eng Data. 2007;52:351–6.
25. Thawornchinsombut S, Park JW. Role of pH in solubility and conformational changes of Pacific whiting muscle proteins. J Food Biochem. 2004;28(2):135–54.
26. Castro BD, Aznar M. Liquid-liquid equilibrium of water+PEG 8000+magnesium sulfate or sodium sulfate aqueous two-phase systems at 35°c: experimental determination and thermodynamic modelling. Braz J Chem Eng. 2005;22(3):463–70.
27. Carolina PC, Jane SRC, Isabele AFC, Luıs AM, Luıs HM, Silva Maria CM. Equilibrium data for PEG 4000+salt+water systems from (278.15 to 318.15) K. J Chem Eng Data. 2007;52:351–6.
28. Bradford MA. Rapid and sensitive method for the quantification of microgram quantities of protein utilizing the principles of protein-dye binding. Anal Biochem. 1976;72:248–54.
29. Murugesan T, Perumalsamy M. Liquid-Liquid equilibria of poly (ethylene glycol) 2000+sodium citrate+water at (25, 30, 35, 40, and 45) °C. J Chem Eng Data. 2005;50:1392–5.
30. Aloui F, Khoufi S, Louki S, Sayadi S. Performances of an activated sludge process for the treatment of fish processing saline wastewater. Desalination. 2009;246:389–96.
31. Thawornchinsombut S, Park JW, Yongsawatdigul J. Conformational changes and dynamic rheological properties of fish sarcoplasmic proteins treated at various pHs. Food Chem. 2010;121:1046–52.
32. Yeung CJ, Jinx-Soo K. Fish protein recovered using pH shifting method and its physicochemical properties. J Ocean Univ China. 2005;4(3):224–8.
33. Hrynets Y, Omana DA, Xu Y, Betti M. Comparative study on the effect of acid- and alkaline-aided extractions on mechanically separated turkey meat (MSTM): chemical characteristics of recovered proteins. Process Biochem. 2011;46:335–43.
34. Madhusudhan MC, Raghavarao KSMS, Sanjay N. Integrated process for extraction and purification of alcohol dehydrogenase from Baker's yeast involving precipitation and aqueous two phase extraction. Biochem Eng J. 2007. doi:10.1016/j.bej.2007.08.007.
35. Gholamreza Pi, Vossoughi M, Taghikhani V. Partitioning of penicillin G acylase in aqueous two-phase systems of poly (ethylene glycol) 20000 or 35000 and potassium dihydrogen phosphate or sodium citrate. J Chem Eng Data. 2010;55:243–8.
36. Perumalsamy M, Batcha MI. Synergistic extraction of bovine serum albumin using polyethylene glycol based aqueous biphasic system. Process Biochem. 2011;46:494–7.
37. Kohler K, Veide A, Enfors SO. Partitioning of β-galactosidase fusion proteins in PEG–phosphate aqueous two-phase systems. Enzyme Microbiol Technol. 1991;13:204–9.
38. Patil G, Raghavarao KSMS. Aqueous two phase extraction for purification of C-phycocyanin. Biochem Eng J. 2007;34:156–64.
39. Yucekan I, Onal S. Partitioning of invertase from tomato in poly (ethylene glycol)/sodium sulphate aqueous two-phase systems. Process Biochem. 2011;46:226–32.
40. Gautam S, Simon L. Partitioning of β-glucosidase from *Trichoderma reesei* in poly(ethylene glycol) and potassium phosphate aqueous two-phase systems: Influence of pH and temperature. Biochem Eng J. 2006;30:104–8.
41. Pandey SK, Banik RM. Extractive fermentation for enhanced production of alkaline phosphatase from Bacillus licheniformis MTCC 1483 using aqueous two-phase Systems. Bioresour Technol. 2010. doi:10.1016/j.biortech.2010.12.066.
42. Ketnawaa S, Rawdkuena S, Chaiwutb P. Two phase partitioning and collagen hydrolysis of bromelain from pineapple peel Nang Lae cultivar. Biochem Eng J. 2010;52:205–11.

Bioremediation of Zinc Using *Bacillus* sp. Isolated from Metal-Contaminated Industrial Zone

2

M.P. Krishna, Rinoy Varghese, V. Arun Babu, S. Jyothy, and A.A. Mohamed Hatha

Abstract

The production of heavy metals has increased quickly since the industrial revolution. Heavy metals frequently form compounds that can be toxic, carcinogenic, or mutagenic, even in very small concentrations. The usual techniques of removing metals from wastewaters are in general expensive and have many restrictions. Alternative methods of metal removal and recovery based on biological materials have been measured. Among various agents, the use of microbes for the removal of metals from industrial and municipal wastewater has been proposed as a promising alternative to conventional heavy metal management strategies in past decades. Thus, the present study aims to isolate and characterize bacteria from soil, sediment, and waters of metal-contaminated industrial area to study the zinc resistance patterns and the zinc bioaccumulation potential of the selected microorganism. Zinc analysis of the samples revealed that concentrations varying from 39.832 μg/L to 310.24 μg/L in water, 12.81 μg/g to 407.53 μg/g in soil, and 81.06 μg/g to 829.54 μg/g in sediment are present. Bacterial zinc resistance study showed that tolerance to Zn was relatively low (<500 μg/ml). Ten bacterial genera were represented in soil and 11 from water, while only 5 bacterial genera were recorded from sediment samples. *Bacillus, Pseudomonas*, and *Enterobacter* were found in soil, sediment, and water samples. Highly zinc-resistant *Bacillus* sp. was selected for zinc removal experiment. Zinc removal studies revealed that at pH 5 about 40% reduction occurs; at pH 7, 25% occurs; and at pH 9, 50% occurs. Relatively an increased removal of Zinc was observed in the first day of the experiment by *Bacillus* sp. The metal bioaccumulative potential of the selected isolates may have possible applications in the removal and recovery of zinc from industrial effluents.

M.P. Krishna • R. Varghese (✉) • V.A. Babu • S. Jyothy
School of Environmental Sciences, Mahatma Gandhi University, Kottayam, Kerala, India
e-mail: renochirackal@gmail.com

A.A.M. Hatha
Department of Marine Biology, Microbiology and Biochemistry, Cochin University of Science and Technology, Cochin, Kerala, India

A. Sabu and A. Augustine (eds.), *Prospects in Bioscience: Addressing the Issues*,
DOI 10.1007/978-81-322-0810-5_2, © Springer India 2013

Keywords
Heavy metal • Bacterial metal resistance • Zinc • Bioaccumulation

Introduction

Heavy metal contamination by industrial process and technological development is posing major threats to the environment and public health because of its toxicity, nonbiodegradability, and bioaccumulation [1]. Zinc and its compounds are found in the earth's crust and are present in most rocks, certain minerals, and some carbonate sediments. As a result of weathering of these materials, soluble compounds of zinc are formed and may be released to water [2]. Zinc is one of the metals found in effluents discharged from industries involved in galvanization, electroplating, manufacturing of batteries, and metallurgical industries. Zinc in its metallic form has limited bioavailability and poses no ecological risk. However, zinc can react with other chemicals like acids and oxygen to form compounds, which can be potentially toxic and can cause serious damage to biological systems [3].

All through the previous two decades, wide consideration has been paid on management of environmental pollution and its control due to hazardous materials like heavy metals. Decontamination of heavy metals in the soil and water around industrial plants has been a challenge for a long time. A lot of physicochemical strategies, such as filtration, electrochemical treatment, oxidation/reduction, ion exchange, membrane technology, and reverse osmosis, have been developed for removing heavy metals from the polluted water [4]. But nearly all of them showed to be expensive, less efficient, labor-intensive operation, or lack of selectivity in the treating process [5].

Some reports have shown that indigenous microbes and plant–microbe symbionts tolerate high heavy metal concentrations in different ways and may play a significant role in the restoration of contaminated soil [6]. It is important to study the indigenous microorganisms in heavy metal-polluted sites. It may provide new insight into bacterial diversity under unfavorable conditions, new isolates, and probably new genetic information on heavy metal resistance, which could be exploited in decontamination process in the future [7].

Bioremediation, which involves the use of microbes to detoxify and degrade environmental contaminants, has received increasing attention in recent times to clean up a polluted environment [8]. Bioremediation, being in situ treatment, offers several advantages over the conventional chemical and physical treatment technologies, especially for diluted and widely spread contaminants [9]. The objective of the present study is to isolate and characterize bacteria from soil, sediment, and waters of metal-contaminated industrial area to study the zinc resistance patterns and the zinc bioremediation potential of the selected microorganism.

Materials and Methods

Study Area

Eloor-Edayar industrial area – largest industrial belt of Kerala – is a part of Ernakulam District of Kerala and lies between 76° 17′ 32.9″ and 076° 18′ 31.8″ E longitudes and 10° 04′ 51.6″ and 10° 04′ 38″ N latitudes and is a chronic polluted area and one of the biggest exporting centers of fertilizers and chemicals. Eloor is the house of Fertilizers and Chemicals Travancore (FACT), Travancore Cochin Chemicals (TCC), Indian Rare Earths (IRE), Hindustan Insecticides Limited (HIL), and many other small and big industries situated in the lower flood plains of the river Periyar, and it is an island of 14.21 km^2 formed between two distributaries of river Periyar.

Collection of Samples

Soil, sediment, and water samples were collected from abandoned paddy fields, canals, and river of the selected industrial area. Totally seven sampling sites were identified, and from each sampling site, soil, sediment, and water samples were collected. Soil samples were collected at a depth of 15–20 cm from the surface after removing the top layer. For each of the sampling sites, subsamples of soil were collected from different locations, pooled together, and homogenized so as to obtain representative sample. Samples were collected using a spade that is thoroughly cleaned and disinfected between sampling so as to prevent cross contamination. Sediment samples were collected using Grab sampler and transferred to sterilized plastic bags. Water samples were collected using sterilized plastic bottles. Soil, sediment, and water samples were transferred to an ice box and transported to the laboratory.

Zinc Analysis

Zinc contamination in water was determined with anodic stripping voltammetry (797 VA Computrace, Metrohm) after acid digestion of sample as per the method described in APHA, 1998. For sediments and soil, samples were air dried first and grounded to fine powder using pestle and mortar. Then the samples were separated into two different granulometric fractions, <200 μm and <63 μm, using stainless steel sieves. An aliquot of 0.25 g of powdered sediments of <63 μm was digested with Selectipur nitric acid using a microwave digester (MARSXpress, CEM, USA) as per USEPA 3051a for heavy metals [10, 11]. The digested solution was filtered through Whatman no. 1 filter paper, and finally the volume was made up to 25 ml with ultrapure water (ELGA ultrapure water system, UK). Cadmium was then determined by voltammetric trace-metal analyzer – 797 VA Computrace, Metrohm [12, 13].

Isolation and Identification of Bacteria

Isolation and enumeration of bacteria were carried out by standard serial dilution plate technique. Serially diluted samples were plated to Nutrient Agar and incubated at 37°C for 24–48 h. Bacterial colonies from Nutrient Agar were isolated, purified, and maintained as a pure culture for further study. Bacterial isolates which are maintained as pure culture on Nutrient Agar were characterized and identified up to genus level by morphological tests as per *Bergey's Manual of Determinative Bacteriology*: 9th edition [14] and 8th edition [15]. Morphological tests carried out for the identification of the isolates are Gram's staining, cell shape and arrangement, pigment production, O/F glucose tests, endospore staining, motility, catalase, oxidase, etc.

Bacterial Zinc Resistance Test

Resistance of the bacterial isolates to varying concentrations of zinc was determined by agar dilution method [16]. Fresh overnight cultures of the isolates grown in peptone water were aseptically inoculated into nutrient agar plates, which were supplemented with increasing concentration of zinc metal ions (500 μg/ml–1.5 mg/ml). The plates were incubated at room temperature and observed for bacterial growth. The lowest concentration of zinc at which no growth occurred when compared with the control plates was considered as the minimal inhibitory concentration (MIC). Metal salts were added to the medium after autoclaving and cooling to 45–50°C, from filter sterilized stock solutions. The metal salt used for this study was zinc sulfate ($ZnSO_4$).

Bioaccumulation Experiment

The bacterial strain TCC51 growing in the most zinc concentration was selected for the study, and it was identified as *Bacillus* sp. Bioaccumulation method at different pH (5, 7, and 9) with living bacterial cells was used for the removal of zinc. To study heavy metal removal with live cells, nutrient broth amended with initial concentration of zinc (20 mg/L) was inoculated from overnight grown cultures of selected bacterial isolates. The inoculated flasks were incubated at room temperature for 72 h in a shaking condition. An aliquot of 5-mL sample was taken daily (24-h interval) from each flask. Control flasks without bacterial biomass were running simultaneously with the experiment flasks. Samples were centrifuged to remove suspended biomass, and concentration of heavy metals was determined in the supernatant.

Results and Discussion

A total of 18 bacterial genera were recorded from the selected industrial area. Ten bacterial genera were represented in soil and 11 from water, while only 5 bacterial genera were recorded from sediment samples (Table 2.1). *Bacillus, Pseudomonas*, and *Enterobacter* were found in soil, sediment, and water samples. Zinc analysis of the samples revealed that Zn concentrations varying from 39.832 μg/L to 310.24 μg/L in water, 12.81 μg/g to 407.53 μg/g in soil, and 81.06 μg/g to 829.54 μg/g in sediment were present. Zinc levels in the soil sediment and water samples of the study area are represented in Table 2.2. Zinc resistance studies of the bacterial isolates showed that out of 164 isolates collected, most of them showed low resistance (<500 μg/ml) and many isolates showed high resistance of >1,500 μg/ml.

Comparatively highly zinc-resistant *Bacillus* sp. was selected for zinc removal. Results of zinc removal study revealed that with increase in time, the biomass of the selected *Pseudomonas* sp. increased (Fig. 2.1). Correspondingly, with increase in biomass, the zinc bioaccumulation was also increased. Zinc removal studies revealed that at pH 5 about 40% reduction occurs; at pH 7, 25% occurs; and at pH 9, 50% occurs. Relatively enhanced removal of zinc was observed in the first day of the experiment by *Bacillus* sp. (Fig. 2.2, 2.3, 2.4, 2.5).

Zinc is commonly found in the earth's crust, and natural release to the environment can be significant. In addition, zinc is one of the most widely used metals in the world. The major industrial sources of zinc include electroplating, smelting and ore processing, and drainage from both active and inactive mining operations [17]. Furthermore, zinc is an important component of

Table 2.1 Bacterial genera found in the soil, sediment, and water samples of the selected industrial area

	Sample		
Sl no.	Soil	Sediment	Water
1	*Bacillus*	*Kurthia*	*Staphylococcus*
2	*Caryophanon*	*Pseudomonas*	*Acinetobacter*
3	*Listeria*	*Bacillus*	*Azotobacter*
4	*Kurthia*	*Enterobacter*	*Bacillus*
5	*Agromyces*	*Escherichia*	*Pseudomonas*
6	*Arthrobacter*		*Xanthobacter*
7	*Cellulomonas*		*Enterobacter*
8	*Deinococcus*		*Escherichia*
9	*Pseudomonas*		*Klebsiella*
10	*Enterobacter*		*Aeromonas*
11			*Thiobacillus*

Table 2.2 Zinc concentration in the soil, water, and sediment samples of selected industrial area

	Zinc concentration		
Sample name	Soil (μg/g)	Water (μg/L)	Sediment (μg/g)
TCC	193.29	86.211	244.48
BPM	378.88	279.836	309.28
KDM	12.81	252.022	829.54
BR	174.76	41.555	280.64
WMH	237.16	39.832	165.49
RFH	117.31	79.381	81.06
WB	407.31	310.246	378.14

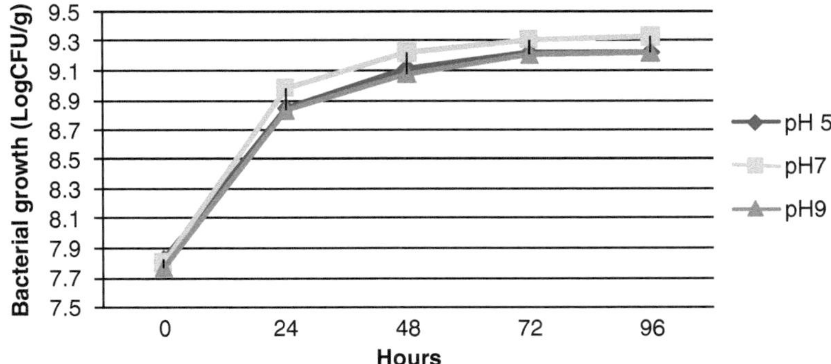

Fig. 2.1 Growth kinetics of *Bacillus* sp. at different pH with 20 mg/L initial concentration of zinc

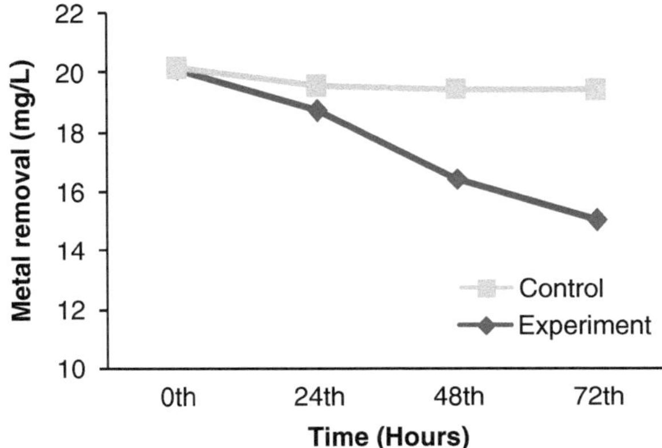

Fig. 2.2 Bioaccumulation of zinc using *Bacillus* sp. at pH 7

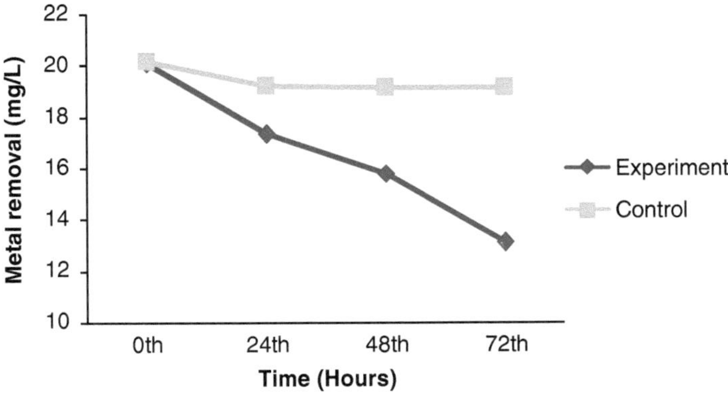

Fig. 2.3 Bioaccumulation of zinc using *Bacillus* sp. at pH 5

brass, bronze, die casting metal, other alloys, rubber, and paints. The environmental releases of zinc from sources of human origin far exceed the releases from natural sources [18]. The permissible concentrations of Zn in water and sediment/soil as per international standards are 5.0 mg/L and 300–600 mg/kg, respectively. In the present study, the values of Zn in water samples are below the permissible limits. In the case of soil and sediment samples, zinc showed enrichment levels exceeding the normally expected distribution. High levels of zinc are observed in several pockets, very nearer to industries which specify that the source of these elements could be the industrial effluents.

Zinc resistance studies of the bacterial isolates showed that out of 164 isolates collected, most of

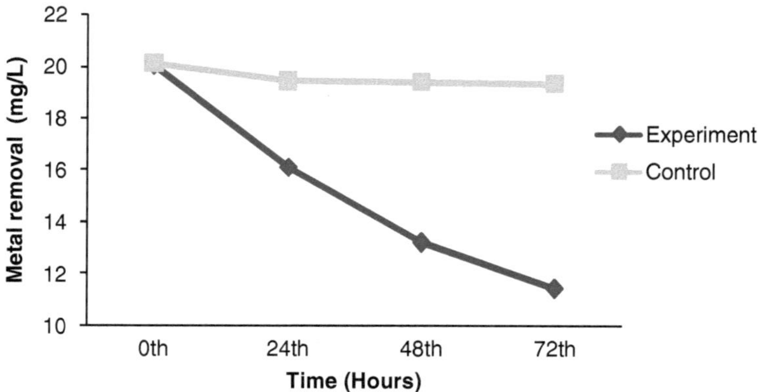

Fig. 2.4 Bioaccumulation of zinc using *Bacillus* sp. at pH 9

Fig. 2.5 Effect of pH on zinc removal using *Bacillus* sp.

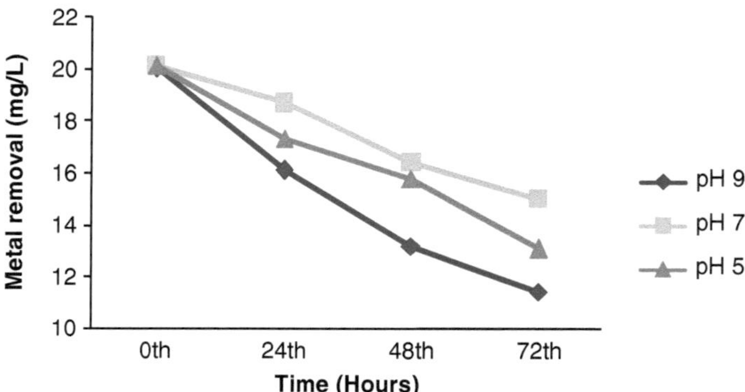

them showed low resistance (<500 μg/ml) and many isolates showed high resistance of >1,500 μg/ml. The high levels of resistance and the widespread tolerance that was found among the isolates are probably attributed to zinc contents in the soil [19]. The level of tolerance among the 164 cultures varied may be due to the difference in the concentration of zinc in the environment. The site from which the samples was taken has been polluted with high levels of heavy metals for many years, perhaps giving a diverse range of bacteria the chance to adapt to the environment, either by convergent evolution of resistance mechanisms or by transferring the resistance genes via a plasmid. Resistance to heavy metals, including cadmium, zinc, copper, chromate, cobalt, arsenic, and nickel, is most often carried by bacteria on plasmids or transposons, and it has been theorized that this allows for lateral transfer in the environment [20].

Results of heavy metal removal studies showed that with increase in time, the biomass of the bacterial strains increased. Likewise, with increase in biomass, zinc bioaccumulation also increased. The increase in surface area that can be due to increase in biomass improves the adsorptive nature or increases the number of active binding sites on cell surface [21]. The selected metal-resistant strains showed that their growth was only slightly affected with different pH. Therefore, it is clear that growth of the newly isolated strains is not inhibited with different pH, and this fact makes them strong candidates for future application in metal bioremediation.

The active mode of metal accumulation by living cells is usually designated as bioaccumulation. This process is dependent on the metabolic activity of the cell referred to its intrinsic biochemical and structural properties, physiological and/or genetic adaptation, environmental modification of metal specification, availability, and toxicity [22]. The capacity of living cells to remove metal ions from aqueous solutions is also significantly influenced by environmental growth conditions, as temperature, pH, and biomass concentrations [23].

Many researchers reported the efficiency and mechanisms of bacteria to remove different metal ions, and many of them are comparable with the present study. Richard et al. [24] reported that Cu^{+2} and Pb^{+2} appear to bind to materials on the cell surface. Lead is precipitated in an insoluble form that is localized to the cell membrane or cell surface [25, 26]. This could be generally explained by the fact that the negatively charged groups (carboxyl, hydroxyl, and phosphoryl) of bacterial cell wall adsorb metal cations through various mechanisms such as electrostatic interaction, van der Waals forces, covalent bonding, or combination of such processes [27].

Since the pH values in metal-containing water and wastewater can vary, it is necessary to use solutions of different pH values to examine the effect of heavy metal removal. In the present work, initial pH was adjusted in the range 5, 7, and 9, before the addition of the biosorbent. The medium pH affects the solubility of metals and the ionization state of the functional groups like carboxylate, phosphate, and amino groups of the cell wall. The inconsistency in literature regarding the influence of pH on biosorption seems to indicate the way pH would alter the adsorption of metal ions to biomass, and it varies with the type of adsorbents (biomass) and also the type of adsorbates (metal ions). The obtained results showed that the selected *Bacillus* sp. is a good bioaccumulation medium for zinc ions and had high adsorption yields for the treatment of wastewater containing zinc. Consequently, bacterial bioaccumulation technologies are still being developed and much more work is required.

References

1. Perez-Marin AB, Ballester A, Gonzalez F, Blazquez ML, Munoz JA, Saez J, Meseguer Zapata V. Study of cadmium, zinc and lead biosorption by orange wastes using the subsequent addition method. Bioresour Technol. 2008;99:8101–6.
2. National Academy of Sciences (NAS). Drinking water and health, vol. 3. Washington, DC: National Academy Press; 1980.
3. Fosmire GJ. Zinc toxicity. Am J Clin Nutr. 1990;51:225–7.
4. Tang L, Zeng GM, Shen GL, Li YP, Zhang Y, Huang DL. Rapid detection of picloram in agricultural field samples using a disposable immunomembrane-based electrochemical sensor. Environ Sci Technol. 2008;42:1207–12.
5. Xiao X, Luo SL, Zeng GM, Wei WZ, Wan Y, Chen L, Guo HJ, Cao Z, Yang LX, Chen JL, Xi Q. Biosorption of cadmium by endophytic fungus (EF) *Microsphaeropsis sp.* LSE10 isolated from cadmium hyperaccumulator *Solanum nigrum* L. Bioresour Technol. 2010;101:1668–74.
6. Ge HW, Lian MF, Wen FZ, Yun YF, Jian FY, Ming T. Isolation and characterization of the heavy metal resistant bacteria CCNWRS33-2 isolated from root nodule of *Lespedeza cuneata* in gold mine tailings in China. J Hazard Mater. 2009;162:50–6.
7. Fabienne G, Antonis C, Hauke H. Comparative 16S rDNA and 16S rRNA sequence analysis indicates that *Actinobacteria* might be a dominant part of the metabolically active bacteria in heavy metal contaminated bulk and rhizosphere soil. Environ Microbiol. 2003;10:896–907.
8. Malik A. Metal bioremediation through growing cells. Environ Int. 2004;30:261–78.
9. Eccles H. Removal of heavy metals from effluent streams: why select a biological process? Int Biodeter Biodegr. 1995;35:5–16.
10. Kingston HM, Walter PJ, Halk SJ, Lorentzen EM, Link D. Environmental microwave sample preparation: fundamentals, sample preparation, and applications, ACS professional reference book series. Washington, DC: American Chemical Society; 1997.
11. Kingston HM, Jassie LB. Safety guidelines for microwave systems in the analytical laboratory. In: Kingston HM, Jassie LB, editors. Introduction to microwave acid decomposition: theory and practice, ACS professional reference book series. Washington, DC: American Chemical Society; 1988.
12. Ireland-Ripert J, Bermond A, Ducauze C. Determination of methyl mercury in the pressure of inorganic mercury by anodic stripping voltammetry. Anal Chim Acta. 1982;143:249–54.
13. Lo J, Lee J. Dithiocarbamate extraction and Au(iii) back extraction for determination of mercury in water and biological samples by anodic stripping voltammetry. Anal Chem. 1994;66:1242–8.
14. Holt JG, Krieg NR, Sneath PHA, Staley JT, Williams ST. Bergey's manual of determinative bacteriology (Ninth edition). London: Williams and Wilkins; 1994.
15. Buchanan RE, Gibbons NE. Bergey's manual of determinative bacteriology. 8th ed. Baltimore: The Williams and Wilkins Co.; 1974.
16. Lui GW, Talnag JW, Strohl WR, Pfister RM. Hexavalent chromium resistant bacteria isolated from river sediments. Appl Environ Microbiol. 1983;46:846–54.

17. Mirenda RJ. Acute toxicity and accumulation of zinc in the crayfish, *Orconectes virilis* (Hagen). B Environ Contam Tox. 1986;37:387–94.
18. Fishbein L. Sources, transport, and alterations of metal compounds: an overview: 1. Arsenic, beryllium, cadmium, chromium, and nickel. Environ Health Persp. 1981;40:43–64.
19. Abou-Shanab RAI, Berkum P, Angle JS. Heavy metal resistance and genotypic analysis of metal resistance genes in gram-positive and gram-negative bacteria present in Ni-rich serpentine soil and in the rhizosphere of *Alyssum murale*. Chemosphere. 2007;68:360–7.
20. Silver S. Plasmid-determined metal resistance mechanisms: range and overview. Plasmid. 1992;27:1–3.
21. Bai S, Abraham R, Emilia T. Studies on enhancement of Cr (VI) biosorption by chemically modified biomass of *Rhizopus nigricans*. Water Res. 2002;36:1224–36.
22. Cha S, Cookseyd A. Copper resistance in *Pseudomonas syringae* mediated by periplasmic and outer membrane proteins. P Natl Acad Sci USA. 1991;88:8915–9.
23. Chen P, Ting YP. Effect of heavy metal uptake on the electrokinetic properties of *Saccharomyces cerevisiae*. Biotechnol Lett. 1995;17:107–12.
24. Richard WG, Krumholz D, Matthew SC, Louis ST. Heavy metal resistance pattern of frankia strains. Appl Environ Microbiol. 2002;68:923–7.
25. Aiking H, Govers H, Vant Riet J. Detoxification of mercury, cadmium, and lead in *Klebsiella aerogenes* NCTC 418 growing in continuous culture. Appl Environ Microbiol. 1985;50:1262–7.
26. Levinson HS, Mahler I, Blackwelde P, Hood T. Lead resistance and sensitivity in *Staphylococcus aureus*. Microbiol Lett. 1996;145:421–5.
27. Chojnacka K, Chojnacki A, Gorecka H. Biosorption of Cr^{+2}, Cd^{+2} and Cu^{+2} ions by blue green alga *Spirulina sp.* kinetics, equilibrium and the mechanism of the processes. Chemosphere. 2005;59:75–84.

3 Evaluation of Bryophyte Protein-Based Defense Against Selected Phytophagous Insects

Remya Krishnan and K. Murugan

Abstract

Insect-resistant proteins have been one of the major successes of applying plant genetic engineering technology to agriculture. Relatively little is known about insect defense mechanisms in the lower plants. The current paradigm is that secondary metabolites and physical barriers are most important in conferring insect resistance in plants. We investigated whether protein-based resistance exists in representatives of bryophytes. We screened a total of 20 bryophytes such as *Porella acutifolia, Fissidens asperifolius, Fissidens crispulus, Hypopterygium tamarisci, Brachymenium nepalense, Brachythecium buchananii, Campylopus pilifer, Marchantia linearis, Leucobryum bowringii, Plagiochila beddomei, Isopterygium albescens, Taxiphyllum taxirameum, Octoblepharum albidum, Bryum argenteum, Riccia frostii, Philonotis fontana, Racopilum cuspidigerum, Funaria hygrometrica, Pallavicinia lyellii,* and *Polytrichum commune* for protein-based insecticidal activity against the two common lepidopteran pests: corn earworm *(Helicoverpa zea)* and armyworm (*Spodoptera litura*). Protein extracts from bryophyte species were compared with those from a lepidopteran-susceptible *Glycine max* cultivar (Cobb) in bioassays for insect resistance. The *Octoblepharum albidum, Fissidens asperifolius, Bryum argenteum,* and *Marchantia linearis* protein extracts caused the greatest decrease in damage in leaf-disk assays and insect larval growth. The results from dietary utilization experiments showed a reduction in efficiency of conversion of ingested food and digested food and an increase in approximate digestibility and metabolic cost. The in vivo effects of proteins of bryophytes showed varied response toward the parameters such as larval weight, growth rate, and proportion of survivors. These bryophyte species are potential candidates for further evaluation.

R. Krishnan (✉) • K. Murugan
Plant Biochemistry and Molecular Biology Lab,
Department of Botany, University College,
Thiruvananthapuram, Kerala, India
e-mail: krishnar07@gmail.com

A. Sabu and A. Augustine (eds.), *Prospects in Bioscience: Addressing the Issues*,
DOI 10.1007/978-81-322-0810-5_3, © Springer India 2013

Keywords
Insect resistance • Proteins • Bryophytes • Lepidoptera • Coleoptera

Introduction

Bryophytes are nonvascular diverse group of plants. In spite of its small stature, the bryophyte plays a significant role in diverse terrestrial ecosystems, yet relatively little is known about their physiology and genetics. In fact, it has been estimated that only 5% of all bryophytes have been studied with regard to any phytochemical characters [1]. Even less is known about the genomics, proteomics, and biochemical pathways of mosses. Although these taxa show rich promise in unlocking physiological properties such as drought tolerance, insect resistance, disease resistance, and tolerance to heavy metals, there is still little gene discovery research being performed on these plants. This may be because these plants are of little economic importance to commercial agriculture. However, one notable exception is *Physcomitrella patens, Tortula ruralis,* which has been the subject of concerted genomic research as the model moss [2].

Although insect species have been estimated to use lower plants as a food source [3], there is a predominant concept that bryophytes are rarely fed on by phytophagous insects in nature [4]. Lepidopterans that eat bryophytes are especially rare [5], yet caterpillars, which possess chewing mouthparts, cause extensive damage to crops. Those insects that do utilize lower plants as a food source often possess piercing-sucking mouthparts that enable them to bypass the high concentration of secondary metabolites found in lower plant cell walls by inserting the mouthparts directly into phloem and sucking sap. Auerbach and Hendrix [6] suggest that the bryophytes being underutilized by insects in comparison to angiosperms might result from the lack of flowers or fruit. Hendrix and Marquis [7] also found that the vegetative portions of bryophyte are fed upon by insects to the same extent as angiosperm vegetative tissue.

Bryophytes produce a pool of secondary metabolites [8]. It has been the general assumption that these secondary compounds are the primary mechanism of lower plants' insect resistance (IR). Ferulic acid, hydrolyzable tannins, terpenes, tannins, and alkaloids have been isolated from lower plants [8]. These same secondary metabolites have been shown to deter insect pests in higher plants and would be expected to demonstrate the same properties in lower plants [9]. Similarly, the other secondary metabolites such as pterosins, phenolic acids, sulfated cinnamic acids, and flavonoids isolated from various lower plants provide effective resistance to feeding insects [10].

Proanthocyanidins are believed to be the most effective broad-spectrum defense against predators [11]. Ecdysones have also been isolated from lower plants and are believed to deter feeding by mimicking insect molting hormones [12]. A literature survey yielded some insight of fern and moss IR proteins. Apparently, thiaminase has been the only fern or moss plant enzyme shown to demonstrate IR activity. Thiaminase is a fern enzyme associated with and the major causative agent of bracken poisoning and B_1 deficiency in cattle and other ruminants [13]. Hendrix [14] showed that thiaminase deterred feeding by the southern armyworm in the fern *Nephrolepis exaltata.*

Our study is an attempt at describing broad protein-based IR in bryophytes. Twenty species were screened for protein-based IR against two common lepidopteran crop pests: *Helicoverpa zea* and *Spodoptera litura.* These insects were chosen because of their economic importance and commercial availability. Protein-based defenses against pest insects on higher plants have been well documented. It would be expected that lower plants might have similar capabilities. Because bryophytes are evolutionarily older than higher plants, it would also be interesting to deduce the evolutionary origins of protein-based insect defenses in land plants. Any findings in this field will increase our knowledge of plant/insect coevolution and add to our understanding of lower plant insect defenses.

Table 3.1 Total protein content (μg) in selected bryophytes and leaf damage ratings for the initial insect bioassays after 72 h by fall armyworm (FAW)

Name of plant	Total protein (μg/g)	Avg. damage rating
Porella acutifolia	1,487.6±0.5	3
*Fissidens asperifolius**	3,389.8±1.2	1.75
Fissidens crispulus	1,399.8±0.09	2.5
Hypopterygium tamarisci	1,378.6±0.2	3
Brachymenium nepalense	1,335.5±2	3.5
Brachythecium buchananii	1,323±0.34	3
Campylopus pilifer	1,299.7±3.3	2.5
*Marchantia linearis**	6,967.4±0.8	0.5
Leucobryum bowringii	1,534.5±0.5	2.75
Plagiochila beddomei	1,517.8±3.7	3
Isopterygium albescens	1,467.8±1.4	3
Taxiphyllum taxirameum	1,434.6±2.4	3
*Octoblepharum albidum**	4,772.8±0.88	1
*Bryum argenteum**	3,471±0.73	1.5
Riccia frostii	1,445±3.6	3
Philonotis Fontana	1,442.4±1.4	3.5
Racopilum cuspidigerum	1,587.5±2	3.5
Funaria hygrometrica	1,521.3±3.5	2.75
Polytrichum commune	1,687.2±3.6	2.75
Pallavicinia lyellii	1,423.9±0.03	3
Glycine max	2,346.3±1.5	3.5

*Four species (asterisks) had statistically significant decreases in FAW damage ratings when compared with control extracts as determined by ANOVA at the significance level of $P=0{:}05$

Materials and Methods

Plant Material

Twenty bryophyte species commonly growing wild in Ponmudi hills were used for a functional screen (Table 3.1). Soybean (*Glycine max* cv. Cobb), which is a suitable food source for lepidopteran larvae, was used as control. Five grams of newly emerged and fully expanded thallus was sampled from plants growing naturally in various habitats, and the tissue was placed into a sterile, prechilled 1.5-ml centrifuge tube and placed immediately on ice for transport back to the laboratory.

Total Protein Extraction

The samples were frozen in liquid nitrogen for 1min and immediately stored at −80°C until homogenization. Tissue was crushed into a fine powder and was resuspended in 1 ml of cold (4°C) HEPES-extraction buffer (20-mM HEPES pH 8, 0.5-mM DTT, 1-mM EDTA, 10% glycerol, 1-mM phenylmethylsulfonyl fluoride, and 1-mM benzamidine). The homogenate was incubated on ice for 1 h with occasional mixing. The samples were then centrifuged at 3,000 g at 4°C for 30 min, and supernatant was concentrated by ammonium sulfate (55%) precipitation. The salt precipitate was centrifuged at 3,000 g for 30 min, and the pellet was resuspended in 0.5ml of extraction buffer. Extracts for each plant tissue was dialyzed using a cellulose membrane with an exclusion limit of <1.2 kD (Sigma). The bags were placed in 10 volumes of cold 20-mM HEPES buffer, kept at 4°C, and gently stirred for 24 h. Three buffer changes were carried out in 8-h intervals. At the end of 24 h, the contents within the dialysis tubing were removed and placed into a Vivaspin 20 concentrator with a molecular mass cutoff of 5 kD (Sartorius). Samples were then centrifuged at 6,000 g for

30 min at 4°C. Extracts were removed from the concentrator and immediately stored at −80°C. Bradford assays were performed to determine protein concentration, and SDS–PAGE analysis was performed to assess the general quality of each sample.

Insect Bioassays: Initial Screen Using *S. litura* and *H. zea*

Soybean leaf disks, 1 cm in diameter, were coated with either control (soybean extract) or bryophyte protein extracts standardized at 100 μM. Five leaf disks were tested for every extract. Individual extract-coated leaf disks were placed in glass petri dishes containing a moist filter paper. Petri dishes were then placed under light for 2 h to facilitate drying and therefore decrease the chance of insect larval death by drowning. Fall armyworm (FAW) (*Spodoptera litura)* and *H. zea* eggs were hatched and deprived of food for 24 h. At the second instar stage of development, 5 larvae were placed onto each treated leaf disk and allowed to feed for 72 h under continuous light at room temperature. The amount of leaf-disk damage by herbivory (or leaf damage) was measured. At 72 h, the plates were placed at −80 °C for 1 h. Surviving FAW head capsules were measured under a dissecting microscope at 20× magnification using a micrometer etched slide. Plates were checked at 24, 48, and 72 h to count and remove dead larvae from each plate. A relative leaf damage scale (0–4) was created as a comparative tool to describe herbivory. A rating of 0 would represent <1% damage, and a rating of 4 represents damage of >50%. Leaf damage was determined at 72 h. The filter paper within the petri dish was kept moist throughout the duration of the experiment.

Midgut Preparation

Fourth instar larvae were cold-immobilized, and the midgut along with its contents was removed in cold 150-mM NaCl and stored frozen (−20°C) until needed. Guts from larvae of *S. litura* and *H. zea* were subsequently homogenized in 150-mM NaCl and centrifuged at 6,000 g for 5 min, and the supernatants pooled and kept on ice for enzymatic assays.

Effects of IR Protein on Endogenous Proteolytic Activity

The effect of IR protein on the proteolytic activity of whole midgut extracts was measured using BAPNA (1 mM) as substrate. The assays were run at pH 8.0 in 100-mM Tris–HCl. IR protein (50–200 μg) was incubated with the midgut extracts (50 μg of protein) at 37°C for 15 min before adding the substrate. The reaction was stopped by adding 30% (v/v) acetic acid. All incubations were done in triplicate and appropriate controls were included. The resulting absorbance was read at 410 nm.

Effects of IR Protein on the Development of *S. litura* and *H. zea* Larvae

The activity of IR protein and soybean Kunitz trypsin inhibitor (SKTI) (as a reference standard) were assessed *in vivo* by feeding the different instars (1st to 5th) of *S. litura* and *H. zea* on diet supplemented with crude IR protein. IR protein and SKTI were incorporated into the larval base diet at concentrations of 50, 100, 150, 200, and 250 μM. A set of controls was maintained where the five different instars were fed on base diet only. In all treatments, the larvae were allowed to develop into pupae. Ten larvae were used for each treatment. For each replicate, survival and larval instar were recorded daily throughout the trail, and from day 3 onward, the larval weights were noted at an interval of 3 days. The efficacy of the inhibitor was determined by measuring the body weight and the number of larvae reaching pupal stage in comparison to the control.

Nutritional Parameters

A number of nutritional parameters were compared among sixth instar larvae exposed to150-μM IR protein-treated or a control diet. The larvae, feces, and remaining uneaten diet were separated in the microscope, dried, and weighted. Nutritional indices of consumption, digestion, and utilization of food were calculated, as described by Waldbauer and Farrar et al. [15, 16]. The nutritional indices, namely, efficiency of conversion of ingested food (ECI), efficiency of conversion of digested food (ECD), and approximate digestibility (AD), were calculated as follows: ECI $(\Delta B/I)\times 100$; ECD $[\Delta B/(I-F)]\times 100$; and AD $[(I-F)/I]\times 100$, where I = weight of food consumed, ΔB is change in body weight, and F = weight of feces produced during the feeding period. Metabolic cost (CM) was calculated as 100 − ECD.

Results and Discussion

Protein Extraction

Protein content varied among the bryophytes from 3389.8 ± 1.2 to 6967.4 ± 0.8 (Table 3.1).

Initial Insect Leaf Damage Assay

Octoblepharum albidum, Fissidens asperifolius, Bryum argenteum, and *Marchantia linearis* are the most effective extract treatments that resulted in significantly decreased leaf damage ratings at $P = 0.05$ level, compared with the control treatment. Control leaf disks had an average damage rating of 3.5 after 72 h of FAW feeding. Treated leaf disks of *Octoblepharum albidum, Fissidens asperifolius, Bryum argenteum,* and *Marchantia linearis* comparatively displayed damage ratings between 0.5 and 1.75, after 72 h (Table 3.1). The remaining extracts resulted in varied damage rates comparable to those of the control (Table 3.1).

Effects of IR Protein on Endogenous Proteolytic Activity

Figure 3.1a and b shows the in vitro specificity of IR protein against enzymes extracted from the larvae of insect pests of economic importance. The larval digestive enzymes extracted from the two insect species were trypsin-like enzymes and were clearly inhibited by IR protein. The inhibitory activity of IR protein against insect proteases, expressed as μmol/min/ml extract against trypsin, ranged from 0.92 to 0.087 for *S. litura* and *H. zea*.

The phytophagous larvae of most lepidopteran species analyzed so far have alkaline midgut fluids, with serine proteases and exopeptidases providing most of the midgut proteolytic activity [17]. IR proteins have been considered as natural control agents against herbivorous insects because they reduce proteolytic enzyme activity in vitro and affect larval development in a number of lepidopteran species. However, not all insect species are susceptible to IR proteins, even though they secrete digestive enzymes that can be inhibited in vitro by these proteins [18, 19].

Bryophyte IR protein was more active against *S. litura* compared to *H. zea*. The substantial inhibition of gut proteases of insect species in the in vitro assay suggested that IR protein may affect the growth and/or survival of these insect pests when incorporated into their diet.

Effects of IR Protein in Diet on Larval Growth and Survival

In vivo effect of IR protein from *Octoblepharum albidum, Fissidens asperifolius, Bryum argenteum,* and *Marchantia linearis* against *S. litura* and *H. zea*, related to the larval growth and development, is of the other interest. Hence, insect feeding assays were performed to evaluate the antibiosis exerted by IR protein toward *S. litura* and *H. zea*. The in vivo effects of IR protein were analyzed toward the parameters such as mass, growth rate, and mortality, respectively, at different developmental stages of *S. litura* and *H. zea* by incorporating different concentrations of the IR proteins using leaf coating assays.

Larvae fed on control diet showed normal development, followed by those consuming 50-μM IR proteins. At moderate (150 μM) and high (200 μM) concentrations of IR proteins, there was a significant reduction in mass of the early instars (1st to 3rd) (Fig. 3.2a, b). Although initially

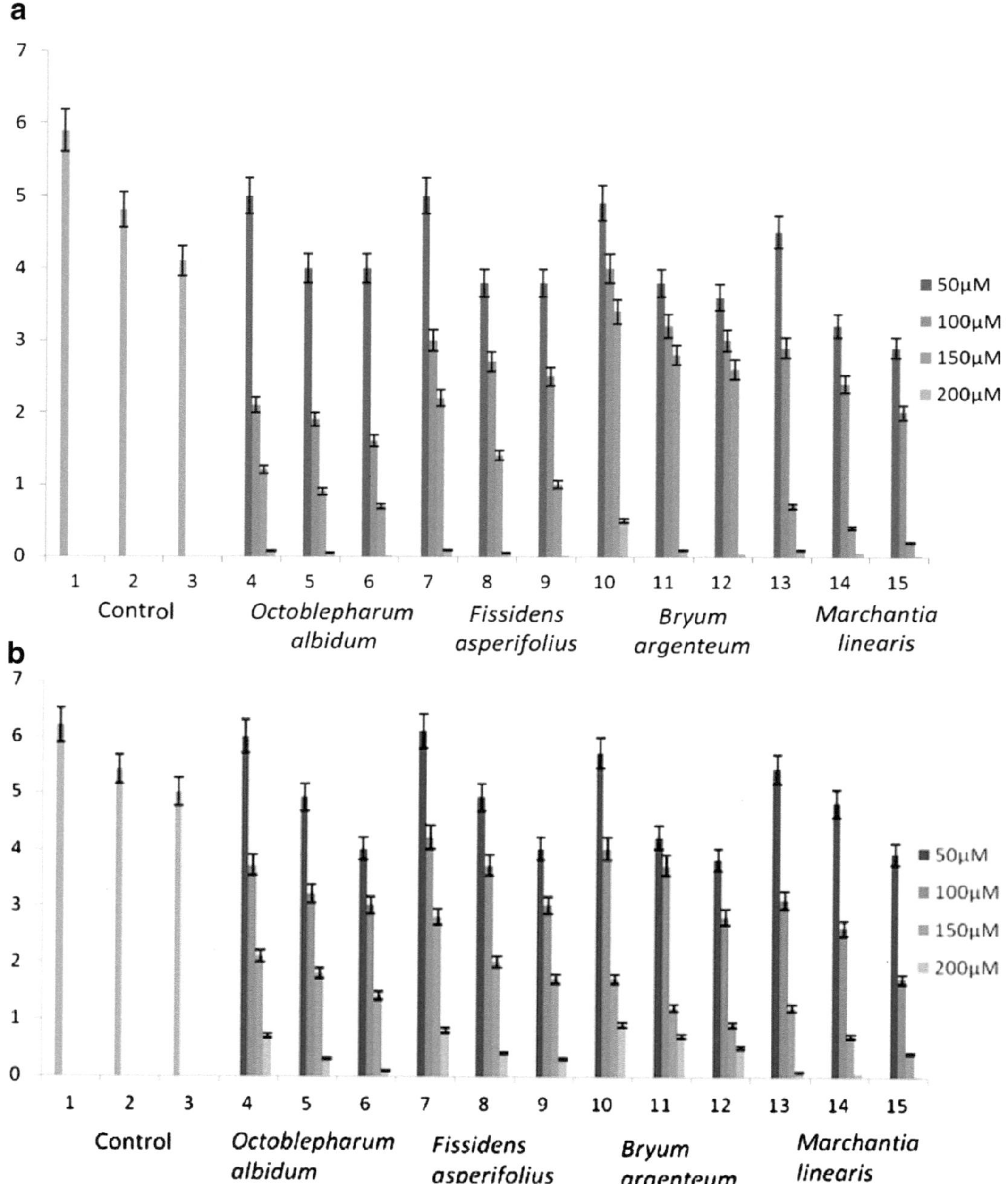

Fig. 3.1 (**a**) IR protein inhibitory activity in digestive tract trypsin extracts from different instars of *Spodoptera litura* larvae. (**b**) IR protein inhibitory activity in digestive tract trypsin extracts from different instars of *Helicoverpa zea* larvae

(0–3 larval development days) there was no change in the difference among means of the various treatments ($P<0.05$), the reduction in larval mass varied significantly ($P<0.001$) between different treatment groups during their remaining growth period (Fig. 3.2a, b). The reduction in larval mass was particularly severe during 3–6 larval development days (F value >10), among all the dosage treatments. Notably, IR protein was comparable to SKTI ($P<0.01$) in lowering of the larval body

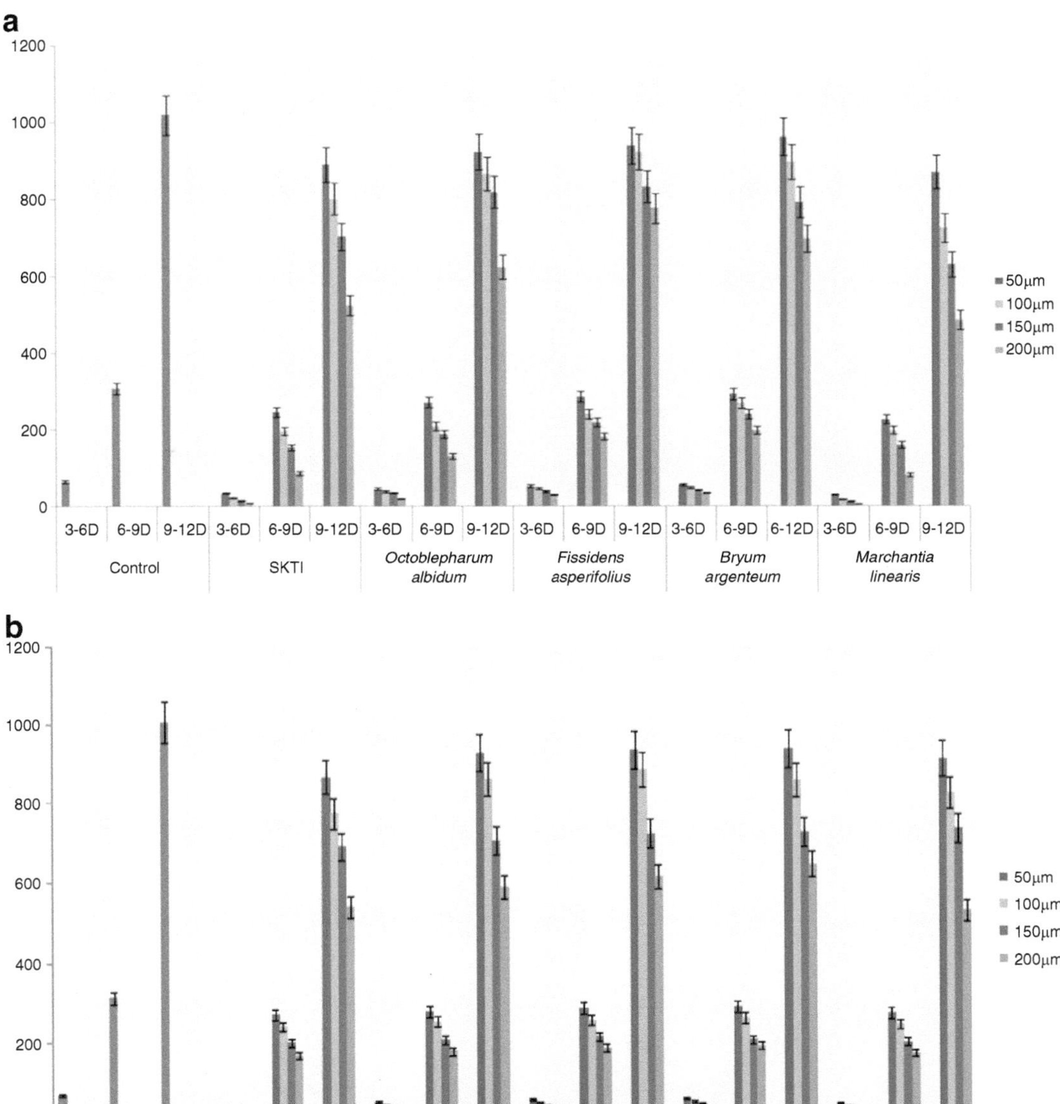

Fig. 3.2 (**a**) Dose-dependent effect of IR proteins on body mass of *Spodoptera litura* larvae. (**b**) Dose-dependent effect of IR proteins on body mass of *Helicoverpa zea* larvae

mass among the early instars. In fact, the decrease in body mass at the later instar stages (4th and 5th) was stabilized for SKTI, unlike IR proteins, whose deleterious effects marginally persisted among the ultimate instars (Fig. 3.2a, b). Similarly, on consumption of high doses of IR protein, there was a marked decrease of larval body mass. Parallel to the larval body mass, retardation in the growth rate with increasing time was also apparent among the control and IR protein-fed larvae (Table 3.2). Notably, growth rates for the control and IR protein-exposed larvae were much conspicuous through the 3–6 days of larval development ($P<0.001$). During this period, growth rate of larvae on experimental diet (200 μM) fell over significantly as compared to larva growing under control conditions (Fig. 3.3a, b). The differences in the mean larval growth rates among the different

Table 3.2 Growth rate of *Spodoptera litura* and *Helicoverpa zea*

Concentration	Control			SKTI			*Octoblepharum albidum*			*Fissidens asperifolius*			*Bryum argenteum*			*Marchantia linearis*		
Spodoptera litura																		
	3–6D	6–9D	9–12D	3–6D	6–9D	9–12D	3–6D	6–9D	9–12D	3–6D	6–9D	9–12D	3–6D	6–9D	9–12D	3–6D	6–9D	9–12D
	0.17± 0.11	0.42± 0.14	0.36± 0.16															
50 μM				0.17± 0.22	0.34± 0.11	0.31± 0.09	0.17± 0.07	0.37± 0.07	0.32± 0.04	0.17± 0.01	0.36± 0.02	0.32± 0.02	0.17± 0.02	0.35± 0.03	0.32± 0.04	0.16± 0.06	0.32± 0.07	0.30± 0.07
100 μM				0.16± 0.03	0.31± 0.01	0.29± 0.01	0.16± 0.02	0.35± 0.02	0.31± 0.03	0.16± 0.01	0.33± 0.03	0.31± 0.04	0.16± 0.03	0.34± 0.02	0.31± 0.01	0.15± 0.03	0.30± 0.03	0.29± 0.03
150 μM				0.15± 0.02	0.29± 0.03	0.25± 0.04	0.11± 0.05	0.26± 0.06	0.30± 0.01	0.14± 0.02	0.23± 0.01	0.30± 0.02	0.15± 0.03	0.25± 0.03	0.31± 0.01	0.13± 0.02	0.22± 0.03	0.28± 0.02
200 μM				0.13± 0.02	0.21± 0.01	0.21± 0.01	0.11± 0.01	0.24± 0.03	0.20± 0.03	0.13± 0.01	0.22± 0.02	0.20± 0.02	0.14± 0.11	0.25± 0.20	0.27± 0.03	0.11± 0.03	0.20± 0.03	0.19± 0.02
Helicoverpa zea																		
	3–6D	6–9D	9–12D	3–6D	6–9D	9–12D	3–6D	6–9D	9–12D	3–6D	6–9D	9–12D	3–6D	6–9D	9–12D	3–6D	6–9D	9–12D
	0.19± 0.01	0.46± 0.06	0.35± 0.08															
50 μM				0.18± 0.02	0.44± 0.01	0.34± 0.01	0.19± 0.02	0.44± 0.03	0.35± 0.06	0.19± 0.04	0.44± 0.06	0.36± 0.07	0.19± 0.06	0.44± 0.08	0.36	0.17± 0.08	0.42± 0.12	0.28± 0.11
100 μM				0.17± 0.04	0.41± 0.01	0.34± 0.04	0.18± 0.04	0.42± 0.06	0.37± 0.09	0.18± 0.06	0.43± 0.09	0.35± 0.04	0.18± 0.08	0.42± 0.08	0.36± 0.07	0.16± 0.06	0.39± 0.08	0.27± 0.07
150 μM				0.17± 0.05	0.40± 0.09	0.32± 0.08	0.16± 0.07	0.41± 0.09	0.40± 0.07	0.16± 0.09	0.40± 0.11	0.34± 0.12	0.16± 0.09	0.41± 0.11	0.35± 0.14	0.14± 0.09	0.39± 0.06	0.26± 0.07
200 μM				0.15± 0.05	0.39± 0.07	0.31± 0.08	0.15± 0.08	0.40± 0.11	0.39± 0.06	0.16± 0.08	0.39± 0.12	0.34± 0.17	0.16± 0.14	0.40± 0.12	0.33± 0.07	0.12± 0.01	0.37± 0.02	0.25± 0.05

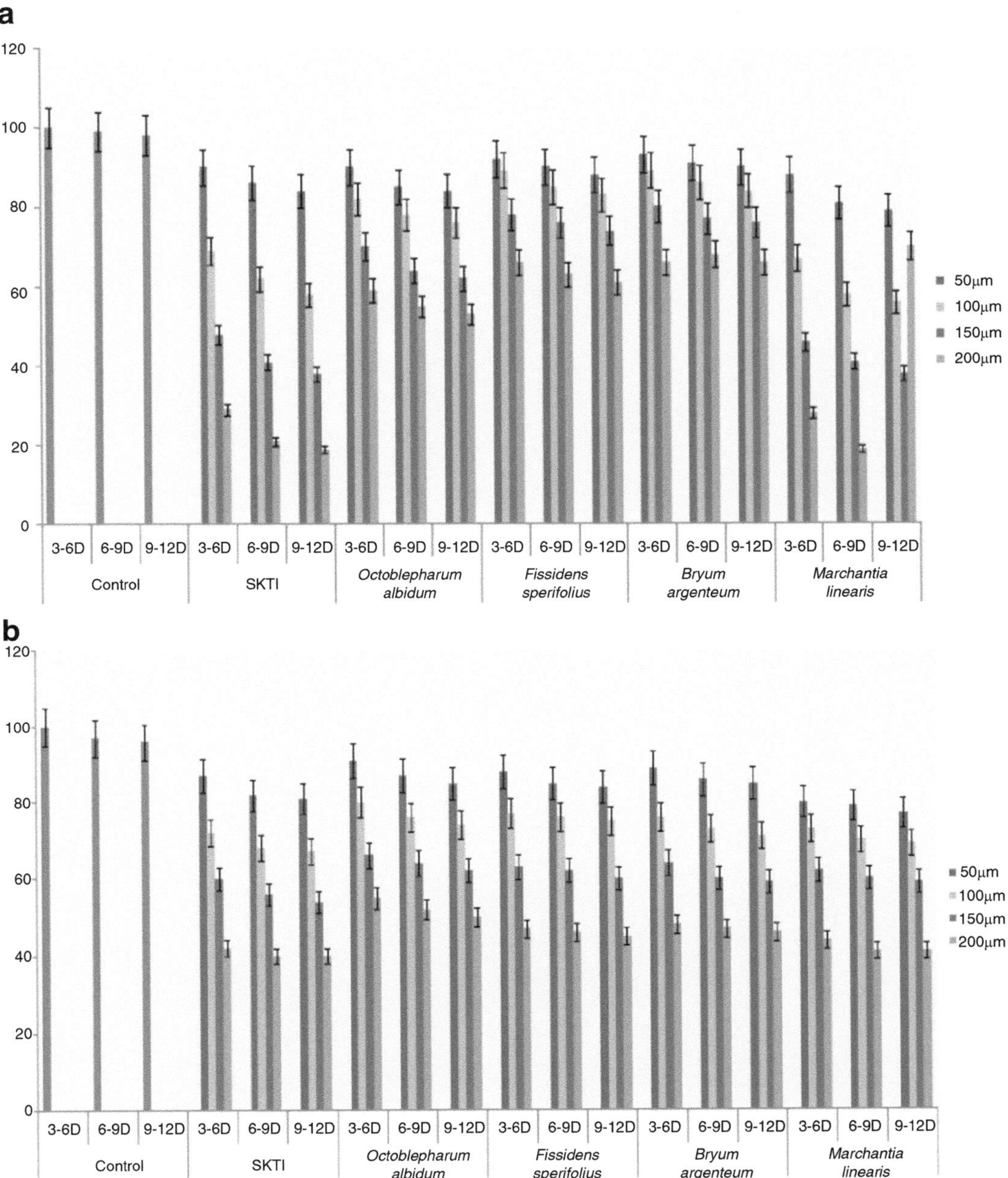

Fig. 3.3 (**a**) Dose-dependent effect of IR proteins on the growth rate of *Spodoptera litura* larvae. (**b**) Dose-dependent effect of IR proteins on the growth rate of *Helicoverpa zea* larvae

treatments thereafter were not significant in any of the following 3-day intervals. Predictably, effect of IR protein on larval growth rate pattern was dose dependent; least exposed recovered the fastest, while those exposed maximum showed a much-delayed response of overcoming the adverse effects. So it seems that impediment of the growth rate rather than any other detrimental effects of IR protein influences the larval body mass.

A trend analogous to the growth rate patterns was also noted on the mortality rate of the larvae of *S. litura* and *H. zea* exposed to IR protein. At minimum dosage (50 μM), IR protein had no adverse effect on larval survival rates (Fig. 3.4a, b).

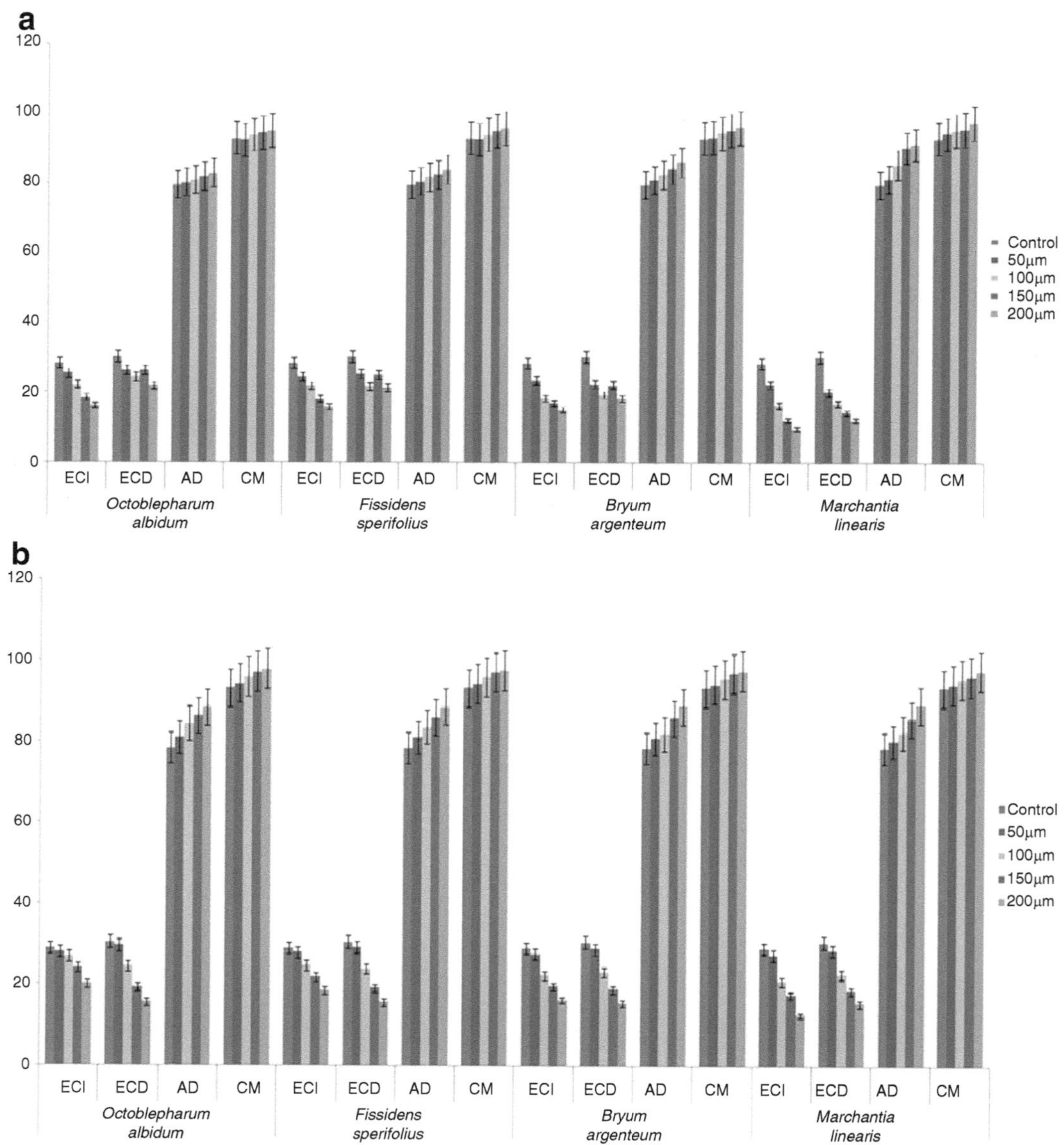

Fig. 3.4 (**a**) Dose-dependent effect of IR proteins on the mortality rate of *Spodoptera litura* larvae. (**b**) Dose-dependent effect of IR proteins on the mortality rate of *Helicoverpa zea* larvae

However, moderate to high doses of IR protein induced high to severe mortality rates. In fact, larva fed on 250-μM IR protein exhibited high mortality by the final instar stage. Additionally, IR protein-containing diet caused moderate mortality among the larval instars compared with a diet supplemented with SKTI (Fig. 3.4a, b).

Our results on the larval growth and development closely follow the studies performed by McManus and Burgess and Gatehouse et al. [20, 21] for SKTI against larval *S. litura* and *Lacanobia oleracea*, respectively. Overall, it is worth mentioning that the deleterious effects of IR protein matched that of SKTI significantly ($P < 0.01$) as exhibited by the differences in mean weight, growth rates, and survival patterns among the larval instars fed on diet with both the SKTI + IR protein, respectively. This can be explained by the fact that lower plants, like bryophytes, provide new proteins; predators are yet to adapt to the inhibitory mechanisms. Though few larvae survived, they did not undergo normal development and transformed either into defective larval–pupal intermediates which either failed to molt or resulted in smaller pupae, which normally develop into short-lived sterile adults.

Nutritional Parameters

Nutritional analyses revealed that IR protein of bryophytes presented a toxic effect when ingested by larvae. IR proteins, when incorporated in larval leaf coating assays at 150-μM concentration, reduced ECI and ECD and increased AD and metabolic cost (CM) for *S. litura* and *H. zea* larvae when compared with the control. The consumption of the basal diet with the incorporation of 150 μM of IR protein by *S. litura* and *H. zea* larvae was not significantly different compared to the consumption of the control diet by larvae. Fecal production reduced proportionally with concentrations of IR protein for *S. litura* and *H. zea* larvae than that of the control group. These results suggest that IR protein acts on the insect's intestinal tract or interferes with digestion. Dietary utilization experiments show that IR proteins, incorporated into basal diet, did not alter the consumption rate, but decreased the fecal production of *S. litura* larvae and *H. zea*. An index of dietary utilization showed that ECI and ECD decreased when 150-μM IR protein diet was employed. In the present study, the AD value for larvae of *S. litura* and *H. zea* increased throughout the feeding period of the experiment; this finding suggests that, during this treatment, the food remained for a greater time in the insect's gut to allow the detoxification of the protein. This increased the exposure of the food to digestive enzymes, probably allowing an increased digestibility of the food in comparison to that seen in the control.

Similar trend was also observed by Morbue and Blackwell [22]. ECI is an overall measurement of an insect's ability to utilize the food that it ingests for growth. A drop in ECI indicates that more food is being metabolized for energy and less is being converted to body mass, i.e., growth of insect [23]. ECD also decreases as the proportion of digested food metabolized for energy increases [24]. Decreased ECI and ECD values indicate that ingested IR protein also exhibits some chronic toxicity. Confirming these results, Table 3.3 demonstrates an increase in CM (metabolic cost) in *S. litura* and *H. zea* larvae, as also described by Morbue and Blackwell [22]. We suggest that the reduction in ECD is likely to result from a reduction in the efficiency to convert food stuffs into growth, perhaps by a diversion of energy from production of biomass into detoxification of IR proteins, i.e., an increase in costs.

The use of IR proteins to protect plants against insect pests is, however, complicated by the ability of insects to circumvent plant defenses. The data corroborates with insects, such as flour moth *Anagasta kuehniella*, whereas *Spodoptera frugiperda* adapts to soybean protease inhibitor by changing the expression of trypsin and chymotrypsin activities [25]. Others selectively induce inhibitor-insensitive proteases, as observed in Colorado potato beetles, *Spodoptera exigua*, and other insect species [26]. Digestive proteolysis mediated by direct proteolytic fragmentation is another strategy used by insects to minimize the adverse effects of plant defense proteins on food digestion and nutrient [27]. Recently, Zhu-Salzman et al. [27] suggested that the cowpea bruchid adapts to the challenge of soybean cysteine protease inhibitor, soyacystatin N (scN), by qualitatively and quantitatively remodeling its digestive protease complement and by activating scN-degrading protease activity. Recent studies suggest that insects can overcome the deleterious effects of protease inhibitors by synthesizing different proteases that are insensitive to particular inhibitors [28].

In conclusion, the action of IR protein on the development of *S. litura* and *H. zea* larvae shows that IR protein influences larval survival, weight, efficiency of conversion food, and approximate digestibility of *S. litura* and *H. zea* larvae, indicating that this protein may have great toxic potential, particularly since IR protein significantly reduced larval weight and increased mortality. Additional studies of IR protein in other insects are required to confirm the biotechnological

Table 3.3 *Nutritional indices of Spodoptera litura and Helicoverpa zea 5th instar larvae on 50, 100 and 150 μM IR Protein treated and control diet*

Nutritional indices (Mean ± S.E.)																
Spodoptera litura																
Treatment (μM)	*Octoblepharum albidum*				*Fissidens asperifolius*				*Bryum argenteum*				*Marchantia linearis*			
	ECI	ECD	AD	CM	ECI	ECD	AD	CM	ECI	ECD	AD	CM	ECI	ECD	AD	CM
Control	36.5 ±0.15	29.2 ±0.14	79.4 ±0.16	94.6 ±0.10	36.5 ±0.15	29.2 ±0.14	79.4 ±0.16	94.6 ±0.10	36.5 ±0.15	29.2 ±0.14	79.4 ±0.16	94.6 ±0.10	36.5 ±0.15	29.2 ±0.14	79.4 ±0.16	94.6 ±0.10
50	29.1 ±0.11	28.4 ±0.20	88.5 ±0.03	95.1 ±0.06	27.3 ±0.03	26.5 ±0.08	87.3 ±0.01	95.1 ±0.06	26.5 ±0.01	21.3 ±0.03	88.7 ±0.04	96.5 ±0.02	28.6 ±0.06	23.2 ±0.01	89.6 ±0.03	96.1 ±0.03
100	25.3 ±0.08	20.5 ±0.09	92.1 ±0.04	96.3 ±0.08	24.5 ±0.01	20.1 ±0.05	90.5 ±0.09	96.1 ±0.07	19.7 ±0.05	17.8 ±0.04	92.5 ±0.02	96.9 ±0.07	17.5 ±0.04	19.4 ±0.03	93.2 ±0.02	96.4 ±0.07
150	19.7 ±0.05	13.5 ±0.03	94.2 ±0.07	97.1 ±0.09	18.8 ±0.05	13.8 ±0.15	93.1 ±0.04	95.2 ±0.01	15.1 ±0.10	15.9 ±0.10	95.1 ±0.04	97.2 ±0.03	13.6 ±0.10	14.1 ±0.05	96.1 ±0.07	98.1 ±0.09
Helicoverpa zea																
Treatment (μM)	*Octoblepharum albidum*				*Fissidens asperifolius*				*Bryum argenteum*				*Marchantia linearis*			
	ECI	ECD	AD	CM	ECI	ECD	AD	CM	ECI	ECD	AD	CM	ECI	ECD	AD	CM
Control	34.4 ±0.11	28.2 ±0.10	82.4 ±0.13	93.5 ±0.09	34.4 ±0.11	28.2 ±0.10	82.4 ±0.13	93.5 ±0.09	34.4 ±0.11	28.2 ±0.10	82.4 ±0.13	93.5 ±0.09	34.4 ±0.11	28.2 ±0.10	82.4 ±0.13	93.5 ±0.09
50	28.7 ±0.09	26.7 ±0.15	83.5 ±0.01	94.1 ±0.02	30.4 ±0.05	27.1 ±0.03	88.1 ±0.03	94.1 ±0.03	30.8 ±0.03	26.2 ±0.08	84.5 ±0.07	94.1 ±0.04	30.9 ±0.06	25.5 ±0.08	86.6 ±0.07	95.4 ±0.09
100	24.5 ±0.08	25.1 ±0.03	85.1 ±0.09	95.2 ±0.06	26.5 ±0.08	26.5 ±0.09	89.2 ±0.02	96.2 ±0.07	26.5 ±0.01	24.5 ±0.05	90.1 ±0.02	96.5 ±0.08	25.8 ±0.02	23.7 ±0.06	90.7 ±0.02	97.5 ±0.03
150	19.2 ±0.03	20.5 ±0.05	90.3 ±0.04	96.7 ±0.09	19.2 ±0.02	19.7 ±0.04	92.1 ±0.07	97.5 ±0.09	18.1 ±0.05	19.5 ±0.15	94.3 ±0.01	97.1 ±0.01	17.4 ±0.11	18.6 ±0.10	95.6 ±0.09	98.6 ±0.07

ECI- Efficiency of conversion of ingested food, ECD- Efficiency of conversion of digested food, AD- Approximate digestibility, CM – Metabolic cost ($P<0.05$)

potential of this inhibitor as an agent against phytophagous insects.

Overall, these initial assays served as an effective method to screen for lower plant IR extracts with insecticidal activity. The protein extracts that exhibited insecticidal potentialities in this experiment may exist in one of two forms: small polypeptide chains (SPC) or gene-encoded IR protein (GEP).

SPCs have been shown in numerous tests to confer insect resistance in angiosperms [9]. SPCs conferring IR are usually the by-products of complex molecular pathways. In this experiment, no remaining protein, theoretically, should be greater than 5 kD (~45 amino acid residues) from the result of the methods used to exclude small metabolites. As a general rule, SPCs exist in the 1–20-kD range. Therefore, we cannot rule out the possibility that the effects seen on both *S. litura* and *H. zea* larvae were not caused by the presence of SPCs. It is likely, therefore, that observed protein toxicity is GEP based.

If the results with *S. litura* and *H. zea* are indicative, the extracts might have broad-range antibiosis effects. Since there are many bryophyte-specific genes not found in angiosperms, which is to be expected in the 400-million-year lineage split, lower plants seem to be potentially rich sources of IR genes [2].

Acknowledgment We would like to thank Kerala State Council for Science, Technology and Environment for providing financial support for the project.

References

1. Asakawa Y. Recent advances in phytochemistry of bryophytes – acetogenins, terpenoids and bis(bibenzyl) s from selected Japanese, Taiwanese, New Zealand, Argentinean and European liverworts. Phytochemistry. 2001;56:297–312.
2. Nishiyama T, Fugita T, Shin IT, Seki M, Nishide H, Uchiyama I, Kamiya A. Comparative genomics of *Physcomitrella patens* gametophytic transcriptome and *Arabidopsis thaliana:* implications for land plant evolution. Proc Natl Acad Sci USA. 2003;100:8007–12.
3. Cooper-Driver GA. Insect-fern associations. Entomol Exp Appl. 1978;24:110–6.
4. Davidson AJ, Harborne JB, Longton RE. Identification of hydroxycinnamic and phenolic acids in *Mnium hornum* and *Brachythecium rutabulum* and their possible role in protection against herbivory. J Hattori Bot Lab. 1989;67:415–22.
5. Weintraub J. Lithinine moths on ferns: a phylogenetic study of insect-plant interactions. Biol J Linn Soc. 1995;55:239–50.
6. Auerbach MJ, Hendrix SD. Insect-fern interactions: macro-lepidopteran utilization and species area association. Ecol Entomol. 1980;5:99–104.
7. Hendrix SD, Marquis RJ. Herbivore damage to 3 tropical ferns. Biotropica. 1983;15:108–11.
8. Asakawa Y. Biologically active substances from bryophytes. In: Chopra RN, Bhatla SC, editors. Bryophyte development: physiology and biochemistry. Boston: CRC Press; 1990. p. 259–88.
9. Basra A, Basra R. Mechanisms of environmental stress resistance in plants. Annu Rev Biochem. 1997; 59:873–907.
10. Harborne JB. Introduction to ecological biochemistry. San Diego: Academic; 1993.
11. Cooper-Driver GA. Anti-predation strategies in pteridophytes: a biochemical approach. Proc R Soc Edinb B (Biol Sci). 1985;86B:397–402.
12. Lafont R, Horn DHS. Phytoecdysteroids: structure and occurrence. In: Koolman J, editor. Ecdysone: from chemistry to mode of action. Stuttgart: Thieme; 1989. p. 39–64.
13. Fenwick GR. Bracken *Pteridium aquilinum:* toxic effects and toxic constituents. J Sci Food Agric. 1988;46:147–73.
14. Hendrix SD. The resistance of *Pteridium aquilinum* to insect attack by *Trichoplusia ni* (Hubn.). Oecologia. 1977;26:347–61.
15. Waldbauer GP. The consumption and utilization of food by insects. Adv Insect Physiol. 1968;5:229–88.
16. Farrar RR, Barbour JD, Kennedy KG. Quantifying food consumption and growth in insects. Ann Entomol Soc Am. 1989;82:593–8.
17. Terra WR, Ferreira C. Insect digestive enzymes: properties, compartmentalization and function. Comp Biochem Physiol B. 1994;109:1–62.
18. Broadway RM. Dietary proteinase inhibitors alter the complement of midgut proteinases. Arch Insect Biochem. 1996;41:107–16.
19. De Leo F, Bonade'-Bottino M, Ceci R, Gallerani R, Joaunin L. Effects of a mustard trypsin inhibitor expressed in different plants on three lepidopteran pests. Insect Biochem Mol Biol. 2001;31:593–602.
20. McManus MT, Burgess EPJ. Effects of soybean (Kunitz) trypsin inhibitor on growth and digestive proteases of larvae *Spodoptera litura.* J Insect Physiol. 1995;41:731–8.
21. Gatehouse A, Norton MR, Davison E, Babbe GM, Newell SM, Gatehouse JA. Digestive proteolytic activity in larvae of tomato moth, *Lacanobia oleracea,* effects of plant proteinase inhibitor *in vitro* & *in vivo.* J Insect Physiol. 1999;45:545–58.

22. Morbue A, Blackwell AJ. Azadirachtin: an update. J Insect Physiol. 1993;39:903–24.
23. Koul O, Multani JS, Singh G, Daniewski WM, Berlozecki S. 6-β- Hydroxygedunin from *Azadirachta indica*. Its potentiating effects with some non-azadirachtin limonoids in neem against lepidopteran larvae. J Agric Food Chem. 2003;51:2937–42.
24. Wheeler DA, Isman MB. Antifeedant and toxic activity of *Trichilia americana* extract against the larvae of *Spodoptera litura*. Entomol Exp et Appl. 2001;98:9–16.
25. Paulillo Cesar LMS, Adriana LR, Plinio TC, Jose PRP, Walter TR, Marcio SFC. Changes in midgut endopeptidase activity of *Spodoptera frugiperda* (Lepidoptera: Noctuidae) are responsible for adaptation to soybean proteinase inhibitors. J Econ Entomol. 2000;93:892–6.
26. Cloutier Jean C, Fournier M, Yelle S, Michaud D. Adult Colorado potato beetles, *Leptinotarsa decemlineata* compensate for nutritional stress on Oryzacystatin 1-transgenic potato plants by hypertrophic behavior and over production of incentive proteases. Arch Insect Biochem Physiol. 2000; 44:69–81.
27. Zhu-Salzman K, Koiwa H, Salzman RA, Shade RE, Ahn JE. Cowpea bruchid *Callosobruchus maculatus* uses a three-component strategy to overcome a plant defensive cysteine protease inhibitor. Insect Mol Biol. 2003;12:135–45.
28. Volpicella M, Ceci LR, Cordewener J, America T, Gallerani R, Bode W, Jongsma MA, Beekwilder. Properties of purified gut trypsin from *Helicoverpa zea*, adapted to proteinase inhibitors. Eur J Biochem. 2003;270:10–9.

4 A Novel Algorithm for Prediction of Hub Proteins from Primary Structure in Eukaryotic Proteome Using Dipeptide Compositional Skew Information and Amino Acid Sequence Likeness

B.L. Aswathi, Baharak Goli, Renganayaki Govindarajan, and Achuthsankar S. Nair

Abstract

We propose a novel hub-finding algorithm which relies on the use of dipeptide composition and amino acid sequence likeness. For extracting the most prominent features in hub identification, two feature selection techniques are widely used in data preprocessing for machine learning problems: fast correlation-based feature selection (FCBFS) and correlation-based feature selection (CFS) algorithms. The performance of two types of classifiers such as random forest classifier (RFC) and RBF network was evaluated with these filter approaches. Our proposed model led to successful prediction of hub proteins from primary structure with 92.52 and 91.28% accuracy for RFC and RBF network, respectively, in case of FCBFS and 90.92 and 93.76% accuracy for RFC and RBF network, respectively, in case of CFS algorithm.

Keywords

Protein-protein interaction networks • Hub proteins • Feature vectors • Machine learning • Protein hubness

Introduction

Research in molecular biology is budging essentially toward the study of complex, multicomponent systems, or networks, which underlie the living cell. These bionetworks are described and modeled in terms of their components, interactions, regulatory features, pathways, and system dynamics. For better realization of cellular level processes, the study of the complex networks they define is vital. Majority of cellular mechanisms are mediated by physical interactions among proteins, including signal transduction, enzyme activity, and post-translational modification [1]. The elucidation of large networks of protein-protein interactions has contributed to the identification of biochemical and signaling pathways and to functional annotation of genes [1]. A group of proteins and their

B.L. Aswathi (✉) • B. Goli • R. Govindarajan • A.S. Nair
Department of Computational Biology and Bioinformatics, University of Kerala, Thiruvananthapuram 695581, India
e-mail: aswathi.bl@gmail.com

A. Sabu and A. Augustine (eds.), *Prospects in Bioscience: Addressing the Issues*,
DOI 10.1007/978-81-322-0810-5_4, © Springer India 2013

interactions form a protein-protein interaction network (PPIN). In a PPIN, a node denotes a protein and a connecting edge represents a protein-protein interaction. Usually, proteins build complex networks that have been shown to have a scale-free topology [2]. Based on this topological property, each protein node is characterized by a term connectivity, which represents the total number of interactions a protein has. Generally, most of the nodes in a protein-protein interaction network are sparsely connected while a few nodes are highly interactive, like any other scale-free network, which follows power law [2]. The interacting proteins with high degree of connectivity are termed as "hubs." It has been observed that most of the network remains fully connected even though the sparingly connected nodes are removed. Instead, the removal of the most central hub nodes may cause the collapse of the system [2]. In fact, it has been established that connectivity and essentiality in the yeast protein interaction network are positively correlated [2]. Thus, hub proteins which literally "hold the protein interaction networks together" are more likely to be essential than sparsely connected proteins [3].

This work focuses on identification of highly interactive hub proteins which are the central actors in protein-protein interaction networks. Hub protein analysis assumes vital importance, since the possibility of their involvement in multiple pathways is higher due to their multiple binding sites [4]. Characterization of hub proteins may help in better realization of cellular functions as well as discovering novel drug targets and predicting the side effects in drug discovery by understanding the pathways, topologies, and dynamics of them. Many of the well-known and widely examined proteins associated with many diseases are hubs. Information about these interactions helps in understanding various diseases and also helps to develop new therapeutic approaches [4–6].

Various attempts have been reported in the identification of highly connected hubs in protein-protein interaction networks using various data including gene ontology [7], gene proximity [8, 9], gene fusion events [10, 11], and gene co-expression data [12, 13]. Lack of availability of the above-mentioned data for the entire protein interaction data of any organism is one of the challenging factors for using them. Structural information is also severely limited as PDB structures are not available for many of the proteins [5]. Hence, in this study, we took only amino acid sequences and sequence-based features for the prediction of hub proteins.

We have developed a statistical approach to classify hub proteins from amino acid sequence information using soft computational algorithms such as random forest which is one of the widely used ensemble classifiers and RBF networks, most commonly using supervised neural networks in pattern recognition.

Materials and Methods

We selected *H. sapiens* as the model organism for this study, which is well annotated and has modest protein interaction information. The features that we used for hub identification include dipeptide composition and sequence similarity. We examined all the derived features for their feasibility in predicting hubs using fast correlation-based feature selection (FCBFS) and correlation-based feature selection (CFS) algorithms, two powerful techniques in purging uncorrelated and redundant features. With the pruned dataset as the backbone, we applied two different algorithms for hub classification: random forest classifier (RFC) and RBF network.

Dataset

Protein interaction data of *H. sapiens* was generated from IntAct [14] database released on January 15, 2012, and corresponding amino acid sequences of varying lengths were compiled from UniProt [15]. Using CD-HIT-2D [16] web server, we removed all redundant sequences. Protein sequences having amino acid sequence length less than 60 were also excluded to avoid any bias. The curated dataset included 5,318 proteins and the total number of protein interactions was 37,281 with an average degree of interaction of 9.641. We constructed training and testing sets in such a way that no pair of proteins from either set had significant sequence similarity.

Identifying Hub Ad Hoc Threshold

Defining an ad hoc threshold or degree cutoff has been a common practice in the analysis of protein interaction networks, where all proteins that have a degree higher than this threshold are considered as "hub" nodes. The perception of a hub protein is a riveting one because by definition hub proteins often do have special biological properties and bind to a large number of proteins [17]. According to previous researches, a steady increase in connectivity with protein age was observed, and from this it is clear that the degree of interaction or connectivity of any protein has a direct correlation with the age of that protein [18]. However, because of the "scale-free" nature of protein interaction networks, there is no consensus in the literature on the degree threshold that defines a hub [17]. Also it is perceived that the connectivity of any hub protein is species specific [7]. So far it is not clear whether the biological significance of hubs is relatively insensitive to their precise definition [17]. Examples of varying criteria used to define hubs in previous studies include the following: in Han et al. [19], nodes with degree greater than 5 were labeled as hubs; in Ekman et al. [17], nodes with degree greater than 8 were labeled as hubs; and in Jin et al. [20], a connectivity cutoff of 20 was used to define hub proteins. In Batada et al. [21], the top 95 and 50% of the high-degree nodes were defined as hubs.

In our *H. sapiens* PPI dataset, the degree of connectivity of proteins ranged from 1 to 148. We had to choose a degree threshold for determining a protein as hub or non-hub. For that the fold-change definition was followed by taking the ratio of connectivity and average connectivity to determine appropriate cutoff to identify hubs in *H. sapiens* PPI network [3]. A node with connectivity fold change greater than or equal to 10 was considered as hub (cutoff, P-value <0.001, using distribution of standard normalized fold-change values in *H. sapiens*). Thus, our final dataset included 342 highly connected proteins and 4,976 sparsely connected proteins.

Fig. 4.1 Dipeptide sequential count

Deriving Feature Vectors

For classification, many of the machine learning techniques require property vectors as input. For that each amino acid sequence should be replaced by a set of numeric values representing its properties. Extracting eloquent biological information from the amino acid sequences of varying length is one of the important and most crucial steps in any classification problem. In this work, dipeptide composition and sequence likeness were used as the feature vectors for hub classification.

Dipeptide Composition

Molecules comprising two amino acids joined by a single peptide bond are referred as dipeptides. We can derive 400 different dipeptide combinations from 20 different amino acids. What makes dipeptide composition important is that it can provide information about residue pairing preferences as well as amino acid distributions [22]. Here we computed the dipeptide composition in a sequential way. Consider a sample sequence C C C P A C Q C C A C. Dipeptide composition of the CC pair, DP_{CC}, is calculated as 3 (Fig. 4.1).

If a protein sequence P is given with *N* amino acids, we can map the sequence with the feature vector as FP = {F_1, F_2, F_3,...F_{400}}. Thus, we obtain 400 different binary residue combinations from 20 amino acids.

Sequence Likeness

In this approach, each protein is described by the vector of its similarities with respect to a predetermined set of hub proteins. This benchmark hub set includes 10 well-known hub proteins in *H. sapiens* and is listed in Table 4.1.

Table 4.1 Predefined hub set for similarity check

Sl. No.	Protein name	Function	Sequence length
1	p53	Tumor suppressor	387
2	p21	Cell cycle regulation	154
3	p27	Cell cycle regulation	921
4	BRCA1	Transcriptional regulation	1,354
5	Kalirin	Nerve growth and axonal development	2,985
6	Ubiquitin	Protein recycling and regulation	404
7	Calmodulin	Messenger protein	149
8	ELAV-like protein 2	Cellular differentiation	359
9	Zinc finger protein 282	Specific interactions with DNA	671
10	Leptin	Receptor	1,165

We compared the sequence similarity of whole proteins in our dataset with the predefined protein set using stand-alone BLAST [23]. Since there were 10 hub proteins in the predefined set, we got a feature vector of 10 corresponding similarity percentages for each protein in our dataset.

Feature Screening

Most commonly the performance of any classifier depends on the reliability of the features chosen and the size of the training set [24]. Faster classification models and smallest subset of important and prominent features should be retained, in order to attain maximal classification performance. Feature selection is one of the significant techniques in data preprocessing for machine learning and data mining problems, which trashes out irrelevant and redundant features and speeds up the data mining algorithm and improves classification accuracy [25–27]. Our feature vector contains 410 values for each protein in the dataset. In order to extract the most consistent features, we adopted fast correlation-based feature selection (FCBFS) [28] and correlation-based feature selection (CFS) algorithms [29], which are two well-known feature selection techniques. In this study, 410 features were generated from the transformation step explained above, 400 from dipeptide composition and 10 features from sequence likeness. After feature pruning, a total of 156 features remained. Of these, 149 features represent dipeptide composition. Seven features from sequence likeness were also yielded well. We briefly describe these feature selection algorithms below.

Fast Correlation-Based Feature Selection (FCBFS)

FCBFS is the most commonly used feature selection algorithm for extraction of high-dimensional data. It can efficiently achieve high degree of dimensionality reduction and enhance classification accuracy with predominant features [30]. If there are two features that are found to be redundant to each other and one of them needs to be removed, this heuristic algorithm will remove the one that is less relevant to the class concept. This approach demonstrates its efficiency and effectiveness in dealing with high-dimensional data for classification [31].

Correlation-Based Feature Selection (CFS)

This is one of the powerful techniques in removing uncorrelated and redundant features. Using a best first-search heuristic approach, it estimates the prominence of subsets of features [32]. This heuristic algorithm contemplates the importance of individual features for predicting the class as well as the level of correlation among them. The basic logic in CFS is that good feature subsets include those features that are highly correlated with the target class and uncorrelated with each other.

Machine Learning Algorithms for Classification

Keeping the extracted feature vectors as the backbone, we have modeled two classifiers, RBF network [32] and an ensemble classifier random forest [33]. Artificial neural network is one of the supervised learning algorithms used commonly to solve classification problems. Recently there has been a lot of interest in ensemble learning methods, which incorporate many classifiers and aggregate their results [34]. The efficiency of each classifier was compared in terms of their classification accuracy, and the time taken by each classifier to build the model was also noted. For the implementation we used Weka suite, a machine learning workbench developed in Java programming language [35].

Construction of Neural Network Classifier

RBF networks are supervised neural networks which are popular substitute to multilayer perceptions which employ reasonably lesser number of locally tuned units and are adaptive in nature [32]. They are widely used for classification and pattern recognition problems. In this study, the training set consisting of 200 hub and 1,500 non-hub elements was given to each network in the tenfold cross validation scheme.

Ensemble Classifier

Two well-known methods are boosting and bagging. Breiman [33] proposed random forests, which add an additional layer of randomness to bagging. In addition to constructing each tree using a different bootstrap sample of the data, random forests change how the classification or regression trees are constructed. In standard trees, each node is split using the best split among all variables. In a random forest, each node is split using the best among a subset of predictors randomly chosen at that node. This somewhat counterintuitive strategy turns out to perform very well compared to many other classifiers, including discriminant analysis, support vector machines, and neural networks, and is robust against overfitting [34].

Results and Discussion

Our results show that there is a strong correlation between the extracted amino acid features and the protein hubness. With feature selection algorithms, FCBFS and the CFS algorithms followed by two classification algorithms, random forest and RBF networks, we could effectively classify hub proteins using the pruned dipeptide composition and sequence likeness features.

Feature-Vector Inference

Dipeptide composition exhibited a remarkable skew in the binary combination of similar peptides like AA and CC in hub proteins, whereas in non-hubs the dipeptide combination of similar and dissimilar residues appeared randomly without showing any specific prominence. Same-residue repeats were rich in hubs compared to non-hubs. By using FCBFS and the CFS algorithms, we could obtain 74 dipeptide compositions from hub train dataset. Of these 53 features were same-residue repeats. Similarly, we obtained 96 dipeptide compositions form non-hub train dataset where 21 features were overlapped with the hub features. When we combined the screened dipeptide compositions of both hubs and non-hubs, a total of 149 non-redundant features were attained, of which 97 were same-residue repeats: 68 from hub dataset and 29 from non-hub dataset.

Likewise, sequence likeness between the predefined hub dataset and the hub as well as the non-hub train dataset showed significant differences in similarity percentage. For large proportion of hub dataset, we obtained similarity percentages greater than 60 with nine predefined hub proteins, P53, P21, P27, kalirin, calmodulin, zinc finger

Table 4.2 Jackknife performance of different hub prediction algorithms

Classification method	Sensitivity (%)	Specificity (%)	Accuracy (%)	Time taken to build the model (s)
RBF network + CFS	92.10	94.18	93.14	1.87
RFC + CFS	89.56	86.36	87.96	2.69
RBF network + FCBFS	91.24	90.04	90.64	1.65
RFC + FCBFS	90.16	93.16	91.66	3.05

protein 282, BRCA1, ubiquitin, and leptin. Using FCBFS and the CFS algorithms, we checked the feasibility of the above-mentioned similarity scores and finally the lopped similarity features excluded BRCA1 and ubiquitin. A large proportion of non-hub data exhibited similarity percentages less than 40% to the entire predefined set.

Table 4.3 Accuracy of each classifier for self-consistency and independent data test

Classification method	Self-consistency (%)	Independent test (%)
RBF network + CFS	94.21	91.28
RFC + CFS	92.73	92.52
RBF network + FCBFS	95.32	93.76
RFC + FCBFS	96.81	90.92

Performance Evaluation

In order to estimate the performance of the classification models, a standard tenfold cross validation was applied where the whole dataset is randomly partitioned into ten evenly sized subsets. Performance is measured for each test set, and the mean is reported as overall accuracy. Several measures were used to evaluate the performance of the classifiers. These measures include specificity = TN/(TN + FP)*100, sensitivity = TP/(TP + FN)*100, accuracy = TP + TN/(TP + TN + FP + TN), and time taken to build the models in seconds, where TP = true positive, TN = true negative, FP = false positive, and FN = false negative. Table 4.2 summarizes the performance of different classifiers and the jackknife test performance of different classifiers. In jackknife test, each protein in the training set is pulled out to make classification using the rest of the training set. Jackknife is considered as more objective and exhaustive than other tests.

Self-consistency test and independent data test were also performed to evaluate the classification models. The results are shown in Table 4.3. Self-consistency test checks the consistency of the developed model. If a classification model produces good performance in self-consistency test, then that classifier can be considered as a good one. In self-consistency test, observations of training datasets are predicted with decision rules acquired from the same dataset. Since the prediction system parameters obtained by the self-consistency test are from the training dataset itself, the success rate is high. The accuracy of self-consistency determines the fitting ability of the rules obtained from the features of training sets.

In this study, a novel hub prediction algorithm which relies only on the use of two features extracted from amino acid sequences, dipeptide composition and sequence likeness, is proposed. FCBFS and the CFS algorithms were used for feature trimming followed by two classification algorithms, random forest and RBF networks for hub prediction. We could effectively classify hub proteins using the extracted dipeptide composition and sequence likeness features.

It is not a novel approach to analyze structural and functional phenomena from sequence information solely. This approach has been widely used with the advent of bioinformatics approaches in genomics and proteomics studies. There have been many computational biology tools which apply this approach to various problems including protein subcellular localization [36] and protein allostery prediction [37]. The biological importance of the chosen amino acid properties in this work is yet to be explained. Investigating the significance of these properties in the formation of PPINs would be remarkable.

References

1. Vallabhajosyula RR, Chakravarti D, Lutfeali S, Ray A, Raval A. Identifying hubs in protein interaction networks. PLoS One. 2009;4(4):e5344.
2. Albert R, Jeong H, Barabási AL. Error and attack tolerance of complex networks. Nature. 2000;406: 378–82.
3. Tun K, Rao RK, Samavedham L, Tanaka H, Dhar PK. Rich can get poor: conversion of hub to non-hub proteins. Syst Synth Biol. 2009;2:75–82.
4. Patil A, Kinoshita K, Nakamura H. Hub promiscuity in protein-protein interaction networks. Int J Mol Sci. 2006;11:1930–43.
5. Aswathi BL, Nair AN, Atmaja S, Pawan KD. Identification of hub proteins from sequence. Bioinformation. 2011;7(4):163–8.
6. He X, Zhang J. Why do hubs tend to be essential in protein networks? PLoS Genet. 2006;2:e88.
7. Hsing M, Byler KG, Cherkasov A. The use of Gene Ontology terms for predicting highly-connected "hub" nodes in protein-protein interaction networks. BMC Syst Biol. 2006;2:80.
8. Dandekar T, Snel B, Huynen M, Bork P. Conservation of gene order: a fingerprint of proteins that physically interact. Trends Biochem Sci. 1998;23:324–8.
9. Overbeek R, Fonstein M, D'Souza M, Pusch GD, Maltsev N. The use of gene clusters to infer functional coupling. Proc Natl Acad Sci USA. 1999;96:2896–901.
10. Marcotte EM, Pellegrini M, Ng HL, Rice DW, Yeates TO, Eisenberg D. Detecting protein function and protein-protein interactions from genome sequences. Science. 1999;285:751–3.
11. Enright J, Iliopoulos I, Kyrpides NC, Ouzounis A. Protein interaction maps for complete genomes based on gene fusion events. Nature. 1999;402:86–90.
12. Ge H, Liu Z, Church GM, Vidal M. Correlation between transcriptome and interactome mapping data from Saccharomyces cerevisiae. Nat Genet. 2001;29: 482–6.
13. Pellegrini M, Marcotte EM, Thompson MJ, Eisenberg D, Yeates TO. Assigning protein functions by comparative genome analysis: protein phylogenetic profiles. Proc Natl Acad Sci USA. 1999;96:4285–8.
14. Kerrien S, Alam-Faruque Y, Aranda B, Bancarz I, Bridge A, Derow C, et al. IntAct –open source resource for molecular interaction data. Nucleic Acids Res. 2007;35:D561–5. http://www.ebi.ac.uk/intact/main.xhtml
15. Apweiler R, Bairoch A, Wu CH, Barker WC, Boeckmann B, Ferro S, et al. UniProt: the Universal Protein knowledgebase. Nucleic Acids Res. 2004;32: D115–9. http://www.uniprot.org
16. Weizhong Li, Adam Godzik. Cd-hit: a fast program for clustering and comparing large sets of protein or nucleotide sequences. Bioinformatics. 2006;22(13): 1658–9.
17. Ekman D, Light S, Björklund ÅK, Elofsson A. What properties characterize the hub proteins of the protein-protein interaction network of *Saccharomyces cerevisiae*? Genome Biol. 2006;7:R45.
18. Prachumwat A, Wen-Hsiung Li. Protein function, connectivity, and duplicability in yeast. Mol Biol Evol. 2006;23(1):30–9.
19. Han JD, Bertin N, Hao T, Goldberg DS, Berriz GF, et al. Evidence for dynamically organized modularity in the yeast protein-protein interaction network. Nature. 2004;430:88–93.
20. Jin G, Zhang S, Zhang XS, Chen L. Hubs with network motifs organize modularity dynamically in the protein-protein interaction network of yeast. PLoS One. 2007;2:e1207.
21. Batada NN, Reguly T, Breitkreutz A, Boucher L, Breitkreutz BJ, et al. Stratus not altocumulus: a new view of the yeast protein interaction network. PLoS Biol. 2006;4:1720–31.
22. Garg A, Gupta D. VirulentPred: a SVM based prediction method for virulent proteins in bacterial pathogens. BMC Bioinform. 2008;9:62.
23. Altschul SF, Madden TL, Schaffer AA, Zhang JH, Zhang Z, Miller W, Lipman DJ. Gapped BLAST and PSI-BLAST: a new generation of protein database search programs. Nucleic Acid Res. 1997;25: 3389–402.
24. Goli B, Aswathi BL, Nair AS. A novel algorithm for prediction of protein coding DNA from non-coding DNA in microbial genomes using genomic composition and dinucleotide compositional skew, advances in computer science and engineering lecture notes of the institute for computer sciences, social informatics and telecommunications engineering. 2012;85: 535–42
25. Hall M, Holmes G. Benchmarking attribute selection techniques for discrete class data mining. IEEE Trans Knowl Data Eng. 2003;15:1–16.
26. Wang C, Ding C, Meraz RF, Holbrook SR. PSoL: a positive sample only learning algorithm for finding non-coding RNA genes. Bioinformatics. 2006;22: 2590–6.
27. Liu H, Yu L. Towards integrating feature selection algorithms for classification and clustering. IEEE Trans Knowl Data Eng. 2005;17(3):1–12.
28. Huan Liu, Lei Yu. Feature selection for high-dimensional data a fast correlation-based filter solution. IEEE Trans Knowl Data Eng. 2005;17(4): 491–502.
29. Hall MA. Correlation based feature selection for machine learning. Doctoral dissertation, The University of Waikato, Department of Computer Science; 1999.
30. Werbos PJ. Beyond regression: new tools for prediction and analysis in the behavioral sciences. PhD thesis, Harvard University; 1974.

31. Parker DB. Learning-logic. Technical report, TR-47, Sloan School of Management, MIT, Cambridge, MA; 1985.
32. Rumelhart DE, Hinton GE, Williams RJ. Learning internal representations by error propagation in Parallel distributed processing: explorations in the microstructure of cognition, vol. I. Cambridge: Bradford Books; 1986.
33. Breiman L. Random forests. Mach Learn. 2001;45(1): 5–32, 18.
34. Kira K, Rendell LA. A practical approach to feature selection. In:Proceedings of the ninth international workshop on machine learning. Morgan Kaufmann Publishers Inc; 1992. p. 249–56.
35. Hall M, Frank E, Holmes G, Pfahringer B, Reutemann P, Witten IH. The WEKA data mining software: an update. SIGKDD Explor. 2009;11(1).
36. Cherian BS, Nair AS. Protein location prediction using atomic composition and global features of the amino acid sequence. Biochem Biophys Res Commun. 2010;391:1670–4.
37. Namboodiri S, Verma C, Dhar PK, Giuliani A, Nair AS. Sequence signatures of allosteric proteins towards rational design. Syst Synth Biol. 2011;4:271–80.

Naïve Bayes-Based Classification for Short Microbial Genes Using Chaos Game Representation

5

Baharak Goli, B.L. Aswathi, and Achuthsankar S. Nair

Abstract

Studies focusing on recognition of short genes encoding small proteins will provide new essential biological insights. This chapter presents a novel method for prediction of short genes based on chaos game representation (CGR). CGR is a graphical representation of biological sequences such as DNAs and proteins. CGR uniquely represents DNA sequences and reveals hidden patterns in it. In this study, genomic feature extraction is implemented by computing the frequency chaos game representation (FCGR) matrix. The order 2, 3 and 4 FCGR matrices are considered here, which consist of 16, 64 and 256 elements, respectively. These element matrices act as the feature descriptor for classification. We utilized principal component analysis (PCA) as a preprocessing step to reduce the feature vector dimensionality and to improve the classification performance. A novel method for classification based on the combination of FCGR and state-of-the-art pattern recognition algorithm, Naïve Bayes classifier, is proposed. The results of the experiment reveal the potential of this representation for discrimination of short genes from noncoding DNA.

Keywords

Frequency chaos game representation • Principal component analysis • Naïve Bayes classifier • Short genes

Introduction

Short proteins play critical roles in many biological processes including stabilizing factors for larger protein complexes [1], regulation of amino acid metabolism [2] and spore development [3]. Short proteins are mediator in other functions such as regulation of innate immunity, cell communication and homeostasis signal transduction.

B. Goli (✉) • B.L. Aswathi • A.S. Nair
Department of Computational Biology and Bioinformatics, University of Kerala, Thiruvananthapuram 695581, India
e-mail: baharak_goli@yahoo.com

A. Sabu and A. Augustine (eds.), *Prospects in Bioscience: Addressing the Issues*,
DOI 10.1007/978-81-322-0810-5_5, © Springer India 2013

The performance of gene-finding tools depends on the coding statistics that are adopted to annotate the sequence. Coding statistic describes the probability that a DNA sequence is coding for a protein. The currently available gene-prediction algorithms in prokaryotic genomes often fail to identify short protein-coding DNA due to lack of prominent coding statistics in short nucleotide sequences, resulting in incomplete or wrong annotations [4]. Hence, the study of coding statistics in short DNA sequences is of major importance in gene prediction and genome annotation. During the past two decades, various gene-finding algorithms have been developed and several review papers have been published [5–8].

We defined the short genes as the fragments ranging from 60 to 400 bp since prediction of significant fraction of short genes lying in this range is beyond the detection ability of many of the topmost gene-prediction tools. Any classification problem deals with selection of feature vectors as its major initial concern. In this study, genomic feature extraction is implemented by computing the Frequency Chaos Game Representation (FCGR) matrix. The order 2, 3 and 4 FCGR matrices are considered here, which consist of 16, 64 and 256 elements, respectively. These element matrices act as the feature descriptor for classification. The proposed feature vector is extracted from six prokaryotic organisms. A straightforward Naïve Bayes classifier in conjunction with principal component analysis was employed. PCA was used to reduce the feature vector dimensionality and to improve the classification performance.

Materials and Methods

The Dataset

In this study, dataset was formed obtaining coding and noncoding sequences of two unique *Escherichia coli* strain, *E. coli* K12-MG1655 and *E. coli* UTI89 (UPEC), and 4 Enterobacteriaceae: *Buchnera aphidicola* 5A, Enterobacter 638, *Klebsiella enteric* a 342 and *Yersinia pestis* KIM 10 from IMG database [9]. Short coding and noncoding sequences falling within the range of 60–400 bp were extracted. Training dataset comprised of 3,270 coding and 2,489 noncoding sequences. Testing dataset comprised of 1,637 coding and 1,247 noncoding sequences.

Feature Vector Formation Using FCGR

Chaos game representation (CGR) is a scale-independent and unique graphical representation for genomic sequences proposed by H.J. Jeffrey [10]. Later on, it was generalized for sequences of arbitrary symbols [11] and other biological sequences such as amino acid sequence [12]. A CGR for a genomic sequence is plotted in a square where the four vertices are assigned to the four nucleotides as A = (0, 0), C = (0, 1), G = (1, 1) and T = (1, 0). This process begins by plotting the first nucleotide of the genomic sequence midway between the centre of the square and the corner of the square corresponding to this nucleotide. The process continues by selecting the consecutive nucleotides in the sequence which is plotted midway between the preceding plotted point and the vertex corresponding to this nucleotide. This continues until one reaches the last nucleotide in the genomic sequence. Mathematically, coordinates of the successive points in the CGR of a DNA sequence are defined as $X_i = 0.5(X_{i-1} + g_{ix})$ and $Y_i = 0.5(Y_{i-1} + g_{iy})$. The plot produced then is referred to as chaos game representation (CGR) of the genomic sequence. This remarkable fractal information content in the CGR has been left almost unexplored.

Most applications of CGR are based on frequency matrix of different oligonucleotides obtained from various grid resolutions. In order to acquire this frequency matrix, CGR space is divided into grids of proper size and counting the occurrence in each square. These frequency matrices are then termed as FCGR. If A is an mth order FCGR matrix containing $2^m \times 2^m$ elements, then $a_{i,j}$ $(1 < i < 2^m, 1 < j < 2^m)$ indicate the elements in the matrix.

In this study, we used frequency matrices extracted from CGR obtained by a $2^m \times 2^m$ grid where $m = 2$ (4×4 FCGR), $m = 3$ (8×8 FCGR) and $m=4$ (16×16 FCGR) of each DNA sequence. The frequency matrices were then reordered by concatenating the rows of each matrix in order to obtain one-dimensional vector for each sequence. Therefore, three different matrices of sizes [4,907×16], [4,907×64] and [4,907×256] were obtained for protein-coding sequences, and three matrices of sizes [3,736×16], [3,736×64] and [3,736×256] were obtained for noncoding sequences. These element matrices act as the feature descriptor for classification. FCGR of order 2, 3 and 4 of the short protein-coding and noncoding DNA are given in Fig. 5.1.

Dimensionality Reduction

In scientific databases, it is usual to come across large amounts of high-dimensional data. In such cases, query processing performance detracts. This occurrence is known as the curse of dimensionality [13] and, as a result, typical data-mining operations, such as clustering or classification, become unproductive [14]. To address this issue, data-mining algorithms are dependent on dimensionality reduction (DR) techniques as a preprocessing step. Dimensionality reduction algorithms solve this challenge by projecting data from the high-dimensional space R^x to a lower-dimensional space R^p (where $p \ll x$). The intention of the dimensionality reduction methods is to preserve the distances between points or other statistical properties in the lower-dimensional space.

In this chapter, well-known PCA-based dimensionality reduction algorithm was employed as a preprocessing step for classification algorithm [15, 16]. Principal component analysis (PCA) is also termed as the Karhunen-Loève transform (KLT), the singular value decomposition (SVD), the empirical orthogonal function (EOF), the eigenvector transform, the Hotelling transform and the proper orthogonal decomposition (POD) method. PCA is an efficient nonparametric linear dimensionality reduction method which extracts the essential structure of a dataset without the need for modeling of the data. It reduces the dimension of the data by projecting the original features into others called principal components and then applies linear combinations of those features. It retains the original features with the largest variance.

Naïve Bayes Classification

In this study, we made use of statistical classifier called Naïve Bayes classifier which is based on the Bayes theorem [17] which relates the probability of the occurrence of an event to the occurrence or nonoccurrence of an associated event. It assumes that the effect of a variable value on a given class is independent of the values of other variable. This theory is known as class-conditional independence. Naïve Bayes classifier requires a small amount of training data to estimate the parameters (means and variances of the variables) essential for classification. Because independent variables are assumed, only the variances of the variables for each class required are being determined.

Algorithm adopts the Bayes rule to express *P* (class | features) in terms of *P* (class) and *P* (features | class):

$$P\ \text{class}|\text{features} = \frac{p(\text{class}) \times p\left(\text{features}|\text{class}\right)}{p(\text{features})} \quad (5.1)$$

It then makes the 'naïve' assumption that all features are independent of one another. This means that

$$P\ \text{class}|\text{features} = p(\text{class}) \times p(f_1 / \text{class}) \times p(f_2 / \text{class}) \times \cdots \times p(f_n / \text{class}) / p(\text{features}) \quad (5.2)$$

The above equation can be written as

$$\text{Posterior} = \text{prior} \times \frac{\text{liklihood}}{\text{evidence}} \quad (5.3)$$

Suppose the data sample is represented by n-dimensional feature vector and there are m classes. Given an unknown data sample Y, the classifier will predict that Y belongs to the class with highest posterior probability, conditional on Y. To classify an unknown sample Y, $P\ (Y|C_i) \times P\ (C_i)$ is calculated for each class C_i. Sample Y is

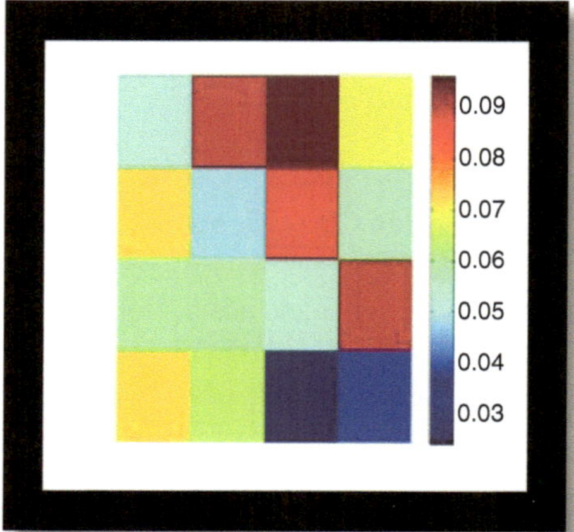

Di-nucleotide frequency matrix for coding region
id | 646314245|
for *E.coli* K-12 MG1655

Di-nucleotide frequency matrix for non-coding region
id | NC_000913|
for *E.coli* K-12 MG1655

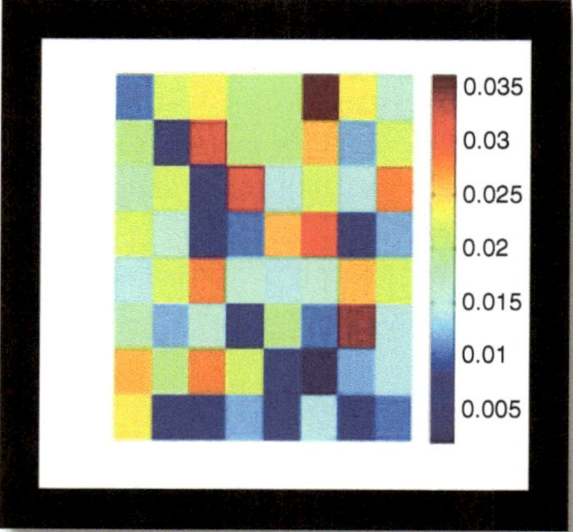

Tri-nucleotide frequency matrix for coding region
id | 646314245|
for *E.coli* K-12 MG1655

0.06
0.05
0.04
0.03
0.02
0.01

Tri-nucleotide frequency matrix for non-coding region
id | NC_000913|
for *E.coli* K-12 MG1655

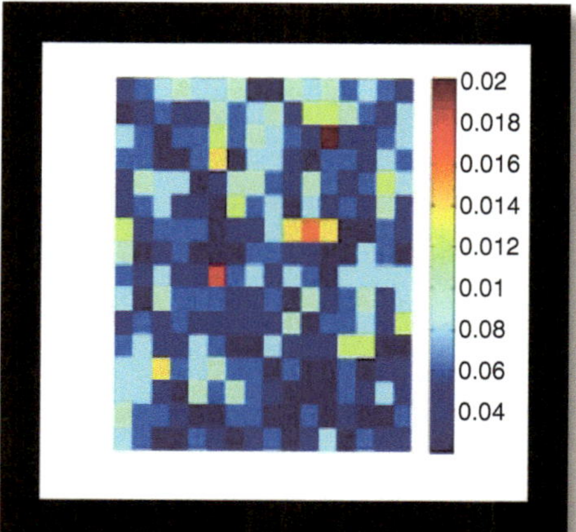

Tetra-nucleotide frequency matrix for coding region
id | 646314245|
for *E.coli* K-12 MG1655

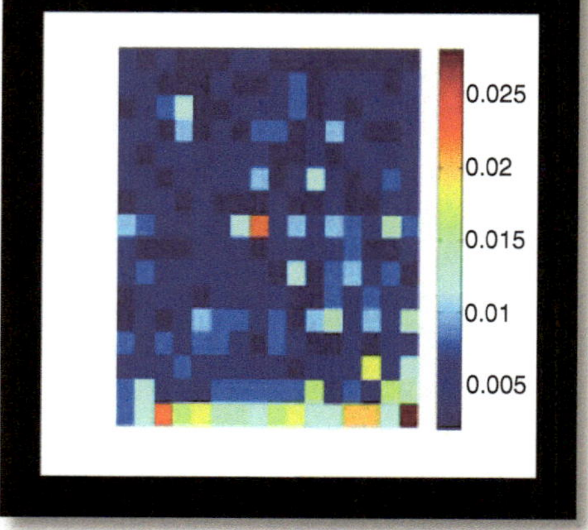

Tetra-nucleotide frequency matrix fornon-coding region
id | NC_000913|
for *E.coli* K-12 MG1655

Fig. 5.1 FCGR of order 2, 3 and 4 of the protein-coding and noncoding DNA

Table 5.1 Performance of Naïve Bayes classifier

Features	Sn	Sp	Pc	MCC	5-fold CV
FCGR obtained by a $2^m \times 2^m$ grid where $m = 2$	91.86	93.04	94.70	0.845	92.36
FCGR obtained by a $2^m \times 2^m$ grid where $m = 3$	94.41	96.65	97.45	0.906	95.37
FCGR obtained by a $2^m \times 2^m$ grid where $m = 4$	94.01	96.56	97.37	0.901	95.10

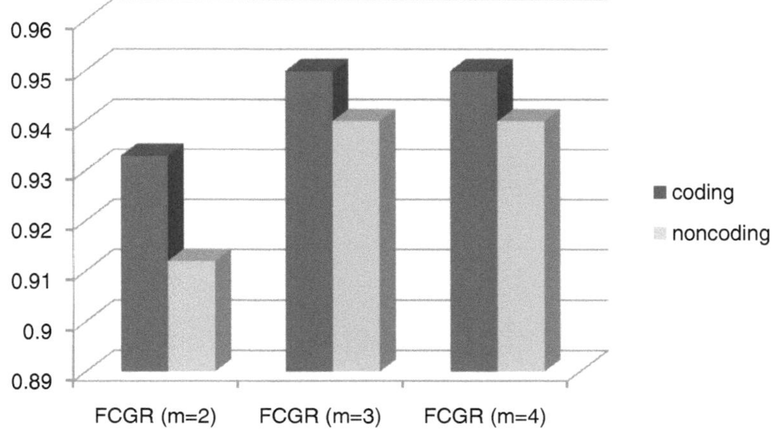

Fig. 5.2 F-measures of different classifiers for coding and noncoding regions

assigned to the class C_i if $P(Y|C_i) \times P(C_i) > P(Y|C_j) \times P(C_j)$ for $1 <= j <= m$, where j is different from i.

Results and Discussion

The CGR images for DNA sequences were plotted using C-GREx software, developed by the Department of Computational Biology and Bioinformatics, University of Kerala, India. The tool is freely available at https://sites.google.com/site/cgrexonline. Conversion of CGR image to FCGR was done using Matlab. We performed the rest of experiments using PCA and Naïve Bayes classifier implemented in data-mining toolkit Weka version 3.6.6 [18] on a PC with a 2.13 GHz Intel CPU and 4 GB RAM, using Windows 7. Default setting was used for all methods. The capability of the propose model was evaluated using basic performance measurements such as accuracy, sensitivity and specificity (TP, TN, FP and FN representing true positive, true negative, false positive and false negative, respectively).

$$\text{Specificity} = \frac{\text{TN}}{(\text{TN} + \text{FP})} * 100$$

$$\text{Sensitivity} = \frac{\text{TP}}{(\text{TP} + \text{FN})} * 100$$

$$\text{Precision} = \frac{\text{TP}}{(\text{TP} + \text{FP})} * 100$$

$$\text{MCC} = \frac{((\text{TP} * \text{TN})(\text{FP} * \text{FN}))}{\left(\sqrt{(\text{TP} + \text{FP}) * (\text{TP} + \text{FN}) * (\text{TN} + \text{FP}) * (\text{TN} + \text{FN})}\right)}$$

$$\text{Accuracy} = \frac{\text{TP} + \text{TN}}{(\text{TP} + \text{TN} + \text{FP} + \text{TN})}$$

The results of 5-fold cross validation indicate that our proposed model avoids overfitting. The results are shown in Table 5.1.

We also computed F-measure which is the harmonic mean of sensitivity and specificity for both coding and noncoding regions. The results are given in Fig. 5.2.

Several alternative measures were used to evaluate the success of proposed ensemble method including mean absolute error, root mean squared error, relative absolute error and root

Table 5.2 Comparison of features based on error rates in prediction using the features

Features	Mean absolute error	Root mean squared error	Relative absolute error (%)	Root relative squared error (%)
FCGR obtained by a $2^m \times 2^m$ grid where $m=2$	0.076	0.276	15.62	55.90
FCGR obtained by a $2^m \times 2^m$ grid where $m=3$	0.046	0.215	9.47	43.52
FCGR obtained by a $2^m \times 2^m$ grid where $m=4$	0.477	0.218	9.76	44.18

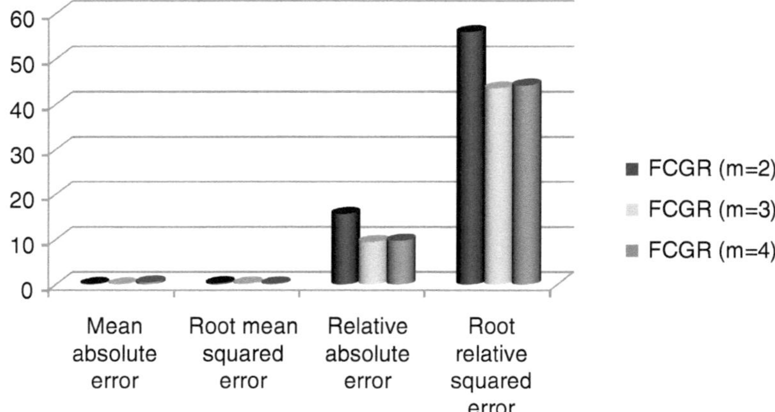

Fig. 5.3 The obtained error rates comparison of features

Table 5.3 Accuracy of proposed classifier for self-consistency test

Features	Accuracy (%)
FCGR obtained by a $2^m \times 2^m$ grid where $m=2$	93.41
FCGR obtained by a $2^m \times 2^m$ grid where $m=3$	96.09
FCGR obtained by a $2^m \times 2^m$ grid where $m=4$	95.57

relative squared error. These measures can be computed using the following formulas. The predicted values are noted by $p_1, p_2, \ldots, p_n$, and the actual values are noted by $a_1, a_2, \ldots, a_n$. The obtained error rates of proposed model are shown in Table 5.2 and Fig. 5.3.

$$\text{Mean absolute error} = \frac{(p_1 - a_1)^2 + \cdots + (p_n - a_n)^2}{n}$$

$$\text{Root mean squared error} = \sqrt{\frac{(p_1 - a_1)^2 + \cdots (p_n - a_n)^2}{n}}$$

$$\text{Relative absolute error} = \frac{|p_1 - a_1| + \cdots + |p_n - a_n|}{|a_1 - \bar{a}| + \cdots + |a_n - \bar{a}|}$$

$$\text{Root relative squared error} = \sqrt{\frac{(p_1 - a_1)^2 + \cdots (p_n - a_n)^2}{(a_1 - \bar{a})^2 + \cdots (a_n - \bar{a})^2}}$$

Self-consistency test (shown in Table 5.3) was also performed to evaluate the proposed algorithm. Self-consistency test reflects the consistency of the developed model. It is an evaluation method to estimate the level of fitness of data in a developed method. In self-consistency test, observations of training datasets are predicted with decision rules acquired from the same dataset. The accuracy of self-consistency reveals the fitting ability of the rules obtained from the features of training sets. Since the prediction system parameters obtained by the self-consistency test are from the training dataset, the success rate is high. Low accuracy of self-consistency test indicates low-efficiency classifier.

To tackle the recognition problem of short genes in microbial genomes, a novel coding statistics is introduced. We extracted genomic features for computing the frequency chaos game representation (FCGR) matrix. The order 2, 3 and 4 FCGR matrices are considered here, which consist of 16, 64 and 256 elements, respectively. These element matrices act as the feature descriptor for classification. We utilized principal component analysis (PCA) as a preprocessing step to reduce the feature vector dimensionality and to improve the classification performance. Classification was done using Naïve Bayes classifier which has very large influence both on theoretical and applied research in classification problems. The results of the experiment reveal the potential of this representation for discrimination of short genes from noncoding DNA. We compared the performance measures of our algorithm using order 2, 3 and 4 FCGR matrices. The experimental results illustrate the superiority of proposed algorithm, when using order 3 FCGR matrix as the feature elements. Naïve Bayes classifier with order 2 FCGR matrix enhances 3.01 and 0.27 % accuracy compared with order 2 and 4 FCGR matrices, respectively. It also has the smallest value for each error measure. Naïve Bayes classifier with order 4 FCGR matrix is the second best classifier. Our work strongly suggests that Naïve Bayes classifier with above-mentioned features of the sequence can be effectively used for classification of short genes.

References

1. Schneider D, Volkmer T, Rogner M. PetG and PetN, but not PetL, are essential subunits of the cytochrome b6f complex from Synechocystis PCC 6803. Res Microbiol. 2007;158:45–50.
2. Yanofsky C. Transcription attenuation: once viewed as a novel regulatory strategy. J Bacteriol. 2000;182:1–8.
3. Cutting S, Anderson M, Lysenko E, Page A, Tomoyasu T, Tatematsu K, Tatsuta T, Kroos L, Ogura T. SpoVM, a small protein essential to development in *Bacillus subtilis*, interacts with the ATP-dependent protease FtsH. J Bacteriol. 1997;179:5534–42.
4. Brent MR, Guigo R. Recent advances in gene structure prediction. Curr Opin Struct Biol. 2004;14:264–72.
5. Fickett JW, Tung CS. Assessment of protein coding measures. Nucleic Acids Res. 1992;20:6441–50.
6. Mathe C, Sagot MF, Schiex T, Rouze P. Current methods of gene prediction, their strengths and weaknesses. Nucleic Acids Res. 2002;30:4103–17.
7. Wang Z, Chen Y, Li Y. A brief review of computational gene prediction methods. Genomic Proteomic Bioinform. 2004;2:216–21.
8. Do JH, Choi DK. Computational approaches to gene prediction. J Microbiol. 2006;44:137–44.
9. Markowitz VM, Korzeniewski F, Palaniappan K, Szeto E, Werner G, Padki A, Zhao X, Dubchak I, Hugenholtz P, Anderson I, Lykidis A, Mavromatis K, Ivanova N, Kyrpides NC. The integrated microbial genomes (IMG) system. Nucleic Acids Res. 2006;34:D344–8.
10. Jeffrey HJ. Chaos game representation of gene structure. Nucleic Acids Res. 1990;18:2163–70.
11. Tino P. Spatial representation of symbolic sequences through iterative function systems. IEEE Trans Syst Man Cybern A Syst Hum. 1998;29(4):386–92.
12. Basu S, Pan A, Dutta C, Das J. Chaos game representation of proteins. J Mol Graph. 1997;15:279–89.
13. Donoho DL. High-dimensional data analysis: the curses and blessings of dimensionality In: American Mathematical Society Conference Math Challenges of the 21st Century; 2000.
14. Beyer K, Goldstein J, Ramakrishnan R, Shaft U. When is "Nearest Neighbor" meaningful? Database theory – ICDT'99. In: Beeri C, Buneman P, editors. Database theory – ICDT'99. Berlin//Heidelberg: Springer; 1999. p. 217–35.
15. Jollie IT. Principal component analysis, Springer series in statistics. New York: Springer; 1986. p. 64–91.
16. Hotelling H. Analysis of a complex of statistical variables into principal components. J Educ Psych. 1933;24
17. Feller W. An introduction to probability theory and its applications, vol. 2. 2nd ed. New York: Wiley; 1971.
18. Hall M, Frank E, Holmes G, Pfahringer B, Reutemann P, Witten IH. The WEKA data mining software: an update, SIGKDD Explor. Newsl. 2009;11:10–8.

PCR-RFLP of 16S rRNA Gene for Identification of Speciation of Meat of Porcine Origin in Processed and Admixed Meat

6

Anjana Sharma, Shiv Kumar, Neha Sharma, and Deepak Sharma

Abstract

The PCR-RFLP using 16S rRNA gene was explored to test its effectiveness in identification of speciation of meat from porcine in raw, cooked as well as admixed meat. Critical perusal of the restriction maps of ~591-bp 16S rRNA fragment sequences from pig as well as other livestocks identified *RsaI* specific to pig, yielding 343-bp and 251-bp fragments in pig, while in other livestock species such as cattle, buffalo, goat and sheep, no restriction enzyme site was found. In all the samples of raw pork, the restriction digestion of 594-bp fragment with *RsaI* yielded the 343-bp and 251-bp fragments, while no restriction digestion in all other livestock species, i.e. cattle, buffalo, goat and sheep. PCR-RFLP with *RsaI* (In cursive) of 594-bp fragment amplified from three different types of cooked meat DNA samples from pig, i.e. cooking at 72°C for 30 min, steam cooking at 90°C for 30 min and autoclaving at 120°C/15 lb/30 min, yielded a similar restriction enzyme digestion profile as with raw meat DNA, suggesting no effect of cooking on PCR-RFLP. Similarly, PCR-RFLP was able to differentiate the origin of meat in mixed meat samples quite consistently up to the level of 25%; however, at the level below 10%, results were not consistent.

Keywords

16S rRNA • PCR-RFLP • Restriction enzyme- *RsaI* • Livestock species

A. Sharma (✉) • S. Kumar
Department of Food Technology, Bundelkhand University, Jhansi, UP, India
e-mail: anjanafst@gmail.com

N. Sharma • D. Sharma
Central Avian Research Institute, Izatnagar, Uttar Pradesh, India

A. Sabu and A. Augustine (eds.), *Prospects in Bioscience: Addressing the Issues*, DOI 10.1007/978-81-322-0810-5_6, © Springer India 2013

Introduction

The identification of species of origin of meat presents considerable problems for food analysts. Correct species identification is important for the consumer for other reasons also, such as medical requirements of individuals who have specific food allergies or religious taboos, apart from possible economic losses. In this regard, pork is a susceptible target due to religious issues associated with adulteration of *halal* food products [1]. With respect to food prescriptions in Islam, Muslims have to follow a set of dietary laws intended to advance their well-being. These dietary laws or prescriptions determine which foods are *halal* (i.e. permitted) for Muslims. Consumption of pork is prohibited according to Islamic rulings [2]. Hence, it is important to develop a reliable technique to ensure the adulteration of meat, especially meat products with meat of porcine origin which is crucial for *halal* authentication.

Identification of species of origin of fresh meat has been achieved using various methods. Anatomical, histological, as well as chemical methods are reliable only in unprocessed raw meat, while the serological methods like precipitation techniques [3] to the more recently developed ELISA [4] and counter immunoelectrophoresis techniques [5] are associated with difficulty in detecting species origin of heat-treated meat samples due to denaturation of the proteins, cross reactivity with other proteins and loss of epitopes in processed meat.

Advent of DNA-based techniques popularly known as molecular techniques like polymerase chain reaction (PCR), restriction fragment length polymorphism (RFLP), random amplified polymorphic DNA (RAPD) fingerprinting, DNA hybridization and DNA sequencing have been very well exploited. In an animal cell, DNA is presentorganoleptic in the nucleus as well as in the mitochondria. Both types of DNA were used for meat differentiation. While some workers targeted nuclear genomic DNA as probes [6, 7], others exploited the polymorphism existing in mitochondrial DNA [8–12].

Among the mitochondrial genes, the most commonly used genes for PCR-RFLP were cytochrome b gene [13] and 12S rRNA gene [10–12]. 16S rRNA is also an important mitochondrial gene, exhibiting sizable polymorphism between livestock species, but has not been exploited much except in few reports such as that from [14] in hairtails, i.e. *Trichiurus lepturus*, *T. japonicus* and *Trichiurus* sp. Hence, PCR-RFLP using 16S rRNA gene was explored to test its effectiveness in identification of speciation of meat from porcine in raw, cooked as well as admixed meat.

Material and Methods

PCR-RFLP Assay

Samples

Raw meat: Pig meat samples were collected from the local pig slaughterhouse, while sample of buffalo meat, goat meat and sheep meat were collected from the municipal abattoirs. Cattle meat samples were collected from Rampur, Uttar Pradesh.

Cooked meat: The fresh meat samples were procured as above and were cooked. Cooking was done in three ways, i.e. cooking at 72°C for 30 min, steam cooking at 90°C for 30 min and autoclaving at 120°C/15 lb/30 min.

Admixed meat: The steam-cooked meat from pig was mixed with buffalo meat in the ratio of 75:25, 50:50, 25:75, 10:90 and 1:99 proportions. From the mixed meat, 10 mg of meat each from 10 different places was taken. A total of 100 mg meat obtained was used for DNA extraction.

Isolation of DNA and Amplification of Partial 16S rRNA

The DNA was isolated using the protocol described by [6]. The universal primers (forward 5′-CGC CTG TTT ACC AAA ACA T-3′ and reverse 5′-CCG GTC TGA ACT CAG ATC ACG T-3′) designed for amplification of partial mt 16S rRNA were synthesized from MBI Fermentas (M/S Genetix) and used.

The PCR was set up in 25-μl reaction volume, and the reaction mixture was comprised of 2.5 μl of

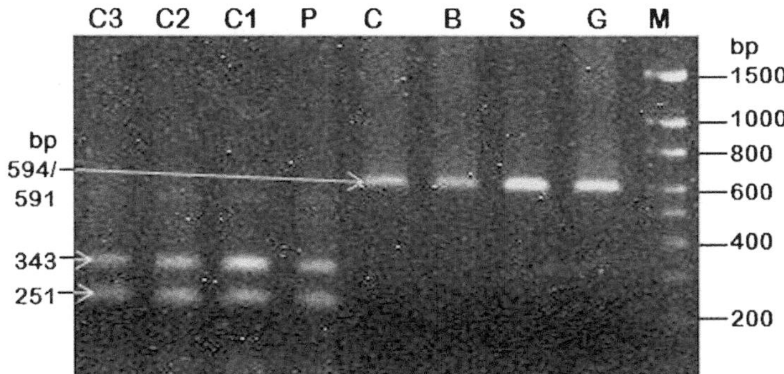

Fig. 6.1 *RsaI* restriction enzyme profile of ~591-bp 16S rRNA fragment amplified from genomic DNA extracted from pork meat cooked under different temperatures (*C3* autoclaved at 120°C/15 lb/30 min, *C2* steam cooked at 90°C for 30 min, *C1* cooked at 72°C for 30 min) and from raw meat of different livestock species (*P* pig, *C* cattle, *B* buffalo, *S* sheep, *G* goat, *M* 100-bp–1,500-bp DNA ladder (Biobasics))

10× assay buffer (100-mM Tris-HC1, pH 9.0, 15-mM MgC12, 500-mM KC1 and 0.1% gelatin), 0.20 mM of dNTP mix, 10 pm of forward and reverse primers, 1U Taq DNA polymerase, 50-ng genomic DNA and autoclaved milliQ water to make up the volume. The reactions were performed in a Mastercycler Gradient Thermocylcer (Eppendorf, Germany), and the cycling conditions were initial denaturation at 94°C for 5 min followed by 30 cycles of 45-s denaturation at 94°C, 45-s annealing at 60°C and 1-min elongation at 72°C. Amplified products were resolved on 1.4% agarose gel.

Restriction Enzyme Mapping of Selective 16S rRNA Fragment

The nucleotide sequences of selective 16S rRNA fragments of pig (AY337045, AY574046, DQ207753, DQ466081 and DQ867012) from database were used for restriction enzyme mapping using GENETOOL software. Restriction enzyme *RsaI* was identified as pig specific after critical perusal of restriction enzyme maps. Sequences of selective 16S rRNA fragment from cattle (AF547270, AY488491, AY702618 and DQ867009), buffalo (AY676871, AB074967, AY526085, AY676856 and DQ867010), goat (AF533441, M55541, DQ867011 and DQ334809) and sheep (AF010406, AY858379, EF490455 and DQ867013) were used to test the species specificity of the identified restriction enzyme.

Restriction Digestion of Selective 16S rRNA Gene

PCR-amplified product of mitochondrial 16S rRNA genes amplified from all the meat samples were subjected to digestion with each of the restriction enzyme *RsaI* separately. Digested product was subjected to electrophoresis in 1.6% agarose gel along with DNA size marker.

Results and Discussion

Critical perusal of the restriction maps developed using the partial 16S rRNA sequences from pig identified *RsaI* specific to pig. *RsaI* restriction enzyme was expected to yield a 343-bp and 251-bp fragments after RE digestion in pig, while in other livestock species such as cattle, buffalo, goat and sheep no restriction enzyme site was found.

In all the samples of raw pork, the restriction digestion of 594-bp fragment with *RsaI* yielded the 343-bp and 251-bp fragments, and this profile was specific to pig only as it yielded no restriction digestion in all other livestock species, i.e. cattle, buffalo, goat and sheep (Fig. 6.1). Various workers have used PCR-RFLP of different mitochondrial genes for differentiation of raw meat from livestock species including sheep and goat [15] could effectively identify the pig, cattle, wild boar, buffalo, sheep, goat, horse, chicken and

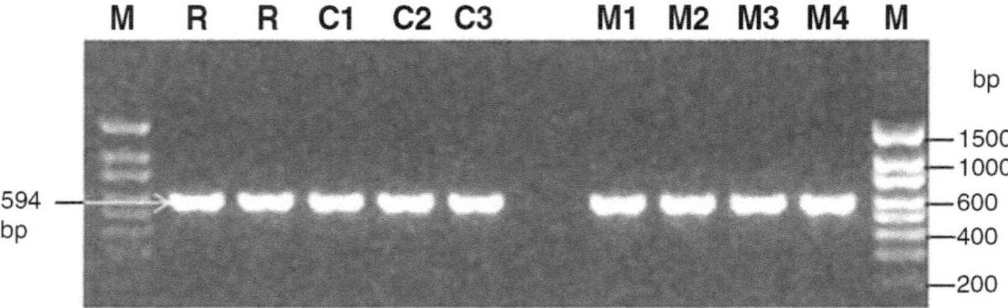

Fig. 6.2 Amplification of ~591-bp 16S rRNA fragment using genomic DNA as template, extracted different type of pork meat. *R* raw meat, *C1* cooked at 72°C for 30 min, *C2* steam cooked at 90°C for 30 min, *C3* autoclaved at 120°C/15 lb/30 min, *M1* mixture of pork and beef in 50:50 ratio, *M2* mixture of pork and beef in 25:75 ratio, *M3* mixture of pork and beef in 10:90 ratio, *M4* mixture of pork and beef in 1:99 ratio, *M* 100-bp–1,500-bp DNA ladder (Biobasics)

turkey meat using PCR-RFLP of 359-bp fragment of cytochrome b gene with restriction digestion by *lul*, *Rsal*, *Taql* and *Hinfl*. A similar study [11] also reported that PCR-RFLP of 12S rRNA gene could differentiate closely related meat species, such as cattle-buffalo, sheep-goat, chicken-quail and turkey-guinea fowl. Fajardo et al. (2007) [16] reported the unequivocal identification of chamois, pyrenean ibex and mouflon/sheep and their differentiation from meat of domestic species such as cattle, goat and swine using the PCR-RFLP of 720 base pairs (bp) from the mitochondrial 12S rRNA gene with *Mse*I and *Apo*I endonucleases.

PCR-RFLP Assay with Cooked Meat

The cooked meat samples from pig were prepared by cooking the meat in three different way, i.e. cooking at 72°C for 30 min, steam cooking at 90°C for 30 min and autoclaving at 120°C/15 lb/30 min. The resolution of extracted genomic DNA from cooked meat showed that cooking of the meat has affected the quality of extracted genomic DNA and the degree of shearing of DNA has increased with the severity of processing; however, the amplified products from all the cooked meat were of similar intensity as that from the raw meat except the intensity of the bands amplified from autoclaved meat (Fig. 6.2), which was comparatively less. These amplified 594-bp 16S rRNA fragments were subjected to restriction enzyme digestion with *Rsal*, which yielded similar species-specific profiles, as that from the PCR-RFLP with raw meat (Fig. 6.1). Earlier workers also found that though the processing of meat affect the quality of genomic DNA, in general, it does not affect the amplification of mitochondrial genes [10, 11, 17].

PCR-RFLP Assay for Mixed Meat Speciation

The efficiency of PCR-RFLP of 16S rRNA gene in detecting the admixing of the meat from pig with meat from other livestock species was evaluated. The admixtures of pig meat with buffalo meat were prepared in 50:50, 25:75, 10:90 and 1:99 ratio separately. Using the genomic DNA extracted from admixture meat samples as template, ~590-bp fragments were amplified in all the mixed meat samples (Fig. 6.2). These ~590-bp PCR product was digested with the species-specific restriction enzymes (i.e. *Rsal*, Fig. 6.3). For testing the repeatability of results, five samples were tested in all the proportions of admixing. RE profile with species-specific restriction enzyme was able to differentiate the origin of meat in mixed meat samples quite consistently up to the level of 25% as the PCR-RFLP of the ~590-bp product amplified from mixed meat with *Rsa* yielded 343-bp and 251-bp in all the replicates. However, at 90:10 and 99:1 ratio of admixing of pig meat, PCR-RFLP of the 590-bp product amplified from these mixed meat with *Hpal* could yield pig-specific RE profile only in 20% replicates. Girish et al. [11] also reported the inconsistency in results in mixed meat in various combinations, i.e. buffalo meat with chevon/mutton,

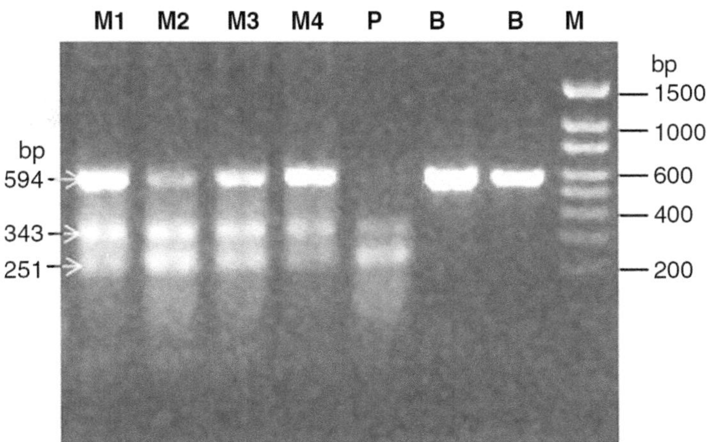

Fig. 6.3 *RsaI* restriction profile of ~591-bp 16S rRNA fragment amplified from genomic DNA, extracted from mixed meat of pig and buffalo meat in different proportions. *M1* 50:50, *M2* 25:75, *M3* 10:90, *M4* 1:99 ratio, *P* 100:0, *B* 0:100, *M* 100-bp–1,500-bp DNA ladder (Biobasics)

beef with chevon/mutton, beef with buffalo meat and mutton with chevon combinations at the inclusion levels of 10% or less than that. This could be due to the disparity in the quantity of PCR product amplified and the amounts of target DNA present [18]. However, many workers were able to identify the admixing of meat even at the level below 10%, even up to 1% or less by using PCR-RFLP of different mitochondrial genes [17, 19, 20].

References

1. Yahya Ishmael R. When ignorance is not bliss. *Halal* J. 2005;8:28–29.
2. Bonne K, Verbeke W. Muslim consumer trust in *halal* meat status and control in Belgium. Meat Sci. 2008; 79:113–23.
3. Allsup TN. Initial progress in the identification of species by testing imported red meat and offal. In: Patterson RLS, editor. Biochemical identification of meat species. London/New York: Elsevier Applied Science Publishers; 1987. p. 176–86.
4. Martin N, Wardale RJ, Jones SJ, Hernandez PE, Patterson RLS. Monoclonal antibody sandwich ELISA for the potential detection of chicken meat in mixtures of raw beef and pork. Meat Sci. 1991;30:23–31.
5. Sherikar AT, Karkare DD, Khot JB, Jayrao BM, Bhilegaonkar KN. Studies on thermostable antigens, production of species-specific anti adrenal sera and comparison of immunological techniques in meat speciation. Meat Sci. 1993;33:121–36.
6. Chikuni K, Ozutsume K, Hoishi Kawa T, Kato S. Species identification of cooked meat by DNA hybridization assay. Meat Sci. 1990;27:119–28.
7. Kirsten F, Ebbehy TPD. Species differentiation of heated meat products by DNA hybridization. Meat Sci. 1991;30:221–34.
8. Matsunaga T, Shibata K, Yamada J, Shinmura Y. Effects of processing condition on species identification of meat products. J Jpn Soc Food Sci Technol. 1999;46:187–94.
9. Verkaar ELC, Nijman IJ, Boutaga K, Lenstra JA. Differentiation of cattle species in beef by PCR-RFLP of mitochondrial and satellite DNA. Meat Sci. 2002;60:365–9.
10. Girish PS, Anjaneyulu ASR, Viswas KN, Shivakumar BM, Anand M, Patel M, Sharma B. Meat species identification by polymerase chain reaction-restriction fragment length polymorphism (PCR-RFLP) of mitochondrial 12S rRNA gene. Meat Sci. 2005;70(1):107–12.
11. Girish PS, Anjaneyulu ASR, Viswas KN, Anand M, Rajkumar N, Shivakumar BM. Sequence analysis of mitochondrial 12S rRNA gene can identify meat species. Meat Sci. 2004;66:551–6.
12. Neha S, Third SS, Deepak S. Effect of meat processing on genomic DNA quality and specific gene amplification. J Appl Anim Res. 2005;28:69–72.
13. Tanabe S, Miyauchi E, Muneshige A, Mio K, Sato C, Sato M. PCR method of detecting pork in foods for verifying allergen labeling and for identifying hidden pork ingredients in processed foods. Biosci Biotechnol Biochem. 2007;71:1663–7.
14. Chakraborty A, Futoshi A, Yukio I. Molecular identification of hairtail species (Pisces: *Trichiuridae*) based on PCR-RFLP analysis of the mitochondrial 16S rRNA gene. J Appl Genet. 2005;46:381–5.
15. Meyer R, Hoefelein C, Luethy J, Candrian U. Polymerase chain reaction-restriction fragment length polymorphism analysis: a simple method for species identification in food. J AOAC Int. 1995;78:1542–51.
16. Fajardo V, Gonzalez I, Lopez-Calleja I, Martin I, Rojas M, Pavon M, Pablo H, Garcia T, Martin R. Analysis of mitochondrial DNA for authentication of meat from Chamois (*Rupicapra rupicapra*), Pyrenean Ibex (*Capra pyrenaica*), and Mouflon (*Ovis ammon*) by polymerase chain reaction-restriction fragment length polymorphism. J AOAC Int. 2007;90:179–86.
17. Sun YL, Lin CS. Establishment and application of a fluorescent Polymerase Chain reaction-restriction

fragment length polymorphism (PCR-RFLP) method for identifying porcine, caprine, and bovine meat. J Agric Food Chem. 2003;7:1771–6.

18. Partis L, Croan D, Guo Z, Clark R, Coldham T, Murby L. Evaluation of a DNA fingerprinting method for determining the species origin of meat. Meat Sci. 2000; 54:369–76.

19. Rodriguez MA, Garcia T, Gonzalez I, Asensio L, Hernandez PR, Martin R. PCR identification of beef, sheep, goat and pork in raw and heat-treated meat mixtures. J Food Prot. 2004;67:172–7.

20. Ilhak IO, Arslan A. Identification of meat species by polymerase chain reaction (PCR) technique. Turk J Vet Anim Sci. 2007;31:159–63.

A Novel HMG-CoA Reductase Inhibitor from Methanolic Extract of *Holoptelea integrifolia* (Roxb.) Planch

7

K.S. Arun and Anu Augustine

Abstract

Holoptelea integrifolia is a large deciduous tree included in lekhaneyagana, a pharmacological classification remarked in *Charaka Samhita*, which means "reduce excess fat". The present study was to check the effect of methanolic fraction of *H. integrifolia* on serum lipids, faecal lipids and HMG-CoA reductase (HMGR), an enzyme which catalyses the committed step in cholesterol biosynthesis. *H. integrifolia* extracts were administered orally at doses (200 and 400 mg/kg body weight) to diet-induced obese Sprague Dawley (SD) strain rats. Mevinolin was used as positive control. After 4 weeks of administration, the concentrations of total cholesterol (T-c), triglycerides (TG) and low-density lipoprotein (LDL-c) in serum were significantly lower in *H. integrifolia*-treated groups compared to high-fat group. Both groups showed higher ratio of high-density lipoprotein cholesterol (HDL-c)/(T-c) compared to high-fat diet group. In this study, HMGR activity was enormously reduced, and detailed faecal analysis showed a remarkable increase in faecal T-c and TG compared to high-fat control. It is presumed that a higher hypolipidemic effect of *H. integrifolia* might be accounted by the combined effect by HMGR inhibition which results in suppression of endogenous cholesterol biosynthesis and blocking the intestinal fat absorption.

The methanolic fraction of *H. integrifolia* on LCMS and tandem mass spectrometry analysis showed the presence of a compound, 2-[(3,4-dimethyl-2-oxo-2H-chromen-7-yl)oxy] propanoate (C1). The compound C1 was subjected to molecular docking studies with human HMGR (1HW8). The results indicate that the compound C1 can bind strongly to free HMGR in its active site and support the in vivo results. Based on this observation, we propose that the compound C1 is a potent inhibitor of HMGR and hence potential anticholesterol agent.

K.S. Arun • A. Augustine (✉)
Department of Biotechnology and Microbiology, Kannur University, Thalassery Campus, Kannur 670661, India

Inter University Centre for Bioscience, Kannur University, Kannur 670661, India
e-mail: anuaugus@rediffmail.com

A. Sabu and A. Augustine (eds.), *Prospects in Bioscience: Addressing the Issues*,
DOI 10.1007/978-81-322-0810-5_7, © Springer India 2013

Keywords
H. integrifolia • HMG-CoA reductase • LCMS • AutoDock

Abbreviations

HDL-c	High-density lipoprotein cholesterol
HMGR	HMG-CoA reductase
LCMS	Liquid chromatography mass spectrometry
LDL-c	Low-density lipoprotein cholesterol
MSMS	Tandem mass spectrometry
RT	Retention time
T-c	Total cholesterol
TG	Triglycerides

Introduction

Overweight and obesity are defined as abnormal or excessive fat accumulation that may impair health. At least 2.8 million adults die each year as a result of being overweight or obese. The medical expenses and cost of lost productivity due to obesity are greater than $100 billion per year and is increasing drastically [1–3]. The elevated plasma cholesterol is a major risk factor for the development of obesity, which led to the search for drugs that could reduce blood cholesterol. One possibility is to reduce cholesterol biosynthesis which is achieved by inhibiting the rate-limiting enzyme HMG-CoA reductase (HMGR) in the cholesterol biosynthetic pathway [4, 5]. HMGR (EC 1.1.1.34) is a transmembrane protein anchored in the membrane of the endoplasmic reticulum that catalyses a key step in the mevalonate pathway, which is involved in the synthesis of sterols, isoprenoids and other lipids [6, 7].

Holoptelea integrifolia is a large deciduous tree, commonly known as Indian Elm. Ethno medically, the leaves and stem bark of this plant were used by local people for skin diseases, obesity, cancer and for wound healing in the form of paste [8–10]. This plant is included in "lekhaneyagana" which means "reduce excess fat", a pharmacological classification mentioned by Charaka in *Charaka Samhita* [11]. The present study aims to check the effect of *H. integrifolia* on HMG-CoA reductase activity in diet-induced obese rats. The fraction with HMGR-inhibiting activity is further analysed using LCMS and MSMS, and the compounds obtained are subjected to molecular docking analysis with human HMGR receptor (1HW8) using AutoDock Version 4.0 [12].

Materials and Methods

Preparation of the Bark Extract of *H. integrifolia*

Fresh bark of *H. integrifolia* was collected from the hill area of Kallanode (Calicut district) (coordinates: 11°31′49″N 75°52′21″E). The plant was identified and authenticated by Centre for Medicinal Plants Research (CMPR), Kottakkal, India. Voucher specimen was processed and deposited (No: CMPR 3949). Methanol extract was prepared using soxhlet apparatus for 48 h.

In Vivo Study: Establishment of High-Fat Model and Drug Treatment in Rats

Sprague Dawley (SD) strain rats, body weight 160 ± 10 g, were used for experiments. The experiment was approved by the Institutional Animal Ethics Committee (IAEC No-KULS/IAEC 2011–04). Rats were randomly divided into two groups: control group ($n = 10$) and high-fat group ($n = 40$). Animals in control group were fed with a basic diet for 4 weeks. All animals in high-fat group were fed with a high-fat diet for 4 weeks. After 4 weeks, blood samples were taken for the analysis of T-c, TG, HDL-c and LDL-c for confirmation of hyperlipidemia [13].

The rats in high-fat model group were randomly divided into four groups with ten rats per group: high-fat model control group, *H. integrifolia* I group (200 mg/kg body weight (BW)), *H. integrifolia* II group (400 mg/kg BW), and mevinolin-treated positive control group (3.0-mg/kg body weight) [13, 14]. At the end of the experiments, the overnight-fasted rats were sacrificed and blood samples were taken for the measurement of serum T-c, TG, HDL-c [15, 16], LDL-c [17] and liver for HMG-CoA reductase assay [18]. Faeces of the last 2 days of the treatment period were collected. They were freeze-dried for 48 h. The faeces were ground into fine powder and extracted with chloroform–methanol (2:1) mixture in accordance with the method described by Folch et al. [19]. Faecal cholesterol was determined using enzymatic kit.

LCMS Analysis of Methanolic Extract of *H. integrifolia*

Methanolic extract of *H. integrifolia* was used to perform LCMS and MSMS analysis (Waters ESI QToF System), and structural elucidation of molecules was performed using MassFragment (MassLynx Mass Spectrometry Software Waters Corporation, Milford, USA).

The MS^E mode of acquisition was used to acquire the data with low and high collision energy. The compound obtained is detailed in Fig. 7.2.

Docking of Compounds with Human HMG-CoA Reductase Enzyme

The compound obtained in LCMS data and compactin were docked with a crystal structure of human HMGR (PDB ID: 1HW8) using AutoDock Version 4.0 [12, 20]. Since the active site is bipartite, A and B chains are selected for docking. Flexible docking strategy was applied in which all the binding site residues were set flexible. ASP690, ARG590, SER684 and LYS691 of A chain and ASN755 and LYS 735 of B chain were selected as flexible residues. The grid dimensions were 30×34×44 Å with points separated by 0.375 Å. The Lamarckian genetic algorithm was applied for minimization, using default parameters to find the lowest energy and its structure prediction accuracy [21]. The complexes of ligands with HMGR resulting from molecular docking were analysed using Python Molecular Viewer (MGL Tools). The pose generated were viewed using PyMOL Molecular Graphics System [22].

Results and Discussions

Effect of *H. integrifolia* on Serum Lipids in Diet-Induced Obese Rats

Oral administration of *H. integrifolia*, 200 and 400 mg/kg BW, to high-fat animals resulted in significant decreases in serum T-c, TG and LDL-c levels ($p<0.05$). To note the decrease of serum TG level is a promising result. Recent studies showed that triglycerides are independently related to coronary heart disease and most of the antihypercholesterolemic drugs were not able to reduce triglyceride levels [4, 23]. Furthermore, LDL-c carries cholesterol to the arteries, infiltrates their walls and may trigger plaque buildup. The treatment shows a significant increase ($p<0.05$) in plasma HDL-c levels (Table 7.1). The hypolipidemic effect of HDL-c is largely attributed to its central function in the reverse cholesterol transport, a process where excess cell cholesterol is taken up and processed by HDL-c particles for further delivery to the liver for metabolism. Thus, the decrease in T-c, TG and LDL-c level and increase in HDL-c show the ability of *H. integrifolia* extract to reduce cardiovascular disease risk.

Effect of *H. integrifolia* on Faecal Lipids in Diet-Induced Obese Rats

Detailed faecal analysis of *H. integrifolia* extract-treated rats shows a remarkable increment in faecal T-c and TG. Therefore, a probable mechanism for the greater hypolipidemic effect of *H. integrifolia* extract-treated rats can be explained on the basis of blocking of intestinal fat absorption.

Table 7.1 Effect of *H. integrifolia* on serum lipids, faecal lipids and HMGR activity in diet-induced obese rats

	Normal control	High-fat model	*H. integrifolia* 200 mg/kg	*H. integrifolia* 400 mg/kg	Standard mevinolin
Serum lipids (mg/dL)					
T-c	72.34 ± 7.6	111.8 ± 6.6[a]	92.85 ± 4.79[b]	86.3 ± 2.76[b]	83.7 ± 1.93
TG	78.10 ± 4.1	107.1 ± 6.6[a]	93.41 ± 5.76[b]	84.7 ± 3.32[b]	81.81 ± 1.77
HDL-c	35.1 ± 3.49	22.4 ± 4.08[a]	25.98 ± 1.86[b]	29.90 ± 1.90[b]	31.6 ± 1.65
LDL-c	21.6 ± 8.02	67.9 ± 9.88[a]	48.18 ± 5.32[b]	39.46 ± 2.17[b]	35.75 ± 1.29
HMGR[c]	2.68 ± 0.17	3.23 ± 0.28[a]	4.93 ± 0.23[b]	5.50 ± 0.32[b]	4.05 ± 0.18
Faecal lipids (mg/dL)					
T-c	52.31 ± 1.57	85.98 ± 3.89	95.75 ± 1.72[b]	88.38 ± 2.96[b]	88.38 ± 2.96
TG	12.75 ± 1.20	15.03 ± 1.21	21.0 ± 1.72[b]	16.3 ± 1.38[b]	16.3 ± 1.38

T-c total cholesterol, *TG* triglycerides, *LDL-c* low-density lipoprotein cholesterol, *HDL-c* high-density lipoprotein cholesterol, *HMGR* HMG-CoA reductase; normal control, rat fed with normal diet; high-fat model, rat fed with high-fat diet; *H. integrifolia*-200 mg/kg and *H. integrifolia*-400 mg/kg, high-fat diet rats treated with 200-mg/kg and 400-mg/kg body weight of *H. integrifolia* bark methanolic extract, respectively; standard mevinolin, high-fat diet rats treated with mevinolin (3.0 mg/kg). The values are mean ± s.d. for ten rats

[a]$p < 0.05$ compared with control group

[b]$p < 0.05$ compared with untreated model group

[c]HMG-CoA reductase activity is expressed in terms of HMG-CoA/mevalonate ratio. The ratio between HMG-CoA and mevalonate is inversely proportional to HMG-CoA reductase activity

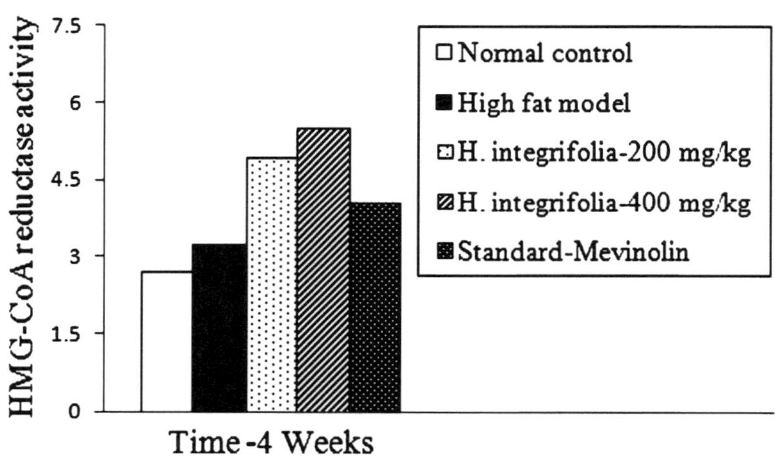

Fig. 7.1 Effect of *H. integrifolia* on HMGR activity in diet-induced obese rats

Effect of *H. integrifolia* on HMGR Activity in Diet-Induced Obese Rats

Estimation of hepatic HMGR activity showed a significant decrease in extract-treated hyperlipidemic rats. In particular, hepatic HMGR significantly went down by 70.27% in the 400-mg/kg *H. integrifolia* extract group than untreated model group (Fig. 7.1) Mevinolin treatment also shows significant reduction on HMGR activity (Table 7.1). HMGR catalyses the four-electron reduction of HMG-CoA to coenzyme A and mevalonate. Moreover, HMGR is the rate-limiting enzyme in the cholesterol pathway; HMG-CoA is water soluble and has an alternative metabolic pathway for its breakdown. Thus, no potential toxic precursors are formed when the pathway is blocked. The above factors make HMGR a promising target for treating hyperlipidemia.

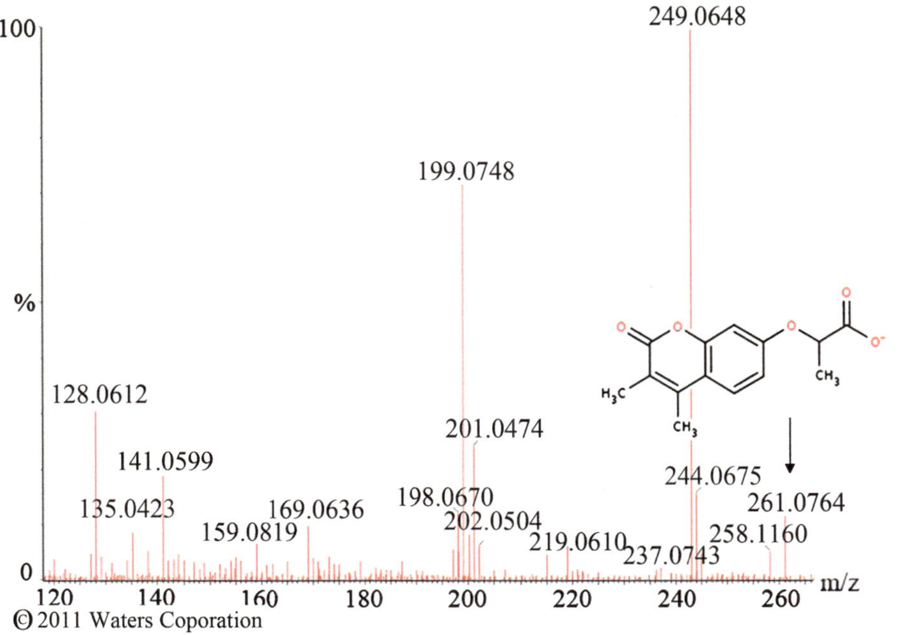

Fig. 7.2 ESI-MSMS and mass fragment analysis of mass 261.0755 m/z at RT 9.1 min and the structure with elemental composition $C_{14}H_{13}O_5$

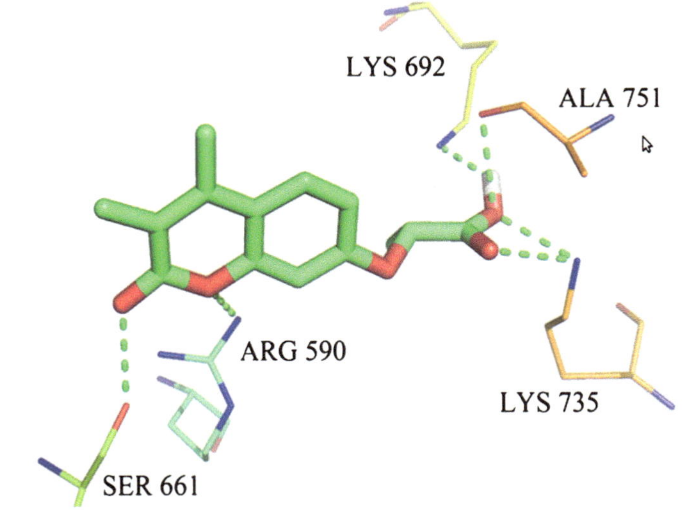

Fig. 7.3 The optimized complex of 2-[(3,4-dimethyl-2-oxo-2H-chromen-7-yl)oxy] propanoate (CSID 322828) with the binding pocket of the catalytic portion of human HMGR substrate: hydrogen bonds are displayed as *green dashes*. Nonpolar hydrogens are omitted for clarity. Figure generated using PyMOL

LCMS Analysis of Methanolic Extract of *H. integrifolia*

LCMS and MSMS analysis done using *H. integrifolia extract* showed the presence of (2R)-2-[(3,4-dimethyl-2-oxo-2H-chromen-7-yl)oxy] propanoate (C1). The MSMS spectrum, library search and structure are detailed in Fig. 7.2.

Docking of Compounds with Human *HMG-CoA Reductase* Enzyme

AutoDock result entails that the compound C1 binds favourably at the active site of IHW8 by 6 hydrogen bonds with the active site residue, ARG590, LYS735 (2 bonds), ALA751 and SER661 (Fig. 7.3). The binding energy −13.68 KCal and

Table 7.2 Results of docking of compounds with the catalytic portion of human HMGR

Compound name	Binding energy	Inhibition constant	Interacting residue
2-[(3,4-dimethyl-2-oxo-2H-chromen-7-yl)oxy] propanoate	−13.68	94.46 pM	SER 661, ARG 590, LYS 735 (2), LYS692, ALA751
Compactin	−12.9	348.66 pM	ARG590, ASN755, LYS691, LYS692, ALA751

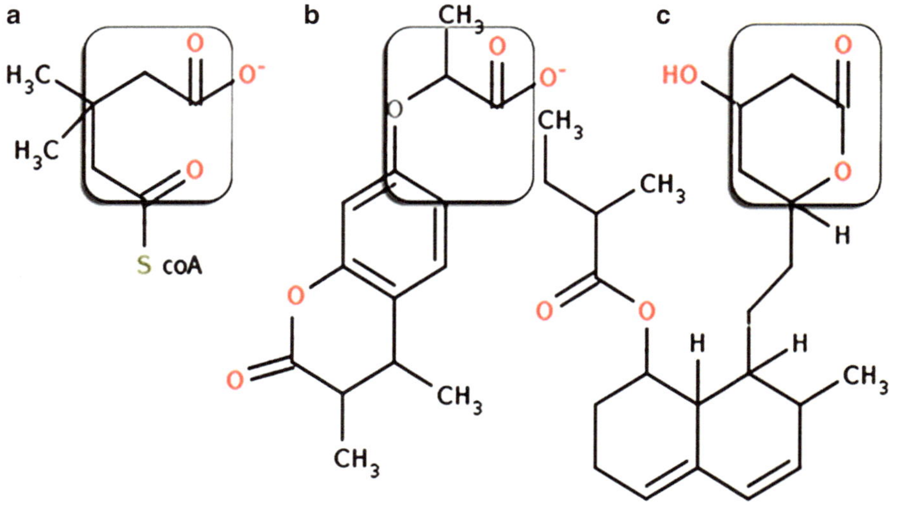

Fig. 7.4 Structural formulas of (**a**) HMG-CoA, (**b**) 2[3,4-dimethyl-2-oxo-2H-chromen-7-yl)oxy] propanoate and (**c**) compactin. The similar portions are shown in *square boxes*

inhibition constant 94.46 pM substantiate that ligand forms a favourable interaction with HMGR receptor (Table 7.2). The HMG-binding pocket of HMGR is characterized by a loop (residues 682–694, referred to as "cis loop")]. C1 shows a hydrogen bond with cis loop residue LYS692. Moreover, the compound shares a structural similarity with HMG-CoA, which is substrate of HMGR. Structural formulae of HMG-CoA, compound C1 and compactin are shown in Fig. 7.4. Thus, the docking results support the in vivo experiment results.

Conclusions

Oral administration of *H. integrifolia* extract (both 200 and 400 mg/kg BW) was able to reduce serum HMGR activity. The treatment was able to reduce serum T-c, TG, LDL-c and VLDL-c levels. Moreover, *H. integrifolia* facilitates agreeable improvement in cardioprotective lipid HDL-c. Docking studies show that C1 can bind to free HMGR in its active site and may prevent its activity. It has the same mode of binding and interactions as that of known direct statins (HMGR inhibitors) [9, 24–27]. Their contacts are also sufficient in number, suggesting strong interactions with the target molecule [28–31]. The above results affirm the in vivo studies in diet-induced obese rats and validate the "lekhaneyagana" remarked in *Charaka Samhita* (700 BC).

Acknowledgements DBT-BIF, Govt. of India, is gratefully acknowledged for the support in the form of Bioinformatics Infrastructure Facility (BIF) at the Department of Biotechnology and Microbiology, Kannur University, for computational and other allied facilities. Arun KS is thankful to UGC-BSR, New Delhi, for providing the fellowship to carry out the research work.

References

1. WHO. Obesity: preventing and managing the global epidemic, Report of a WHO consultation. World Health Organ Tech Rep Ser 894, i-xii; 2000. pp. 1–253.
2. Allison DB, Fontaine KR, Manson JE, Stevens J, Van Itallie TB. Annual deaths attributable to obesity in the United States. J Am Med Assoc. 1999;282:1530–8.
3. Wolf AM, Colditz GA. Current estimates of the economic cost of obesity in the United States. Obes Res. 1998;6:97–106.
4. Kee EA, Livengood MC, Carter EE, McKenna M, Cafiero M. Aromatic interactions in the binding of ligands to HMG-CoA reductase. J Phys Chem B. 2009;113:14810–5.
5. Menge T, Hartung HP, Stüve O. Statins – a cure-all for the brain? Nat Rev Neurosci. 2005;6:325–31.
6. DeBose-Boyd RA. Feedback regulation of cholesterol synthesis: sterol-accelerated ubiquitination and degradation of HMG-CoA reductase. Cell Res. 2008;18:609–21.
7. Istvan ES, Deisenhofer J. Structural mechanism for statin inhibition of HMG-CoA reductase. Science. 2001;292:1160–4.
8. Bambhole VD, Jiddewar GG. Anti obesity effect of *Iris versicolor* and *Holoptelea integrifolia* in rats. Sachitra Ayurved. 1985;37:557–61.
9. Dinesh K, Karunesh K, Jyoti G, Navita B, Sunil K. A mini review on chemistry and biology of *Holoptelea integrifolia* (Roxb). Planch (Ulmaceae). Asian Pac J Trop Biomed. 2012;2:1–6.
10. Gokaraju GR, Gokaraju RR, Golakoti T, Sengupta K, Bhupathiraju K. Anti-obese compositions containing *Holoptelea integrifolia* extracts, US patent 20100203078; 2010.
11. Sharma PV. Charaka samhita. Varanasi: Chaukhambha Orientalia; 2000.
12. Morris GM, Huey R, Lindstrom W, Sanner MF, Belew RK, Goodsell DS, Olson AJ. AutoDock4 and AutoDockTools4: automated docking with selective receptor flexibility. J Comput Chem. 2009;30: 2785–91.
13. Xu C, Haiyan Z, Hua Z, Jianhong Z, Pin D. Effect of *Curcuma kwangsiensis* polysaccharides on blood lipid profiles and oxidative stress in high-fat rats. Int J Biol Macromol. 2009;44:138–42.
14. Tajuddin Ahmad S, Latif A, Qasmi IA, Amin KMY. An experimental study of sexual function improving effect of *Myristica fragrans* Houtt. (nutmeg). BMC Complement Altern Med. 2005;5:16–23.
15. Rubinstein A. National cholesterol education program, second report of the Expert Panel on detection, evaluation, and treatment of high blood cholesterol in adults. Circulation. 1995;91:908–9.
16. Stein EA, Myers GL. National cholesterol education program recommendations for triglyceride measurement: executive summary. The national cholesterol education program working group on lipoprotein measurement. Clin Chem. 1995;41:1421–6.
17. Friedewald WT, Levy RI, Fredrickson DS. Estimation of the concentration of low-density lipoprotein cholesterol in plasma, without use of the preparative ultracentrifuge. Clin Chem. 1972;18:499–502.
18. Rao AV, Ramakrishnan S. Indirect assessment of hydroxymethylglutaryl-CoA reductase (NADPH) activity in liver tissue. Clin Chem. 1975;21:1523–5.
19. Folch J, Lees M, Stanley GHS. A simple method for the isolation and purification of total lipids from animal tissues. J Biol Chem. 1957;226:497–509.
20. Gasteiger J, Marsili M. Interactive partial equalization of orbital electro negativity-a rapid access to atomic charges. Tetrahedron. 1980;36:3219–28.
21. Delano W. The PyMOL molecular graphics system. 2002. URL http://www.pymol.org
22. Goodsell DS. Computational docking of biomolecular complexes with AutoDock. Cold Spring Harb Protoc. 2009;2009:pdb.prot5200.
23. Bainton D, Miller NE, Bolton CH, Yarnell JW, Sweetnam PM, Baker IA, Lewis B, Elwood PC. Plasma triglyceride and high density lipoprotein cholesterol as predictors of ischaemic heart disease in British men. The caerphilly and speedwell collaborative heart disease studies. Br Heart J. 1992;68:60–6.
24. Istvan ES. Structural mechanism for statin inhibition of 3-hydroxy-3-methylglutaryl coenzyme A reductase. Am Heart J. 2002;144:S27–32.
25. Argüelles N, Sánchez-Sandoval E, Mendieta A, Villa-Tanaca L, Garduño-Siciliano L, Jiménez F, Cruz MDC, Medina-Franco JL, Chamorro-Cevallos G, Tamariz J. Design, synthesis, and docking of highly hypolipidemic agents: *Schizosaccharomyces pombe* as a new model for evaluating alpha-asarone-based HMG-CoA reductase inhibitors. Bioorg Med Chem. 2010;18:4238–48.
26. Wilson PW. High-density lipoprotein, low-density lipoprotein and coronary artery disease. Am J Cardiol. 1990;66:7A–10.
27. Williams P, Robinson D, Bailey A. High-density lipoprotein and coronary risk factors in normal men. Lancet. 1979;1:72–5.
28. Pyo YH, Seong KS. Hypolipidemic effects of Monascus-fermented soybean extracts in rats fed a high-fat and -cholesterol diet. J Agric Food Chem. 2009;57:8617–22.
29. Tobert JA. Lovastatin and beyond: the history of the HMG-CoA reductase inhibitors. Nat Rev Drug Dis. 2003;2:517–26.
30. Sarver RW, Bills E, Bolton G, Bratton LD, Caspers NL, Dunbar JB, Harris MS, Hutchings RH, Kennedy RM, Larsen SD, Pavlovsky A, Pfefferkorn JA, Bainbridge G. Thermodynamic and structure guided design of statin based inhibitors of 3-hydroxy-3-methylglutaryl coenzyme A reductase. J Med Chem. 2008;51:3804–13.
31. Medina-Franco JL, López-Vallejo F, Rodríguez-Morales S, Castillo R, Chamorro G, Tamariz J. Molecular docking of the highly hypolipidemic agent alpha-asarone with the catalytic portion of HMG-CoA reductase. Bioorg Med Chem Lett. 2005;15:989–94.

Optimisation of Cultural and Nutritional Parameters for the Production of Protease from Newly Isolated Bacterial Strain *Bacillus* SDR 10

8

Dinta Daniel, B.S. Rakhi, S. Subramaniyan, and G.S. Sandhia

Abstract

The newly isolated bacterial strain SDR-10 showing growth-associated protease production was a gram-positive rod-shaped bacterium producing endospores. The partial gene sequence of the 16S rRNA gene of the strain followed by sequence alignment confirmed the genus as *Bacillus*. *Bacillus* SDR-10 showed a maximum protease production in 24 h. The enzyme showed its optimum temperature of activity at 45°C and optimum pH at 10 and stability at this temperature and pH. Production of protease from the strain was enhanced by optimising the cultural and nutritional parameters. Wheat bran resulted in the highest protease production at a concentration of 0.6%. Inorganic nitrogen sources were poor in inducing enzyme production from the culture. Among the carbon sources studied, glucose was the best for protease production at a level of 1.0%. Factors affecting the enzyme production were maximum at growth pH 6.5 and inoculum size of 5%. Optimisation of the various factors resulted in enhancing the production from 56.51 units to 132.48 units (2.35%).

Keywords

Alkaline protease • *Bacillus subtilis SDR-10* • Optimisation • Submerged fermentation • Characterisation

Introduction

Microbial enzymes are part of various industries. Among these enzymes, proteases account for nearly 60% of the total industrial enzyme market. A major proportion of commercially available proteases are currently derived from *Bacillus* strains [1]. Microbial proteases have major application in detergent formulations, food, leather, diagnostics, agrochemicals, waste management, silver recovery and pharmaceutical industries [2, 3]. The industrial use of proteases was first proposed by Röhm (German Patent GP283923, 1913), who incorporated pancreatic enzymes into a detergent. It was only in the 1960s that the first detergent containing a bacterial

D. Daniel • B.S. Rakhi • S. Subramaniyan
• G.S. Sandhia (✉)
Department of Botany and Research Centre,
University College,
Thiruvananthapuram 695034, Kerala, India
e-mail: sandhiags@rediffmail.com

A. Sabu and A. Augustine (eds.), *Prospects in Bioscience: Addressing the Issues*,
DOI 10.1007/978-81-322-0810-5_8, © Springer India 2013

protease appeared in the market. Here the enzyme used was subtilisin Carlsberg isolated from *Bacillus licheniformis* [4].

Alkaline proteases are produced by a wide range of microorganisms including bacteria, moulds and yeasts. Among bacteria, *Bacillus* strains are the most important producer of commercial proteases [2, 5], and these are the specific producers of extracellular proteases [6]. About 35% of the total microbial enzymes used in detergent industry are derived from bacterial sources and most of them produced by *Bacillus* sp. It is a very good source for protease production because it is a nonpathogenic, gram-positive bacteria and capable of synthesising various types of proteases. Protease production is influenced by nutrients like carbon and nitrogen sources and vitamins and metal ions and environmental factors such as pH and temperature.

Present study is focused on the bacterial strain with high proteolytic activity isolated from forest soil samples. For enhancing the enzyme production from this strain, the cultural and nutritional parameters have been optimised, and characterisation of the crude enzyme has also been attempted.

Materials and Methods

Microorganism and Culture Condition

Forest soil samples were screened on casein agar plates for isolating proteolytic microorganisms. The culture showing highest proteolytic activity was selected for further studies and was maintained at 4°C.

Identification of Isolate

Bacterial strain was identified following standard microbiological and biochemical procedures. For confirming the identity of the strain, genomic DNA was extracted through CTAB method [7] and 16S rRNA gene was amplified under standard PCR conditions. Amplified gene was subjected to partial sequencing. BLAST search and multiple sequence alignment were done for strain confirmation.

Pre-inoculum Preparation

The pre-inoculum was prepared by transferring a loop full of 24-h old culture into 50 ml of inoculums with composition as follows. The inoculated medium was incubated in a shaker for 24 h for raising the pre-inoculums.

Submerged Fermentation

The fermentation study was performed in triplicate in 250-ml Erlenmeyer flasks containing 50 ml of the medium with the following composition (g/l): peptone 10.0 g, NaCl 1.0 g, $CaCl_2.H_2O$ 0.5 g, KH_2PO_4 0.2 g and $MgSO_4.7H_2O$ 0.5 g. The medium was inoculated with 5% inoculum, and the initial pH of the medium was maintained at 7.0 using 1% Na_2CO_3. The inoculated medium was kept in a shaker for a period of 24 h, and samples were taken in a regular interval of 24 h for 4 days. The cell-free culture supernatant obtained after centrifugation at 10,000 rpm at 4°C for 15 min served as the enzyme source and was assayed for protease activity.

Analytical Methods

Protease Enzyme Assay

The enzyme activity was determined by using the procedure of Tsuchida et al. [8] with the following modifications. 0.5 ml 2.0% casein in phosphate buffer (pH 7) was taken as the substrate to which properly diluted enzyme samples were added. After incubation at 40°C for 10 min, 1-ml TCA was added. Blank was prepared in the same procedure, but TCA was added prior to enzyme addition. The mixture was centrifuged at 2,000 rpm for 3 min. To the supernatant was added 5 ml of 0.44 M Na_2CO_3, and after 10 min of incubation, 0.5-ml twofold diluted Folin–Ciocalteau reagent was added. After 20 min read the absorbance at 660 nm in a spectrophotometer. One unit enzyme activity is defined as the amount of enzyme that releases 1 μg of tyrosine per ml per minute under the above assay conditions.

Cell Protein

0.5 ml of uniformly mixed sample was made up to 5 ml by adding 4.5 ml distilled water and was centrifuged. The pellet was separated and added with 1 ml 1 N NaOH. Protein in the sample was estimated as per Lowry's method [9] using BSA as standard.

Characterisation of Protease

Optimum Temperature of Protease Activity

Optimum temperature for enzyme activity was determined using the standard assay procedure in the temperature range of 30–60°C. Thermal stability was monitored by measuring the residual activity after 2 h of incubation at a temperature of 45°C for a period of 2 h.

Optimum pH of Protease Activity

Optimum pH for enzyme activity was identified using the standard assay procedure in the pH range of 5.0–11. pH stability was monitored by measuring the residual activity after 2-h incubation at a pH value of 10.0 for a period of 2 h. After incubation, the activity was measured at pH 7.0 at 40°C.

Optimisation of Protease Production

For enhancing the production of protease from the culture, cultural and nutritional parameters have to be optimised. The optimisation was done by sequentially varying one factor at a time, the selected optimal level serving as the control medium for further studies. Thus, a set of conditions could be achieved which can enhance protease production from SDR-10.

Optimisation of Different Cultural Conditions

Effect of Inoculum on Protease Production

Effect of inoculum on protease production was studied by inoculating the media with different inoculum concentrations such as 1.0, 2.5, 5.0 and 10.0.

Effect of Medium pH on Protease Production

The effect of pH on protease production and growth of organism in fermentation media was determined by growing the culture in the control medium at different initial values ranging from 5 to 9. Suitable buffer systems were used for this study.

Optimisation of Nutritional Parameters

Effect of Nitrogen Source on Protease Production

To test the effect of different nitrogen sources on protease production, peptone in the basal medium was substituted with 1% of different organic/complex, inorganic and low-cost agrowastes. Selected complex/organic substrates include urea, casein and gelatine. The different inorganic nitrogen sources studied include NH_4Cl, (NH_4) NO_3, $NaNO_3$ and $(NH4)_2SO_4$. The low-cost substrates selected were wheat bran, ground nut meal, rice bran and black gram bran. Their effects on enzyme production were evaluated against control medium used initially with peptone as the nitrogen source.

Effect of Different Levels of the Selected Substrate on Protease Production

The selected substrates from the above were evaluated at different levels for understanding their influence on protease production.

Effect of Carbon Sources on Protease Production

To test the effect of different carbon sources on protease production, different carbon sources such as starch, sucrose, xylose, glucose and glycerol were added to the basal medium at 1.0% and assayed the activity. Analytical procedures were carried out as described earlier.

Effect of Different Concentration of Selected Carbon Sources

Different concentrations of the selected carbon source were used to examine the production of proteases. The medium contained different concentrations such as 0.1, 0.25, 0.5, 1.0 and 1.5 (%w/v) of carbon source and assayed the activity.

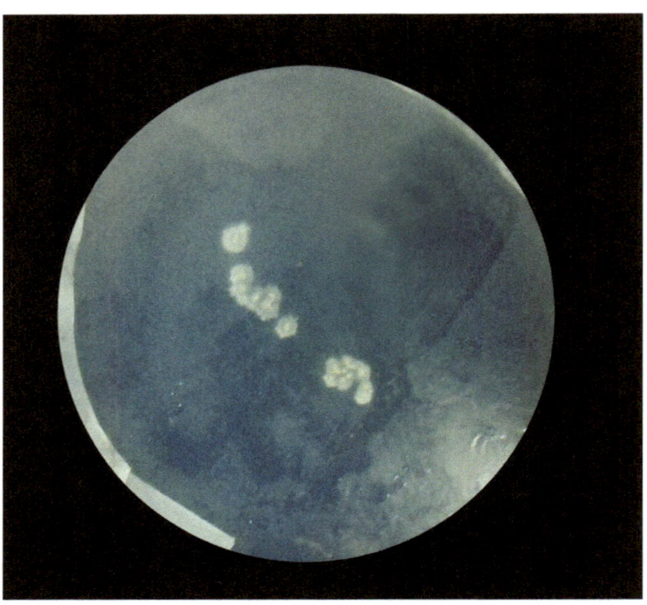

Fig. 8.1 Morphology of the individual colony

Optimised Medium Against Control

All the optimised conditions were combined together and tested against a control under the fermentation conditions already mentioned. This was to evaluate the effect of the chosen conditions on enzyme production.

Results and Discussion

The production and characterisation of proteases from different microorganisms are ever increasing [4]. However, new organisms producing novel protease suitable for industrial uses are always rewarding because of the emergence of new industries producing unique products like detergent preparation with enzyme additives. In the present investigation, a bacterial strain isolated from soil was selected for the production of high levels of protease activity. Attempts were made to identify the bacterium and characterisation of the enzymes synthesised. Subsequent studies involve the enzyme production using different nutritional and cultural parameters.

Isolation of Proteolytic Bacteria

The bacterium used for the present study was isolated from soil samples from Kerala, India, using the clear zone formation method. Colonies that are showing good clear zones of 1 cm or above are selected for further studies. The cultures which were positive in the clear zone production were subjected to liquid medium growth studies resulting in the selection of culture SDR-10 as the potent microorganism with protease production efficiency (Fig. 8.1).

Newly isolated bacterium is a gram-positive rod producing endospores. Biochemical and physiological tests performed according to the *Bergey's Manual of Determinative Bacteriology* identified the strain number SDR-10 as *Bacillus* sp. On the basis of 16S rRNA gene BLAST studies, the isolate was found to be having 100% sequence similarity with species of *Bacillus* (Figs. 8.2 and 8.3). The phylogenetic tree (Fig. 8.3) constructed by the neighbour-joining method indicated that the isolate SDR-10 was having maximum affinity to *B. subtilis*.

Fermentation Profile of *Bacillus* SDR-10

The present investigation was aimed at optimisation of medium components which have been predicted to play a significant role in enhancing the production of alkaline proteases [3]. Proper combinations of various nutritional parameters such as carbon and nitrogen sources, growth pH

>SDR10_Protease_producer_

AGTGATGAAGGTTTTCGGATCGTAAAGCTCTGTTGTTAGGGAAGAACAAGTGCCG
TTCAAATAGGGCGGCACCTTGACGGTACCTAACCAGAAAGCCACGGCTAACTACG
TGCCAGCAGCCGCGGTAATACGTAGGTGGCAAGCGTTGTCCGGAATTATTGGGCG
TAAAGGGCTCGCAGGCGGTTTCTTAAGTCTGATGTGAAAGCCCCCGGCTCAACCG
GGGAGGGTCATTGGAAACTGGGGAACTTGAGTGCAGAAGAGGAGAGTGGAATTC
CACGTGTAGCGGTGAAATGCGTAGAGATGTGGAGGAACACCAGTGGCGAAGGCG
ACTCTCTGGTCTGTAACTGACGCTGAGGAGCGAAAGCGTGGGGAGCGAACAGGAT
TAGATACCCTGGTAGTCCACGCCGTAAACGATGAGTGCTAAGTGTTAGGGGGTTT
CCGCCCCTTAGTGCTGCAGCTAACGCATTAAGCACTCCGCCTGGGGGAGTACGGT
CGCAAGACTGAAACTC

Fig. 8.2 Sequence of the 16S rRNA gene of SDR-10 in 'FAST A' format

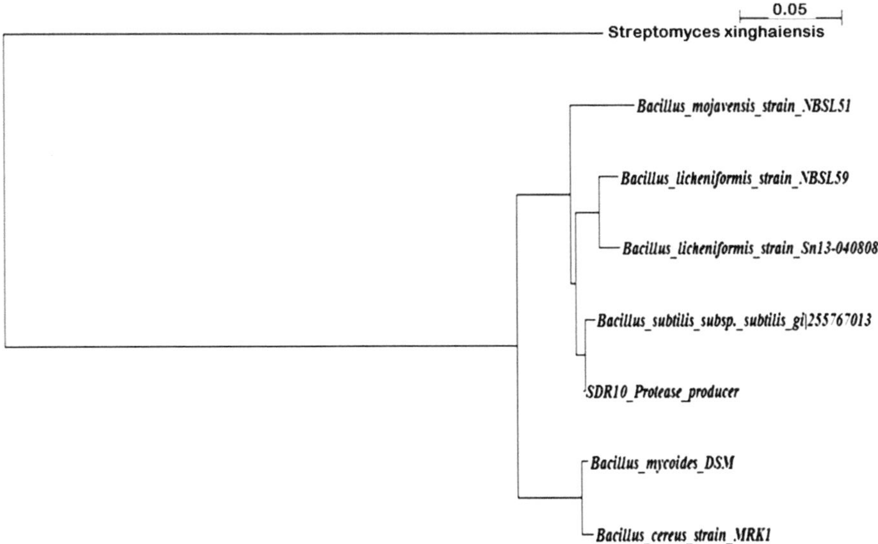

Fig. 8.3 Neighbour-joining tree showing the position of isolate *Bacillus* SDR-10 to a selected number of members of genus *Bacillus*. Branch lengths reflecting the actual distances between the sequences could be estimated using the value provided

and inoculum size can be established in order to suit the bacterial isolate for high secretion of alkaline protease. Protease production was evaluated along with change in medium pH and growth for a period of 96 h, and the values were recorded at every 24-h interval.

It was observed that the highest protease production was obtained after 24 h. Protease production started in the early logarithmic phase, which along with growth reached the maximum after 24 h (Fig. 8.4). Similar growth-associated enzyme production has been reported in *Bacillus* sp. SSR1 with a peak after 15 h by Kezia et al. [10]. Haddar et al. [1] reported maximum protease production after 24 h in *Bacillus mojavensis A21*. Growth-associated protease production could be correlated with the onset of a high rate of protein turnover during sporulation [11]. Protease production was studied at the initial pH value of 7.0 which gradually shifted towards alkaline side as the growth proceeded (Fig. 8.4).

Characterisation of Crude Protease

Effect of Temperature on Enzyme Activity

Optimum temperature for protease activity was determined at different temperatures ranging from 35 to 60°C using casein as the substrate at pH 7.0. Protease activity was optimum at the

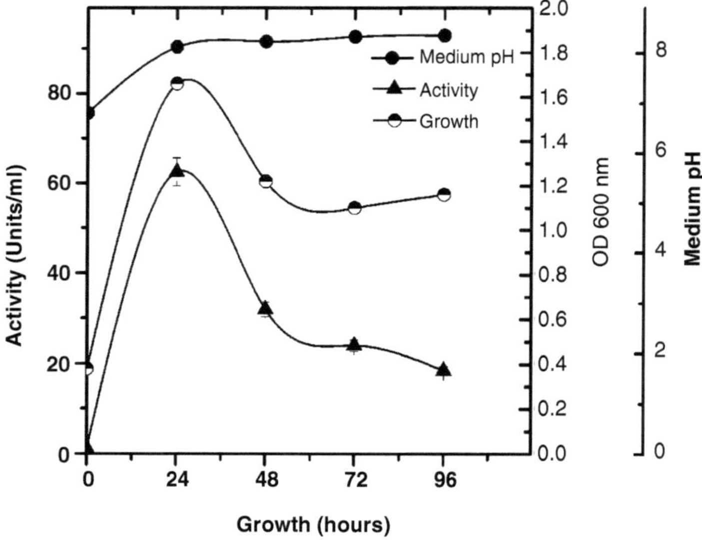

Fig. 8.4 Growth and protease production from *Bacillus* SDR-10. Change in the medium pH is also monitored

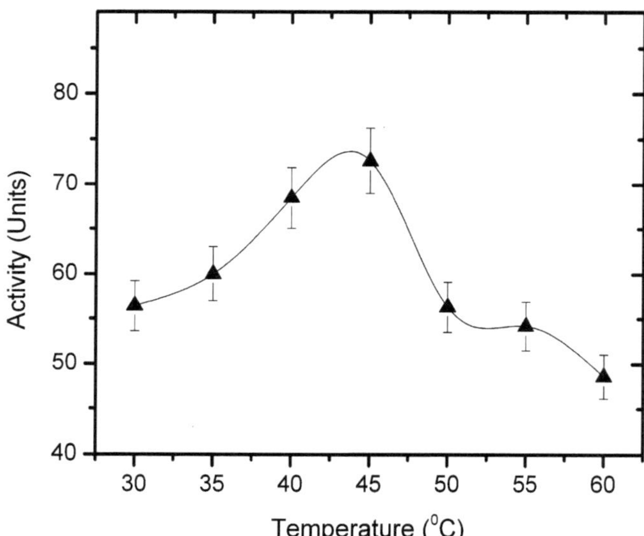

Fig. 8.5 Activity of SDR-10 protease at different temperatures

temperature of 45°C. The activity gradually declined both at higher and lower temperatures (Fig. 8.5). The optimum temperature of alkaline proteases ranged from 50 to 70°C [12]. This slightly elevated optimum temperature by the *Bacillus* SDR-10 is in accordance with reports for proteases from other *Bacillus* spp. The optimum temperature of protease activity at 60°C has been reported in *Bacillus* sp. SMIA-216 [13] and at 70°C in *Bacillus* sp. GUS1 [14].

Effect of Temperature on the Stability of the Enzyme

The protease enzyme was incubated at the optimum temperature of 45°C for a period of 4 h. The result indicated that the enzyme was completely stable at 45°C for 120 min; thereafter, there was reduction in activity. Chomsri [15] reported that protease produced by *Bacillus* sp. was stable at 60°C for 90 min. *Bacillus* sp. SMIA produced protease stable at 30°C for 2 h [16]. The enzyme

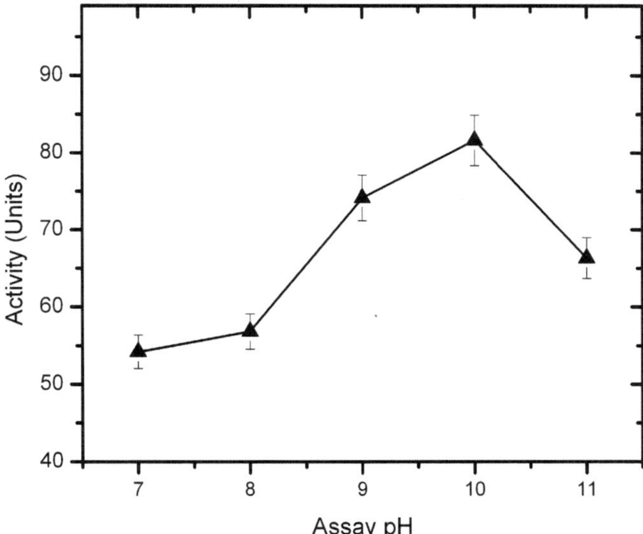

Fig. 8.6 Activity of SDR-10 protease at different pH values

retained 84% of its original activity after incubation at 50 and 60°C for 2 h [17]. Stability of protease at 55°C for 3 h was reported by Sevinc and Demirkan [18].

Effect of pH on Enzyme Activity

Bacillus species are known to produce proteases belonging to three categories: acidic, neutral and alkaline based on their pH of activity [19]. In the present study we tried to characterise the pH optimum of the protease produced by *Bacillus* SDR-10. The optimum pH for protease activity was at pH 10.0. In the pH values 11.0 and 9.0, 81.3% and 90.8% of activity at pH 10.0 was observed (Fig. 8.6). Higher protease activity under alkaline pH is suggestive that the enzyme under study can be accounted under alkaline protease. Optimum pH of 10.0 was observed in *Bacillus* spp. [20]: *B. subtilis* CN215 [21] and *Bacillus subtilis* PE-11 [22]. The higher pH optimum is always advantageous for application as detergent additive in commercial detergent formulations.

pH Stability

The alkaline protease from *Bacillus* SDR-10 was having optimum pH at 10.0. In order to understand the stability of this protease under harsh alkaline conditions of detergent washing, the stability was studied for 4 h at a temperature of 45°C at pH value 10. The enzyme was stable for 90 min, thereafter started losing its stability. SDR-10 protease is important as this is having high activity in the alkaline range. The optimum temperature of 45°C is also promising because low-temperature washing are preferred due to environment and energy concerns. This is because of the possibilities of saving energy and greater use of synthetic fibres which cannot be washed at higher temperature [23]. In this aspect, SDR-10 protease is of potential application.

Optimisation of Nutritional Parameters for Protease Production

A series of experiments were carried out to study the effects of various carbon and nitrogen sources including both simple and complex ones on the protease production by the newly isolated *Bacillus* spp.

Effect of Nitrogen Sources on Protease Production

Nitrogen is very essential for the growth of microorganisms that could be either as organic or inorganic form influencing the production of various

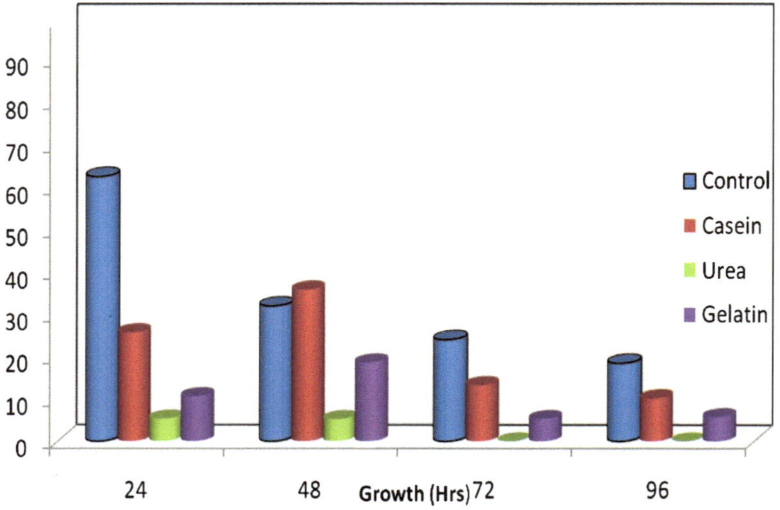

Fig. 8.7 Protease production profile of *Bacillus* SDR-10 using different organic/complex nitrogen sources

biomolecules by the bacteria. It is reported that proteases constitute only 15.6% of the biomolecules [24]. The requirement of the type and amount of nitrogen source varies from one organism to the other.

Effect of Organic/Complex Nitrogen Sources

The production of protease was evaluated in three different media with organic or complex nitrogen sources against the control (peptone medium). The growth of bacteria in the control medium for a period of 24 h was found to be giving the maximum protease activity (62. 5 units) in the cell-free culture supernatant (Fig. 8.7). Casein resulted in 41% of the maximum activity. However, the production pattern showed a difference here; i.e. the peak production was shifted to 48 h.

Urea as a nitrogen source was poor in inducing protease. Similar to the medium containing casein, gelatine also expressed maximum protease production at 48 h (30.8%).

In *Bacillus* sp. PCSIR EA-3 [25], among the various nitrogen sources studied peptone induced the highest production of protease as in *Bacillus* SDR-10. The growth of the strain was good in casein as well as control media. Although gelatine supported growth, the enzyme production was low. Urea was ineffective in inducing enzyme production. Similarly growth was least with respect to urea. The inhibitory effect of urea on enzyme production has been reported in *B. licheniformis* ATCC 21415 by Mabrouk et al. [26].

Irrespective of the initial pH value, the medium pH turned towards 8.0 to 8.5 at the end of fermentation. In *Bacillus polymyxa*, [27] reported that even though the pH initially decreased, it reached alkalinity when secondary metabolites were released and became constant.

Effect of Inorganic Nitrogen Sources

The different inorganic nitrogen sources failed to induce protease production to a significant level. Out of the four different types studied, ammonium sulphate was the most promising. It could induce nearly 35% of the activity compared to the peptone medium. However, sodium nitrate and ammonium nitrate were found to be very poor inducers of proteolytic enzymes in this study (Fig. 8.8).

The growth of the strain was noticed in different inorganic nitrogen sources. Here also the growth was not promoted in inorganic nitrogen sources as compared against control medium. Production of protease by many bacteria was found to be inhibited by ammonium salts [28].

The pH profile in the different inorganic sources showed the following trend. All the media were

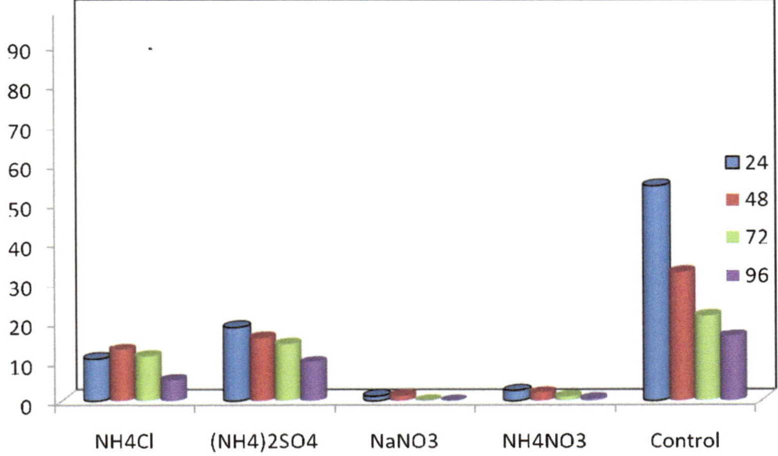

Fig. 8.8 Comparative production profile of *Bacillus* SDR-10 in different inorganic nitrogen sources

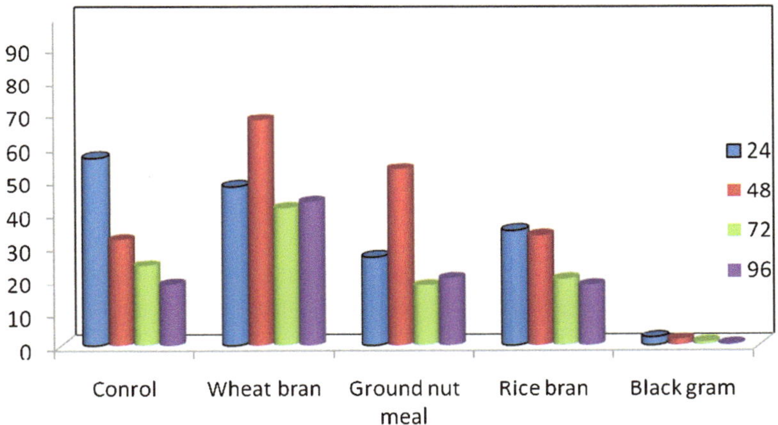

Fig. 8.9 Comparative profile of *Bacillus* SDR-10 in different agrowastes

having a starting pH between 6.5 and 7.0. However, as growth progressed, control medium pH shifted towards alkaline side, while in the inorganic media the pH remained more or less near the neutral range indicating the non-production of enzymes using these substrates.

Effect of Low-Cost Agrowastes

Different low-cost agrowastes have always attracted enzymologists as alternative economical substrates for enzyme production. In the present study, the effect of four different substrates such as wheat bran, ground nut meal, rice bran and black gram husk was evaluated. All these are cheap substrates and are locally available.

Wheat bran showed good protease production with 67.8 units after 48 h of growth (Fig. 8.9), while in the control medium, it was 56.5 units after 24 h. With wheat bran, the production was shifted to 48 h. Here, growth was also maximum after 48 h. *Bacillus* spp. generally secreted alkaline protease in the late exponential and stationary phases of growth [29]. The production of protease in this stage is probably to utilise macromolecules in the environment in response to the readily available nutrients in the medium [30]. In fact, the production pattern of enzyme is largely influenced by the various cultural and nutrient conditions [31].

Production in ground nut meal was also effective in inducing protease. Nearly 95% of the

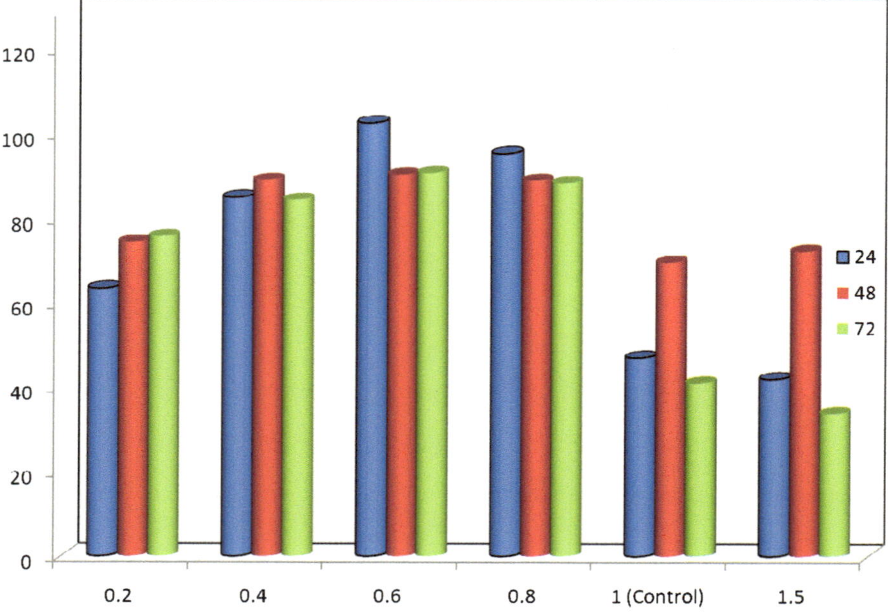

Fig. 8.10 Comparative production profile of *Bacillus* SDR-10 in different concentrations of wheat bran

activity at control medium could be obtained using this as the substrate after 48 h. Growth was also promising in this medium.

Rice bran, a by-product of rice industry, was tried as a substrate in the study. Rice bran as a medium component resulted only 60% enzyme production of that in the control medium. Moreover, the pH of the culture medium dropped to acidic side after the growth. However, there are reports regarding the use of rice bran as substrate in solid state fermentation [32]. When black gram husk was used as the substrate, this failed both in inducing proteolytic activity as well as growth.

In the present investigation, it is noticed that wheat bran is resulting in the highest level of protease activity when compared to other organic and inorganic nitrogen sources. The high level of protease production in the presence of wheat bran was observed at 24 and 48 h of incubation periods. Casein and peptone also showed good enzyme activity, but it is less than wheat bran. It has been previously reported that nitrogen sources have considerable influence on production of alkaline protease. Many bacterial species have different preferences towards the use of either organic or inorganic nitrogen for growth and protease production. The inducing effect of nitrogen sources like peptone and soybean meal on bacterial alkaline protease production has been well documented [27].

Effect of Different Levels of Wheat Bran

The present investigation showed that wheat bran has a significant role in alkaline protease production; 0.6% of wheat bran showed the maximum enzyme activity at 24 h (Fig. 8.10). At higher concentrations of wheat bran, there was reduction in protease production. This may be due to catabolic repression or substrate inhibition [33].

Effect of Different Carbon Sources

Various sources of carbon such as glucose, sucrose, starch, xylose and glycerol were selected to study the production of protease by the newly isolated *Bacillus* SDR-10. Results obtained showed highest protease production in glucose medium (Fig. 8.11). Except glycerol, all other carbon sources could result in significant level of protease production from the culture. Starch caused high level of enzyme expression in *Bacillus* species [34]. Geethanjali and Subash [35] reported similar results in *Bacillus subtilis*. Based on the results of the present study, glucose has the maximum protease production which was selected for further optimisation studies.

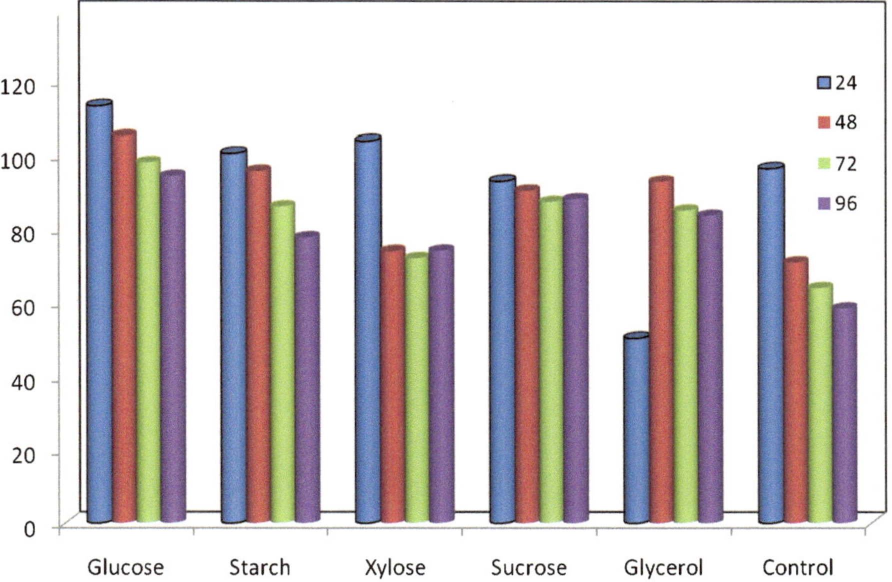

Fig. 8.11 Comparative production profile of *Bacillus* SDR-10 in different carbon sources

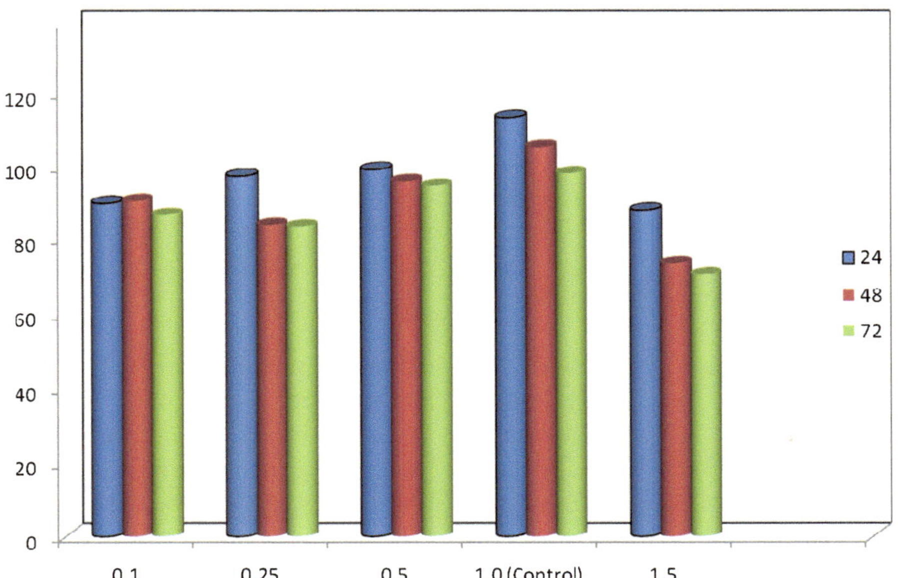

Fig. 8.12 Comparative production profile of *Bacillus* SDR-10 in different concentrations of glucose

Effect of Different Levels of Glucose on Protease Production

Among the different carbon sources tried, glucose resulted in the maximum growth of *Bacillus* SDR-10. The effect of different concentration of glucose on protease production revealed that the maximum amount of enzyme production was observed in 1% of glucose. The concentration of 1% glucose significantly gave the highest yield of protease (Fig. 8.12).

It was equally observed that initially the production of protease increased with increase in the concentration of glucose but declined at 1.5%. The result obtained showed that 1% of glucose showed the maximum enzyme activity at 24 h; the protease activity was reduced at 48 h of

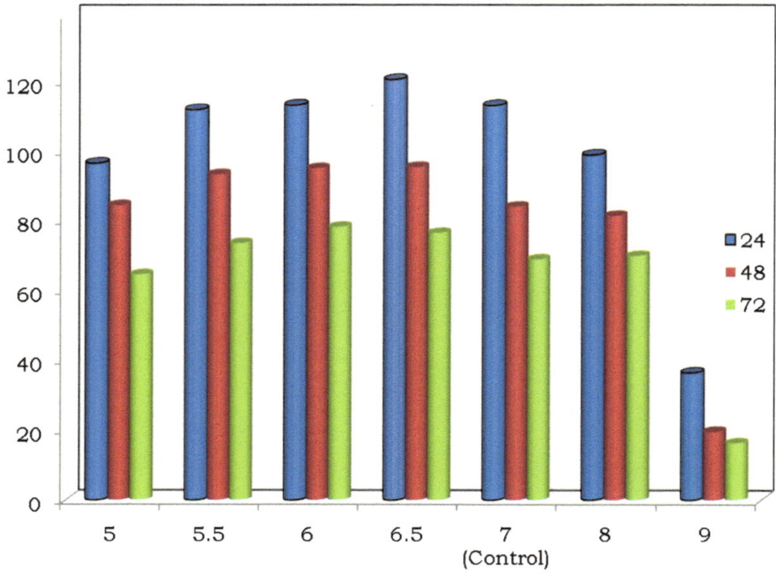

Fig. 8.13 Protease production in different media pH

incubation. The optimum concentration of glucose for the maximum production of alkaline protease by *Bacillus subtilis* is 1%. Francis et al. [36] reported that the concentration of 5% glucose was chosen for the optimum concentration for the protease production by *A. carbonarius*.

Optimisation of Cultural Parameters

Effect of pH on Protease Production

Medium pH is a critical factor in any fermentation process. In order to find the optimum pH for growth and enzyme production, the pH values such as 5, 6, 6.5, 7, 8 and 9 were adjusted in the fermentation medium. Maximum enzyme activity was observed in a medium of pH 6.5 (Fig. 8.13). Growth was optimum in the neutral range. Sarkar et al. [37] reported maximum enzyme activity when pH of the medium was maintained at pH 6.0 in *Bacillus* sp. and pH 6.5 was optimum in *B. subtilis* [38]. Geethanjali et al. [35] reported that the maximum protease production was observed in the medium pH 9.0.

The growth of the bacterium in different pH values is shown in Fig. 8.13. Although fairly good growth could be observed in the higher pH values, in the pH range of 5.0–7.0, growth was rather similar, with optimum growth in the pH range of 6.5–7.0. A similar result has been reported by Sarkar et al. [37].

Irrespective of the initial pH value, the medium pH shifted towards alkaline side in all the media. This trend has been observed in different species of *Bacillus* [27]. Increase in pH can be possibly due to proteolysis and release of metabolites [37].

Effect of Inoculum on Protease Production

The amount of inoculum used to grow the bacteria also affects protease production. Results from the present study showed that optimum inoculum size for protease production in the newly isolated strain was 5.0% (Fig. 8.14). An increase in inoculum size from 1 to 5% showed a progressive increase in enzyme activity reaching the highest at 5% level, while further increase caused overcrowding the spores that decreases the enzyme activity. Such results were also reported by Kezia et al. [10].

The increase in protease production using small inoculum sizes was suggested to be due to the higher surface area to volume ratio resulting in increased protease production [38]. In addition, improved distribution of dissolved oxygen

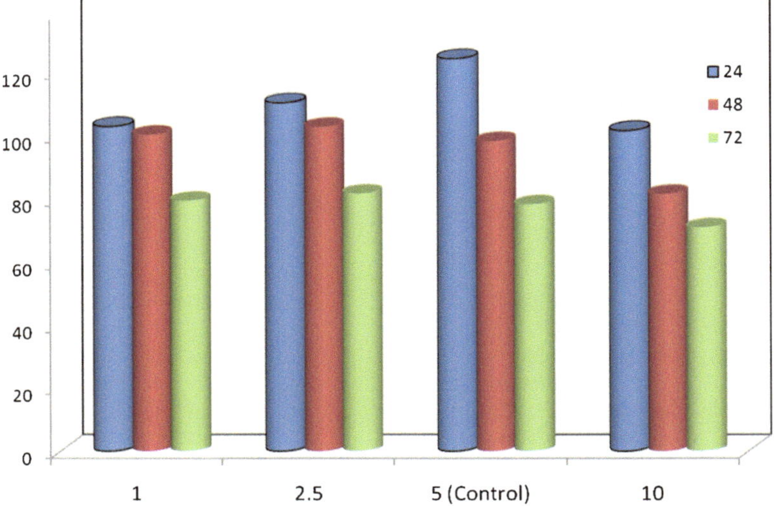

Fig. 8.14 Comparative protease profile of *Bacillus* SDR-10 in different levels of inoculum

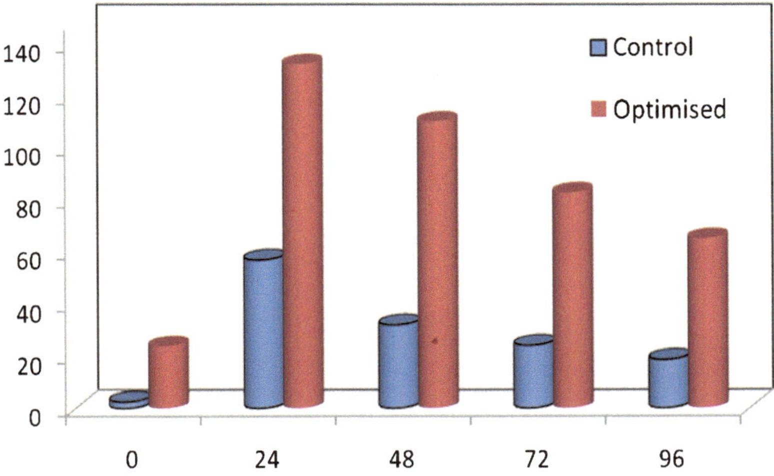

Fig. 8.15 Optimisation against control

and more effective uptake of nutrients also contributed to higher protease production. If the inoculum size is too small (2.0%), insufficient number of bacteria would lead to reduced amount of secreted protease. One study has suggested that higher inoculum sizes resulted in reduced dissolved oxygen and increased competition towards nutrients [38]. Kezia et al. [10] has reported the maximum enzyme production and growth at 5.0% inoculum level. Similar result has been noted in the present study.

Comparative Study of Optimised Medium Against Control

When optimised medium was compared against the control, 2.35-fold increases in activity could be obtained (Fig. 8.15). The alkaline protease from *Bacillus* SDR-10 could produce high amount of protease as a result of optimisation, and the enzyme possesses many of the required characteristics as a detergent additive. The future work will be concentrated on the purification and application aspects of this enzyme.

Acknowledgement The authors are thankful to the Environmental Technology Division, National Institute for Interdisciplinary Science and Technology, Thiruvananthapuram 695019, for the scientific assistance. We are also grateful to the Director, Department of Collegiate Education, Government of Kerala, India, and the Principal, University College, Thiruvananthapuram.

References

1. Haddar A, Fakhfakh-Zouari N, Hmidet N, Frikha F, Nasri M, Kamoun AS. Low-cost fermentation medium for alkaline protease production by *Bacillus mojavensis* A21 using hulled grain of wheat and sardinella peptone. J Biosci Bioeng. 2010;110(3):288–94.
2. Kelly CT, Fogarty WM. Microbial alkaline enzyme. Process Biochem. 1976;11:3–9.
3. Gupta R, Beg QK, Lorenz P. Bacterial alkaline protease molecular approaches and industrial applications. Appl Microbiol Biotechnol. 2002;59:15–32.
4. Saeki K, Ozaki K, Kobayashi IS. Detergent alkaline proteases: enzymatic properties, genes, and crystal structures. J Biosci Bioeng. 2007;6:501–8.
5. Godfrey T, Reichet P. Industrial enzymology, the application of enzyme on industry. London: The Nature Press; 1985.
6. Singh J, Batra N, Sobti CR. Serine alkaline protease from a newly isolated *Bacillus* sp. SSR1. Process Biochem. 2001;36:781–5.
7. Wilson HR, Chan PT, Turnbough Jr CL. Nucleotide sequence and expression of the pyrC gene of E. coli K-1. J Bacteriol. 1987;169(7):3051–8.
8. Tsuchida O, Yamagota Y, Ishizuka J, Arai J, Yamada J, et al. Alkaline protease activity was determined by the modified procedure by *Bacillus* sp. Curr Microbiol. 1986;14:7–9.
9. Lowry OH, Rosebrough NJ, Farr AL, Randall RJ. Protein measurement with the folin phenol reagent. J Biol Chem. 1951;193:265.
10. Kezia D, Narasimha Rao M, Naidu SV. Influence of various environmental parameters on protease secretion from *Bacillus subtilis* DKMNR. Int Res J Pharm. 2011;2(3):178–82.
11. Mehrotra S, Pandey PK, Gaur R. The production of alkaline protease by a *Bacillus* sp. Isol Bioresour Technol. 1999;67:201.
12. Kumar CG, Takagi H. Microbial alkaline proteases: from a bio industrial viewpoint. Biotechnol Adv. 1999;17:561–94.
13. Banerjee R, Bhattacharya BC. Optimization of multiple inducers effect on protease biosynthesis by *Rhizopus oryzae*. Bioprocess Eng. 1992;7:225–9.
14. Seifzadeh S, Sajedi RH, Sariri R. Isolation and characterisation of thermophilic alkaline proteases resistant to SDS and EDTA from *Bacillus* sp. GUS1. Iran J Biotech. 2008;6(4):214–21.
15. Chomsri N. Thermostable protease enzymes. Master thesis in Biotechnology. The Graduate school, Chiangmai University, Chiangmai; 2001.
16. Nascimento WCAD, Martins MLL. Production and properties of an extracellular protease from thermophilic *Bacillus* sp. Braz J Microbiol. 2004;35:91–6.
17. Akel H, Al-Quadan F, Yousef TK. Characterization of a purified thermostable protease from hyper thermophilic *Bacillus* strain HUTBS71. Eur J Sci Res. 2009;31(2):280–8.
18. Sevinc N, Demirkan E. Production of protease by *Bacillus* sp. N-40 isolated from soil and its enzymatic properties. J Biol Environ Sci. 2011;5(14):95–103.
19. Ward OP. Proteolytic enzymes. In: Moo-Young M, editor. Comprehensive biotechnology, vol. 3. Blanch HW, Drew S, Wang DIC, editors. The practice of biotechnology; current commodity products. Oxford: Pergamon; 1985. pp. 789–818.
20. Gupta R, Beg QK, Khan S, Chauhan B. An overview on fermentation, downstream processing and properties of microbial alkaline proteases. Appl Microbiol Biotechnol. 2005;60:381.
21. Uchida H, Kondo D, Yamashita S, Tanaka T, Tran HL, Nagano H, Uwajima T. Purification and properties of a protease produced by *Bacillus subtilis* CN2 isolated from a Vietnamese fish sauce. World J Microbiol Biotechnol. 2004;20:579–82.
22. Adinarayana K, Ellaiah P. Response surface Optimization of the critical medium components for the production of alkaline protease by a newly isolated *Bacillus* sp. J Pharm Pharm Sci. 2002;5(3):272.
23. Griffin HL, Green CMA. Isolation and characterisation of an alkaline protease from a marine ship worm bacterium. Curr Microbiol. 1992;24:111–2.
24. Kole MM, Draper I, Gesson DF. Protease production by *Bacillus subtilis* in oxygen controlled glucose fed batch fermentation. Appl Microbiol Biotechnol. 1988;28:404–8.
25. Qadar SAU, Shireen E, Iqbal S, Anwar A. Optimization of protease production from newly isolated strain of *Bacillus* sp. PCSIR EA-3. Indian J Biotechnol. 2009;8:286–90.
26. Mabrouk SS, Hashem AM, El-Shayeb NMA, Ismail A-MS, Abdel-Fattah AF. Optimization of alkaline protease productivity by *Bacillus licheniformis* ATCC 21415. Bioresour Technol. 1999;69:155–9.
27. Kaur M, Dhillon S, Chaudhary K. Production, purification and characterization of a thermostable alkaline protease from *Bacillus polymyxa*. Indian J Microbiol. 1998;38:63–7.
28. Liu V, Hsieh HC. Inhibition of protease production of various bacteria by ammonium salts: its effect on toxin production and virulence. J Bacteriol. 1969;-99(2):406–13.
29. Priest FG. Extracellular enzyme synthesis in the genus *Bacillus*. Bacteriol Rev. 1977;41:711–53.
30. Hangeman JK, Shankar Weiler GW, Wall PR, Franich K, Mac Cowan GW, Cauble SM, Quinones C. J Bacteriol. Single, chemically defined sporulation medium for

Bacillus subtilis: growth, sporulation, and extracellular protease production. 1984;160:438.
31. Greesham R, Inamine F. Nutritional importance of proteases. In: Manual of industrial microbiology and Biotechnology; 1986. pp. 41–48.
32. Selvamohan T, Sherin S. Optimization of protease production from *Bacillus cereus* using different substrates. Plant Arch. 2010;10(2):651–66.
33. Nadeem M, Javed I, Shah Jahan B. Effect of medium composition on commercially important alkaline protease production by *Bacillus licheniformis* N2. Food Technol Biotechnol. 2008;46(4):388–94.
34. Mahmood AU, Greenman J, Scragg AH. Microbios. Effects of macromolecular growth substracts on production of extracellular enzymes by *Bacillus* species in continous culture. 2000;103:85.
35. Geethanjali S, Subash A. Optimization of protease production by *Bacillus subtilis* isolated from midgut of fresh water fish *Labeo rohita*. World J Fish Mar Sci. 2011;3(1):88–95.
36. Francis IS, Okolo BN, Moneke AN, Odibo FJC. Influence of cultivation conditions on the production of a protease from *Aspergillums carbonarius* using submerged fermentation. Afr J Food Sci. 2011;5(6):353–65.
37. Sarkar PK, Cook PE, Owens JD. *Bacillus* fermentation of soybeans. World J Microbiol Biotechnol. 1993;9:-295–9.
38. Rahman RNZA, Geok LP, Basri M, Salleh AB. Bioresour Technol. Physical factors affecting the production of organic solvent-tolerant protease by *Pseudomonas aeruginosa* strain K. 2005;96:429.

Identification of a PPAR Delta Agonist: An *In Silico* Approach Towards Drug Design for Metabolic Syndrome

9

Serena D'Souza, Uma Maheshwari, Santhosh Rebello, and Asha Abraham

Abstract

Metabolic syndrome is characterized by the clustering of abnormalities such as high blood pressure, low HDL, high LDL, central obesity and raised fasting blood glucose. These abnormalities increase the risk of developing cardiovascular disease and diabetes. The nuclear receptor, peroxisome proliferator activated receptor delta(PPARδ), also known as NR1C2, is known to regulate lipid metabolism. It is modulated by ligand binding. In this study, we have used molecular docking technique to find a candidate agonist for PPARδ. Docking study was performed with 346 compounds (obtained from PubChem) similar to GW501516 and the molecular target PPARδ (crystallographic structure was obtained from RSCB database) using AutoDock Vina 4.2 and ADT 1.5.4 program. The docked ligands were analyzed on the basis of binding free energy and hydrogen bond interactions. The docked ligands showed favourable hydrogen bonding with the receptor at a distance of 3.7 Å. Among the compounds analyzed, two ligands were considered, namely, ethyl-2-[2-methyl-4-[1-[4-methyl-2-[4-(trifluoromethyl)phenyl]-1,3-thiazol-5-yl]-5-phenylpentyl]sulfanylphenoxy] acetate with a binding free energy

S. D'Souza • A. Abraham (✉)
Father George Albuquerque Pai Cell and Molecular Biology Laboratory, PG Department of Biotechnology, St Aloysius College (Autonomous), Mangalore 575003, Karnataka, India
e-mail: drashaabraham@staloysius.edu.in

U. Maheshwari
Department of Bioinformatics, AIMIT, St Aloysius College (Autonomous), Beeri, Kotekar Post, Mangalore 575022, Karnataka, India

S. Rebello
PG Department of IT and Bioinformatics, AIMIT, St Aloysius College (Autonomous), Beeri, Kotekar Post, Mangalore 575022, Karnataka, India

A. Sabu and A. Augustine (eds.), *Prospects in Bioscience: Addressing the Issues*, DOI 10.1007/978-81-322-0810-5_9, © Springer India 2013

of −11.0 kcal/mol and zero hydrogen bonds and ethyl-2-[2-methyl-4-[[4-methyl-2-[4-(trifluoromethyl)phenyl]-1,3-thiazol-5-yl]-(2,3,4,5,6-pentafluorophenyl)methyl]sulfanylphenoxy] acetate with binding free energy of −10.4 kcal/mol and two hydrogen bonds with Glu 471 of the B chain and Asn 312 of the A chain of the receptor. Further wet lab studies need to be carried out to study the functional characteristics of the ligands on PPARδ and its role in the syndrome.

Keywords
Metabolic syndrome • Peroxisome proliferator activated receptor delta • Molecular docking

Introduction

The metabolic syndrome represents a combination of cardiovascular risk determinants such as obesity, insulin resistance, hypertension and lipid abnormalities. Long-term energy excess and/or decrease of physical activity is known to create profound disturbance in lipid homeostasis and possibly lead to metabolic syndrome. The syndrome is associated with a five fold higher risk of developing type II diabetes and two to three fold higher risk of cardiovascular disease. Expected lifespan is reduced by an average of 15 years. The causes for this syndrome are multiple and not totally elucidated. Peroxisome proliferator activated receptors (PPARs) are membrane associated transcription factors belonging to the nuclear hormone family and are implicated in the lipid metabolism. Hence, they are ideal candidates for therapeutic approaches. These receptors are ligand activated and are to date of three subtypes, namely, PPARα (NR1C1), PPARβ/δ (NR1C2) and PPARγ (NR1C3) encoded by different genes. PPARγ has two isoforms, PPARγ1 and PPARγ2 [1]. Each PPAR subtype shows a distinct tissue distribution wherein PPARδ/β has the broadest expression pattern and the levels of expression in certain tissues depend on the extent of cell proliferation and differentiation. PPARδ/β is highly expressed in colon, small intestine, liver and keratinocytes [2]. The role of PPARδ in metabolic syndrome is being investigated lately. It is known to be an important regulator of fuel handling, namely, lipid metabolism [3], accelerate the depletion of intracellular lipid stores and have a beneficial role in hypertriglyceridemia and insulin resistance [4]. It is implicated in the expression of genes involved in lipid uptake such as fatty acid-binding protein and fatty acid translocase [5]. These findings point out the potential of using PPARδ as a promising new target receptor. Down regulation of PPARδ has also been observed in metabolic syndrome [6, 7]. PPARδ agonists like GW501516 [8] are known to increase HDL cholesterol and decrease triglycerides, along with VLDL and LDL cholesterol, affect lesion progression in a mouse model of artherosclerosis and cause weight loss [3]. These studies suggest that PPARδ agonists might be useful in the treatment of metabolic syndrome.

Identification of an agonist can be carried out using the traditional 'trial-and-error method of testing' or through *in silico* approaches such as computer-based drug design. The use of bioinformatics tools requires the combined knowledge of structural biology and computational drug design and can be done either by using ligand based approach (indirect drug design) or receptor based approach (direct drug design). The ligand based approach uses known active and inactive compounds which are used to develop a pharmacophore. The receptor based approach requires the availability of the three dimensional structure of the receptor which is then used to examine the interactions of the ligands. The search can include natural product derived agonists or synthetic ligands. This may include screening libraries of

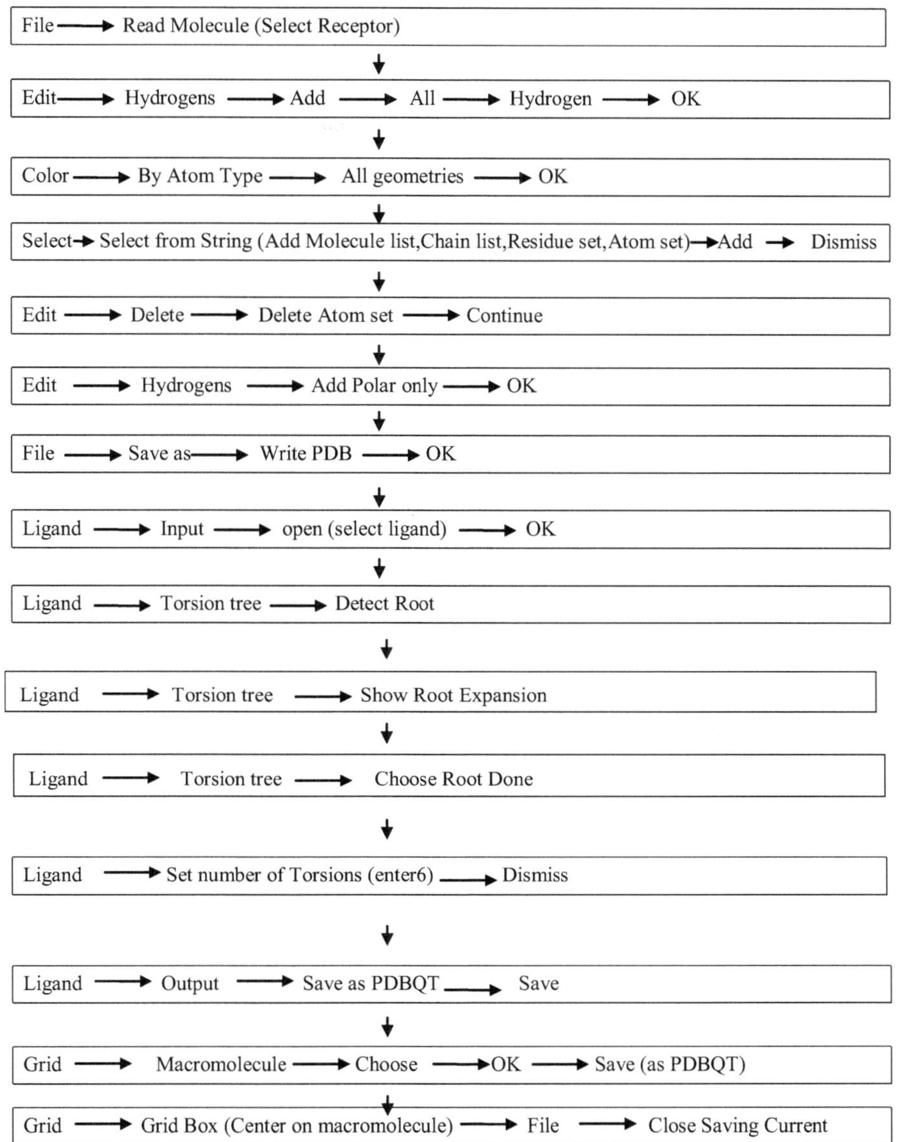

Fig. 9.1 Summary of methodology used during the study

existing ligands (virtual screening), designing suitable ligands using tools like Marvin Sketch or developing a library using structural template. The final step includes molecular docking which predicts both ligand conformation and orientation within a targeted binding site. In this study, a receptor based approach utilizing structural template has been used. The structural template used is GW501516 since it is known to have selective agonism for PPARδ.

Materials and Methods

The methodology adopted is summarized in Fig. 9.1.

Receptor

Briefly, the receptor used for the study is PPARδ. The structure 2ZNP for PPARδ receptor was

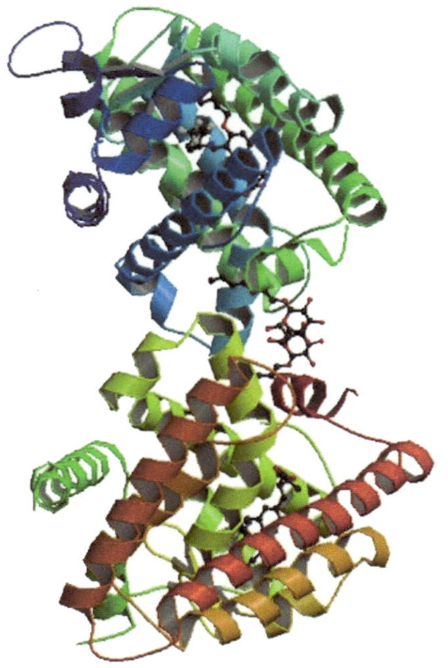

Fig. 9.2 PDB structure of PPAR delta with TIPP204

Fig. 9.3 Parent template GW501516

retrieved from Research Collaboratory for Structural Bioinformatics (RCSB) Protein Data Bank (http://www.rcsb.org/pdb). The crystal structure is a complex of synthetic agonist TIPP204 and the PPARδ receptor (Fig. 9.2). In this crystal structure, the ligand binding domain contains two chains A and B. The interface of the A and B monomer of the homodimer was chosen as the target for docking studies. The ligand was removed from the complex using Swiss-PdbViewer (DeepView).

Ligand

The ligand GW501516 PubChem id 9803963 was considered as the parent template (Fig. 9.3). The 3D conformer of GW501516 was obtained from PubChem (http://pubchem.ncbi.nlm.nih.gov/). Using GW501516 as a parent template, a similar compound search was carried out and 346 similar compounds were obtained.

Docking

The docking analyses of 346 compounds and GW501516 were carried out by means of the AutoDock Vina 4.2 [9]. AutoDock 4.0 was compiled and run under Windows XP operating system. The graphical user interface program 'AutoDock tools', also called 'MGLTools 1.5.4', was used to prepare the receptor and the ligand. The active site of the receptor was considered as a flexible molecule; Kollmann charges and polar hydrogens were added. The grid was centred at the interface of the A and B monomers of the receptor. The grid box size was set at 40, 40, 40 Å, whereas the centre of the grid box was set at 13.962 (x), −18.372 (y) and 23.759 (z). The spacing between the grid points was 0.375 Å. The ligands were treated as being flexible, Gasteiger charges were assigned as the ligands were not peptides, the nonpolar hydrogens were merged, and rigid roots of each ligand were defined automatically. The amide bonds were made non-rotatable. To run AutoDock Vina, command prompt was used, and the name of the protein, ligand in the PDBQT format, the values for the centre of grid box and the size of the grid box were fed.

The docked molecules were viewed in PyMOL. To view the docked molecules, the receptor is required to be present in the PDB format and the docked molecule in the PDBQT format. The interactions were viewed using the

Table 9.1 Top 10 ligands

Receptor (PDB ID)	Ligand no.	Interactions Receptor Chain	Residue	Atom	Ligand Atom	Bond length(Å)	Number of interactions	Energy (Kcal/mol)
2ZNP	134	Nil					0	−11.0
	127	B	Glu471	O	N	3.68	2	−10.4
		A	Asn312	O	N	3.7		
	131	B	Glu47	O	N	3.65	2	−10.3
		A	Asn312	O	N	3.75		
	160	A	Ile472	N	O	3.28	3	−10.0
		A	Gln471	O	O	3.11		
		B	Gln471	O	O	3.66		
	205	A	Leu311	N	O	2.97	5	−10.0
		B	Thr297	O	O	3.18		
		B	Thr297	O	O	3.11		
		A	Thr316	O	O	3.6		
		A	Asn312	O	N	3.67		
	245	B	Glu471	O	N	3.66	1	−9.9
	342	B	Thr316	O	HO1(O)	3.04	8	−9.9
		B	Asn312	O	O	3.02		
		B	Asn312	O	O	3.1		
		A	Thr316	O	HO1(O)	3.52		
		B	Thr316	O	O	3.04		
		B	Gly399	O	HO1(O)	3.33		
		A	Glu471	O	HO1(O)	3.23		
		A	Asn312	O	N	3.65		
	129	B	Glu471	O	N	3.66	2	−9.8
		A	Asn312	O	N	3.74		
	128	A	Asn312	O	N	3.74	2	−9.8
		B	Glu471	O	N	3.71		
	266	A	Asp313	O	O	3.17	10	−9.8
		A	Thr316	O	O	3.05		
		A	Asn312	O	O	3.05		
		A	Gly399	O	O	3.16		
		A	Gly399	O	O	3.31		
		A	Asp313	O	O	3.48		
		A	Asp313	O	O	3.36		
		A	Asp313	O	O	3.05		
		A	Met401	N	O	3.58		
		A	Glu471	O	N	3.11		
	301	A	Glu471	O	O	2.95	5	−9.8
		B	Asn312	O	O	2.93		
		A	Glu471	O	HO1(O)	3.29		
		B	Glu471	O	N	3.79		
		B	Thr316	O	N	3.31		
	323	A	Glu471	O	N	3.06	2	−9.8
		A	Asn312	O	O	3.39		
	326	B	Thr316	O	O	2.71	6	−9.8
		B	Thr316	O	O	2.8		
		A	Asn312	O	N	3.67		
		A	Asn312	O	O	2.97		
		A	Asn312	O	O	3.77		
		B	Thr316	N	O	3.37		

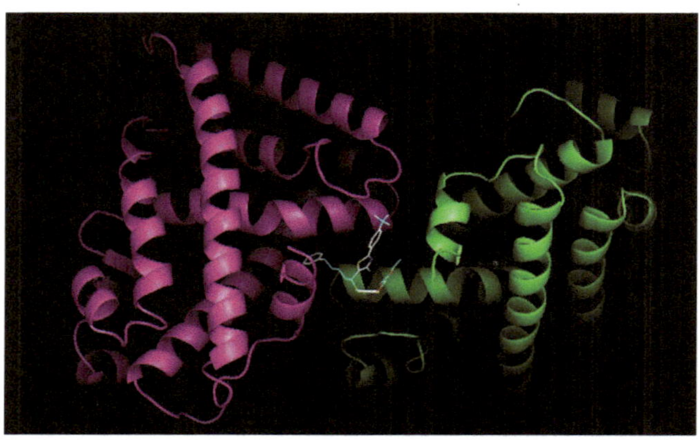

Fig. 9.4 Docked test ligand 134+PPARδ. *Magenta colour* represents A chain of the PPARδ receptor. *Light green colour* represents B chain of PPARδ receptor. *Ligand* represented as line model

command prompt wherein hydrogen was added both to the receptor and the ligand, the donor and the acceptor were selected and the length of the hydrogen bond between the donor and the acceptor was specified.

Results and Discussion

On performing similar compound search using GW501516 as the parent template, 346 compounds were obtained. The docked receptor ligand complexes were analyzed on the basis of binding free energy and number of interactions, namely, hydrogen bonds. Binding free energy gives an idea of the affinity of the ligand to the receptor and also the stability of the docked molecule. Hydrogen bonds provide an insight into the overall flexibility of the molecule. More than five hydrogen bonds in a docked molecule are known to interfere with the permeability of the ligand *in vivo*. It was found that the energy of the docked ligands ranged between −11.0 kcal/mol and −5.4 kcal/mol, and the number of hydrogen bonds varied from 0 to 10. Upon analysis, the top 10 ligands were considered (Table 9.1). On the basis of the energy parameter, test ligand 134, namely, ethyl-2-[2-methyl-4-[1-[4-methyl-2-[4-(trifluoromethyl)phenyl]-1,3-thiazol-5-yl]-5-phenylpentyl] sulfanylphenoxy] acetate (Fig. 9.4) was found to have the lowest binding free energy of −11.0 kcal/mol with zero hydrogen bonds. However, on considering both energy and number of hydrogen bonds, test ligand 127, namely, ethyl-2-[2-methyl-4-[[4-methyl-2-[4-(trifluoromethyl)phenyl]-1,3-thiazol-5-yl]-(2,3,4,5,6-pentafluorophenyl) methyl]sulfanylphenoxy] acetate (Fig. 9.5a, b), was found to have binding free energy of −10.4 kcal/mol and 2 hydrogen bonds with Glu 471 of the B chain and Asn 312 of the A chain of the receptor. A comparison of the obtained test ligands with the parent template showed that both test ligands had lower binding free energy. Also, test ligand 127 showed more hydrogen bonds as compared to the parent template (Fig. 9.6a, b). Thus, the obtained test ligands do show the potential to be used as an agonist for the receptor.

Since PPARδ is still an orphan receptor, *in silico* approaches provide a means of developing/designing drug targets pro-PPARδ. As of now, very few efficacious synthetic agonists have been developed for this receptor. Utilization of structural templates to guide the development of a possible candidate agonist is a very useful approach to accelerate the process of drug design. Such approaches have been put to use only in the near past [10, 11]. However, these approaches have used the structural template to design ligands. In our study, the parent template, namely, GW501516, has been used

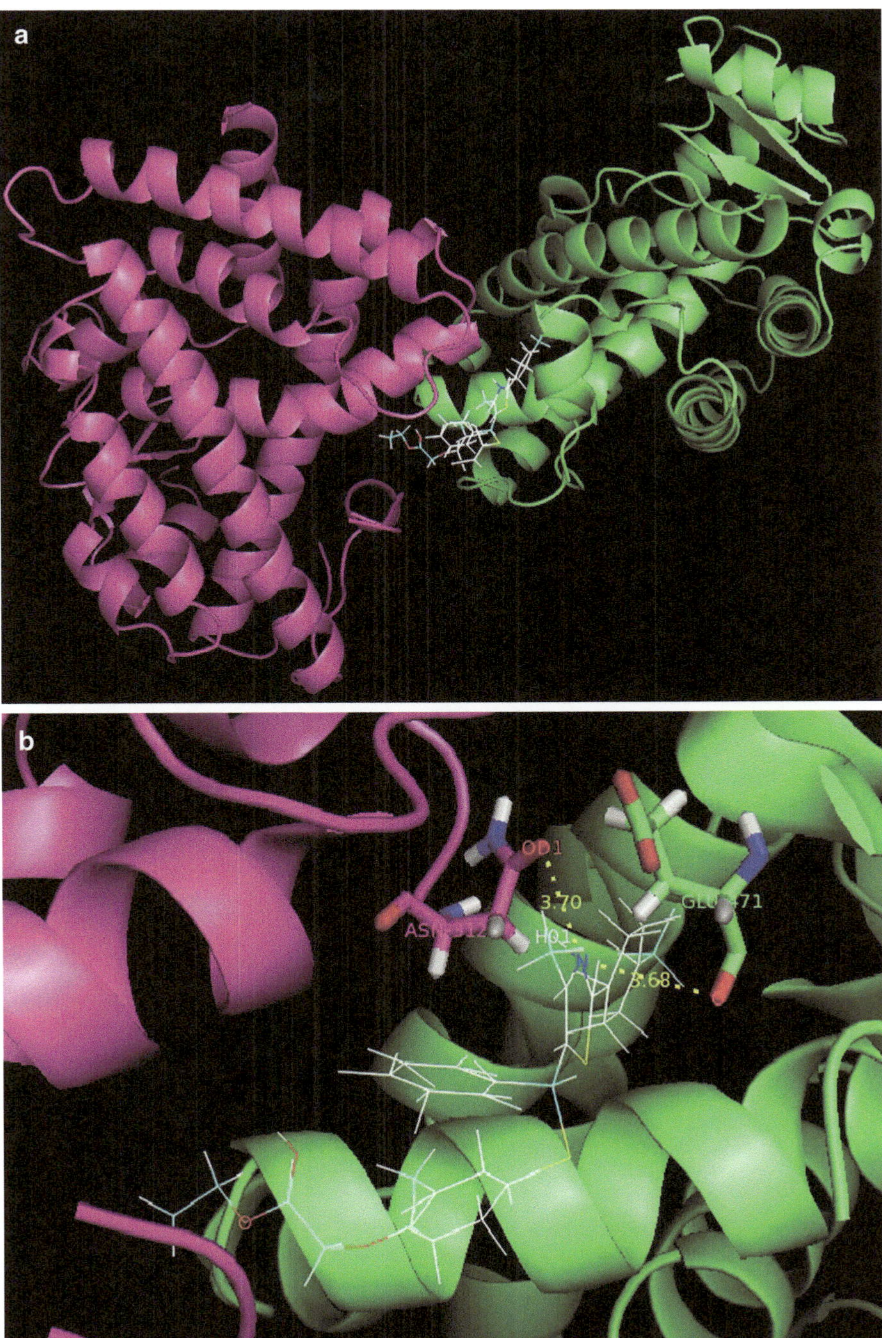

Fig. 9.5 (**a**) Docked test ligand 127 + PPARδ. *Magenta colour* represents A chain of the PPARδ receptor. *Light green colour* represents B chain of PPARδ receptor. *Ligand* represented as line model. (**b**) Docked test ligand 127 + PPARδ with interactions. *Magenta colour* represents A chain of the PPARδ receptor. *Light green colour* represents B chain of PPARδ receptor. *Ligand* represented as line model. *Yellow dotted lines* represent hydrogen bonds between ligand and receptor

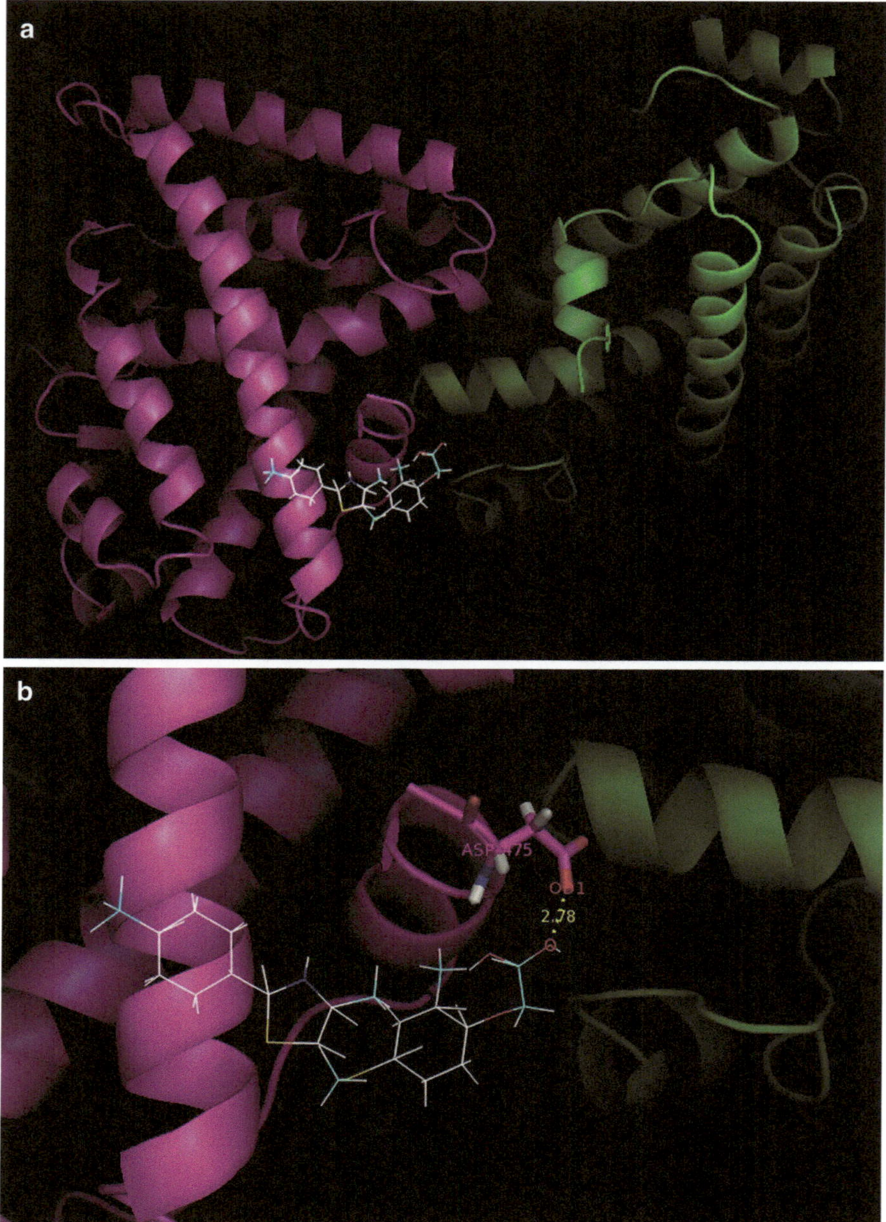

Fig. 9.6 (**a**) Parent ligand + PPARδ. *Magenta colour* represents A chain of the PPARδ receptor. *Light green colour* represents B chain of PPARδ receptor. *Ligand* represented as line model. (**b**) Parent ligand + PPARδ with interactions. *Magenta colour* represents A chain of the PPARδ receptor. *Light green colour* represents B chain of PPARδ receptor. *Ligand* represented as line model. *Yellow dotted lines* represent hydrogen bonds between ligand and receptor

as a guide to search for structures similar to the ligand. The beneficial role played by a PPARδ may be due its mimicking of the actions of physical exercise [12], improving adaptive response to endurance exercise [13] or limiting substrate availability for lipid synthesis and accumulation in adipose, as well as other insulin-sensitive tissues. However, the regulatory role of this receptor in lipid homeostasis, insulin sensitivity and metabolism as a whole remains yet a mystery. Agonists may act as useful tools to elucidate its possible role.

Conclusions

In conclusion, our results show that ethyl-2-[2-methyl-4-[1-[4-methyl-2-[4-(trifluoromethyl) phenyl]-1,3-thiazol-5-yl]-5-phenylpentyl] sulfanylphenoxy] acetate and ethyl-2-[2-methyl-4-[[4-methyl-2-[4-(trifluoromethyl) phenyl]-1,3-thiazol-5-yl]-(2,3,4,5,6-pentafluorophenyl)methyl]sulfanylphenoxy] acetate could be used as agonists for the receptor PPARδ and may also be useful tools in understanding the cascade.

Acknowledgement Serena D'Souza would like to thank CSIR for providing Junior Research Fellowship to pursue doctoral studies. Dr. Asha Abraham would like to thank UGC for providing a major research project. All the authors thank the Principal, St. Aloysius College and Director, AIMIT for providing support and necessary facilities to carry out this work.

References

1. Yu S, Reddy JK. Transcription coactivators for peroxisome proliferator-activated receptors. Biochim Biophys Acta. 2007;1771:936–51.
2. Girroir EE, Hollingshead HE, He P, Zhu B, Perdew GH, Peters JM. Quantitative expression patterns of peroxisome proliferator-activated receptor-β/δ (PPAR-β/δ) protein in mice. Biochem Biophys Res Commun. 2008;371(3):456–61.
3. Lee CH, Olson P, Hevener A, Mehl I, Chong LW, Olefsky JM, Gonzalez FJ, Ham J, Kang H, Peters JM, Evans RM. PPARδ regulates glucose metabolism and insulin sensitivity. Proc Natl Acad Sci USA. 2006;103(9):3444–9.
4. Barish GD, Narkar VA, Evans RM. PPAR delta: a dagger in the heart of the metabolic syndrome. J Clin Invest. 2006;116(3):590–7.
5. Poirier H, Niot J, Monnot MC, Braissant O, Meunier-Durmort C, Costet P, Pineau T, Wahli W, Willson TM, Besnard P. Differential involvement of peroxisome proliferator activated receptors alpha and delta in fibrate and fatty acid mediated inductions of the gene encoding liver fatty acid binding protein in the liver and the small intestine. Biochem J. 2001;355:481–8.
6. Kumar A, Hasamnis A. A clinical update on peroxisome proliferator-activated receptors. Syst Rev Pharm. 2010;1:175–81.
7. Nunn AVW, Bell J, Barter P. The integration of lipid-sensing and anti-inflammatory effects: how the PPARs play a role in metabolic balance. Nucl Recept. 2007;5:1.
8. Oliver WR, Shenk JL, Snaith MR, Russell CS, Plunket KD, Bodkin NL, Lewis MC, Winegar DA, Sznaidman ML, Lambert MH, Xu HE, Sternbach DD, Kleiwer SA, Hansen BC, Willson TM. A selective peroxisome proliferator-activated receptor δ agonist promotes reverse cholesterol transport. Proc Natl Acad Sci USA. 2001;98(9):5306–11.
9. Trott O, Olson AJ. AutoDock Vina: improving the speed and accuracy of docking with a new scoring function, efficient optimization and multithreading. J Comput Chem. 2010;31:455–61.
10. Epple R, Azimioara M, Russo R, Bursulaya B, Tian SS, Gerken A, Iskandar M. 1,3,5-Trisubstituted aryls as highly selective PPARδ agonists. Bioorg Med Chem Lett. 2006;16:2969–73.
11. Mogensen JP, Jeppesen L, Bury PS, Pettersson I, Fleckner J, Nehlin J, Frederiksen KS, Albrektsen T, Din N, Mortensen SB, Svensson LA, Wassermann K, Wulff EM, Ynddal L, Sauerberg P. Design and synthesis of novel PPARα/γ/δ triple activators using a known PPARα/γ dual activator as structural template. Bioorg Med Chem Lett. 2002;13:257–60.
12. Gaudel C, Grimaldi PA. Metabolic functions of peroxisome proliferator-activated receptor β/δ in skeletal muscle. PPAR Res. 2007;2007:86394-400;86394.
13. Luquet S, Lopez-Soriano J, Holst D, Fredenrich A, Melki J, Rassoulzadegan M, Grimaldi PA. Peroxisome proliferator activated receptor δ controls muscle development and oxidative capability. Fed Am Soc Exp Biol. 2003;17:2299–301.

Preliminary Phytochemical and Antimicrobial Activity of Ethanolic Extract of *Pittosporum viridulum* Stem Bark

10

C.T. Sadashiva, Satheesh George, Nitha Balan, U. Prakash Kumar, and Indira Balachandran

Abstract

The ethanolic extract of *Pittosporum viridulum* stem bark was screened for antibacterial activity using disc diffusion method. The phytochemical analysis of the extract was also carried out to detect total phenolic content and total flavonoids. The ethanolic extract of *P. viridulum* showed a good antimicrobial activity against *Staphylococcus aureus*, *Escherichia coli*, *Proteus mirabilis*, *Pseudomonas aeruginosa* and *Klebsiella pneumoniae*. Antibiotic chloramphenicol is used as positive control. Maximum zone of growth inhibition was observed against *Pseudomonas aeruginosa*, with an inhibition zone of 18 mm for ethanol extract. The extract used for screening was prepared by dissolving 1 and 2 mg/100 μl DMSO. Phytochemical screening of *Pittosporum viridulum* showed the presence of carbohydrates, glycosides, amino acids, fixed oils and fats, flavonoids, saponins and phenol.

Keywords

Pittosporum viridulum • Antimicrobial activity • Phytochemical analysis

Introduction

The plant family *Pittosporaceae* consists of 9 genera and 250 mainly subtropical species, 160 of which occur in the genus *Pittosporum Banka* ex. Soland and largely distributed in tropical and subtropical regions of old world, Africa, Asia, Australia, New Zealand and India, where the genus is represented by 11 species [1]. The species of *Pittosporum* genus are widely used as medicinal plants. The flowers, root, bark and leaves of this genus have well-documented uses tic, the main components being essential oils, oleoresin, triterpenoid saponins and stigmasterol [2–4]. In India, bark of *Pittosporum* genus is used as anti-inflammatory, antispasmodic and for narcotic effects [5, 6]. The bark is also useful in chronic bronchitis and is also administered in leprous affections and rheumatic swellings, and the paste of root is applied to dropsical and rheumatic swellings [5]. The oil is applied topically in sprains, bruises, sciatica, rheumatism, chest affection and

C.T. Sadashiva (✉) • S. George • N. Balan • U.P. Kumar • I. Balachandran
Centre for Medicinal Plants Research, Arya Vaidya Sala, Kottakkal, Malappuram, Kerala, India
e-mail: sada1hassan@gmail.com

A. Sabu and A. Augustine (eds.), *Prospects in Bioscience: Addressing the Issues*, DOI 10.1007/978-81-322-0810-5_10, © Springer India 2013

Fig. 10.1 Stem bark of *Pittosporum viridulum*

in certain skin diseases [6]. To our knowledge, this is the first report on preliminary phytochemical screening and antimicrobial activity of ethanolic extract of stem bark of *P. viridulum* (Fig. 10.1).

Materials and Methods

Plant Material and Extraction

The plant, *Pittosporum viridulum*, was collected from Kakkayam forest. The collected plant material was washed, air-dried under shade at room temperature and ground with a grinder into powder and used for further extract. Hundred grams of the dried bark powder of the plant material (100×3) was extracted sequentially by refluxing with ethanol for 3 h using Soxhlet apparatus at a temperature of 400°C. The extract was filtered and concentrated under reduced pressure using rotary evaporator below 600°C and stored at 40°C for further use.

Phytochemical Screening of Crude Extract

The phytochemical components in ethanolic extract were screened using the standard methods [7].

Antimicrobial Activity

Antimicrobial activity was tested by agar disc diffusion method against test microorganisms, which included *Staphylococcus aureus*, *Escherichia coli*, *Proteus mirabilis*, *Pseudomonas aeruginosa* and *Klebsiella pneumonia*. All cultures were obtained from *IMTECH*, *Chandigarh*, India. The antibacterial activity was determined using the agar disc diffusion method [8, 9] with slight modification. Standard antibiotic discs chloramphenicol C30 (Himedia, India), each

Table 10.1 Antibacterial activities in ethanol extract *Pittosporum viridulum* stem bark

Microorganisms	Ethanol extract A	Ethanol extract B	Positive Control (*chloramphenicol*)	MIC (μg/ml)
Staphylococcus aureus	8	10	35	500
Escherichia coli	9	12	40	500
Proteus mirabilis	7	11	35	200
Pseudomonas aeruginosa	12	18	30	100
Klebsiella pneumoniae	8	9	35	250

Values (mean of the three replicates) are diameter of zone of inhibition (mm)
A = 1% solution
B = 2% solution
Positive control = each antibiotic disc contains 30 μg of *chloramphenicol*

disc containing 30 μg, were used as positive control. Each solvent extract for screening was prepared by dissolving 1 mg/100 μl to 5 mg/100 μl in DMSO. Sterilized filter paper discs made from Whatman No. 1 (3 mm in diameter) were immersed in appropriate concentration of each plant extract and allowed to be saturated [9]. Bacterial cultures were aseptically inoculated onto plates of Nutrient Agar. The discs containing the plant extract were transferred onto the surface of the seeded agar plates. Discs dipped in DMSO were used as a negative control. Plates with the organisms were incubated for 24 h, and growth inhibition was quantified by measuring the diameter for the zone of inhibition to the nearest millimetre from the lower surface of the petri dishes.

The minimum inhibitory concentration (MIC) of the plant extract was determined by a broth dilution technique [10] with slight modification. The plant extract was diluted to give 100, 200, 250, 500 and 750 μg/ml final concentration in nutrient broth. Using a standard micropipette, 0.05 ml of the 18-h old bacterial broth (106 cfu/ml) cultures was introduced into each of the test tubes. A set of tubes containing growth medium and each of the test bacteria were used as control. All the tubes were incubated at 35°C for 24 h. The minimum inhibitory concentration was taken as the lowest concentration of the extract that will prevent growth of the bacterial strains. All experiments were repeated thrice.

Results and Discussion

The ethanolic extract of stem bark of *P. viridulum* was evaluated for antimicrobial activity against gram-positive and gram-negative bacterial strains by disc diffusion method. The results of antibacterial activity of ethanolic extract of stem bark of *P. viridulum* are represented in Table 10.1. Maximum zone of inhibition was observed against gram-negative bacteria, *Pseudomonas* sp., which had inhibition zone of 18 mm for ethanol extract. The minimum inhibitory concentration of crude plant extracts varied between 25 and 150 μg/ml, depending on the bacterial strains used. *Pseudomonas* sp., highly sensitive to the extract, has shown minimum inhibitory concentration at 100 μg/ml. The components analysed for are phenol, flavonoids, tannins, steroids, glycosides, saponins, fixed oils and terpenoids represented in Table 10.2. Antibacterial activity observed in the present study may be because of the presence of compounds like carbohydrates, glycosides, flavonoids, saponins, phenol and terpenoids. Our results firmly support the popular use of bark of

Table 10.2 Preliminary phytochemical screening of *Pittosporum viridulum* stem bark

Name of the test	Test for	Present (+) Absent (–)
(a) Dragendroff's test	Alkaloids	–
(b) Hager test		–
(c) Wagner's test		–
(a) Molisch's test	Carbohydrates	+
(b) Fehling's test		+
(a) Legal's test	Glycosides	+
(a) Biuret test	Amino acids	–
(b) Ninhydrin test		–
(a) Spot test	Fixed oils and fats	–
(b) Saponification test		–
(a) Shinoda test	Flavonoids	+
(b) Concentrated sulphuric acid test		+
(c) Aqueous sodium hydroxide test		+
(a) Foam test	Saponins	+
(a) Ferric chloride test	Phenol	+
(b) Lead acetate test		+

P. viridulum in folk medicine as antimicrobial agent. Therefore, this species could be a potential source for antimicrobial compounds.

References

1. Sharma BD, Balakrishnan NP, Rao RR, Hajra PK. Flora of India, vol. 1. Calcutta: Botanical Survey of India; 1993.
2. Gurib-Fakim A, Demarne FE. Constituents of the essential oil of the leaves of *Pittosporum balfourii* growing in rodrigues. Planta Med. 1994;60(5):584–5.
3. Weston RJ. Composition of essential oil from the leaves of seven New Zealand species of *Pittosporum* (Pittosporaceae). J Essent Oil Res. 2004;16(5):453–8.
4. Weston RJ. Aromatic components of the leaves of the New Zealand lemonwood tree *Pittosporum eugenioides*. Z Natforsch. 2004;59C:32–4.
5. Yoganarasimhan SN. Medicinal plants of India, vol. I. Bangalore: Interline Publishing Pvt, Ltd.; 1996.
6. Khare CP. Indian medicinal plants, an illustrated dictionary. Berlin: Springer; 2007. 496.
7. Harborne JB. Phytochemical methods. London: Chapman and Hall, Ltd; 1973. p. 49–188.
8. Xu XH, Lee SF. Activity of plant flavonoids against antibiotic-resistant bacteria. Phytother Res. 2001;15(1):39–43.
9. Mahasneh AM. Screening of some indigenous Qatari medicinal plants for antimicrobial activity. Physiother Res. 2002;16:751–3.
10. Irobi ON, Young M, Anderson WA. Antimicrobial activity of Annatto (*Bixa orellana*) extract. Int J Pharm. 1996;34:87–90.

Biochemical and Nutritional Evaluation of *Monochoria vaginalis* Presl.

11

S. Pradeesh, G. Nair Archana, M. Devi Chinmayee, C.S. Sarika, I. Mini, and T.S. Swapna

Abstract

Monochoria vaginalis of Pontederiaceae is an aquatic herb mostly found in tropical regions of India in places such as ditches, rice fields and on margins of water reservoirs, pool, marshes and swamps. It has a short sub-erect, spongy root stock and long petiolated leaves. This weedy leafy plant commonly known as 'Karim Koovalam' possesses numerous medicinal properties. The root is used to cure toothache, asthma, stomach and liver disorders. The entire plant excluding the root is eaten as vegetable by Paniya and Chetti tribes of Wayanad District in Kerala, and the leaves are used to treat cough, fever, etc. The paste of the tuber is used externally as an antidote to snakebite poisoning among the Kadars in the Anaimalai Hills of Tamil Nadu. Phytochemical and nutritional analysis of this plant with diverse medicinal properties will be helpful to explore untapped potentialities of the plant for effective and sustainable utilization. Extract was taken from young, mature leaves and mature petiole using different solvents like methanol, acetone, chloroform and petroleum ether for biochemical analysis. Various phytochemicals like reducing sugar, glycosides, flavonoids, alkaloids, tannins, steroids, terpenoids, coumarins, saponins, anthraquinones, phlobatannins and iridoids were analysed and found to be present actively. Nutritional factors like total carbohydrates, total proteins, total reducing sugar, different amino acids, free fatty acids, crude fibre, lipids, total moisture content and vitamins were quantified using standard estimation procedures. Antinutritional factors like phytic acid, total phenol, tannic acid and trypsin inhibitor were also estimated. Micronutrients and different pigments were quantified. Result showed that maximum amount of total carbohydrates, total protein, reducing sugar and amino acids like methionine were present in young leaf. This plant can be

S. Pradeesh • G.N. Archana • M.D. Chinmayee • C.S. Sarika • I. Mini • T.S. Swapna (✉)
Department of Botany, University College, Thiruvananthapuram, Kerala, India
e-mail: swapnats@yahoo.com

A. Sabu and A. Augustine (eds.), *Prospects in Bioscience: Addressing the Issues*, DOI 10.1007/978-81-322-0810-5_11, © Springer India 2013

proposed as alternative nutrient rich leafy vegetable and can be utilized in a sustainable way. Phytochemical evaluation along with the sustainable utilization and domestication of edible plants from the wild will definitely help to achieve food security of the nation.

Keywords
Phytochemicals • *Monochoria vaginalis* • Nutritional factors • Antinutritional factors

Introduction

Vegetables are important sources of bioactive molecules, which are highly beneficial for the maintenance of good health and prevention of diseases [1]. Leafy vegetables which originated from an area may or may not be confined to that particular region [2] and often regarded as weeds. But as some leafy vegetables grow in the wild and are readily available in the field, they do not require any formal cultivation. Many of them are resilient and adaptive and tolerate adverse climatic conditions more than the exotic species [3]. Although they can be raised comparatively at lower management cost and on poor marginal soil, they were remained underutilized, due to lack of awareness of their nutritional values in favour of the exotic ones [4]. Crop production system should aim to increase the use of underexploited natural resources such as traditional food crops [5].

Leafy vegetables are rich in carotene, ascorbic acid, riboflavin, folic acid and minerals like calcium, iron and phosphorous [6]. George [7] stated that leafy vegetables represent a veritable natural pharmacy of minerals, vitamins and phytochemicals. He concluded that the potassium content of leafy vegetables is good in the control of diuretic and hypertensive complications, while fibre content of vegetables prevents constipation [8], and the proteins in vegetables are superior to those found in fruits [7]. Leafy vegetables are popular and important in oriental diets [9]. But malnutrition is a very serious health problem to humans, and about 780 million people in the world are chronically undernourished [10]. Green leaves contain significant amount of iron, β-carotene, folic acid and protein, as well as considerable amount of pyridoxine, riboflavin and copper [11]. The phytochemicals in vegetables protect the human body from a large number of ailments and are often known as protective foods [12]. The difference between the world's supply of quality foods and the growth of the global population continues to widen, and the current surge in the search for nutritious foods is therefore not surprising. Several plants exist with very high nutritive value and yet remain unexploited for human and animal benefits [13]. Therefore, in the present investigation, an attempt has been made to understand the phytochemicals, nutritional and antinutritional factors of the wild edible plant *Monochoria vaginalis*, consumed by the Paniyar, Chetti and Kattunaayika tribes of Western Ghats in India. Studies on nutritional value of wild leafy vegetables are of considerable significance since it may help to identify long forgotten resources.

Materials and Methods

Monochoria vaginalis was collected from Kalpetta, Wayanad district of Kerala, India. After shade drying, the mature leaf, young leaf and mature petiole were ground well using mechanical blender to fine powder and extracted in Soxhlet apparatus using different solvents like methanol, acetone, petroleum ether and chloroform. The fresh samples of the mature leaf, young leaf and mature petiole were also used for the analysis. The analysis were performed following standard methods for estimation of total protein, total carbohydrates, reducing sugar, total phenols, pigments, vitamins, amino acids, etc. Dried powder was used for the analysis of phytic acid, free fatty acids, tannic acid and different secondary metabolites. All the analyses were carried out in triplicate.

Analytical Techniques

The moisture content was calculated from the difference in weight of fresh sample and dried sample, after drying the plant for 2 h in 110°C temperature by following the method of IUPAC [14].

Phytochemical Analysis

Reducing sugars present in samples were extracted with distilled water. Total reducing sugar was estimated by dinitrosalicylic acid method and read the absorbance at 540 nm against a blank [15]. The amount of total carbohydrate present in the sample was estimated by Anthrone method [16], total protein by Lowry's method [17] and chlorophyll content by the method of Arnon [18]. For chlorophyll estimation, fresh tissue was homogenized in 80% acetone and read the absorbance at 645, 663 and 652 nm, and the chlorophyll present was calculated by Arnon's formula. The amount of amino acids present in the samples was estimated by Moore and Stein [19] and free fatty acid by following the method of Cox and Pearson [20].

Crude fibre extraction was carried out by petroleum ether to remove the fat and was estimated by Wet method [21]. For lipid estimation, powdered sample was extracted with chloroform/methanol mixture and was estimated by the method of Bligh and Dyer [22]. The amount of β-carotene was determined following the method of IUPAC (1987) [23], vitamin C by the method of Sadasivam and Balasubramanian [24] and vitamin E by the method of Rosenberg [25]. Folin–Ciocalteu method was used for total phenol estimation [26], trypsin inhibitor by Kakade et al. [27], phytic acid content in the sample by Wheeler and Ferrel method [28], and tannic acid was estimated by Folin–Denis method [29].

Phytochemical Screening

Phytochemical analysis of extract was done as described by Harbone [30]. The phytochemicals like reducing sugar (Fehling's test), glycosides (Keller–Killani test), flavonoids (Shinoda test), alkaloids (Dragendorff's method), tannins (ferric salt test), steroids (chloroform–acetic anhydride test), terpenoids (Liebermann–Burchard method), coumarins (alcoholic sodium hydroxide–hydrochloric acid test), saponins (chloroform and H_2SO_4 test), anthraquinones, phlobatannins and iridoids were tested. Dried samples were digested in 4:1 ratio of nitric acid and perchloric acid, and micronutrients like iron, manganese, magnesium, copper, zinc, aluminium, selenium and nickel were analysed by atomic absorption spectroscopy.

Results and Discussion

Present investigation includes quantitative analysis of total moisture content along with fatty acids, crude fibre, lipids, vitamin A, vitamin E, vitamin C, different pigments and antinutritional factors like total phenols, phytic acid, trypsin inhibitor and tannic acid. To understand the proximate and chemical composition of the genus, major primary and secondary metabolites from different parts of plant such as mature leaf, young leaf, and mature petiole were selected and quantified. Studies showed that moisture forms major part of the fresh weight of the plant. The percentage of moisture was high in young leaf (12.56%) compared to other parts that is shown in Fig. 11.1c. Sugars are the product of photosynthesis in plants. In plants having high photosynthetic efficiency, the amount of sugar produced will be high [31]. Reducing sugar from mature leaf, young leaf and mature petiole were extracted in distilled water. The analysis showed high sugar content in the young leaf (0.046 mg g^{-1}) of the plant (Fig. 11.1a) compared to other parts.

Carbohydrates are the important components of storage and structural materials in the plants. They exist as free sugars and polysaccharides.

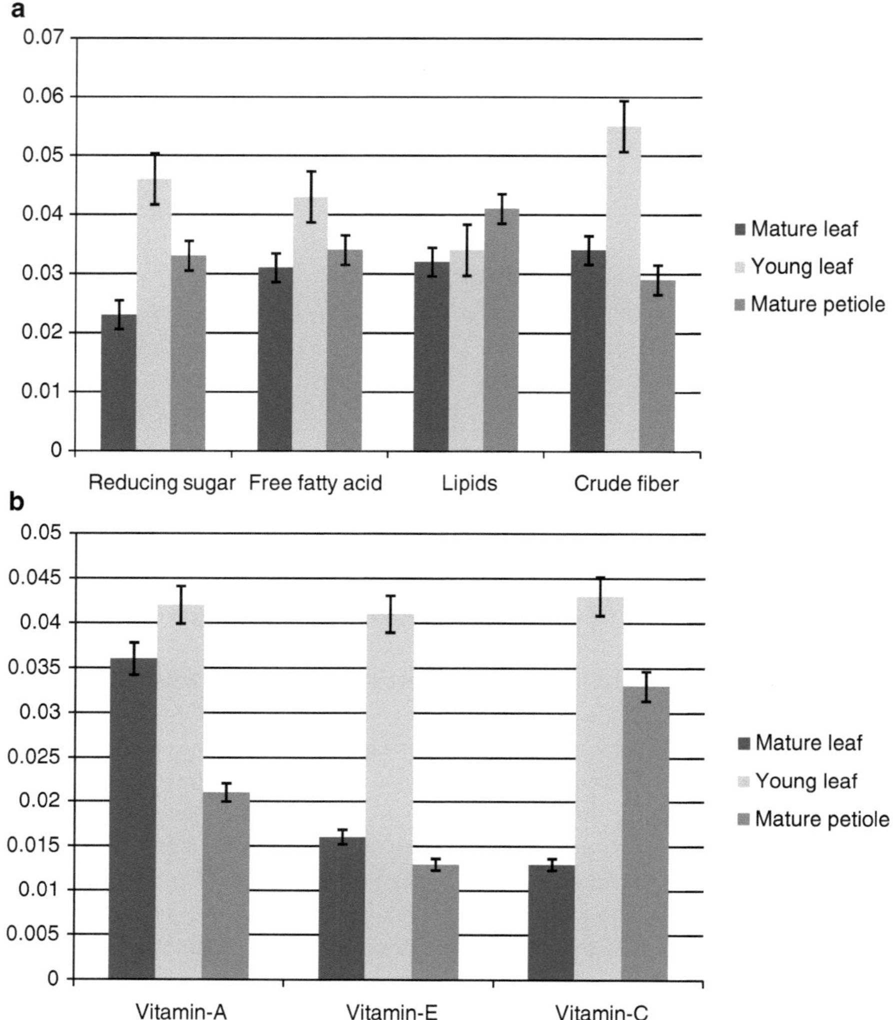

Fig. 11.1 (**a**) Nutritional factors (reducing sugar, free fatty acid, lipids, crude fibre) in *Monochoria vaginalis* (in mg g^{-1}). (**b**) Nutritional factors (vitamins A, E and C) in *Monochoria vaginalis* (in mg g^{-1}). (**c**) Nutritional factors (total protein, total carbohydrate, total moisture) in *Monochoria vaginalis* (in mg g^{-1}). (**d**) Nutritional factors (chlorophyll A, chlorophyll B, carotenoids, total chlorophyll) in *Monochoria vaginalis* (in mg g^{-1})

Maximum amount of total carbohydrates was present in young leaf (24.3 mg g^{-1}) compared to the other parts like mature leaf (16.92 mg g^{-1}) and mature petiole (11.27 mg g^{-1}) (Fig. 11.1c). Analysis of total protein in different parts of the plant was done and results showed maximum amount of protein in young leaf (21.86 mg g^{-1}) in comparison with mature leaf (12.23 mg g^{-1}) and mature petiole (12.47 mg g^{-1}) (Fig. 11.1c). This shows the photosynthetic efficiency of young leaves. Lipid content was also higher in young leaf (Fig. 11.1a). The crude fibre content is commonly used as a measure of the nutritive value of poultry and livestock feeds [21]. Analysis shows that maximum amount of crude fibre was present in young leaf (0.055 mg g^{-1}) than the other plant parts (Fig. 11.1a). In *Monochoria vaginalis*, free fatty acid content was higher in young leaf (0.043 mg g^{-1}) compared to mature leaf and petiole (Fig. 11.1a). Results also showed that the young leaf has higher amount of vitamin A, vitamin E and vita-

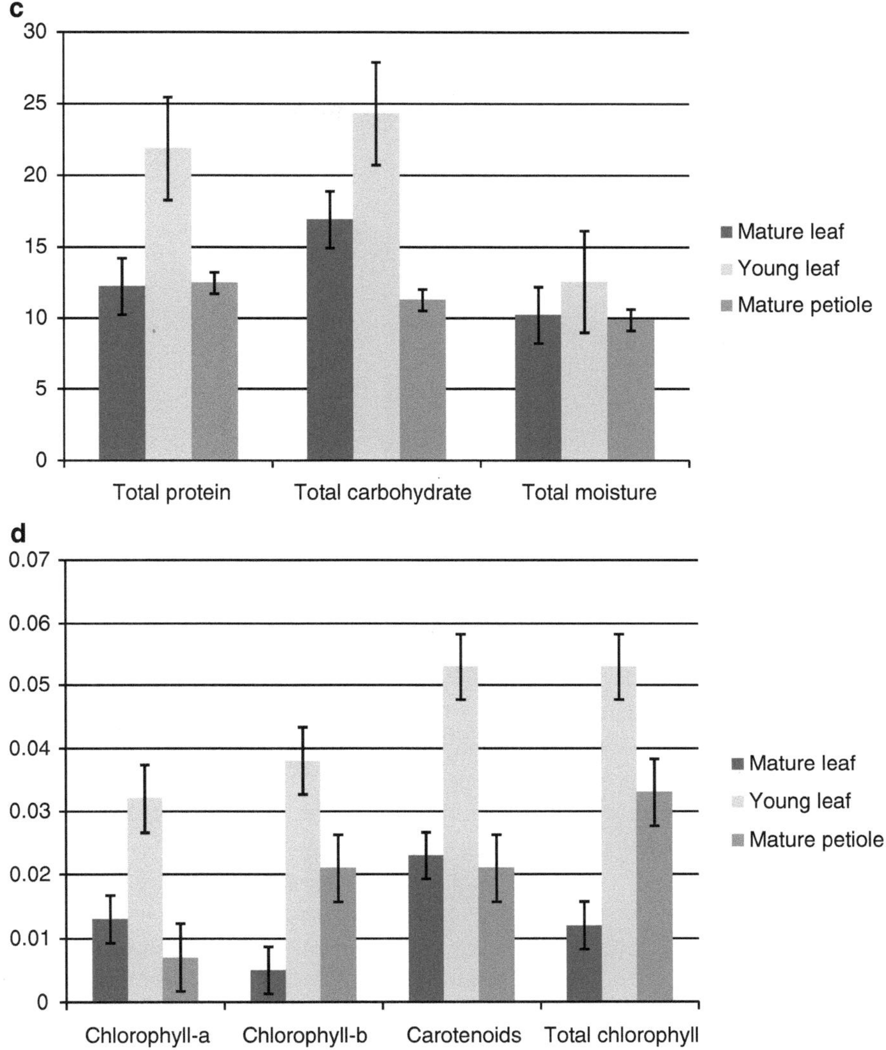

Fig. 11.1 (continued)

min C (Fig. 11.1c). Amount of chlorophyll in a green plant or any crop plant is an index of photosynthetic ability, which in turn is an index of its productivity level, primarily of primary metabolites [18]. The amount of chlorophyll was measured and compared in mature leaf, young leaf and mature petiole of *Monochoria vaginalis*, and it was found that high content of chlorophyll A, chlorophyll B, total chlorophyll and carotenoids was present in young leaf than mature leaf and petiole (Fig. 11.1d).

Total phenol content in different parts of *Monochoria vaginalis* was studied, and lower concentration of phenols was found in young leaf (0.024 mg g^{-1}) compared to petiole (0.035 mg g^{-1}) and mature leaf (0.026 mg g^{-1}) (Fig. 11.2). Trypsin inhibitors, which are antinutritional factors for monogastric animals, do not exert adverse effects in ruminants because they are degraded in the rumen [32]. Results obtained showed that very low amount of trypsin inhibitors was found in the plant parts such as mature leaf (0.015 mg g^{-1}) compared to young leaf (0.019 mg g^{-1}) and mature petiole (0.025 mg g^{-1}) as shown in Fig. 11.2. Phytic acid is a common storage form of phosphorus in seeds and is also considered as an antinutritional factor. The complexing of phytic acid with nutritionally essential elements

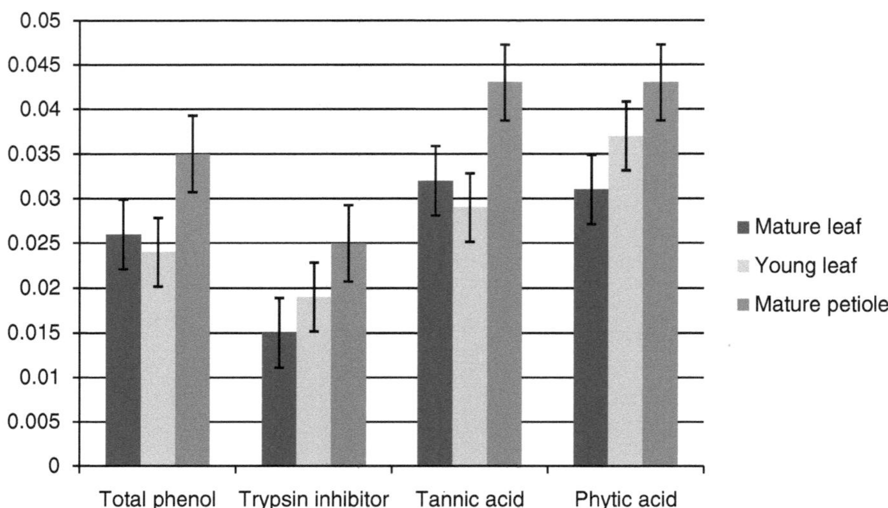

Fig. 11.2 Antinutritional factors in *Monochoria vaginalis* (in mg g^{-1})

and the possibility of interference with proteolytic digestion have been suggested as reasons for antinutritional activity. Phytic acid also interferes with calcium and iron absorption. Hence, estimation of phytic acid in food becomes essential [28]. The result showed that very low amount of phytic acid was found in the plant parts such as mature leaf (0.031 mg g^{-1}), young leaf (0.037 mg g^{-1}) and mature petiole (0.043 mg g^{-1}) as shown in Fig. 11.2. Higher amount of tannins was reported in cereals, and they display impaired nutritional quality, lower digestibility and reduction of food consumption [29]. Analysis showed that only low concentration of tannic acid was present in young leaf (0.029 mg g^{-1}), mature leaf (0.032 mg g^{-1}) and mature petiole (0.043 mg g^{-1}) of *Monochoria vaginalis* (Fig. 11.2).

Present study showed that amino acids content was higher in both young leaf and mature leaf compared to mature petiole (Fig. 11.3a, b). Among the different amino acids, methionine (Fig. 11.3b) was present in higher amounts in young leaf compared to some reported leafy vegetables like *Crassocephalum crepidioides* (methionine 0.58 mg g^{-1}) and *Senecio biafrae* (methionine 0.58 mg g^{-1}) [33].

Phytochemical screening showed that glycosides, alkaloids, steroids and tannins were pre-sent actively, but the presence of anthraquinones, phlobatannins, iridoids, etc. were not detected (Table 11.1).

This nutrient rich leafy vegetable shows sufficient quantities of micronutrients like iron, manganese, magnesium, copper, zinc and aluminium (Table 11.2). Mature leaf possesses higher amount of iron, manganese, magnesium, copper and zinc compared to other plant parts such as young leaf and mature petiole and other reported leafy vegetables [33]. Selenium and nickel were not detected in plant tissues.

The above results show that this wild leafy vegetable is rich in nutritional factors compared to many other wild leafy vegetables reported [34, 35]. Western Ghats of Kerala favours the growth of a variety of useful plants. Plants like *Monochoria vaginalis* contribute in diverse ways to the local people including their daily diet and folk system of medicine. Nutritional factors like total carbohydrates, total proteins and reducing sugars were present in high amount in young leaf than mature leaf and mature petiole. Nutritional evaluation of mature leaf, young leaf and mature petiole of this wild leafy vegetable *Monochoria vaginalis* indicates that this plant possesses high quantities of proteins, total reducing sugar, carbohydrates, vitamins, amino acids and crude fibre along with very low quantities of total phenols, phytic acid, tannic acid, trypsin inhibitor, etc., compared to other reported leafy vegetables.

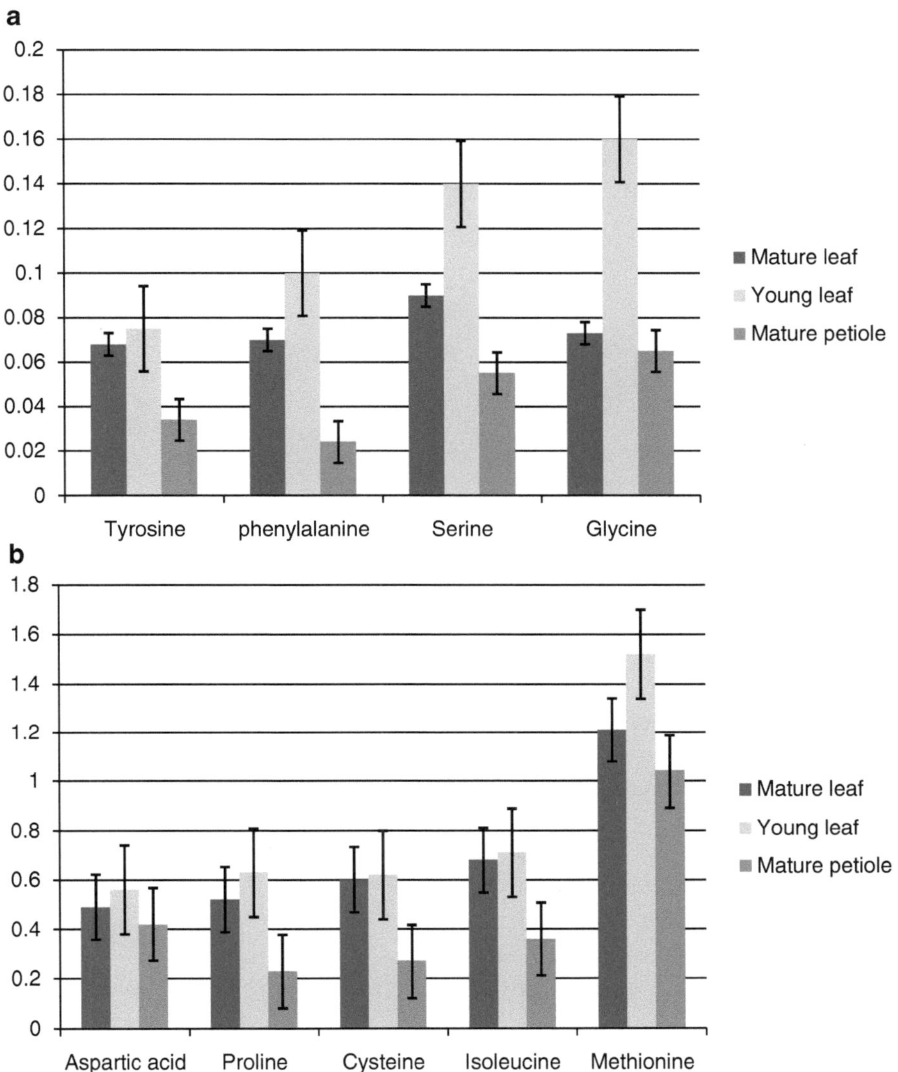

Fig. 11.3 (**a**) and (**b**) Amino acids in *Monochoria vaginalis* (in mg g^{-1})

Table 11.1 Preliminary phytochemical evaluation of mature leaf, young leaf and mature petiole of *Monochoria vaginalis*

	Mature leaf					Young leaf					Mature petiole				
Phytochemicals	M	A	P.E	C	D.W	M	A	P.E	C	D.W	M	A	P.E	C	D.W
Reducing sugar	+	−	+	+	+	+	−	−	+	+	+	+	−	+	+
Glycosides	+	+	−	+	+	+	+	−	+	+	+	+	−	+	+
Flavonoids	+	−	+	−	−	+	+	+	+	+	+	+	−	+	+
Alkaloids	+	+	−	+	+	+	+	−	+	+	+	+	−	+	+
Tannins	+	+	−	+	+	+	+	−	+	+	+	+	−	+	−
Steroids	+	+	+	+	+	+	+	+	+	+	+	+	+	+	−
Terpenoids	+	−	−	−	+	+	−	−	−	+	+	+	−	−	−
Coumarins	−	−	+	+	+	+	−	−	−	+	+	−	−	+	+
Saponins	+	+	−	+	+	+	+	+	+	+	+	−	−	+	+
Anthraquinones	−	−	−	−	−	−	−	−	−	−	−	−	−	−	−
Phlobatannins	−	−	−	−	−	−	−	−	−	−	−	−	−	−	−
Iridoids	−	−	−	−	−	−	−	−	−	−	−	−	−	−	−

M methanol, *A* acetone, *P.E* petroleum ether, *C* chloroform, *D.W* distilled water

Table 11.2 Micronutrient detection of mature leaf, young leaf and mature petiole of *Monochoria vaginalis*

Samples	Iron (mg kg^{-1})	Manganese (mg kg^{-1})	Magnesium (mg kg^{-1})	Copper (mg kg^{-1})	Zinc (mg kg^{-1})	Aluminium (mg kg^{-1})
Mature leaf	0.919	0.325	0.032	0.056	0.367	0.712
Young leaf	0.532	0.151	0.016	0.015	0.289	0.754
Mature petiole	0.149	0.083	0.051	0.071	0.184	0.154

Conclusions

Wild edible plants play a major role in meeting the nutritional requirement of the tribal population in remote parts of the country. Wild edible plants not only provide food quantity but also make significant contribution to the population nutrition throughout the year. The potential of traditional vegetables may help to meet the increasing demands of the growing population. Increased use of traditional vegetables can contribute to enhance people's health and standard living as well as the economic and social status of the food producers themselves. Identifying some of these underutilized leafy vegetables like *Monochoria vaginalis* from Western Ghats and their inculcation in to our diet could potentially address some of the challenges like life expectancy and food security. The present study reveals that *Monochoria vaginalis* is a nutrient rich plant which possesses high quantities of proteins, total reducing sugar, carbohydrates, vitamins, amino acids and crude fibre along with very low quantities of antinutritional factors can be used against nutritional deficiency problems across the globe.

Acknowledgements Authors gratefully acknowledge the help, support and guidance extended by Dr. N Anilkumar, Director, Community Agro Biodiversity Centre, MS Swaminathan Research Foundation, Wayanad. Financial assistance from the Western Ghat Development Cell of Government of Kerala is also gratefully acknowledged.

References

1. Sheela K, Kamal GN, Vijayalakshmi D, Geeta MY, Roopa BP. Proximate analysis of underutilized green leafy vegetables in Southern Karnataka. J Human Ecol. 2004;15(3):227–9.
2. Guarino L. Traditional Africa vegetables. Promoting the conservation and use of underutilized and neglected crops. In: Proceedings of the IPGRI international workshop on genetic resources of traditional vegetables in Africa. Rome: IPGRI; 1997.
3. Raghuvanshi RS, Singh R. Nutritional composition of uncommon foods and their role in meeting in micronutrient needs. Int J Food Sci Nutr. 2001;32:331–5.
4. Chweya JA, Eyzaguilre PB. The biodiversity of traditional leafy vegetables. Rome: IPGRT Publication; 1999. 540.
5. FAO. Agriculture, food and nutrition for Africa. A resource book for teachers of agriculture. Rome: Food and Agricultural Organization of the United Nations; 1997. p. 412.
6. Nnamani CV, Oselebe HO, Okporie EO. Ethno botany of indigenous leafy vegetables of Izzi Clan, in Ebonyi State, Nigeria. In: Proceedings of 20th annual national conference of biotechnology society of Nigeria. Abakaliki, 14–17 Nov, 2007. pp. 111–4.
7. George PM. Encyclopedia of foods, vol. 1. Washington: Humane Press; 2003. p. 526.
8. Noonan SC, Savage GP. Oxalate content of foods and its effect on humans. Asia Pac J Clin Nutr. 1999; 67:64–74.
9. Department of Agriculture and Forestry, Taiwan Provincial Government. Taiwan agricultural yearbook, 1997 edition. Taichung: Department of Agriculture and Forestry, Taiwan Provincial Government; 1997.
10. FAO. Nutrition et development: Une evaluation d'ensemble. In: Conference internationale sur la nutrition. Rome. 1992. p. 2.
11. Mathur B, Joshi RM, Bray WJ. Impact of supplementing leaf protein concentrate on hemoglobin level of children. Indian J Nutr Diet. 1989;26:267–71.
12. Mathura R, Jagdish S, Pandey AK. Vegetables: a source of nutritional security. Indian Hort. 2004;48(4):14–7.
13. Oladele PE, Oshodi AA. Nutritional potential of Berlandier nettle spurge (*Jatropha cathartica*) seed. Pak J Nutr. 2007;6(4):345–8.
14. IUPAC. Standard methods for the analysis of oils, fats and derivatives. Oxford: Blackwell Scientific publication; 1987.
15. Miller GL. Use of dinitrosalicylic acid reagent for determination of reducing sugars. Anal Chem. 1972;31:426–8.
16. Hedge JE, Hofreiter BT. In: Whistler RL, Be Miller JN, editors. Carbohydrate chemistry, vol. 17. New York: Academic; 1962.

17. Lowry OH, Rosebrough NJ, Farr AL, Randall RJ. Protein measurement with the Folin phenol reagent. J Biol Chem. 1951;193:265–75.
18. Witham FH, Blaydes DF, Devlin RM. Experiments in plant physiology. New York: Van Nostrand; 1971. p. 245.
19. Moore S, Stein WH. In: Colowick SP, Kaplan ND, editors. Methods in enzymol. New York: Academic Press; 1948. pp. 3, 468.
20. Cox HE, Pearson D. The chemical analysis of foods. New York: Chemical Publishing Co Inc; 1962. p. 420.
21. Maynard AJ. Method in food analysis. New York: Academic; 1970. p. 176.
22. Bligh EG, Dyer WJ. Rapid method of total lipid extraction and purification. Can J Biochem Physiol. 1959;37:911–7.
23. Bayfield RF, Cole ER. Colorimetric estimation of Vitamin A with trichloroacetic acid. Meth Enzymol. 1980;67:189–95.
24. Sadasivam S, Balasubramanian T. Practical manual in biochemistry. Coimbatore: Tamil Nadu Agricultural University; 1987. p. 14.
25. Rosenberg HR. Chemistry and physiology of vitamins. New York: Inter Science Publishers Inc; 1992. p. 452–3.
26. Mayer V, Treutter D, Santos-Buelga C, Bauer H, Feuchat W. Study of carbohydrate influence on protein tannin aggregation by nephelometry. J Agric Food Chem. 1995;38:1151–5.
27. Kakade ML, Rackie JJ, McGhee JE, Puski G. Determination of trypsin inhibitor activity of soy products: a collaborative analysis of an improved procedure. Cereal Chem. 1974;51:376.
28. Wheeler EL, Ferrel RE. Method for Phytic acid determination in wheat and wheat fractions. Cereal Chem. 1971;48:312–6.
29. Schanderl SH. In: Method – 48 312. In Food analysis. New York: Academic; 1970. p. 709.
30. Harbone JB. Phenolic glycosides and their natural distribution in the biochemistry of Phenolic compounds. New York/London: Academic; 1977. p. 152–62.
31. Somogyi M. Estimation of sugars by colorimetric method. J Biol Chem. 1952;200:245.
32. Cheeke PR, Shull LR. Natural toxicants in feeds and livestock. West Port: AVI Publishing Inc; 1985.
33. Dairo FAS, Adanlawo IG. Nutritional quality of *Crassocephalum crepidioides* and *Senecio biafrae*. Pak J Nutr. 2007;6(1):35–9.
34. Nnamani CV, Oselebe HO, Agbatutu A. Assessment of nutritional values of three underutilized indigenous leafy vegetables of Ebonyi State, Nigeria. Afr J Biotechnol. 2009;8(9):2321–4.
35. Kubmarawa D, Andenyang IFH, Magomya AM. Amino acid profile of two non-conventional leafy vegetables, *Sesamum indicum* and *Balanites aegyptiaca*. Afr J Biotechnol. 2008;7(19):3502–4.

12 Assessment of Genotypic Variability for Growth, Biophysical Parameters, Yield and Yield-Attributing Characters Under Drought Stress in Cotton

B.C. Patil, A.G. Babu, and K.N. Pawar

Abstract

Increasing scarcity of irrigational water is a major threat to sustainable production of cotton (*Gossypium hirsutum* L.). Drought stress is a complex phenomenon affecting the physiology of cotton plant, in turn reducing crop growth, yield and yield components. An experiment was conducted using 20 *Gossypium hirsutum* genotypes grown in RBD design at Agricultural Research Station, University of Agricultural Sciences, Dharwad, Karnataka, and evaluated for genotypic variability for growth phenomena, biophysical parameters, yield and yield-attributing characters under irrigated and water-deficit condition. Under both irrigated and rainfed conditions, ARBH-813 recorded highest numbers of bolls per plant, while GSHV-97/612 recorded the least. Maximum boll weight was recorded by AKH-0205 followed by H-1452/10 and the least by NH-635 in irrigated condition. Under rainfed conditions, H-1452/10 recorded the highest boll weight. Under both irrigated and rainfed conditions, BS-30 and ARBH-813 recorded the highest water potential, while GJHV-358, GISV-218, H-1452/10 and HBB-101 recorded the least, and the genotype BS-279 recorded highest photosynthesis rate, while LRA-5166 recorded the least SPAD value. The genotypes BS-279, ARBH-813 and Sahana proved better as they recorded highest seed cotton yield under both conditions and least to moderate drought susceptibility indices. These are found to be desirable genotypes for drought situations.

Keywords

Cotton • Growth • Photosynthesis rate • Yield • Water deficit • Water potential

---

B.C. Patil (✉) • A.G. Babu • K.N. Pawar
Agricultural Research Station-Dharwad,
University of Agricultural Sciences, Dharwad
580005, Karnataka, India
e-mail: bc_patil@yahoo.com

A. Sabu and A. Augustine (eds.), *Prospects in Bioscience: Addressing the Issues*,
DOI 10.1007/978-81-322-0810-5_12, © Springer India 2013

Introduction

Cotton (*Gossypium hirsutum* L.) is the world's leading natural fibre crop, grown in arid and semiarid regions of the world. Sustainability and improvement of cotton yield are major challenges to meet the upcoming threats of increasing volume of world population, deterioration of arable land, depletion of water resources and environmental stresses. However, drought is one of the major yield-limiting factors among these stresses [1]. Growth and yield of a crop plant is drastically affected directly or indirectly by altering metabolism, growth and development [2]. Well-focused research work to understand drought tolerance mechanism was conducted on cereals, especially rice [3, 4]. However, reports on drought tolerance of cotton crop are limited. The detrimental impact of drought on cotton crop can be minimized by developing drought-tolerant cultivars. Drought tolerance is a complex mechanism that is influenced by a wide range of physiological traits which have some relationship with productivity under water-deficit conditions. It is known that the quantity and quality of the fibre produced in cotton plant crops are directly related to water availability during the different physiological phases of development. Regardless of whether it is irrigated or not, cotton is often exposed to drought, which adversely affects both yield and lint quality [5]. Genetically equivalent cotton plant populations, when submitted to water deficits, show reductions in yield of up to 50% if compared to those that have been irrigated, especially when the stress factor is imposed during the period between flowering and fructification [6]. Cotton has a C3 carbon metabolism; however, its photosynthetic potential is relatively high [7, 8]. Reduction of photosynthetic rate in cotton under water-limited environment is documented [9]. This reduction may be attributed to stomatal [10] and non-stomatal factors [11]. A higher photosynthetic rate under drought is a decisive factor for higher cotton production [12], and the empirically determined differences in drought tolerance of cotton cultivars can be related to some physiological parameters such as photosynthetic rate and stomatal conductance [13]. There is a considerable diversity of growth parameters, namely, plant height, number of leaves, specific leaf area, plant dry weight, plant dry weight, leaf dry weight and productivity traits, namely, no. of bolls per plant, boll diameter, staple length, fibre strength and ginning out percentage in relation to varying moisture-deficit periods. And the juvenile growth stages were more susceptible to water deficit [14]. Hence, in this direction, we studied the genotypic variability among the cotton cultivars for growth, biophysical parameters, yield and yield-attributing characters, and we focused to identify a suitable screening index for drought tolerance.

Materials and Methods

An experiment was conducted at Agricultural Research Station, University of Agricultural Sciences, Dharwad, Karnataka, India. The experimental material consisted of 20 upland cotton (*G. hirsutum* L.) cultivars selected on the basis of putative differences in yield under drought conditions. Seeds of the cultivars were obtained from Agricultural Research Station, Dharwad, and cotton research stations located at different ecological regions of India.

Experimental Design

Twenty cotton cultivars were evaluated under irrigation regime and in rained regime in the field during 2010–2011 at the research area of the Agricultural Research Station, University of Agricultural Sciences, Dharwad, Karnataka, India. Daily rainfall during each growing season was recorded. The experimental plot was laid out in RBD design. Cotton seed was delinted with sulphuric acid and soaked in water for 12 h before planting. Sowing was completed during the first week of June. A commercial chemical fertilizer was applied at the rate of 100:50:50 kg/ha of $N:P_2O_5:K_2O$ at the time of seedbed preparation. Plant population was maintained at 4 plants/m^2 by hand-thinning 25 days after germination.

Appropriate control measures were adopted for insect pest and weed infestation and applied evenly to all the plots.

Measurement of Physiological Attributes

Net assimilation rate was determined with a LICOR-6400 Photosynthesis System. This portable, battery-operated system consisted of a clamp-on leaf chamber (6 cm^2 area) connected to an infrared gas analyzer. Different gas exchange components change with environmental conditions and developmental stages of the plant. Photosynthetic rate increased to a maximum in young leaves (13–20 days after unfolding), remained at that level after 12 days and then gradually declined [15].

All physiological measurements were performed between 1,000 and 1,300 h at PARX1700 mmol/m^2/s during cloud-free days. The youngest fully expanded main stem leaf (16–18 days) exposed to direct sunlight was used to determine the gas exchange parameters. Leaves were tagged on the day they unfold which was designated as day one. Measurements were recorded on four randomly selected plants per plot 75–78 DAS each year when all the cultivars were at least 50% flowering. A mean of four observations per plot was used for the statistical analysis. Seed cotton was hand-picked from all the plots 180 DAS and was sun-dried for 1 day after removing trash and dry carpels before weighing.

Results and Discussion

Yield Parameters

Seed Cotton Yield

Seed cotton was hand-picked from all the plots 180 DAS. In irrigated condition, the variety BS-279 has recorded highest yield (1750.1 kg/ha) followed by BS-30 (1737.2 kg/ha) and ARBH-813 (1452.7 kg/ha), whereas GISV-218 (691.1 kg/ha) and LRA-5166 (691.7 kg/ha) recorded the least yield. In rainfed situation, BS-279 (1526.7 kg/ha) followed by ARBH-813 (1096.9 kg/ha) recorded the highest yield, while GISV-218 (481.1 kg/ha) followed by LRA-5166 (581.0 kg/ha) recorded the least yield. The percentage of yield reduction in rainfed condition compared to irrigated condition was highest in the genotype H-1353/10 (43.2) followed by GJHV-358 (38.6 %) indicating their sensitivity to moisture regimes, whereas the least reduction was observed in CPD-168 (0.7 %) and F-2228 (4.9 %) (Table 12.1). The biplot of yield vs. DSI shows that the genotypes BS-279, ARBH-813 and Sahana are having the most desirable traits like high yield as well as low drought susceptibility indices, while GJHV-358 and GISV-218 recorded the low yield and more DSI (Fig. 12.1).

Number of Bolls Per Plant

In irrigated condition, the highest numbers of bolls were recorded in ARBH-813 (15.6) followed by H-1353/10 (14.53) and the least in HBB-101 (7.43) and GSHV-97/612 (7.73), respectively. In rainfed condition, highest number of bolls were observed in ARBH-813 (14.2) followed by H-1353/10 (13.33), while GSHV-97/612 (6.60) followed by HBB-101 (6.47) recorded the least number of bolls per plant (Table 12.1).

Boll weight (g/boll)

Maximum boll weight was recorded by AKH-0205 (5.57 g/boll) followed by H-1452/10 (5.04 g/boll) and the least by NH-635 (3.19 g/boll) in irrigated condition. In rainfed condition, it was H-1452/10 which recorded the highest boll weight (4.51 g/boll), and HBB-101 recorded the least (2.63 g/boll). Maximum percent reduction was observed in MR-786 (47.21 %) followed by AKH-0205 (43.80 %) and minimum in RCR-102 (0.20 %) (Table 12.1).

Phenological Observations

There was significant difference between the genotypes for all phenological parameters in both irrigated and rainfed condition.

Table 12.1 Seed cotton yield and yield components in cotton genotypes under irrigation and rainfed condition

		Yield (kg/ha)				No. of bolls/plant			Boll wt. (g/boll)		
Sl. no.	Genotypes	(Irr)	(Rf)	% Redn.	DSI	(Irr)	(Rf)	% Redn.	(Irr)	(Rf)	% Redn.
1	ARBH-2004	1,049.4	944.8	0.4	0.15	13.93	11.33	18.7	4.60	4.10	10.9
2	H-1452/10	1,180.8	961.1	18.6	0.74	14.40	12.67	12.0	5.04	4.51	10.5
3	CPD-231	1,135.1	937.3	17.4	0.69	11.87	10.47	11.8	4.40	3.51	20.2
4	F-2228	860.8	818.9	4.9	0.19	10.27	8.20	20.2	4.17	3.10	25.7
5	H-1353/10	1442.8	819.4	43.2	1.72	14.53	13.33	8.3	3.47	3.15	9.2
6	GJHV-358	793.8	487.2	38.6	1.54	10.97	7.60	30.7	5.03	3.83	23.9
7	CPD-168	843.4	837.4	0.7	0.03	12.33	8.80	28.6	4.31	3.32	23.0
8	NH-635	789.4	716.2	9.3	0.37	8.80	6.93	21.3	3.19	3.04	4.7
9	GBHV-164	819.7	657.2	19.8	0.79	11.67	9.60	17.7	4.77	2.94	38.4
10	GISV-218	691.1	481.1	30.4	1.21	9.43	8.40	21.5	4.07	3.13	23.1
11	LRA-5166(NC)	691.7	581	16	0.64	11.60	8.07	30.4	3.18	3.04	4.4
12	BS-30	1,737.2	933.8	46.2	1.84	15.40	13.00	15.6	3.96	3.50	11.6
13	ARBH-813	1,452.7	1096.9	24.5	0.97	15.60	14.20	9.0	3.81	2.84	25.5
14	BS-279	1,750.1	1526.7	12.8	0.51	13.73	10.47	23.7	4.51	3.31	26.6
15	MR-786	945.6	848	10.3	0.41	11.40	10.33	9.4	4.63	2.65	47.1
16	AKH-0205	989.3	706.7	28.6	1.14	12.40	10.20	17.7	5.57	3.13	43.8
17	GSHV-97/612	792.2	558.3	29.5	1.17	7.73	6.60	14.6	4.90	3.10	36.7
18	Sahana (CC)	1,452.3	1084.9	25.3	1.01	15.57	11.23	27.9	3.78	2.83	25.1
19	RCR-102	1,164.8	549	52.9	2.1	9.93	5.97	17.7	4.42	4.41	0.2
20	HBB-101	1,060.6	658	38	1.51	7.43	6.47	12.9	4.59	2.63	42.7
Mean		**1,082.1**	**810.20**	**23.85**	**0.95**	**0.95**	**11.6**	**9.708**	**4.32**	**4.34**	**3.25**
SEm ±		**109.7**	**104.3**	–	–	**1.31**	**1.21**	–	**0.473**	**0.44**	–
CD @ 5 %		**314.0**	**298.6**	–	–	**3.76**	**3.48**	–	**1.355**	**1.262**	–
CV %		**19.49**	**19.53**	–	–	**19.4**	**19.6**	–	**18.91**	**19.66**	–

Note: *Irr* irrigation, *Rf* rainfed, *Redn* reduction

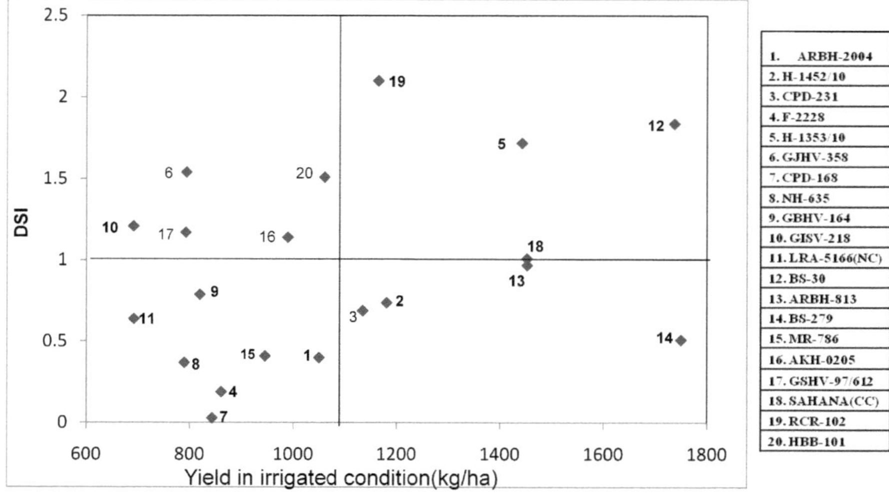

Fig. 12.1 Biplot of yield and drought susceptibility index (DSI) in cotton genotypes (2010–2011)

Table 12.2 No. of days to squaring, flowering and boll opening in cotton genotypes under irrigation and rainfed condition

		50 % Flowering			50 % Squaring			50 % Boll opening		
Sl. no.	Genotypes	(Irr)	(Rf)	Diff	(Irr)	(Rf)	Diff	(Irr)	(Rf)	Diff
1	ARBH-2004	74.0	71.7	2.3	55.00	50.33	4.7	148.00	145.00	3.0
2	H-1452/10	74.7	69.7	5.0	55.00	51.33	3.7	149.33	147.00	2.3
3	CPD-231	74.3	71.3	3.0	55.00	53.33	1.7	150.00	148.00	2.0
4	F-2228	72.7	71.3	1.4	54.67	49.67	5.0	149.00	147.00	2.0
5	H-1353/10	70.0	69.3	0.7	55.67	47.67	8.0	147.00	143.00	4.0
6	GJHV-358	73.0	69.7	3.3	54.00	48.33	5.7	149.67	148.67	1.0
7	CPD-168	74.0	72.3	1.7	54.33	51.00	3.3	151.33	150.67	0.7
8	NH-635	73.3	70.7	2.6	52.67	50.00	2.7	150.67	147.33	3.3
9	GBHV-164	72.7	71.3	1.4	53.67	49.33	4.3	151.00	150.67	0.3
10	GISV-218	72.7	70.3	2.4	54.67	53.33	1.3	152.00	151.00	1.0
11	LRA-5166(NC)	74.7	69.0	5.7	54.33	53.00	1.3	151.67	149.33	2.3
12	BS-30	72.0	69.0	3.0	54.33	54.00	0.3	150.00	149.00	1.0
13	ARBH-813	72.3	68.3	4.0	54.67	51.33	3.3	150.33	149.00	1.3
14	BS-279	73.7	71.7	2.0	54.00	52.33	1.7	150.00	149.00	1.0
15	MR-786	73.3	70.7	2.6	54.33	52.33	2.0	151.00	149.67	1.3
16	AKH-0205	74.3	71.3	3.0	53.33	52.00	1.3	151.33	150.00	1.3
17	GSHV-97/612	72.3	69.7	2.6	56.33	52.67	3.7	152.00	151.67	0.3
18	Sahana (CC)	71.7	69.7	2.0	55.00	52.00	3.0	149.33	149.00	0.3
19	RCR-102	71.3	67.7	3.6	52.00	52.00	0.0	152.33	150.00	2.3
20	HBB-101	71.7	69.3	2.4	54.00	51.00	3.0	152.33	150.33	2.0
Mean		**72.93**	**70.20**	**3.75**	**54.35**	**51.350**	**5.49**	**150.42**	**148.77**	**1.10**
SEm ±		**0.54**	**0.53**	–	**0.74**	**0.73**	–	**2.302**	**0.520**	–
CD @ 5 %		**1.53**	**1.51**	–	**NS**	**2.10**	–	**6.592**	**1.488**	–
CV %		**0.26**	**1.25**	–	**2.36**	**2.47**	–	**2.67**	**0.60**	–

Number of Days to 50 % Squaring

In irrigated condition, the genotypes differed significantly for phenology (Table 12.2). Number of days to 50% squaring was early in RCR-102 (52.00) and NH-635 (52.7), and more number of days was taken by GSHV-97/612 (56.33). In rainfed condition, H-1353/10 (47.7) followed by GBHV-164 (49.3) recorded early squaring, and it was late in BS-30 (54.0).

Number of Days to 50 % Flowering

More number of days to 50% flowering was taken by LRA-5166 (74.7) and H-1452/10 (74.7) and least by H-1353/10 (69.3) in irrigated conditions. Whereas in rainfed condition, LRA-5166 (74.7) took more number of days for 50% flowering, but the genotype RCR-102 (67.7) took the least number of days for 50% flowering (Table 12.2).

Number of Days to 50 % Boll Opening

The genotypes RCR-102 and HBB-101 took more number of days for 50% boll opening (152.3 days) and H-1353/10 the least in irrigated condition (147 days). In rainfed condition, GSHV-97/612 (151.7) and GISV-218 (152) took highest number of days, and ARBH-2004 (145.0) followed by H-1353/10 (143) took least number of days for 50 % boll opening (Table 12.2).

Plant Height

The genotypes differed significantly for plant height at harvest both in irrigated and rainfed condition (Table 12.3). At harvest, H-1353/10 (146.3) recorded more plant height and NH-635 (98.1) recorded the least in irrigated condition. In rainfed condition, H-1353/10 (134.7) and NH-635 (93.1) recorded highest and lowest plant height, respectively. The mean percent reduction in plant

Table 12.3 Plant height and number of monopodia and sympodia plant in cotton genotypes at 120 DAS

		Plant height			TDM		
Sl. no.	Genotypes	(Irr)	(Rf)	% Redn.	(Irr)	(Rf)	% Redn.
1	ARBH-2004	140.3	127.6	9.03	100.7	74.2	35.64
2	H-1452/10	141.2	125.8	10.91	75.7	127.0	40.39
3	CPD-231	126.8	122.6	3.31	77.1	71.4	5.71
4	F-2228	113.8	97.7	14.12	68.0	95.8	28.99
5	H-1353/10	146.3	144.7	1.09	78.9	80.9	2.39
6	GJHV-358	122.1	113.4	7.15	44.2	127.7	65.38
7	CPD-168	132.9	130.7	1.63	78.1	71.7	8.85
8	NH-635	98.3	98.1	0.14	102.8	46.8	119.58
9	GBHV-164	107.6	105.5	1.98	59.7	150.0	60.18
10	GISV-218	136.0	133.2	2.06	64.8	93.5	30.73
11	LRA-5166(NC)	123.2	122.1	0.87	75.1	77.7	3.36
12	BS-30	125.1	123.8	1.01	62.3	96.9	35.68
13	ARBH-813	124.2	122.1	1.72	89.0	99.9	10.87
14	BS-279	129.3	125.5	2.94	100.7	70.0	43.84
15	MR-786	137.9	131.9	4.40	82.6	97.2	15.04
16	AKH-0205	132.1	127.5	3.48	64.0	49.1	30.37
17	GSHV-97/612	143.1	139.3	2.66	49.7	57.5	13.59
18	Sahana (CC)	138.2	134.2	2.89	87.9	94.3	6.84
19	RCR-102	122.9	120.4	2.06	74.5	84.1	11.44
20	HBB-101	119.7	111.2	7.08	102.8	86.9	18.35
Mean		**128.05**	**122.86**	**4.05**	**76.9**	**87.6**	**35.48**
SEm ±		2.24	5.61	–	**1.4**	**1.48**	–
CD @ 5 %		6.41	6.10	–	**4.0**	**4.24**	–
CV %		3.06	7.84	–	**19.8**	**14.77**	–

height was highest in F-2228 (14.12 %) followed by H-1452/10 (10.91 %), while the least reduction was observed in LRA-5166 (2.87 %) followed by CPD-168 (1.63 %).

Total Dry Matter (g/Plant)

The total dry matter at harvest showed significant differences for genotypes in both irrigated and rainfed condition (Table 12.3). The mean TDM in irrigated condition was 95.43 g/plants, and in rainfed condition it was 68.8 g/plants. Thus, there was a mild stress experienced by the crop under rainfed condition than it would have been. Among the genotypes, CPD-231 (7.40 %) followed by Sahana (6.84 %) recorded the least reduction in TDM in rainfed condition compared to irrigated. But the percent reduction was maximum in GJHV-358 (65.38 %) and GBHV-164 (60.20 %).

Biophysical Parameters

The observation on biophysical parameters, namely, water potential and rate of photosynthesis and chlorophyll, was recorded at regular intervals.

Water Potential

The mean water potential measured at 120 DAS showed significant differences among the genotypes and moisture regimes (irrigated and rainfed condition). In irrigated condition, the mean water potential was −16.4 bars, while in rainfed condition it was −24.1 bars indicating that the cumulative effect of water stress exists which shows a difference of more than −7.71 bars of difference between irrigated and rainfed conditions. In irrigated condition, BS-30 (−15.2 bars) followed by ARBH-813 (−15.8 bars) recorded the highest water potential, while GJHV-358 (−18.2) and

Table 12.4 Photosynthetic rate, leaf area and SPAD readings (chlorophyll content) in irrigated and rainfed condition

		Photosynthetic rate (μ mol co_2 m^2/s)			SPAD reading (for chlorophyll content)			Water potential at 90 DAS (–bars)		
Sl. no.	Genotypes	IR	RF	% Reduction	IR	RF	% Reduction	IR	RF	Diff
1	ARBH-2004	13.3	10.3	22.56	40.6	43.9	8.13	16.4	23.4	7.0
2	H-1452/10	13.1	9.1	30.53	45.7	39.7	13.13	17.3	26.8	11.9
3	CPD-231	13.0	13.0	0.0	42.4	41.4	2.36	17.1	22.8	5.7
4	F-2228	12.9	11.6	10.08	45.2	43.6	3.54	16.5	23.5	7.0
5	H-1353/10	13.3	13.7	3.01	40.6	36.3	10.59	17.5	26.1	8.6
6	GJHV-358	15.0	13.1	17.56	44.9	43.8	2.1	18.2	26.4	8.2
7	CPD-168	13.0	11.6	10.77	37.3	37.3	27.35	17.2	23.1	5.5
8	NH-635	14.6	13.4	8.96	43.9	40.5	8.4	15.9	24.8	8.9
9	GBHV-164	13.1	11.2	14.5	45.0	40.7	9.56	16.4	24.2	7.8
10	GISV-218	14.5	13.0	19.23	44.1	38.2	15.45	17.4	24.6	7.2
11	LRA-5166(NC)	13.4	13.2	0.75	34.3	34.9	8.02	17.3	23.8	6.5
12	BS-30	13.3	11.3	15.04	42.3	39.0	1.53	15.2	21.3	6.1
13	ARBH-813	13.9	11.5	17.27	40.8	40.0	0.25	15.8	21.8	6.0
14	BS-279	16.1	14.5	11.03	44.0	42.8	9.74	16.3	20.9	5.6
15	MR-786	11.8	11.8	0.0	44.3	41.7	6.24	16.4	23.4	7.0
16	AKH-0205	14.5	13.3	9.02	41.5	39.3	5.6	16.1	24.1	8.0
17	GSHV-97/612	15.1	12.1	19.87	41.0	37.9	8.18	16.5	25.5	9.0
18	Sahana (CC)	13.8	13.2	4.35	39.2	37.2	5.38	16.4	23.9	8.5
19	RCR-102	12.8	11.0	14.06	42.5	37.8	12.43	16.2	24.4	9.2
20	HBB-101	13.1	11.8	5.34	40.7	40.0	0.98	16.7	27.2	10.5
Mean		**13.8**	**12.19**	**11.69**	**42.69**	**39.8**	**7.94**	**16.39**	**24.1**	**7.71**
SEm ±		**0.509**	**1.618**	–	**2.448**	**2.769**	–			–
CD @ 5 %		**1.456**	**4.632**	–	**7.009**	**7.926**	–			–
CV %		**6.59**	**19.4**	–	**10.16**	**11.88**	–			–

GISV-218 (−17.4 bars) and LRA-5166 (−17.63) recorded the least water potential in irrigated condition. In rainfed condition, BS-279 (–20.9 bars) followed by BS-30 (−21.3 %) recorded high water potential, while H-1452/10 (–26.8 bars) followed by GJHV-358 (−16.4 %) and HBB-101 (–27.2 %) recorded by least water potential. The difference in water potential (the reduction in water potential) in rainfed condition compared to irrigated was least in BS-279 (5.6) and CPD-168 (5.5), while the maximum difference recorded was in H-1452/10 (11.9) and HBB-101 (10.5) (Table 12.4).

Photosynthesis Rate

The mean rate of photosynthesis was 13.8 μ mol/ m^2/s in irrigated condition, while it was 12.19 in rainfed condition. The highest photosynthesis rate was recorded in BS-279 (16.1) followed by GSHV-97/612 (15.1), while the least rate was recorded by MR-786 (11.8) and RCR-102 (12.8). In rainfed condition, the highest photosynthetic rate was recorded by BS-279 (14.6) followed by AKH-0205 (14.5), while the least rate was recorded by H-1452/10 (9.1), ARBH-2004 (10.3) and RCR-102 (11.9). The percentage of reduction in photosynthesis rate in rainfed condition over irrigated condition was least in MR-786 (0.21), while it was more in H-1452/10 (30.5) (Table 12.4).

Total Chlorophyll Content

The chlorophyll content as measured by SPAD readings at 110 DAS. The mean SPAD reading (chlorophyll content) was decreased from 42.7 in irrigated to 39.8 in rainfed condition. Among the genotypes, H-1452/10 (45.7) followed by F-2228 (45.2) recorded the highest SPAD reading, while

LRA-5166 (34.3) followed by CPD-168 (37.3) recorded least chlorophyll content. In rainfed condition, more or less same trend was observed (Table 12.4).

Conclusions

The genotypes BS-279, ARBH-813 and Sahana proved better as they recorded highest seed cotton yield both in irrigated as well as rainfed condition and least to moderate drought susceptibility indices. This may be because the genotypes BS-279 and ARBH-813 maintained good water potential and photosynthesis rate. Sahana showed least reduction in TDM in rainfed condition. ARBH-813 recorded highest numbers of bolls in both irrigated as well as in rainfed condition. These are desirable characters for drought situations. But GISV-218, GJHV-358 and GSHV-97/612 recorded least yield and highest drought susceptibility indices. These are non-desirable characters for moisture stress condition. These conclusions are supported by yield components and physiological characters. These lines can be used in breeding programs for moisture stress tolerance as well as yield stability.

References

1. Boyer JS. Plant productivity and environment. Science. 1982;218:443–8.
2. Garg AK, Kim JK, Owens TG, Ranwala AP, Choi YD, Kochian LV, Wu RJ. Trehalose accumulation in rice plants confers high tolerance levels to different abiotic stresses. Proc Natl Acad Sci USA. 2002;99: 15898–903.
3. Babu RC, Nguyen BD, Chamarerk V, Shanmugasundaram P, Chezhian P, Jayaprakash P, Ganesh SK, Palchamy A, Sadasivam S, Sarkarung S, Wade LJ, Nguyen HT. Genetic analysis of drought resistance in rice by molecular markers: association between secondary traits and field performance. Crop Sci. 2003;43:1457–69.
4. Tester M, Bacic A. Abiotic stress tolerance in grasses: from model plants to crop plants. Plant Physiol. 2005;137:791–3.
5. Pettigrew WT. Physiological consequences of moisture deficit stress in cotton. Crop Sci. 2004;44:1265–72.
6. Araújo AE, Silva CAD, Azevedo DMP, Freire EC, Amalho FS, Andrade FP, Ferreira GB, Santana JCF, Amaral JAB, Medeiros JC, Bezerra JRC, Pereira JR, Silva KL, Silva LC, Barros MAL, Carvalho MCS, Luz MJS, Beltrão NEM, Suassuna ND, Ferreira PF, Santos RF, Fonsêca RG. Cultivo do algodão irrigado. Série Documentos, Sistemas de Produção, 3, Versão eletrônica, Jan 2003. ISSN 1678–8710.
7. Ephrath J, Marani A, Bravdo BA. Effect of moisture stress on stomatal resistance and photosynthetic rate in cotton (*Gossypium hirsutum* L.): controlled level of stress. Field Crops Res. 1990;23:117–31.
8. Faver KL, Gerik TJ, Thaxton PM, El-Zik KM. Late season water stress in cotton: II. Leaf gas exchange and assimilation capacity. Crop Sci. 1996;36:922–8.
9. Ephrath JE, Marani A, Bravdo BA. Photosynthetic rate, stomatal resistance and leaf water potential in cotton (*Gossypium hirsutum* L.) as affected by soil moisture and irradiance. Photosynthetica. 1993;29:63–71.
10. Flexas J, Bota J, Loreto F, Cornic G, Sharkey DT. Diffusive and metabolic limitations to photosynthesis under drought and salinity in C3 plants. Plant Biol. 2004;6:269–79.
11. Ennahli S, Earl HJ. Physiological limitations to photosynthetic carbon assimilation in cotton under water stress. Crop Sci. 2005;45:2374–82.
12. Lopez M, Gutierrez JC, Leidi EO. Selection and characterization of cotton cultivars for dry land production in the south west of Spain. Eur J Agron. 1995;4: 119–24.
13. Nepomuceno AL, Oosterhuis DM, Stewart JM. Physiological responses of cotton leaves and roots to water deficit induced by polyethylene glycol. Environ Exp Bot. 1998;40:29–41.
14. Mahmood S, Irfan M, Raheel F, Hussain A. Characterization of cotton (*Gossypium hirsutum* L.) varieties for growth and productivity traits under water deficit conditions. Int J Agric Biol. 2006;8(6): 796–800.
15. Wullschleger SD, Oosterhuis DM. Photosynthesis of individual field grown cotton leaves during ontogeny. Photosynth Res. 1990;23:163–70.

Study of Serum Immunoglobulin Levels: Complements C3 and C4 in Vitiligo Patients in Northern India

13

Usha Singh, Suman Singh, and S.S. Pandey

Abstract

Vitiligo is an acquired, specific, common, often heritable, depigmenting skin disorder characterized by progressive, well-circumscribed milky cutaneous macules devoid of identifiable melanocytes. It is multifactorial and polygenic. The precise pathology remains elusive; however, several theories have been proposed to explain the loss of epidermal melanocytes in this disorder. None of these theories can completely explain all the clinical and experimental observations made in this disorder. The present study was aimed to find out the role of humoral immunity in the pathogenesis of vitiligo. The serum immunoglobulin profiles of vitiligo patients were compared with that of normal healthy controls. Two hundred vitiligo patients and 75 controls were recruited randomly from the Department of Dermatology and Venereology, Sir Sunderlal Hospital, and Banaras Hindu University Hospital, Varanasi, UP, India, for this study. Complements C3 and C4 and immunoglobulins IgG, IgA, and IgM were estimated by turbidometric method using immunoglobulin kit.

Serum C3 levels in 76% controls and 60.5% vitiligo patients were between 80 and 160 mg/dl, and 39.5% vitiligo patients had below 160 mg/dl and significantly decreased in vitiligo patients (86.99 ± 24.97 mg/dl) as compared to controls (95.09 ± 23.96 mg/dl). Serum C4 level is also decreased in vitiligo patients as compared to controls, but the decrease is nonsignificant. Significant increase in serum IgM was noted in vitiligo patients (148.36 ± 54.05 mg/dl) as compared to controls (125.06 ± 42.48 mg/dl). On the contrary, serum IgG and IgA were significantly decreased in

U. Singh • S. Singh (✉)
Department of pathology, Division of immunopathology, Institute of Medical Sciences, BHU, Varanasi, India
e-mail: sumanbiotechster@gmail.com

S.S. Pandey
Department of dermatology & venereology, Institute of Medical Sciences, BHU, Varanasi, India

A. Sabu and A. Augustine (eds.), *Prospects in Bioscience: Addressing the Issues*, DOI 10.1007/978-81-322-0810-5_13, © Springer India 2013

vitiligo patients (1399.63 ± 241.33 and 183.80 ± 60.80 mg/dl) as compared to controls (1,460 ± 89.93 and 212.46 ± 67.33 mg/dl), respectively. These findings tend to suggest that aberrations in cell-mediated immunity and humoral immunity may be operative in the pathogenesis of vitiligo. Whether the change in immunoglobulin concentrations is a cause or an effect needs further research in large-scale prospective studies.

Keywords
Vitiligo • Complement • Immunoglobulins • Humoral immunity

Introduction

Vitiligo is an acquired, specific, common, often heritable, depigmenting skin disorder characterized by progressive, well-circumscribed milky cutaneous macules devoid of identifiable melanocytes. The worldwide incidence is 1% [1]. It is multifactorial and polygenic. The human immune system is extremely complex and interacts bidirectionally with environmental, metabolic, and endocrine factors as well as with infectious agents. It is well arranged and controlled by genetics. In a previous study, it was demonstrated that pathological skin synthesized immunoglobulins in different patterns as compared to normal skin [2]. Despite much research, the precise pathology and etiology of vitiligo remains elusive. Several theories have been proposed to explain the loss of epidermal melanocytes in this disorder, and the autoimmunity is the most popular hypothesis [3]. The association of vitiligo with autoimmune disorders is well established [4, 5]. There are numerous publications that describe the involvement of both cellular and humoral immunity in the pathogenesis of vitiligo [6]. Antibodies against melanocyte antigens are found in the sera of vitiligo patients mainly belonging to IgG class [7, 8]. Histopathological investigations have shown that inflammatory cells are prominent in perilesional areas of vitiligo patients which increase the expression of activation molecules such as interleukin-2 receptor (IL-2R) HLA-DR and major histocompatibility complex (MHC) II [9, 10]. The present study was aimed to find out the role of humoral immunity in the pathogenesis of vitiligo.

Materials and Methods

A total of 200 vitiligo patients and 75 controls were recruited randomly from the Department of Dermatology and Venereology, Sir Sunderlal Hospital, and Banaras Hindu University Hospital, Varanasi, UP, India, for this study. Complements C3 and C4 and immunoglobulins IgG, IgA, and IgM were estimated by turbidometric method using Quantia kit and the turbidometer of Tulip Diagnostics, Goa, India.

Statistical analysis: The statistical analysis of the data was done using student's *t*-test for difference of mean on SPSS for windows (version 16.0) statistical package (SPSS Inc., Chicago, IL) computer statistics program. *P* value less than 0.05 was taken as significant.

Result and Discussion

The mean ages of vitiligo patients and controls were 33.23 ± 16.67 and 31.45 ± 11.02 year old, respectively. Age in vitiligo varied from 12 to 48 years. The 200 patients included 82 (41.0) men with age range 13–48 years and 118 (59.0) women with age range 13–42 years.

Serum C3 in 39.5% patients was below 80 mg/dl, while in controls only 24% patients had low C3 level. Most of the controls (76%) and cases (60.5%) had C3 level between 80 and 160 mg/dl. None of the either vitiligo cases or control group had C3 level above 160 mg/dl. The mean value of C3 is decreased in vitiligo cases as compared to controls, and this decrease was statistically significant (Table 13.1).

Table 13.1 Serum C3 level in patients with vitiligo and control group

Groups	Range (mg/dl)	<80 mg/dl No. (%)	80–160 mg/dl No. (%)	>160 mg/dl No. (%)	Mean ± SD mg/dl	*P* value
Vitiligo (200)	27.50–158	79 (39.5)	121 (60.5)	0 (0.0)	86.99 ± 24.97	0.016*
Control (75)	56.20–155.40	18 (24.0)	57 (76.0)	0 (0.0)	95.09 ± 23.96	

*Statistically significant ($P < 0.05$)

Table 13.2 Serum C4 level in patients with vitiligo and control group

Groups	Range (mg/dl)	<20 mg/dl No. (%)	20–40 mg/dl No. (%)	>40 mg/dl No. (%)	Mean ± SD mg/dl	*P* value
Vitiligo (200)	11–44.40	61 (30.5)	129 (64.5)	10 (5.0)	23.71 ± 7.39	0.056
Control (75)	11.71–38.40	17 (22.7)	55 (73.3)	3 (4.0)	25.62 ± 7.28	

Table 13.3 Serum IgG level in patients with vitiligo and control group

Groups	Range (mg/dl)	<1279.72 mg/dl No. (%)	1279.72–1674.19 mg/dl No. (%)	>1674.19 mg/dl No. (%)	Mean ± SD mg/dl	*P* value
Vitiligo (200)	542–1,788	57 (28.5)	140 (70.0)	3 (1.5)	1399.63 ± 241.33	0.036*
Control (75)	1279.72–1678.21	0 (0.0)	72 (96.0)	3 (4.0)	1,460 ± 89.93	

*Statistically significant ($P < 0.05$)

Table 13.4 Serum IgM level in patients with vitiligo and control group

Groups	Range (mg/dl)	<61.79 mg/dl No. (%)	61.79–285.67 mg/dl No. (%)	>285.67 mg/dl No. (%)	Mean ± SD mg/dl	*P* value
Vitiligo (200)	18–305	6 (3.0)	165 (82.5)	29 (14.5)	138.36 ± 54.05	0.065*
Control (75)	60–290	5 (6.7)	69 (92.0)	1 (1.3)	125.06 ± 42.48	

*Statistically significant ($P < 0.05$)

Out of 200 cases, 30.5% cases were below the normal range, 64.5% within the normal range, and rest 5% had elevated serum C4 value, while in controls, 73.3% were within the normal range, 22.7% had low C4 value, and only 4% showed increased C4 value. The mean value of C4 is also decreased in vitiligo patients as compared to controls, but this decrease is not statistically significant (Table 13.2).

About 70% patients had serum IgG within this range while 28.5% had IgG level below this range, and rest 1.5% patients had elevated IgG value. None of the controls had decreased IgG level. The mean value of serum IgG is found to be decreased in vitiligo patients as compared to controls, and this decrease is statistically significant (Table 13.3).

Out of 200 cases, 82.5% cases were within the normal range, 3% had decreased, and rest 14.5% had increased serum IgM levels. In controls, 92% were between 60 and 290 mg/dl, i.e., within normal range. Only 6.7% showed increased and 1.3% showed decreased IgM value. The mean value of IgM is increased in vitiligo patients as compared to controls, but this increase was statistically nonsignificant (Table 13.4).

Serum IgA in control group was between 101.11 and 362.20 mg/dl. About 91% patients had IgA level within this range. In 7.5% patients' serum, IgA level was below this range, while 1.5% patients had serum IgA level above this range. The mean value of IgA was decreased in vitiligo patients as compared to control which was statistically significant (Table 13.5).

Table 13.5 Serum IgA level in patients with vitiligo and control group

Groups	Range (mg/dl)	<101.11 mg/dl No. (%)	101.11–362.20 mg/dl No. (%)	>362.20 mg/dl No. (%)	Mean ± SD mg/dl	*P* value
Vitiligo (200)	70–407	15 (7.5)	182 (91.0)	3 (1.5)	183.80 ± 60.80	0.001*
Control (75)	102–369	0 (0.0)	75 (100.0)	0 (0.0)	212.46 ± 67.33	

*Statistically significant ($P < 0.05$)

In a study by some workers, abnormal serum immunoglobulins level had been reported in vitiligo patients [11]. Studies have found that patients with vitiligo and simultaneous atopy presented lower IgE and higher IgA values as compared to patients with atopy alone [12]. Some investigators had reported an increase of IgE count in 20% vitiligo patients [13]. According to our study, the mean value of serum IgG and IgA decreased significantly as compared to controls, while serum IgM level was found to be increased, but the increase is statistically nonsignificant. The mean serum C3 and C4 decreased significantly in vitiligo patients as compared to controls. This was in accordance with the study held on 30 Bangladeshi vitiligo patients. The workers had found that concentration of IgG and IgA in vitiligo patients decreased significantly as compared to normal healthy controls, but the change of IgM was not significant [14]. They also found that there was significant correlation between immunoglobulin (IgG and IgA only). Furthermore, a statistically significant decrease in helper cells and helper/suppressor ratios in comparison with control subjects had been previously reported [15] which supports the above findings and suggested that aberration in cell-mediated immunity may be operative in the pathogenesis of vitiligo. This may also indicate that IgG anti-melanocyte antibodies could have an important role in this disorder, as they can induce melanocyte damage in vitro by a complement-mediated mechanism and antibody-dependent cellular cytotoxicity. In a case study, decrease in serum IgA level had been reported, while the other immunoglobulins (IgG, IgM) and C3 and C4 were in normal range [16]. In vitiligo patients, an increased frequency of heterozygous C4B gene encoding the fourth component of complement and C2 deficiency has been reported, further supporting the concept that vitiligo is associated with abnormalities in the complement system [17]. In another study in Iran on 55 vitiligo patients, the workers determined C3 and C4 values with single radial immunodiffusion (SRID) and reported statistically significant decreased C3 and C4 values in 25.5% patients [18]. Thus, the change in immunoglobulin concentration may be a cause or an effect in pathogenesis of vitiligo.

References

1. Wolff K, Goldsmith LA, Katz SI, Gilchrest BA, Paller AS, Leffell DJ. Fitzpatrick's dermatology in general medicine, vol. I. 7th ed. New York: McGraw-Hill; 2007. p. 616–21.
2. Lai A, Fat RFM, Suurmond D, Furth RV. In vitro synthesis of immunoglobulin secretory component and complement in normal and pathological skin and the adjacent mucous membranes. Clin Exp Immunol. 1973;14:377–95.
3. Daneshpazhooh M, Mostofizdeh GM, Behjati J, Akhyani M, Robati RM. Antithyroid peroxidase antibody and vitiligo: a controlled study. BMC Dermatol. 2006;10(6):3.
4. Alkahateeb A, Fain PR, Thody A, Bennett DC, Spritz RA. Epidemiology of vitiligo and associated autoimmune disease in Caucasian probands and their families. Pigment Cell Res. 2003;16:208–14.
5. Pichler R, Sfetsos K, Badics B, Gutenbrunner S, Berg J, Aubo CKJ. Lymphocyte imbalance in vitiligo patients indicated by elevated CD4+/CD8+ T-cell ratio. Wien Med Wochenschr. 2009;159:337–41.
6. Ogg GS, Rod Dunbar P, Romero P, Chen JL, Cerundolo V. High frequency of skin-homing melanocyte-specific cytotoxic T lymphocytes in autoimmune vitiligo. J Exp Med. 1998;188:1203–8.
7. Kemp EH, Gawkrodger DJ, MacNeil S, Watson PF, Weetman AP. Detection of tyrosinase autoantibodies in vitiligo patients using 35S-labelled recombinant human tyrosinase in radioimmunoassay. J Invest Dermatol. 1997;109:69–73.
8. Kovacs SO. Vitiligo. J Am Acad Dermatol. 1998; 38:647–66.

9. Al Badri AM, Foulis AK, Todd PM, Gariouch JJ, Gudgeon JE, Stewart DG, Gracie JA, Goudie RB. Abnormal expression of MHC class II and ICAM-1 by melanocytes in vitiligo. J Pathol. 1993;169:203–6.
10. Van den Wijngaard R, Wankowicz-Kalinska A, Le Poole C, Tigges B, Westerhof W, Das P. Local immune response in skin of generalized vitiligo patients. Destruction of melanocytes is associated with the prominent presence of CLA+T cells at the perilesional site. Lab Invest. 2000;80:1299–309.
11. Mittal R, Popli R, Jassal J, Kumar D. Immunological study in vitiligo and contact depigmentation. Indian J Dermatol Venereol Leprol. 1994;60:92–4.
12. Chatain C, Ring J, Schallreuter KU. Total serum immunoglobulins and atopic symptoms in patients with vitiligo. Dermatology. 1994;189:27–31.
13. Perfetti L, Cespa M, Nume A, Orecchia G. Prevalence of atopy in vitiligo. Dermatologica. 1991;182:218–20.
14. Ali R, Ahsan MS, Azad MA, Ullah MA, Bari W, Islam SN, Yeasmin S, Hasnat A. Immunoglobulin levels of vitiligo patients. Pak J Pharm Sci. 2010;23:97–102.
15. Grimes PE, Ghoneum M, Stockton T, Payne C, Kelly AP, Alfred L. T cell profiles in vitiligo. J Am Acad Dermatol. 1986;14:196–201.
16. Gul U, Soylu S, Demiriz M. Colocalization of lichen planus and vitiligo associated with selective IgA deficiency. SKINmed Dermatol Clin. 2007; 6:202–3.
17. Venneker GT, Westerhof W, de Vries IJ, Drayer NM, Wolthers BG, de Waal LP, Bos JD, Asghar SS. Molecular heterogeneity of the fourth component of complement (C4) and its genes in vitiligo. J Invest Dermatol. 1992;99:853–8.
18. Farsangi MH, Farokhi SA, Nouhpisheh MK, Tahmasbi R. The role of humoral immune system in the pathogenesis of vitiligo. Iran S Med J. 2003;5:124–8.

14 Evaluation of Antimicrobial Potential of an Invasive Weed *Amaranthus spinosus* L.

A. Mary Sheeba, S.R. Deepthi, and I. Mini

Abstract

Amaranthus spinosus L. growing luxuriantly as an invasive weed on road sides and abandoned land was screened for antimicrobial activity. The bioparts, namely, root, stem, leaves and flower, extracted in distilled water, hexane and methanol were assayed against five bacterial strains such as *Staphylococcus* sp., *Escherichia coli*, *Pseudomonas* sp., *Klebsiella* sp., *Paracoccus* sp. and three fungal strains, namely, *Fusarium* sp., *Aspergillus* sp. and *Alternaria* sp. Preliminary qualitative analysis was also done. Stem and flower of *Amaranthus spinosus* displayed more antibacterial activity among the bioparts. Maximum zones of inhibition in *E. coli* (14 mm) and *Pseudomonas* (13 mm) were noticed in stem extract in distilled water (4.7 mg/disc) and methanol (3.8 mg/disc), respectively. Distilled water fraction of flower inhibited all strains under investigation. Maximum zones of inhibition in respect of *Staphylococcus* (10 mm), *Paracoccus* (9 mm) and *Klebsiella* (15 mm) were evident in flower extracted in various solvents. The methanol fraction of the bioparts exhibited total inhibition of fungal growth. Root, stem and leaves of *A. spinosus* extracted in hexane imparted partial inhibition on all fungal strains. In general, the bioparts of the weed *A. spinosus* displayed antimicrobial activity which can be employed as a natural antimicrobial agent leading to its sustainable utilization.

Keywords

Amaranthus spinosus • Invasive weed • Antimicrobial activity

Introduction

Nature has been a source of medicinal agents for thousands of years, and an impressive number of modern drugs have been isolated from natural sources. Long before mankind discovered the

A.M. Sheeba • S.R. Deepthi • I. Mini (✉)
Department of Botany, University College, Thiruvananthapuram 695 034, Kerala, India
e-mail: dr.mini08@gmail.com

A. Sabu and A. Augustine (eds.), *Prospects in Bioscience: Addressing the Issues*,
DOI 10.1007/978-81-322-0810-5_14, © Springer India 2013

existence of microbes, the idea that certain plants had healing potential as they contained antimicrobial principles was well accepted. This plant-based, traditional medicine system continues to play an essential role in health care, with about 80% of the world's inhabitants relying mainly on traditional medicines for their primary health care [1]. According to the WHO, the plants would be the best source to obtain a variety of drugs, and such plants should be investigated to better understand their properties, safety and efficacy [2]. The investigation of certain indigenous plants for their antimicrobial properties may yield useful results. Many studies indicated that in some plants, there are many substances such as peptides, unsaturated long-chain aldehydes, alkaloid constituents, some essential oils, phenols and water, ethanol, chloroform, methanol and butanol-soluble compounds. These plant products then emerged as compounds with potentially significant therapeutic application against human pathogens, including bacteria, fungi or virus.

There are many reports on the antimicrobial activity of plants belonging to Amaranthaceae family such as *Blutaparon portulacoides* [3], *Alternanthera maritima* [4], *Achyranthus aspera* [5], *Alternanthera brasiliana* [6], *Alternanthera tenella* [7], *Alternanthera sessilis* and *Guilleminea densa* [8], Amaranthus *tricolor* [9], *Amaranthus spinosus* [10] and *Amaranthus viridis* [11] against strains of gram-positive and gram-negative bacteria, along with the yeasts and dermatophytes. *Amaranthus spinosus* L. is an invasive weed commonly found on the wastelands, road sides and abandoned lands. *Amaranthus* is from the Greek "amarantos" which means "unfading", commonly known as spiny amaranth, prickly amaranth and spiny pigweed. Plant is nitrophilous and prefers soils with high organic matter content, but is also able to grow on sandy soils. Plant, especially the young leaves, can be used as vegetable and has medicinal properties like refrigerant, antidote, diaphoretic, febrifuge, astringent, emmenagogue, sudorific, lactagogue, diuretic and emollient. Eradication of weeds from our environment is rather difficult as it consumes much money, labour and time. Hence, combating the problem of weeds by exploiting their beneficial potentials is receiving increasing attention among the scientific community around the globe. Antimicrobial studies on *Amaranthus spinosus* remain few and fragmentary that prompted the present investigation.

Materials and Methods

Whole plant was collected from its natural habitat washed thoroughly to remove adhering particles. Root, stem, leaves and flower were separated, dried under shade followed by oven drying. The dried samples were powdered and were extracted in solvents, namely, distilled water, hexane and methanol in a Soxhlet apparatus for 8 h continuously. The extracts were concentrated, weighed and made up to a known volume and was used for the detection of phytochemicals and antimicrobial assay.

Phytochemical Screening

The extracts were subjected to phytochemical screening to detect the presence of phytoconstituents such as saponins, alkaloids, tannins, coumarins, iridoids, flavonoids, steroids, terpenoids, phlobatannin, reducing sugar, anthraquinone and glycosides.

Antibacterial Assay

The extracts were tested to detect antibacterial properties against five strains of bacteria (*Staphylococcus* sp., *Escherichia coli*, *Pseudomonas* sp., *Klebsiella* sp. and *Paracoccus* sp.) by using disc diffusion method [12]. The bacterial strains were subcultured by inoculating spores of the bacteria from the isolated colonies in a test tube containing peptone water (13.6 g/L). Nutrient agar medium was prepared (37.5 g/L). Discs of 6 mm diameter were cut out of Whatman No.1 filter paper. The prepared paper disc was impregnated with known amounts of various extracts from the bioparts of *A. spinosus*. Control and standards were also prepared accordingly.

Table 14.1 Phytochemical screening of *A. spinosus* L.

Bioparts	R			S			L			F		
Extracts	DW	H	M	DW	H	M	DW	H	M	DW	H	M
Reducing sugar	+	–	+	+	–	+	+	+	+	+	+	+
Glycosides	–	–	–	–	–	–	–	–	–	–	–	–
Flavonoids	–	–	–	–	+	+	–	–	–	–	+	+
Tannins	+	–	+	+	–	+	+	+	+	+	+	+
Alkaloids	+	+	–	+	+	–	+	+	–	+	+	–
Terpenoids	+	–	–	+	–	–	+	–	+	–	–	+
Steroids	–	+	–	–	+	–	–	+	–	+	+	–
Saponins	+	+	+	+	+	+	+	+	+	+	+	+
Phlobatannins	–	–	–	–	–	–	–	–	–	–	–	–
Coumarins	+	+	+	+	+	+	+	+	+	+	+	+
Iridoids	–	+	+	–	+	+	–	+	+	–	+	+
Anthraquinones	–	–	–	–	–	–	–	–	–	–	–	–

R root, *S* stem, *L* leaf, *F* flower, *DW* distilled water, *H* hexane, *M* methanol, + present, – absent

Pure cultures of each bacterial strain were spread evenly on the agar plate. Prepared disc with the extracts was placed over each bacterial colony; the inoculated petri dishes were labelled and incubated at 37°C for 24 h. The plates were observed for zone of inhibition surrounding the discs, which were measured to detect and quantify the antibacterial activity. The experiments performed in duplicate.

Antifungal Assay

Antifungal activity against three fungal strains (*Fusarium* sp., *Aspergillus* sp. and *Alternaria* sp.) by incorporating various extracts in the media used for fungal cultures [13]. The test was conducted by incorporating crude extract in the Sabouraud dextrose agar medium. SDA slants were prepared by dissolving SDA (65 g/L) and agar (10 g/L) in distilled water by heating and autoclaved at 121°C for 20 min. The media (4–5 mL) were poured in test tubes, mixed well with known volume of the crude extracts and allowed to set in to slants. Pure cultures of fungal strains were inoculated in to the slants using inoculation loop under aseptic condition. The tubes were labelled and kept at room temperature and the observations recorded after 3 days on incubation. The control was also run in the same manner. The experiments performed in duplicate to confirm the results.

Results and Discussion

Phytochemical Screening

Phytochemicals are secondary products of metabolism, occur in small quantities and represent a centrepiece of ecological biochemistry [14]. Many of these compounds act to defend the host from the harmful effects of toxins, carcinogens or mutagens found in the plant or plant environment or from the attack of external predators. The preliminary phytochemical screening of the extracts in different solvents of various bioparts of *A. spinosus* (Table 14.1) revealed the presence of important secondary metabolites in the plant body. Maximum numbers of secondary metabolites were eluted by methanol in the present study. The results were more or less similar in various bioparts. Saponins and coumarins were eluted irrespective of the bioparts and solvents. Among the bioparts, the flower contained most of the phytochemicals under investigation.

Numerous plants are reported to possess phytochemicals, and most of the currently identified

Table 14.2 Antibacterial analysis of *A. spinosus* L.

Biopart	Extract	Concentration mg/disc	Zone of inhibition (mm diameter) *E. coli*	*Pseudomonas*	*Staphylococcus*	*Paracoccus*	*Klebsiella*
Root	Methanol	2.4	7	12	–	6	–
	Hexane	1.7	–	6	–	7	9
	Dis. water	3.2	10	6	6	7	8
Stem	Methanol	2.5	8	13	–	6	7
	Hexane	3.6	6	–	–	6	8
	Dis. water	4.7	14	8	–	7	11
Leaves	Methanol	3.4	9	8	–	–	7
	Hexane	1.5	–	–	–	–	–
	Dis. water	2.3	9	–	7	7	7
Flower	Methanol	3.8	–	7	10	–	7
	Hexane	4.2	–	8	7	6	15
	Dis. water	3.7	9	7	6	9	7
Control	Methanol	–	–	–	–	–	–
	Hexane	–	–	–	–	–	–
	Dis. water	–	–	–	–	–	–

compounds are products of the shikimic acid and acetate pathway. The common ones include phenolics like cinnamic and benzoic acids, coumarins, tannins, terpenoid and a few alkaloids, steroids and quinines [15]. Phytochemicals are very important with regard to their potential pharmacological effects [16]. Many alkaloids are toxic to other organisms. They often have pharmacological effects and are used as medications and as drugs [17]. Saponins are a class of chemical compounds which may serve as antifeedants and to protect the plant against microbes and fungi. The antifungal and antibacterial properties of saponins are important in cosmetic applications in addition to their emollient effects [18]. They help to prevent cancer and increase the efficiency of vaccines. Terpenoids have a large and diverse role in antibacterial activity. They defend many species of plants, animals and microorganisms against predators, pathogens and competitors [19].

Tannins are polyphenolic compounds and are widely distributed in many species of plants, where they play a role in protection from predation, and perhaps also as pesticides, and in plant growth regulation [20]. Anthraquinone is an organic compound which acts as an effective repellent. The activity reported for coumarins includes anti-HIV, antitumour, anti-hypertension, anti-inflammatory, antiseptic, analgesic and is also used in the treatment of asthma [21]. Flavonoids extent antimicrobial activity in the healing of wounds and in the treatment of skin diseases [22]. They also play significant role as hypoglycaemic, antioxidant, anti-inflammatory and anticarcinogenic activity [23].

Antibacterial Assay

The bioparts of *A. spinosus* exhibited antibacterial activity (Table 14.2) and varied with bacterial strains, bioparts and solvents. Flower of *A. spinosus* extracted in distilled water inhibited all bacterial strains under study. The inhibition zone of *E. coli* varied from 6 to 14 mm in stem extracted in hexane and methanol, respectively. The growth of *E. coli* was inhibited more effectively in root extracted in distilled water (10 mm) and stem extracted in methanol (14 mm). Most *E. coli* strains are harmless, but some serotypes can cause serious food poisoning in humans.

Root and flower of *A. spinosus* extracted in all solvents inhibited the growth of *Pseudomonas*, while the activity was restricted to methanolic extract of stem and leaf. Highest inhibition zone was noticed in methanolic extract of stem (13 mm) followed by root (12 mm). Flower

Table 14.3 Antifungal assay of *A. spinosus* L.

Biopart	Extract	Concentration mg/mL	Inhibitory efficacy *Alternaria* sp.	*Aspergillus* sp.	*Fusarium* sp.
R	M	24	+	+	+
	H	17	–/+	–/+	–/+
	DW	32	–	–	–
S	M	25	+	+	+
	H	36	–/+	–/+	–/+
	DW	47	–	–	–
L	M	34	+	+	+
	H	15	–/+	–/+	–/+
	DW	23	–	–	–
F	M	38	+	+	+
	H	42	–	–	–/+
	DW	37	–	–	–
C	M	–	+	+	+
	H	–	+	–	–
	DW	–	–	–	–

R root, *S* stem, *L* leaf, *F* flower, *C* control, *M* methanol, *H* hexane, *DW* distilled water, + total inhibition, –/+ partial inhibition, – no inhibition

extract of *A. spinosus* irrespective of the solvent inhibited *Staphylococcus* with highest inhibition zone of 10 mm in methanol. Distilled water extract of root and leaf exhibited an inhibition zone of 6 and 7 mm, respectively. *Staphylococcus* can cause a wide variety of diseases through either toxin production or penetration. A number of strains have developed antibiotic resistance, which means that the infection may not always respond to treatment.

Root and stem extracted in all solvents displayed inhibitory activity against the soil bacteria *Paracoccus*. The activity was restricted to leaves and flower extracted in distilled water. The denitrification properties of *Paracoccus* sp. are an important cause for the loss of nitrogen fertilizers in agricultural soil due to the chemical process called denitrification. *Klebsiella* sp. can cause the disease *Klebsiella pneumonia*, and they are often resistant to multiple antibiotics. The growth of *Klebsiella* was inhibited with stem and flower extracted in all solvents. Distilled water extract of all bioparts inhibited *Klebsiella*. The inhibition zone of *Klebsiella* fluctuated from 7 to 15 mm. Highest activity was noticed in flower extracted in hexane (15 mm) and stem in distilled water (11 mm).

Antifungal Assay

Antifungal assay of the extracts of *A. spinosus* (Table 14.3) in various solvents exhibited varied activity against the fungal strains under investigation. Fungi are a group of achlorophyllous plants which include many thousand members, and most of them are poisonous. The methanol extracts showed complete antifungal activity compared with the extracts of hexane and distilled water. Methanol extracts of the root (24 mg/g), stem (25 mg/g), leaf (34 mg/g) and flower (38 mg/g) exhibited inhibitory efficacy against *Aspergillus* sp., *Alternaria* sp. and *Fusarium* sp. The extracts showed varying degrees of antifungal activity which may be due to the active principles that can be extracted differently depending on the ability of the extracting solvent [22, 24].

The results of the present study on the efficacy of the weed *Amaranthus spinosus* against strains of bacteria and fungi corroborated with various reports of plants belonging to Amaranthaceae. The antibacterial activity of chloroform, n-hexane and ethyl acetate extracts of *Amaranthus spinosus* were evaluated [10] against gram-positive and gram-negative human pathogenic bacteria, and *all* extracts showed good activity against all bacteria with an average zone of inhibition 8–15 mm that strongly supported the present study. The antibacterial activity of the isolated compounds in *Blutaparon portulacoides* was assessed [3] against 15 strains of gram-positive and gram-negative bacteria, along with certain fungus. The ethanolic extracts (aerial parts and roots) and callus generated in vitro of *Alternanthera maritima* exhibited activity against 30 strains including gram-positive and gram-negative bacteria and yeast with inhibition halo between 6 and 20 mm [4] and inferred that all the organic crude extracts and plant cell callus culture extracts obtained from two culture conditions were bioactive and the minimum inhibitory concentrations (MIC) values for isolated constituents varied from 25 to 500 μg/mL.

There are reports [5] on the antibacterial analysis of *Achyranthus aspera* against *Staphylococcus aureus*, *S. haemolyticus* and *Bacillus typhosus*. The aqueous ethanol (70%) crude extract of *Alternanthera brasiliana* leaves and dichloromethane, ethyl acetate and butanolic fractions of the crude extract were tested against a panel of microorganisms and demonstrated that the crude extract and some fractions showed moderate activities against *Staphylococcus aureus*, *Pseudomonas aeruginosa*, *Saccharomyces cerevisiae* and *Prototheca zopfii* [6]. The efficacy of *Amaranthus viridis* was analysed [11] through the screening of the chloroform and aqueous leaf extract of this plant against the bacterial species, namely, *Xanthomonas axonopodis*, *Pseudomonas syringae* (gram-negative bacteria), *Corynebacterium minutissimum*, *Clostridium difficile* (gram-positive bacteria) and major seed-borne fungi *Aspergillus niger*, *Alternaria alternata*, *Drechslera biseptata* and *Fusarium solani*.

Another study was conducted [7] on crude extracts of a callus (two culture media) and adult plants of *Alternanthera tenella* and evaluated their antibacterial and antifungal activity, in vivo and in vitro against 30 strains of microorganisms including gram-positive and gram-negative bacteria, yeasts and dermatophytes with inhibition halos between 6 and 20 mm and inferred that ethanolic and hexane extracts of adult plants inhibited the growth of bacteria, yeasts and dermatophytes. The antimicrobial activity of *Achyranthes aspera*, *Alternanthera sessilis* and *Guilleminea densa* of Amaranthaceae family was tested for bactericidal and fungicidal effect [8].

A study to evaluate the antimicrobial properties of ethanolic extracts of the leaves of *Amaranthus tricolor* against 15 strains of different gram-positive and gram-negative pathogenic bacteria, including *Bacillus cereus*, *Listeria monocytogenes*, *Staphylococcus aureus*, *Vibrio para* and *Escherichia coli*, was conducted [9] and inferred that the plant exhibited specific inhibition of one of the tested bacteria *Bacillus cereus*. Resinous exudates and diterpenoids were known to inhibit bacterial growth with inhibition zone exceeding 10 mm [25]. Secondary metabolites such as alkaloids and flavonoids possess antimicrobial potential [26]. The results in the present study also corroborated with the reports that flowering plants contain antimicrobial substances [27].

Conclusions

The present study help assess the characterization of the weed to elucidate data on the preliminary phytochemical screening for secondary metabolites as well as antibacterial and antifungal activity. *Amaranthus spinosus* is a widespread weed, and its control is rather difficult due to its perennial nature. The identification of antimicrobial activity of the weed opens up new vistas in the sustainable management of these "out of place plants". In general, the invasive weed *Amaranthus spinosus* was rich in secondary metabolites which exhibited antibacterial and antifungal activity and can be effectively exploited for controlling the strains under investigation and as a biocontrol agent against pathogenic microbes. The present study warrants further detailed investigation on to the isolation, purification, characterization as well as biosynthetic pathways of the phytochemicals and its dynamics in medical field.

References

1. Owolabi JO, Obasuyi O. Antifungal and antibacterial activities of the ethanolic and aqueous extract of *Kigelia africana* (Bignoniaceae) stem bark. Afr J Biotechnol. 2007;6(14):882–5.
2. Nascimento GF, Juliana L, Paulo CF, Giuliana LS. Antibacterial activity of plant extracts and phytochemicals on antibiotic resistant bacteria. Braz J Microbiol. 2000;31:247–56.
3. Salvador MJ, Ferreira EO, Pral EM, Alfieri SC, Albuquerque S, Ito IY, Dias DA. Bioactivity of crude extracts and some constituents of *Blutaparon portulacoides* (Amaranthaceae). Phytomedicine. 2002;6: 566–71.
4. Salvador MJ, Zucchi OLAD, Candido RC, Ito IY, Dias DA. *In vitro* antimicrobial activity of crude extracts and isolated constituents of *Alternanthera maritime*. Pharm Biol. 2004;42:138–48.
5. Goyal BR, Goyal RK, Mehta AA. Phyto pharmacology of *Achyranthus aspera*: s review. Pharmacog Rev. 2007;1(1):6.
6. Pereira DF, Mariane DS, Sydney HA, Marli MA, Margareth LA. Antimicrobial activity of a crude extract and fractions from *Alternanthera brasiliana* (L.) *O. Kuntze* leaves. Lat Am J Pharm. 2007;26(6): 893–6.
7. Salvador MJ, Pereira PS, França SC, Candido RC, Ito IY, Dias DA. Bioactive chemical constituents and comparative antimicrobial activity of callus culture and adult plant extracts from *Alternanthera tenella*. Z Naturforsch C. 2009;64(56):373–81.
8. Singh A. Bioactivity of famine food plants from the family: Amaranthaceae. 2009. URI: http://hdl.handle.net/10321/450
9. Aziz SM, Low CN, Chai LC, Razak ASSN, Selamat J, Son R, Sarker MZI, Khatib A. Screening of selected Malaysian plants against several food borne pathogen bacteria. Int Food Res J. 2011;18(3):1195–201.
10. Bulbul IJ, Nahar L, Ripa FA, Haque O. Antibacterial, cytotoxic and antioxidant activity of chloroform, n-hexane and ethyl acetate extract of plant *Amaranthus spinosus*. Int J PharmTech Res. 2011; 3(3):1675–80.
11. Mushatq S, Haider MS, Ali A, Javed S, Khokhar I, Mukhtar I. *In vitro* comparative screening of antibacterial and antifungal activities of some common weeds extracts. Pak J Weed Sci Res. 2012;18(1): 15–25.
12. Kirby WM, Sherris JC, Turck M. Antibiotic susceptibility testing by a standardized single disc method. Am J Clin Pathol. 1966;45:493–6.
13. Collins CH, Lynn PN. Microbiological methods. 3rd ed. Butter Worth and Co. Pvt. Ltd. London; 1970. p. 414–27.
14. Harborne JB. Phytochemical methods: a guide to modern techniques of plant analysis. 3rd ed. New Delhi: Springer; 1998. p. 41–4.
15. Einhelling FA, Leather GR. Potentials for exploiting allelopathy to enhance crop production. J Chem Ecol. 1988;14(10):1829–42.
16. Pascual ME, Carretero ME, Slowing KV, Villar A. Simplified screening by TLC of plant drugs. Pharm Biol. 2002;40(2):139–43.
17. Manske RHF. The alkaloid. Chemistry and physiology, vol. VIII. New York: Academic; 1965. p. 673.
18. Cheeke PR. Saponins: surprising benefits of desert plants. The Linus Pauling Institute. 1998. http://www.orst.edu
19. Gershenzon J, Dudareva N. The function of terpene natural products in the natural world. Nat Chem Biol. 2007;3(7):408–14.
20. Ferrell KE, Thorington RW. Squirrels: the animal answer guide. Baltimore: Johns Hopkins University Press; 2006. p. 91.
21. Moreland DE, Novitzky WP. Effects of phenolic acids, coumarins and flavonoids on isolated chloroplast and mitochondria. In: Waller GR, editor. Allelochemicals: role in agriculture and forestry, American Chemical Society symposium series 330. Washington, DC: American Chemical Society; 1985. p. 247–61.
22. Barnabas CG, Nagarajan S. Antimicrobial activities of flavonoids of some medicinal plants. Fitotherapia. 1988;3:508–10.
23. Anila L, Vijayalekshmi NR. Flavonoids from *Emblica officinalis* and *Mangifera indica*: effectiveness for dyslipidemia. J Ethnopharmacol. 2002;79(1):81–7.
24. Kafaru E. Uses of Bitter leaf. Natl Assoc Altern Med Pract Herb Rememb. 1996;2:29.
25. Urzua A, Caroli M, Vasquez L, Mendoza L, Wilkens M, Tojo E. Antimicrobial study of the resinous exudates and diterpenoids isolated from *Eupatorium salvia* (Asteraceae). J Ethnopharmacol. 2002;62(3):251–4.
26. Okunade AL. *Ageratum conyzoides* L. Asteraceae. Fitoterapia. 2002;73(1):1–16.
27. Ibekwe VI, Nnanyere NF, Akyobi CO. Antimicrobial activities and phytochemical qualities of extracts of orange peels. Int J Environ Heath Human Dev. 2001;2(1):41–6.

Evolutionary Genomics of Avian MHC BLB2 Gene by Molecular Phylogenetic Analysis

15

Sudina Kizhakkayil, Arun Raveendran, and E. Jayadevi Variyar

Abstract

The major histocompatibility complex (MHC) molecules play crucial role in the regulation of the immune response by communicating among different cellular components of the immune system (Lamont, Poult Sci 77:1111–1118, 1998). Chicken (*Gallus gallus*) has been developed into an important experimental animal model for implicating MHC genes in resistance to infectious diseases and economic traits (Schat et al., Poult Sci 73:502–508, 1994). As the process of evolution in this region takes place over long periods of time that cannot be observed directly, phylogenetic analysis with present-day species and breeds could help. The evolutionary links between various domesticated birds were analysed using tree construction distance methods such as neighbour-joining (N-J) and UPGMA (unweighted pair group method with arithmetic mean) and character-based methods like maximum likelihood and maximum parsimony by MEGA version 5.05. The genetic distance was analysed using the same software. The phylogenetic analyses using various methods produced near identical topologies. The genetic distance analysis was in accordance with the tree studies providing a clear overview upon the evolutionary changes of the MHC gene under analysis, i.e. MHC B-L Beta 2.

Keywords

MHC • Chicken • Evolution • Phylogeny

S. Kizhakkayil • E.J. Variyar (✉)
Department of Biotechnology and Microbiology, School of Life Science, Kannur University, Thalassery Campus, Palayad P.O, Kannur, Kerala 670661, India

Inter University Center for Biosciences, Kannur University, Thalassery Campus, Palayad, Kannur, Kerala 670661, India
e-mail: drjayadevi@rediffmail.com

A. Raveendran
Biozon Scientifics, Anchal, Kollam, Kerala 691306, India

Introduction

Speciation and domestication were the stages of extensive evolution in immunity-related genes such as major histocompatibility complex (MHC). The importance of MHC lies in its role in determining the life and death in response to many infectious pathogens, both simple and complicated ones [1]. The simple and compact structure

A. Sabu and A. Augustine (eds.), *Prospects in Bioscience: Addressing the Issues*,
DOI 10.1007/978-81-322-0810-5_15, © Springer India 2013

of chicken MHC as described by 'minimal essential MHC model of the chicken' made it a potent experimental model for MHC studies [2]. These key receptor molecules recognize and bind foreign peptides for presentation to specialist immune cells and subsequent initiation of an immune response [3]. A highly polymorphic multigene family encodes these versatile molecules. They represent the most polymorphic genetic system known in vertebrates to date. In chicken (*Gallus gallus*), MHC genes are located on chromosome 16 [4]. Chicken B or B-F/B-L region has been fully sequenced and found to contain 19 genes within 92 kb of genomic sequence, including two class I (B-F) genes and two class II (B-L) genes [5]. The chicken MHC is associated with resistance to bacterial, parasitic, and viral diseases and other important economic traits in chickens [1]. There are significant numbers of studies which undoubtedly prove the association of MHC genes with diseases such as Marek's disease, Rous sarcoma, Newcastle disease and salmonellosis [6].

Among immune genes, several major histocompatibility complex (MHC) genes are suggested to be under strong pathogen-driven selective forces [7]. As generations undergo various kinds of ecologic adaptations, evolutionary novelties arise following genetic and genomic duplications. This results in vast rearrangements and pronounced variation in gene number and genomic organization among organisms [8]. Studies in the MHC gene organization, diversity, polymorphisms and expression suggest that various avian species have a more complex MHC structure than that described in chickens. Due to the fact that evolution takes place over long periods of time that cannot be observed directly, we must reconstruct phylogenies by inferring the evolutionary relationships among present-day organisms. Though studies have been reached far ahead on the evolutionary aspect of this multigene family at species level, scope of in-depth analysis up to breed level persists. The geographical isolation, pathogen encounters and other environmental factors account for large diversity and rapid evolutionary changes of these loci. Such variations may be responsible for the drastic difference in the immune response of various breeds of birds.

It is commonly observed that modern day breeds are highly susceptible to infectious disease outbreaks such as avian influenza and Newcastle disease compared to indigenous and wild counterparts. The present study aimed at revealing evolutionary background of the gene MHC BLB2 (MHC B-L Beta 2) which is a potential region for modifying a variety of pathogens.

Materials and Methods

Sequences of MHC BLB2 gene of Indian red jungle fowl (progenitor of domestic chicken) and domesticated birds like turkey (*Meleagris gallopavo*), guinea fowl (*Numida meleagris*), Japanese quail (*Coturnix japonica*) and Muscovy duck (*Cairina moschata*) were compared with that of important breeds like white leghorn (an egg-type breed), Cornish (a well-known meat breed) and Kadaknath (a prominent native chicken of India). The sequences were extracted from DNA databases such as NCBI (National Center for Biotechnology Information), DDBJ (DNA Database of Japan) and EMBL (European Molecular Biology Laboratory) nucleotide sequence database. The accession numbers of the sequences analysed were given in Table 15.1. Sequences in FASTA format were aligned by progressive method of multiple sequence alignment using Molecular Evolutionary Genetic Analysis (MEGA Version 5.05) software [9]. Molecular phylogenies based on sequence alignments are only as accurate as the alignment data from which they are produced; consequently, it is important that the alignment data quality is preserved. Sequence contributions to the multiple sequence alignment are weighted according to their relationships on the predicted evolutionary tree. Weights are based on the distance of each sequence from the root in case of distance method. Several statistical methods exist that can be used for reconstruction of the phylogenetic trees. Using tree construction distance methods, N-J (neighbour-joining) and UPGMA (unweighted pair group method with arithmetic mean), the evolutionary connections between organisms were analysed. Then the same sequences were

Table 15.1 Details of the nucleotide sequences of MHC BLB2 sequences from different poultry species used in present study

Sl no.	Category	Accession no.	Length
1	Indian red jungle fowl (*Gallus gallus*)	DQ885564	277 bp
2	Kadaknath (*Gallus gallus domesticus*)	EF579807	277 bp
3	White Cornish (*Gallus gallus domesticus*)	HQ218317	773 bp
4	White leghorn (*Gallus gallus domesticus*)	DQ885562	276 bp
5	Turkey (*Meleagris gallopavo*)	AM233486	277 bp
6	Guinea fowl (*Numida meleagris*)	EU826064	754 bp
7	Muscovy duck (*Cairina moschata*)	HQ909102	1188 bp
8	Japanese quail (*Coturnix japonica*)	AB110482	830 bp

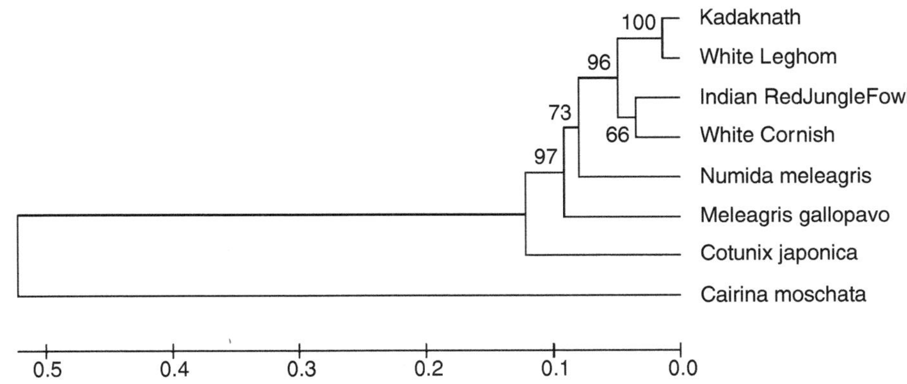

Fig. 15.1 Phylogeny of MHC BLB2 gene by UPGMA method

subjected to tree construction by character-based methods such as maximum likelihood and maximum parsimony. Support of the clusters was evaluated by bootstrap (Kimura 2-parameter), as percentage recurrence of clusters based on 100 bootstrapped replications. The overall mean genetic distance and pairwise distances among the sequences were estimated with Kimura 2-parameter distances by uniform rates among sites using MEGA software.

Results and Discussion

Upon UPGMA (Fig. 15.1) method, MHC BLB2 gene of Muscovy duck diverged into a separate branch from the remaining members. The next member to diverge was Japanese quail. Turkey and guinea fowl moved apart one after the other in accordance with the expected species tree. Various breeds of chicken evolved into a different clade where Kadaknath and white leghorn moved together, while Indian red jungle fowl and white Cornish formed a distinct clade. The positioning of white Cornish is not as well supported by neighbour-joining (Fig. 15.2) as it is using UPGMA, which places it basal to the turkey (*Meleagris gallopavo*). Upon N-J analysis, turkey is forming a common clade with Japanese quail (*Coturnix japonica*).

Upon maximum likelihood analysis with MEGA, Kadaknath and white leghorn, and Indian red jungle fowl were found to have a common ancestor where Cornish formed a separate branch distal to other chicken breeds (Fig. 15.3). The overall mean genetic distance (Kimura 2-parameter) estimated using the bootstrap (100 replicates) was obtained to be 0.3855 ± 0.035. Between

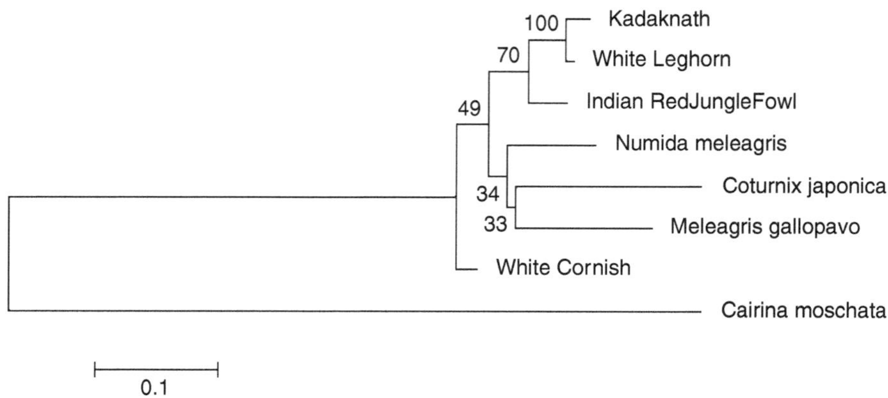

Fig. 15.2 Phylogeny of MHC BLB2 gene by neighbour-joining method

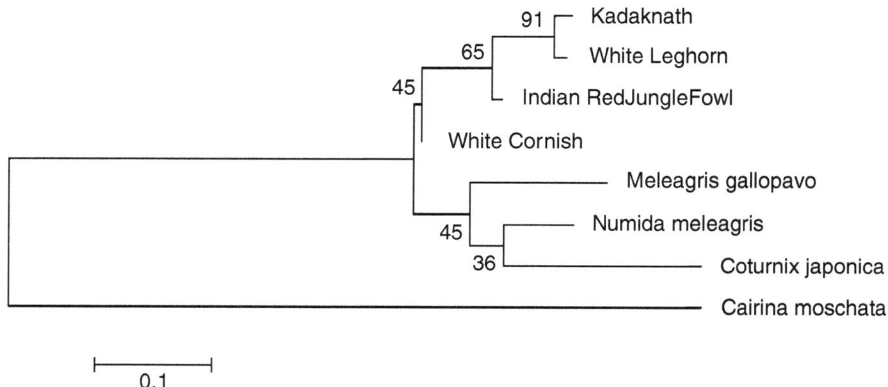

Fig. 15.3 Phylogeny of MHC BLB2 gene by maximum likelihood method

species genetic distances ranged from 0.026 between white leghorn and Kadaknath to 1.128 between turkey and Muscovy duck. Indian red jungle fowl showed lower genetic distances with Cornish, Kadaknath and white leghorn (0.068–0.076), while with other poultry species, it showed higher genetic distances (0.179–1.02). The maximum likelihood analysis (Fig. 15.3) presented the Japanese quail and guinea fowl forming a common clade, whereas the maximum parsimony analysis (Fig. 15.4) placed them in separate adjacent branches. Both analyses proved their relative resemblance of genetic make-up.

Evolutionary studies have shown that the formation of avian MHC genes has been driven by ancient gene duplication events after the bird-mammal split. The Indian red jungle fowl is considered to be the ancestor of domestic chicken of Indian continent. But almost all the analyses produced a varied result where the present-day Indian red jungle fowl and the other chicken breeds having a common ancestor. All types of analyses proved the far earlier divergence of Muscovy duck from all other members under analysis. The genetic distance analysis also proved the same showing maximum distances (0.951–1.128) with all other species. Upon all types of analyses, Kadaknath and white leghorn showed high similarity with respect to this locus where between species genetic distance was found to be minimum (0.026). The within species

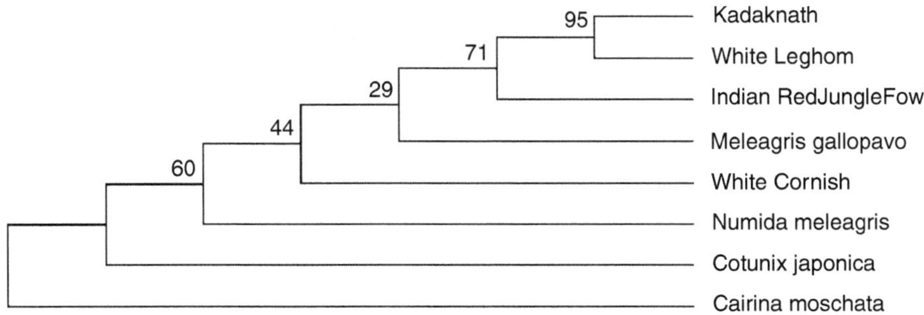

Fig. 15.4 Phylogeny of MHC BLB2 gene by maximum parsimony method

Table 15.2 Between and within species genetic distance based on nucleotide sequence diversity in MHC BLB2 gene

Sl no.	Category	1	2	3	4	5	6	7	8
1	Kadaknath	0	0.03	0.08	0.17	0.24	0.12	0.18	1.1
2	White leghorn		0	0.07	0.15	0.23	0.12	0.18	1.06
3	Indian red jungle fowl			0	0.18	0.26	0.07	0.18	1.02
4	*Numida meleagris*				0	0.23	0.14	0.19	1.02
5	*Coturnix japonica*					0	0.24	0.26	1.11
6	White Cornish						0	0.17	0.95
7	*Meleagris gallopavo*							0	1.13
8	*Cairina moschata*								0

or between-breed genetic distance was found to be considerably low compared to between-species distance as expected. Among the poultry species under analysis, MHC BLB2 nucleotide sequence of Cornish is highly complex. It showed minimum genetic distance with Indian red jungle fowl (0.68) at the same time minimum distance with Muscovy duck (0.951). The UPGMA and N-J analyses confirm these results where Fig. 15.1 showed close proximity of Cornish and Indian red jungle fowl and Fig. 15.2 showed its relative closeness with Muscovy duck.

The MHC gene family is a typical example of trans-species polymorphism [10]. The present analysis also revealed a mixture of alleles among the various species under study proving this effect. More data from other exons, UTRs, other MHC genes and ultimately genomic structure are awaited to get deeper insights into the long-term evolutionary history of the avian MHC.

Acknowledgement The authors of nucleotide sequence accession numbers from data bank utilized in the analysis are acknowledged.

References

1. Lamont SJ. Impact of genetics on disease resistance. Poult Sci. 1998;77:1111–8.
2. Schat KA, Taylor RL, Briles WE. Resistance to Marek's Disease in chickens with recombinant haplotypes of the Major Histocompatibility (B) complex. Poult Sci. 1994;73:502–8.
3. Klein J. Origin of major histocompatibility complex polymorphism: the trans-species hypothesis. Hum Immunol. 1987;19:155–62.
4. Jacob JP, Milne S, Beck S, Kaufman J. The major and a minor class II beta-chain (B-LB) gene flank the tapasin gene in the B-F/B-L region of the chicken major histocompatibility complex. Immunogenetics. 2000;51:138–47.
5. Kaufman J, Milne S, Gobel TW, Walker BA, Jacob JP, Auffray C, Zoorob R, Beck S. The chicken B locus is a minimal-essential major histocompatibility complex. Nature. 1999;401:923–5.
6. Boonyanuwat K, Thummabutra S, Sookmanee N, Vatchavalkhu V, Siripholvat V. Influences of major histocompatibility complex class I haplotypes on avian influenza virus disease traits in Thai indigenous chickens. Anim Sci J. 2006;77:285–9.
7. Hughes AL, Yeager M. Natural selection at major histocompatibility complex loci of vertebrates. Annu Rev Genet. 1998;32:415–35.

8. Kulski JK, Shiina T, Anzai T, Kohara S, Inoko H. Comparative genomic analysis of the MHC: the evolution of class duplication blocks, diversity and complexity from shark to man. Immunol Rev. 2002;190:95–100.
9. Tamura K, Peterson D, Peterson N, Stecher G, Nei M, Kumar S. MEGA5: molecular evolutionary genetics analysis using maximum likelihood, evolutionary distance, and maximum parsimony methods. Mol Biol Evol. 2011;28:2731–9.
10. Edwards SV, Hedrick PW. Evolution and ecology of MHC molecules: from genomics to sexual selection. Trends Ecol Evol. 1998;13:305–10.

16 Phytoremediation of Cadmium Using *Pteris* sp.

Sudha Sajeev, P.V. Ramya, B.C. Sunitha, J.S. Melo, and Smitha Hegde

Abstract

Pteris spp., a member of *Pteridaceae* family, was investigated for its efficacy in remediation of cadmium through foliar absorption. Fern species were selected for phytoremediation particularly for their limited role in food chain of herbivores and humans. The fern fronds of uniform size and mass were cultured in Knop's medium containing different concentrations of cadmium salt (50, 100, 150, and 200 mM) under controlled conditions. Parameters such as biomass, hydration, chlorophyll content, protein content, and proline content were monitored at regular intervals of 24 h for a period of 1 week. The samples were tested for cadmium accumulation at the end of 1 week using atomic absorption spectroscopy. Results have shown that the total protein content has increased significantly up to 72 h. Increased proline concentration was observed at 150 mM of Cd salt treatment up to 48 h irrespective of the position of the frond. Maximum foliar absorption of 70 mg kg^{-1} dry weight was found just after 24 h of Cd salt treatment in the tip region of the frond. Our studies indicate that local *Pteris* sp. is a potential candidate plant for phytoremediation.

Keywords

Pteris sp. • Foliar absorption • Heavy metal • Cadmium • Phytoremediation

Introduction

Cadmium is an important widespread trace pollutant with high toxicity to plants, animals, and humans. It is a nonessential heavy metal having long biological persistence [1]. Cadmium (Cd) is released into the environment by power stations, heating systems, metal working industries, nickel-cadmium batteries, and phosphate fertilizers [1, 2] as well as from geochemical weathering of rocks. Cadmium causes various phytotoxic symptoms

S. Sajeev • P.V. Ramya • B.C. Sunitha • S. Hegde (✉)
Center for Research and Postgraduate studies in Biotechnology, St. Aloysius College, Mangalore 575 003, India
e-mail: shegde@staloysius.edu.in

J.S. Melo
Nuclear Agriculture and Biotechnology Division, Bhabha Atomic Research Centre, Trombay, Mumbai 400 085, India

A. Sabu and A. Augustine (eds.), *Prospects in Bioscience: Addressing the Issues*,
DOI 10.1007/978-81-322-0810-5_16, © Springer India 2013

including chlorosis, growth inhibition, water imbalance, phosphorus and nitrogen deficiency reduced manganese transport, and accelerated senescence [2, 3]. Cadmium causes oxidative stress either by inducing oxygen free radical production or by decreasing enzymatic and nonenzymatic antioxidants due to its high affinity toward sulfur-containing peptides and proteins [3–5].

Despite the toxicity of cadmium, high tolerance levels have evolved in a number of plant species, mainly through mechanisms of exclusion. Some plant species, belonging to the class of hyperaccumulators, can accumulate exceptional concentrations of Cd (above 0.01%) in their shoot dry weight without toxicity symptoms. Cadmium hyperaccumulation is present only in some populations of *T. caerulescens*, *T. praecox*, and *Arabidopsis halleri*, all belonging to *Brassicaceae* family and *Sedum alfredii* (*Crassulaceae*). The primary target of Cd toxicity is unknown. An important source of Cd toxicity is its chemical similarity with essential elements, in particular Zn, but also Ca and Fe, deregulating the homeostasis of latter elements or causing their displacement from proteins. Toxicity of cadmium is primarily due to high sulfhydryl reactivity. Cd causes oxidative stress and can deplete reduced glutathione, an important cellular antioxidant, through the formation of Cd-glutathione complexes or Cd (II)-GS_2 or Cd-induced phytochelatin (PC) synthesis [6]. The mitochondrial electron transfer chain of plant cells is thought to be one of the major targets of Cd toxicity and is the site of the most rapid Cd-induced ROS production [7]. Increased ROS production induces lipid peroxidation. Cd also inhibits water transport, causing proline accumulation as a consequence of water stress [8].

In nature, certain plants have been identified which have the potential to uptake heavy metals. At least 45 families have been identified to hyperaccumulate heavy metal; some of the families are *Brassicaceae*, *Fabaceae*, *Euphorbiaceae*, *Asteraceae*, *Lamiaceae*, and *Scrophulariaceae*. Indian mustard (*B. juncea*) is a high biomass, rapidly growing plant that has an ability to accumulate Ni and Cd in its shoots. It is a promising plant for phytoremediation [9]. Aquatic plants such as the floating *Eichhornia crassipes* (water hyacinth), *Lemna minor* (duckweed), and Pistia have been investigated for use in rhizofiltration [10]. Recently, a fern *Pteris vittata* (Chinese brake fern) has been shown to accumulate as much as 14,500 mg kg^{-1} arsenic in fronds without showing symptoms of toxicity [11]. Corn, sunflower, and sorghum [12] were found to be effective due to their fast growth rate and large amount of biomass [13–15]. Gardea-Torresdey et al. [16] have shown that alfalfa is a potential source of biomaterials for the removal and recovery of heavy metal ions. Pollution of the biosphere with the toxic metal has accelerated dramatically since the beginning of the industrial revolution [17]. The high cost of existing cleanup technologies led to the search for new cleanup strategies that have the potential to be low cost, low impact, visually benign, and environmentally sound. Phytoremediation is a new cleanup concept that involves the use of plants to clean or stabilize contaminated environments. This natural and environmental friendly technology is an effective, aesthetically pleasant, soil organism friendly, diversity enhancer [18–20], and more importantly, it is able to retain the fertility status of the soil even after the removal of heavy metals [21]. Cd accumulation in soil and water now poses a major environmental and human health problem which is in need of an effective and affordable solution. In our present study, we have evaluated *Pteris* sp., a member of *Pteridaceae* family, for hyperaccumulation. Fern species were selected for phytoremediation particularly for their limited role in food chain of herbivores and humans. The objectives of our study were to screen local ferns for the purpose of phytoremediation; to study the time taken by the plants to efficiently uptake the toxic heavy metal cadmium and the effects of cadmium on the plant physiology by assessing fresh weight, dry weight, percentage hydration, chlorophyll, and protein content; and to determine the levels of heavy metal tolerance and distribution in the plants tested using foliar absorption.

Materials and Methods

Pteris spp. were collected from the campus of St. Aloysius College, Mangalore, India, and washed under tap water to remove surface dirt and dead tissue. Fronds of approximate weight (80 g) were chosen for the experiment. Knop's liquid media (KNO_3 – 200 mg l^{-1}; $MgSO_4.7H_2O$ – 200 mg l^{-1}; KH_2PO_4 – 200 mg l^{-1}; Ca (NO_3). $4H_2O$ – 800 mg l^{-1}) with pH 5.6 were prepared; 50 ml of the medium was dispensed to each of the culture bottles and autoclaved. Test concentrations of cadmium chloride ($CdCl_2$), i.e., 50, 100, 150, and 200 mM, were added to the autoclaved media. The fronds were grown under controlled conditions on the media for a period of 7 days. Controls were maintained on Knop's liquid media without the cadmium salt. Duplicates were maintained for each treatment. Two centimeter of the sample tissue was taken in time intervals of 24 h from the basal and tip portion of the treated and control fronds to study the physiological parameters affected as well as spatial distribution of cadmium salts in the fronds.

Determination of Fresh Weight

Fresh sample of uniform length of 2 cm from the treated and control fronds of *Pteris* sp. was weighed before and after the cadmium salt treatment at regular time intervals. This was considered as the fresh weight of the sample.

Determination of Dry Weight and Estimation of Percentage Hydration

Weighed samples from the treated as well as control fronds were dried in an oven at 50°C. The weight was determined after complete drying for 48 h.

Percentage hydration was calculated from the following formula:

$$\%\text{Hydration} = \frac{(\text{fresh weight} - \text{dry weight})}{(\text{fresh weight})} \times 100$$

Estimation of Chlorophyll

Total chlorophyll content of the control- and stress-induced fern fronds was determined by *Arnon's* method [22]. Two centimeter of fresh leaf sample was crushed in a microfuge tube with 1 ml of 80% acetone. The sample was centrifuged at 10,000 rpm for 10 min and the supernatant was separated. The pellet was resuspended in acetone, crushed, and centrifuged and the supernatants were pooled. The procedure was repeated till the residue was colorless. The volume was made up to 2 ml and the absorbance was taken at 663 nm and 645 nm for chlorophyll a and b respectively. Chlorophyll content was calculated from the formulae given below:

$$\text{mg chl a / g tissue} = 12.7\ (A_{663}) - 2.69\ (A_{645})\ (V/1{,}000 \times w)$$

$$\text{mg chl b / g tissue} = 22.9\ (A_{645}) - 4.68\ (A_{663})\ (V/1{,}000 \times w)$$

$$\text{Total chlorophyll} = 20.2\ (A_{645}) + 8.02\ (A_{663})\ (V/1{,}000 \times w)$$

where V = total volume of the sample and w = weight of the sample

Extraction and Estimation of Protein Content

Fresh leaf samples were homogenized in 1.5 ml of buffer I (50 mM Tris–HCl (pH 7.5), 1 mM Na_2EDTA (pH 8), 8 mM $MgCl_2$, and 2% W/V 2-mercaptoethanol) and centrifuged at a speed of 12,000 rpm for 10 min at 4°C. The supernatant was collected and the precipitate was re-extracted with the same buffer and centrifuged at similar conditions. The supernatants were pooled together, and this formed the soluble protein extract. The precipitate was washed with 1 ml of buffer II (50 mM Tris–HCl (pH 6.8)). Then, 0.5 ml of buffer III (50 mM Tris HCl buffer (pH 6.8), 8 M urea, and 3.2% (w/v) SDS) was added to the pellet and incubated at 40 °C for 1 h. The sample was centrifuged at 12,000 rpm for 10 min at 4 °C and the supernatant was collected. This formed the insoluble protein extract. Absorbance for the extracted proteins was taken at 700 nm using colorimeter. The concentration of the protein was estimated using Folin-Lowry's method [23].

Estimation of Proline

Hundred milligram of fresh tissue from each treatment was subjected to proline estimation as per Bates et al. [24]. Sample was extracted by homogenizing in 2 ml of 3% aqueous sulfosalicylic acid and filtered through Whatman No. 2 filter paper. One milliliter each of glacial acetic acid and acid-ninhydrin was added to 1 ml of the filtrate and was kept in the boiling water bath for 1 h. The reaction was terminated by placing the tube in ice bath. Two milliliter of toluene was added to the reaction mixture and stirred well for 20–30 s. The toluene layer was separated and brought to room temperature and the red color intensity was read at 520 nm. The amount of proline in the control and treated samples was estimated from the standard curve plotted using pure proline.

Sample Preparation for Heavy Metal Detection Using AAS

Two centimeter of leaf samples from the control and treated fronds was kept in an oven at 50 °C for drying till constant dry weight was obtained. The dried leaves were crushed with (1:1, v/v) HCl and HNO_3. Crushed sample was kept in Borosil glass tubes for 2–3 days for acid digestion. Digested sample was filtered using Whatman filter paper and the metal was estimated using atomic absorption spectroscopy (GBC 6000).

Statistical Analysis

The experiment was done as randomized block design. The data of all the experiments were subjected to the analysis of variance (ANOVA) and subsequent least significant difference (LSD) to determine the significant difference between the treatments, position of the frond, and time durations and their interactions. A correlation analysis ($P \leq 0.05$) or 95% confidence level was performed to study all the parameters.

Results and Discussion

Frond Growth Parameters

The fronds of *Pteris* sp. treated with different concentrations of cadmium did not show a consistent pattern of increase in its fresh weight. The frond showed a slight increase in the fresh weight in 50-mM concentration and then gradually decreased and once again showed slight variation at 150-mM concentration (Table 16.1). The R-square value for the fresh weight is 0.568. The response of the tip region of the frond is higher compared to the basal portion in gaining fresh weight. Maximum increase in the fresh weight was observed after 24 h of the treatment (Table 16.2). As the time duration increased, there was a significant decrease in the fresh weight. There is a significant increase (R-squared = 0.775) in the dry weight up to 150 mM of the Cd concentration, then slowly the dry weight has decreased (Table 16.2). The tip region has an increased dry weight than the basal region (Fig. 16.1); overall it has maintained a constant dry weight up to 7th day of the experiment. There

Table 16.1 Influence of the different concentrations of cadmium on the fern frond's growth

Treatment	Fresh weight[a]	Dry weight[a]	% Hydration[a, b]
Control	10.800 ± 1.000	2.500 ± .238	62.570 ± 2.446
50 mM	8.750 ± 1.000	2.600 ± .238	58.389 ± 2.446
100 mM	10.400 ± 1.000	3.750 ± .238	50.707 ± 2.446
150 mM	6.200 ± 1.000	1.550 ± .238	58.542 ± 2.446
200 mM	9.750 ± 1.000	2.750 ± .238	55.686 ± 2.446

[a]Data shown is the mean of two replicates ± standard error (SE)
[b]Values have been adjusted as per arcsine transformation

Table 16.2 Influence of the time duration on the fresh weight, dry weight, and hydration of the fern fronds

Time duration	Fresh weight[a]	Dry weight[a]	% Hydration[a, b]
1st day	11.800±1.000	3.100±.238	58.832±2.446
2nd day	7.700±1.000	2.000±.238	58.342±2.446
3rd day	8.450±1.000	3.150±.238	52.111±2.446
5th day	8.350±1.000	2.000±.238	61.234±2.446
7th day	9.600±1.000	2.900±.238	55.375±2.446

[a]Data shown is the mean of two replicates±standard error (SE)
[b]Values have been adjusted as per arcsine transformation

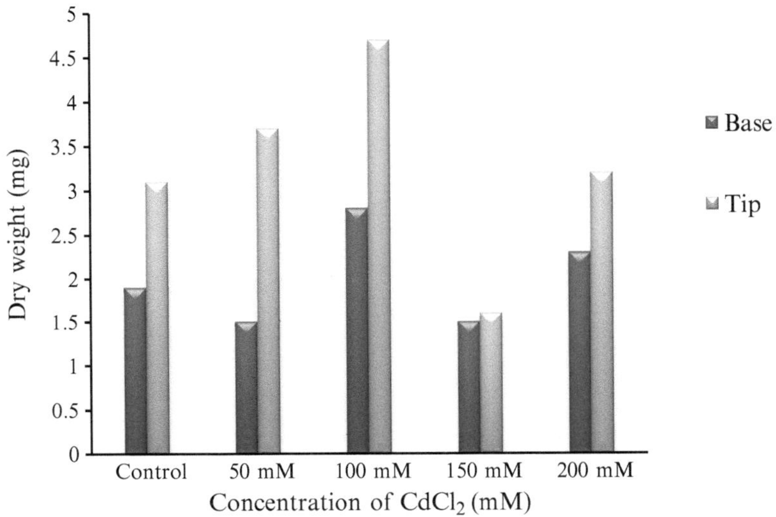

Fig. 16.1 Effect of treatment and position on dry weight

is a significant effect of treatment and position on the percentage hydration of the fronds ($P=0.021$ and $P=0.000$, respectively), whereas there is no substantial change with respect to the time duration. The tip region (55%) of the frond has less hydration than the basal region (65%).

Physiological Parameters

Chlorophyll Content

Among the photosynthetic pigments, different concentrations of Cd have significant effect on chlorophyll a, while there was no significant effect with respect to position of the frond and time duration. Chlorophyll b content has increased at 50 mM, then decreased, and gradually increased to maximum value (9.413±0.679) at 150 mM. There was no differential response of chlorophyll b content in the basal and tip region of the frond. Maximum response of chlorophyll b content was after 72 h of the treatment (6.978±0.679). Different concentration of Cd salt has significant effect on the total chlorophyll content. Total chlorophyll content is higher at a concentration of 150 mM on the 3rd day of the treatment (Fig. 16.2). The chlorophyll content was observed to decrease resulting in the wilting and fading of the fronds.

Protein Content

The study showed that soluble protein concentration increased progressively at all the concentrations of Cd salt up to 7th day of the treatment both in the basal and tip regions of the frond. Insoluble protein was higher in the tip region and

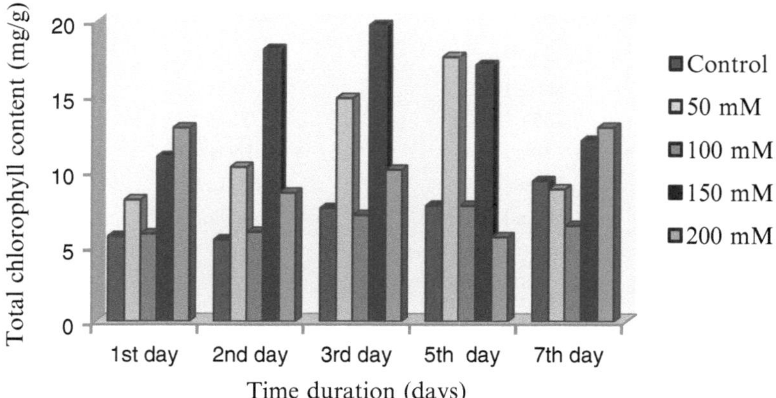

Fig. 16.2 Effect of treatment and time on total chlorophyll content

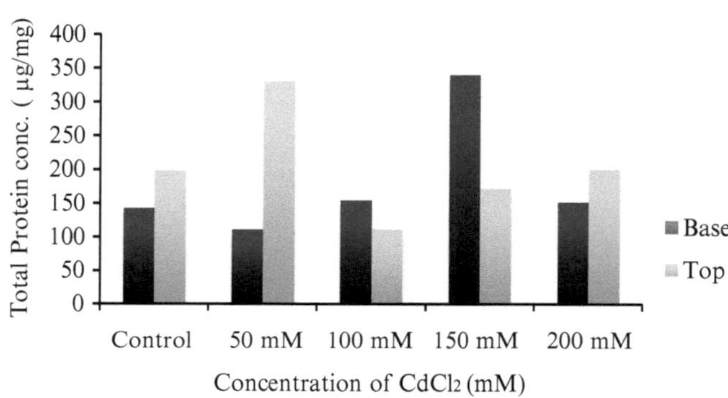

Fig. 16.3 Effect of treatment and position on total protein concentration

there was no uniform pattern with respect to the treatment of $CdCl_2$. But the concentration of insoluble protein increased with time with a maximum value after 72 h of treatment. Total protein content increased significantly up to 72 h, and basal region of the frond has higher protein content at 150 mM (Fig. 16.3). Both the treatment and the time duration have a significant effect on the total protein content of the fern fronds (Fig. 16.4).

Proline Content

Increased proline concentration was observed at 150 mM of Cd salt treatment up to 48 h irrespective of the position of the frond (Fig. 16.5). After 72 h proline content decreased. On the 7th day, it showed a higher proline content indicating the tolerance level of the fronds to cadmium.

Cadmium Accumulation

Cadmium accumulation by the *Pteris* sp. frond was found to be concentration, position, and duration dependent. Maximum absorption was found just after 24 h of Cd salt treatment in the tip region of the frond at 200 mM. *Pteris* sp. can accumulate cadmium up to 90 mg kg^{-1} of the dry weight (dw) in the initial 24 h of the treatment. The tip region of the frond showed a higher accumulation of 70 mg kg^{-1} dry weight of Cd than the basal region (Fig. 16.6). As the time progresses, there is a substantial decrease in the foliar absorption of Cd.

Huang et al. [19] justified that for an economical phyto-extraction process, plants should be able to accumulate at least 1% of the total heavy metal content present in the soil into their dry shoot biomass. Much progress has been made in the recent years in the mechanism, distribution,

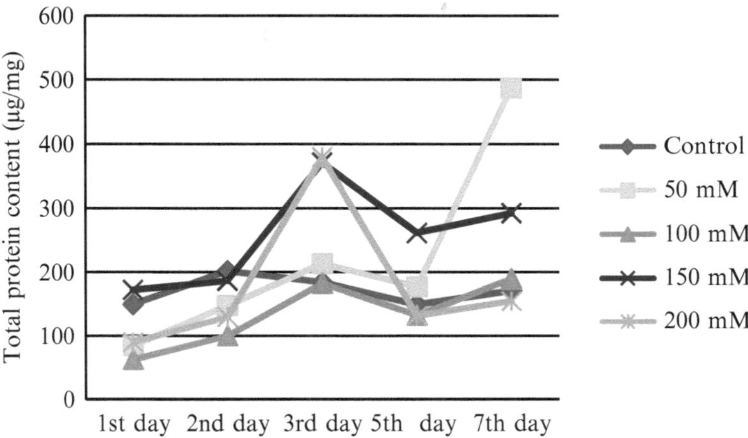

Fig. 16.4 Effect of treatment and time on total protein concentration

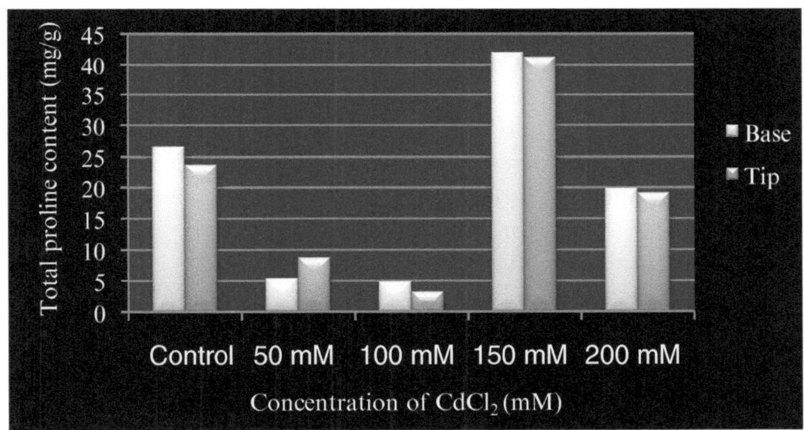

Fig. 16.5 Effect of treatment and position on total proline concentration

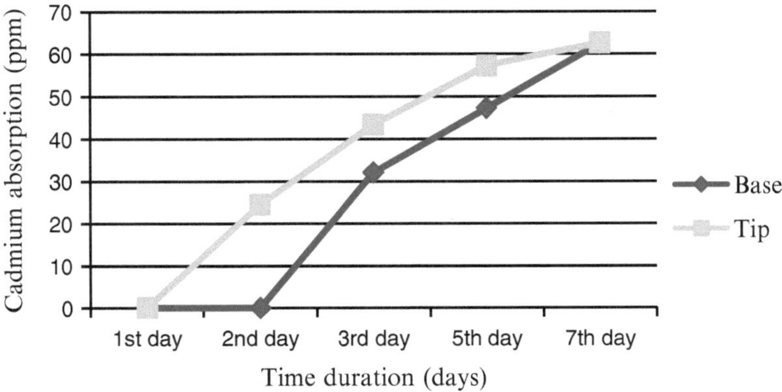

Fig. 16.6 Preferential foliar accumulation of cadmium

and detoxification of heavy toxic metals from the hyper accumulators [6]. This forms an exceptional research material for further study of the genes involved in the phytoremediation process. Tissue-specific gene expression, differential distribution of the heavy metal within the system, and regulation of protein levels and activities need to be studied extensively for efficient application of these hyperaccumulators on field trials.

It is evident from our study that *Pteris* sp. is responding to the abiotic stress given in the form of cadmium. There is no significant increase in the total chlorophyll content compared to the control. Significant reductions were observed at higher concentrations of Cd and also at longer durations. This may be attributed to strong oxidation of the photochemical apparatus in the plant [25].

The increased levels of soluble proteins up to 7th day at all concentrations of Cd reflect that the fern fronds are in continuous metabolic state. Similar kind of results is observed in *Eleusine coracana* under water stress [26]. Total protein content was higher at low doses of Cd (Fig. 16.3). This initial increase in protein level could be attributed to the enzymatic antioxidants that help the fern to recover from the oxidative stress. Stress proteins constitute various antioxidant enzymes and other enzymes involved in glutathione (GSH) and phytochelatin (PC) biosynthesis including some heat shock proteins [27]. But the high doses of Cd were found to decrease the total protein content after 72 h, and this may be due to Cd-induced oxidation of proteins, mediated by H_2O_2 and due to proteolytic activity [27]. Protein degradation and proteolytic activity have been proposed as an index of oxidative stress [28]. It is clear from our study (Fig. 16.5) that proline levels are high at a concentration of 150 mM of Cd even up to 48 h, indicating that the ferns' fronds can withstand up to 150 mM of Cd salt. This tolerance level could be higher when the whole fern is subjected to Cd stress. The accumulation and protective effect of proline have been observed in many higher plants and bacteria as well as protozoa, algae, and marine invertebrates [29]. The total amount of Cd accumulated is very high, i.e., 90 mg kg^{-1} dw at 24 h, and the tip region of the frond has stored maximum Cd (n). This could be due to the actively growing cells. From our study we have found that *Pteris* sp. is showing all the physiological adaptations of a plant under stress. Foliar absorption of around 90 mg kg^{-1} dry weight of cadmium is a positive indication that *Pteris* sp. could be utilized for phytoremediation of cadmium. Chelating agents like EDTA will further enhance the accumulation of Cd by these ferns.

Conclusions

In the present study, foliar absorption of cadmium by *Pteris* spp., was established. As the fronds tolerated 150 mM of Cadmium up to 3rd day efficiently, our future research will be focused on the response of the entire fern system in hydroponics and also on the enzymatic and nonenzymatic defense mechanism under cadmium stress. Further, the effect of chelating agents in enhancing the hyperaccumulating property will be studied.

Acknowledgments We are grateful to the Board of Research in Nuclear Sciences (BRNS), Mumbai, for the financial assistance. The authors are also thankful to Rev. Fr. Swebert D'Silva, Principal, St. Aloysius College (Autonomous), and Rev. Fr. Dr. Leo, Director of Research, for providing the facilities to undertake this study.

References

1. Wagner RC. Accumulation of cadmium in crop plants and its consequences to human health. Adv Agron. 1993;51:173–212.
2. Toppi SDL, Gabbrielli R. Response to cadmium in higher plants. Environ Exp Bot. 1999;41:105–30.
3. Benavides MP, Gallego SM, Tomaro ML. Cadmium toxicity in plants. Braz J Plant Physiol. 2005;17:21–34.
4. Sandalio LM, Dalurzo HC, Gomez M, Romero-Peurtas MC, del-Rio LA. Cadmium induces changes in the growth and oxidative metabolism of pea plants. J Exp Bot. 2001;52:2115–26.
5. Romero-Peurtas MC, Rodriguez-Serrano M, Corpas FJ, Gomez M, Del-Rio LA, Sandalio LM. Cadmium induced sub cellular accumulation of O2- and H2O2 in pea leaves. Plant Cell Environ. 2004;27:1122–34.
6. Verbruggen N, Hermans C, Schat H. Mechanisms to cope with arsenic or cadmium excess in plants. Curr Opin Plant Biol. 2009;12:1–9.

7. Heyno E, Klose C, Kreiger-Liszkay A. Origin of cadmium – induced reactive oxygen species production: mitochondrial electron transfer versus plasma membrane NADPH oxidase. New Phytol. 2008;179:687–99.
8. Schat H, Sharma SS, Vooijs R. Heavy metal-induced accumulation of free proline in a metal-tolerant and a non tolerant ecotype of Silene vulgaris. Physiol Plant. 1997;101:477–82.
9. Terry N, Carlson C, Raab TK, Zayed A. Rates of Se volatilization among crop species. J Environ Qual. 1992;21:341–4.
10. Karkhanis M, Jadia CD, Fulekar MH. Rhizofiltration of metals from coal ash leachate. Asian J Water Environ Pollut. 2005;3(1):91–4.
11. Ma LQ, Komar KM, Tu C, Zhang WH, Cai Y, Kennelley ED. A fern that hyperaccumulates arsenic – a hardy, versatile, fast-growing plant helps to remove arsenic from contaminated soils. Nature. 2001;409:579–9.
12. Jadia CD, Fulekar MH. Phytotoxicity and remediation of heavy metals by fibrous root grass (sorghum). J Appl Biosci. 2008;10:491–9.
13. Schmidt U. Enhancing Phytoextraction: the effects of chemical soil manipulation on mobility, plant accumulation, and leaching of heavy metals. J Environ Qual. 2003;32:1939–54.
14. Tang S, Xi L, Zheng J, Li H. Response to elevated CO_2 of Indian Mustard and Sunflower growing on copper contaminated soil. Bull Environ Contam Toxicol. 2003;71:988–97.
15. Pilon-Smits E. Phytoremediation. Annu Rev Plant Biol. 2005;56:15–39.
16. Gardea-Torresdey JL, Tiemann KJ, Gamez G, Dokken K, Cano-Aguilera I, Renner MW, Furenlid LR. Reduction and accumulation of gold (III) by *Medicago sativa alfalfa* biomass: X-ray absorption spectroscopy, pH, and temperature dependence. Environ Sci Technol. 2000;34:4392–6.
17. Nriagu JO. Global metal pollution. Environment. 1990;32:7–33.
18. Chaney RL, Angle JS, McIntosh MS, Reeves RD, Li YM, Brewer EP, Chen KY, Roseberg RJ, Perner H, Synkowski EC, Broadhurst CL, Wang S, Baker AJ. Using hyperaccumulator plants to phytoextract soil Ni and Cd. Z Naturforsch [C]. 2005;60:190–8.
19. Huang XD, El-Alawi Y, Penrose DM, Glick BR, Greenberg BM. A multi-process phytoremediation system for removal of polycyclic aromatic hydrocarbons from contaminated soils. Environ Pollut. 2004;130:465–76.
20. Susarla S, Medina VF, McCutcheon SC. Phytoremediation: an ecological solution to organic chemical contamination. Ecol Eng. 2002;18:647–58.
21. Kirkham MB. Cadmium in plants on polluted soils: effects of soil factors, hyper accumulation, and amendments. Geoderma. 2006;137:19–32.
22. Arnon DI. Copper enzymes in isolated chloroplasts. Polyphenoloxidase in *Beta vulgaris*. Plant Physiol. 1949;24:1–15.
23. Lowry OH, Rosenbrough NJ, Farr AL, Randall RJ. Protein measurement with folin phenol reagent. J Biol Chem. 1951;193:265–75.
24. Bates LS, Waldren RP, Teare ID. Rapid determination of free proline for water-stress studies. Plant Soil. 1973;39(1):205.
25. Somashekaraiah BV, Padmaja K, Prasad ARK. Phytotoxicity of cadmium ions on germinating seedlings of mungbean (*Phaseolus vulgaris*): involvement of lipid peroxides in chlorophyll degradation. Physiol Plant. 1992;85:85–9.
26. Rajendra P, Vaidyanathan CS, Udayakumar M, Krishna Sastry KS, Appaji Rao N. Alterations in the captivities of the enzymes of proline metabolism in Ragi (*Eleusine coracana*) leaves during water stress. J Biosci. 1981;3(4):361–70.
27. Mishra S, Srivastava S, Tripathi RD, Govindarajan R, Kuriakose SV, Prasad MNV. Phytochelatin synthesis and response of antioxidants during cadmium stress in *Bacopa monnieri L.* Plant Physiol Biochem. 2006;44:25–37.
28. Romero-Puertas MC, Palma JM, Gomez M, Del Rio A, Sandalio LM. Cadmium causes the oxidative modification of proteins in pea plants. Plant Cell Environ. 2002;27:677–86.
29. Cechin I, Rossi SC, Oliveira VC, Fumis TF. Photosynthetic responses and proline content of mature and young leaves of sunflower plants under water deficit. Photosynthetica. 2006;44(1):143–6.

Seasonal Variation of Spore Density and Root Colonisation of Arbuscular Mycorrhizae in Crop Plants in Relation to Soil Edaphic Factors

17

Deepake Ujwala Sheshrao and G. Gyananath

Abstract

Arbuscular mycorrhizal (AM) fungi are recognised as an essential component of sustainable agricultural ecosystem. Keeping in view the importance of AM fungi, the present study was undertaken for assessing the AM fungal spore population dynamics in the rhizosphere soil and its colonisation as a function of seasonal variation in relation to soil physico-chemical factors associated with eight crop plants collected from Nanded district, Maharashtra state, India. The major crop plants grown in these region includes *Glycine max*, *Saccharum officinarum*, *Zea mays*, *Cajanus cajan*, *Gossypium sativum*, *Triticum aestivum*, *Sorghum vulgare* and *Cicer arietinum*. Therefore, these crop plants were selected for the present study. The rhizospheric soil physico-chemical parameters included in this study are temperature, pH, electrical conductivity, alkalinity, % moisture, water holding capacity, organic matter, C, N, P, K, Na, Ca, Mg, Fe, Zn and Cu. The pH of soil samples was found to be alkaline. The electrical conductivity and alkalinity was observed to be high in summer and rainy season, respectively. The high amount of organic matter, organic C, N, P and Ca was recorded in rainy season. In all the three seasons, the percentage of root colonised by arbuscular mycorrhizal AM fungal propagules was found to be more in *Zea mays* as compared to other crop plants. The Pearson correlation coefficient data showed that the AM fungal spore density and percentage of root colonisation by native AM fungi varied significantly in crop plants and influenced by soil edaphic factors. Significant negative correlation of mean spore density with temperature, electrical conductivity and K was recorded during winter and rainy season, and it was positive with Cu in winter season. The high degree of positive correlation of % root colonisation with pH, Zn and Cu was

Deepake U.S. (✉) • G. Gyananath
School of Life Sciences, Swami Ramanand Teerth Marathwada University, Nanded-431606,
Maharashtra, India
e-mail: dreamujwala@gmail.com

A. Sabu and A. Augustine (eds.), *Prospects in Bioscience: Addressing the Issues*,
DOI 10.1007/978-81-322-0810-5_17, © Springer India 2013

observed during winter and rainy season. However, it was negative with N and Ca content of soil. The regression analysis showed a significant positive correlation ($R^2=0.57$, $y=7.69x+6.47$) between spore density and % root colonisation during rainy season. Therefore, this study suggests that status of AM fungi in agricultural crop plants may be influenced by season and soil edaphic factors.

Keywords
Crop plants • AM fungi • Soil parameters • Seasonal variation

Introduction

Mycorrhizal fungi are ubiquitous, associated with almost all plants in nature and are important in agriculture and forestry for bidirectional nutrient transfer [1, 2]. The external mycelium of arbuscular mycorrhizal (AM) fungi acts as an extension of host plant roots and serves as a direct link between roots and soil nutrient reserves [3]. AM fungi play a major role in soil fertility and nutrient acquisition [4], especially the uptake of phosphorus from the soil, and thereby enhance plant growth [5] and yield, provide greater resistance to plant diseases [6] and increased tolerance to drought [7]. The rhizosphere is a site of complex interactions between plants and microorganisms, where environmental factors such as soil physico-chemical parameters as well as fertilisers or cultivation practices may have large effect on microbial communities.

It is now widely accepted that climatic and edaphic factors can substantially influence AM fungi and their populations [8]. Rapid changes in soil nutrients may affect AM association and spore numbers [9]. Several edaphic factors, namely, textural class, soil pH, organic matter, soil moisture and nutrient levels, have been shown to affect spore germination, root colonisation and efficiency of AM fungi [10, 11].

AM colonisation varies with change of season. Seasonal effects also influence the establishment of plants under field conditions, depending on the efficiency of indigenous fungi. Therefore, it is essential to understand the nature of natural processes that help to increase the crop yield and quality with more efficient use of nutrient inputs and reduced need for pesticides. Despite the importance of mycorrhiza in agriculture and forestry, little work has been done regarding their seasonal variation with respect to soil edapho-climatic factors associated with the crop plants in Nanded district, Maharashtra, India.

Keeping these facts in mind, the present study was undertaken to elucidate the seasonal variation in AM root colonisation and spore population in relation to soil edaphic factors associated with eight crop plants in Nanded district.

Materials and Methods

Study Site

The present study sites are situated in Nanded district, Maharashtra, India. All the sampling sites are agricultural lands of conventional farming types. Geographically, Nanded district is lying between 18,015′ to 19,055′ North latitudes and 7,707′ to 78,010′ East longitudes. The total geographical area of Nanded district is 10,760 km^2 that is 3.41% of the total area of the state. The district has 16 talukas. The average rainfall in the district is about 897.8 mm (35.34 in.). Physiography of the study region is uneven. Physiographically, Nanded district is divided into three broad divisions, namely, hilly region, plateau region and plain region. This region has black, lateritic and alluvial soils. Godavari is the main river of the study region which flows west to east in the central part of the district. The climate in the district is hot and dry except the monsoon season. In the field survey, visit to the nearest agricultural fields in Nanded district includes villages such as Wadepuri, Nageli, Shemboli, Barad, Pandharwadi and Wakad. The major crop plants grown in these region include *Glycine max*, *Saccharum officinarum*, *Zea mays*, *Cajanus cajan*, *Gossypium sativum*, *Triticum aestivum*, *Sorghum vulgare* and *Cicer arietinum*. Therefore, these crop plants were selected for the present study.

Sample Collection

The rhizospheric soil samples and fresh roots were collected from agricultural crop plants (*Glycine max*, *Saccharum officinarum*, *Zea mays*, *Cajanus cajan*, *Gossypium sativum*, *Triticum aestivum*, *Sorghum vulgare* and *Cicer arietinum*) during three seasons from various fields. The soil samples were collected from 10 to 30 cm depth. The temperature was recorded at the site of collection. All the samples were transported to the laboratory in polythene bags [12].

Soil Analysis

The rhizospheric soil samples were analysed for pH, EC (electrical conductivity), moisture (%), alkalinity and water holding capacity [13]. Available phosphorus (mg/100 g) in soil was determined by the method of Ball-Coelho et al. (1993). Total carbon and organic matter was determined [14]. The level of mineralisable N was determined by Kjeldahl method [15]. Aqua regia (HNO_3: HCl = 1:3) was used for the extraction of macronutrients (K, Ca and Mg) and micronutrients (Na, Fe, Zn and Cu). Amount of K and Na were measured by flame photometer and Ca, Mg, Fe, Zn and Cu were measured by atomic absorption spectrophotometer from the soil extracts.

Estimation of VAM Infection

The collected root samples were properly washed in running tap water and cleared in a near-boiling 10% aqueous solution of KOH for 48 h. Roots were stained in lactophenol cotton blue following several washings in distilled water to drain out the KOH [16]. The stained roots were cut into 1-cm segments, and 80–100 such segments were picked up and examined under the microscope. AM colonisation was determined by Nicolson's formula [17]: percent root colonisation = no. of root segments colonised/total no. of root segments examined × 100.

Extraction and Enumeration of AM Fungal Spores

Extraction of AM fungal spores from soil was done by following the method outlined by Gerdmann and Nicolson [18]. Fifty gram of soil was dispersed in 500 ml water and decanted through a series of 710- to 38-μm sieves. Residues were filtered through gridded filter papers and all intact spores (noncollapsed spores with cytoplasmic contents free from parasitic attack) were counted using a dissection microscope at ×40 magnification. Sporocarps and spore clusters were considered as one unit.

Statistical Analysis

All the values were given as mean ± standard deviation (SD) of three replicates. Pearson's correlation coefficients were employed to determine the relationships of soil edaphic factors with AM colonisation and spore density. Furthermore, regression analyses were done to find out the variation of soil endomycorrhizal spore populations and AM root colonisation during three seasons. These tests were performed using Microsoft Office Excel 2010 software.

Results

Soil Edaphic Factors

The mean values of soil physico-chemical characteristics are presented in Table 17.1, which vary significantly in all three seasons. The soil analyses revealed edaphic factors such as pH (8.18 ± 0.29), alkalinity (2.45 ± 1.23 meq/100 g), organic matter (1.59 ± 0.51%), water holding capacity (72 ± 3.93%),

Table 17.1 Seasonal variations in physico-chemical parameters of agricultural rhizospheric soil

		Seasons		
Sr. no.	Soil parameters	Winter	Monsoon	Summer
1.	Temperature (°C)	23.86±2.69	23.38±5.53	**33.13±4.39**
2.	pH	8.09±0.19	**8.18±0.29**	7.77±1.35
3.	EC (μs)	35.69±12.14	31.78±16.91	**502.17±56.25**
4.	Alkalinity (meq/100 g)	1.6±0.31	**2.45±1.23**	0.87±0.112
5.	Organic matter (%)	1.25±0.64	**1.59±0.51**	1.37±0.39
6.	Water holding capacity (%)	61.73±10.67	**72±3.93**	65.21±4.4
7.	Moisture (%)	**27.59±7.37**	20.46±8.98	10.3±4.18
8.	C (%)	0.72±0.37	**0.925±0.297**	0.8±0.21
9.	N (kg/ha)	27.83±1.413	**35.67±1.85**	27.44±1.294
10.	Available P (%)	4.4±1.2	**20.14±1.8**	4±0.52
11.	K (mg/kg)	1876.9±15.68	1952.31±12.72	**2257.3±37.94**
12.	Na (mg/kg)	**616.6±21.54**	173.34±10.32	83.245±8.833
13.	Ca (mg/kg)	1874.2±63.51	**2061.8±78.86**	1078.8±7.046
14.	Mg (mg/kg)	ND	ND	ND
15.	Fe (mg/kg)	17916.8±49.6	5666.5±12.46	**54,833**±117.7
16.	Zn (mg/kg)	**166.6±5.59**	133.31±7.2	141.64±5.241
17.	Cu (mg/kg)	**112.5±1.24**	87.49±1.0213	79.16±1.849

Values are given as mean±SD *ND*-not detected. Highest values are emboldened among three seasons for each parameter.

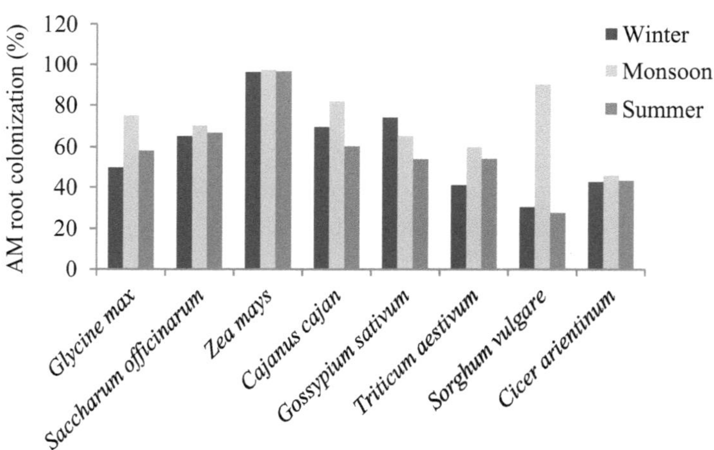

Fig. 17.1 Seasonal variation in percentage of AM root colonisation in eight crop plants

C (0.925±0.297%), N (35.67±1.85 kg/ha), P (20.14±1.8%) and Ca (2061.8±78.86 mg/kg) were found to be elevated during monsoon season as compared to other seasons. However, during winter season, soil parameters, namely, moisture (27.59±7.37%), Na (616.6±21.54 mg/kg), Zn (166.6±5.59 mg/kg) and Cu (112.5±1.24 mg/kg), were higher. In summer season, higher values were obtained for temperature (33.13±4.39°C), EC (502.17±56.25 μs), K (2257.3±37.94 mg/kg) and Fe (54,833±117.7 mg/kg).

Seasonal Variations of Percentage of Root Colonisation and AMF Spore Density

AM root colonisation levels varied according to the crop plants and season (Fig. 17.1). The colonisation rates were more in all crop plants during monsoon season except *Gossypium sativum* (74% in winter season). During monsoon season, highest root colonisation was observed in *Zea mays* (97.03%), followed by *Sorghum*

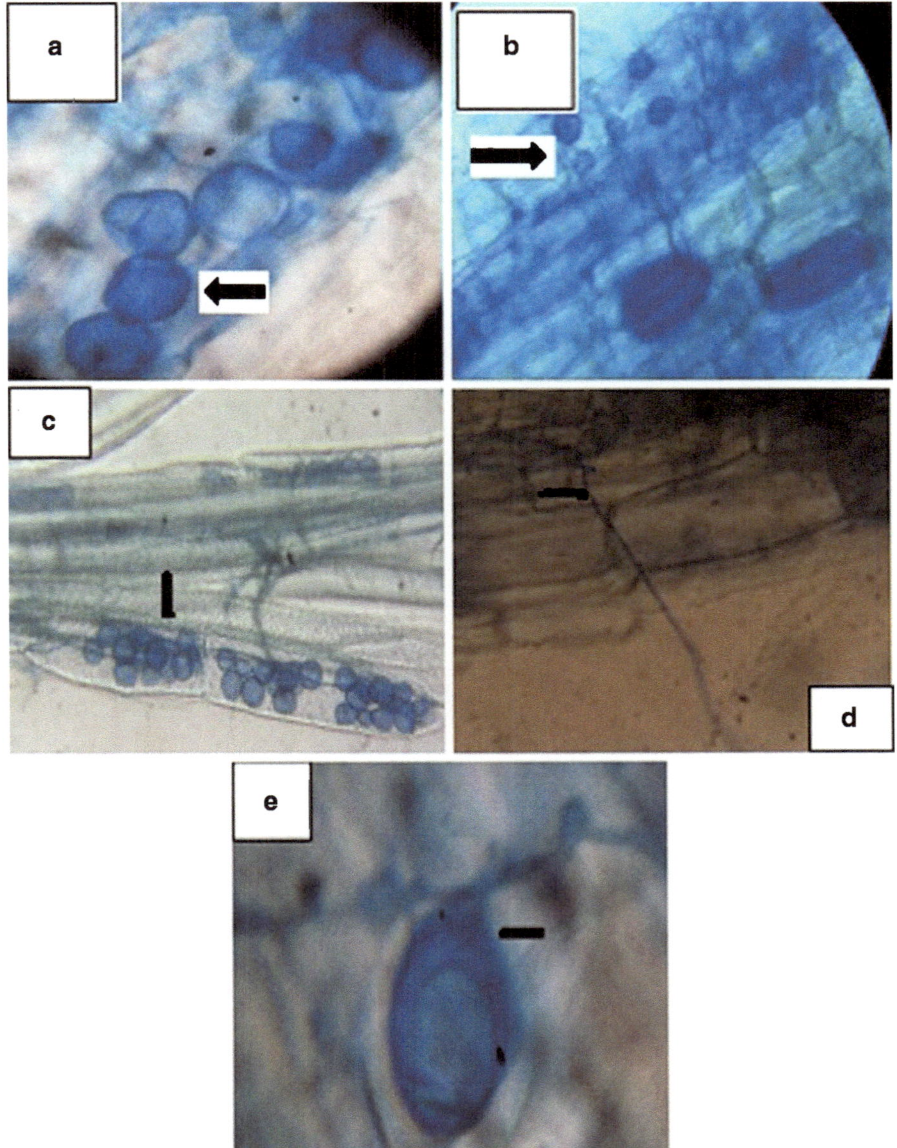

Fig. 17.2 Root colonisation (40×) of various crop plants by AM fungi: (**a**) vesicular colonisation in *Zea mays*; (**b**) arbuscular and vesicular colonisation in *Gossypium sativum*; (**c**) vesicular colonisation in *Saccharum officinarum*; (**d**) internal hyphal colonisation in *Sorghum vulgare*; (**e**) AM fungal spore attached to hyphae in *Cajanus cajan*

vulgare (90%), *Cajanus cajan* (81.82%), *Glycine max* (75%), *Saccharum officinarum* (70.12%), *Gossypium sativum* (65.4%), *Triticum aestivum* (60%) and *Cicer arietinum* (45.69%). AM root colonisation in the form of arbuscule, vesicle, hyphae and occasionally intraradical spores was observed (Fig. 17.2). The density of AM fungal spores was estimated during all the three seasons in rhizospheric soil samples of crop plants (Fig. 17.3). It was observed from the data that more number of spores (50 g^{-1} soil) recovered from all soil samples during winter season except *Saccharum officinarum* (330 during summer season) and *Cicer arietinum* (643 during

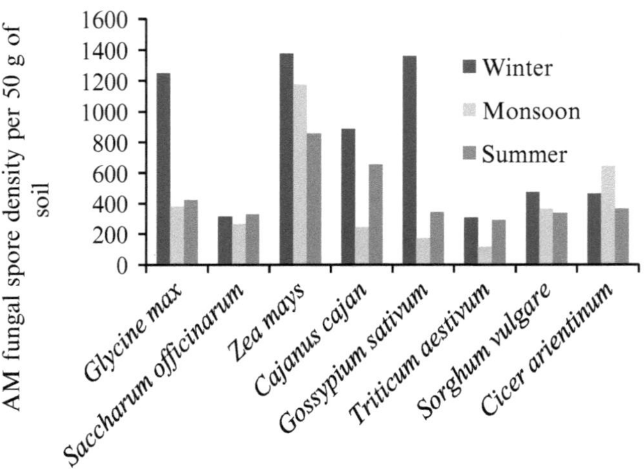

Fig. 17.3 Seasonal variation of AM fungal spore density per 50 g of rhizospheric soils of eight crop plants

Table 17.2 The correlation coefficient between soil parameters, percentage of root colonisation and AM fungal spore density

		r Values					
		AM root colonisation			AM fungal spore density		
Sr no.	Soil parameters	Winter	Monsoon	Summer	Winter	Monsoon	Summer
1.	Temperature	0.297	0.051	0.076	0.6166	**–0.761****	–0.1117
2.	pH	0.582	**0.86*****	–0.298	0.1277	0.1727	0.0843
3.	EC	–0.142	–0.121	–0.182	**–0.667***	–0.6053	–0.3415
4.	Alkalinity	–0.068	–0.115	0.272	–0.3541	0.43014	–0.1757
5.	Organic matter	–0.22	–0.156	0.103	–0.1967	0.31722	0.0063
6.	Water holding capacity	–0.046	–0.205	–0.256	–0.1663	0.01883	0.31677
7.	Moisture	0.135	–0.045	–0.431	0.5275	0.1302	–0.2785
8.	C	–0.192	–0.157	–0.286	–0.1963	0.3146	0.16
9.	N	0.05868	0.34559	**–0.68***	0.4644	0.24345	–0.489
10.	P	–0.046	–0.205	–0.256	–0.2563	0.7193	0.3676
11.	K	–0.5643	–0.3218	–0.4171	**–0.669***	**–0.866*****	0.0514
12.	Na	0.6244	0.4528	0.3704	0.0821	0.5553	0.6554
13.	Ca	–0.4438	**–0.825*****	–0.1033	–0.5903	–0.5393	–0.1619
14.	Mg	ND	ND	ND	–	–	–
15.	Fe	0.2036	–0.1887	0.2742	0.2031	–0.0197	0.2472
16.	Zn	**0.814*****	–0.3545	–0.0723	0.2968	–0.6148	–0.3221
17.	Cu	**0.772****	0.0342	–0.0642	**0.787****	0.4632	–0.3199

ND not detected. Emboldened values indicates strong correlation.
Statistical significance: *$p<0.05$; **$p<0.02$ and ***$p<0.01$

monsoon season). The high spore density of AM fungi was observed in *Zea mays* (1,377), followed by *Gossypium sativum* (1,359), *Glycine max* (1,250), *Cajanus cajan* (887), *Sorghum vulgare* (472) and *Triticum aestivum* (305) during winter season.

Relation Between AM Root Colonisation, AM Fungal Spore Density and Soil Parameters

Table 17.2 reveals seasonal variation in relationship between AM root colonisation, spore density

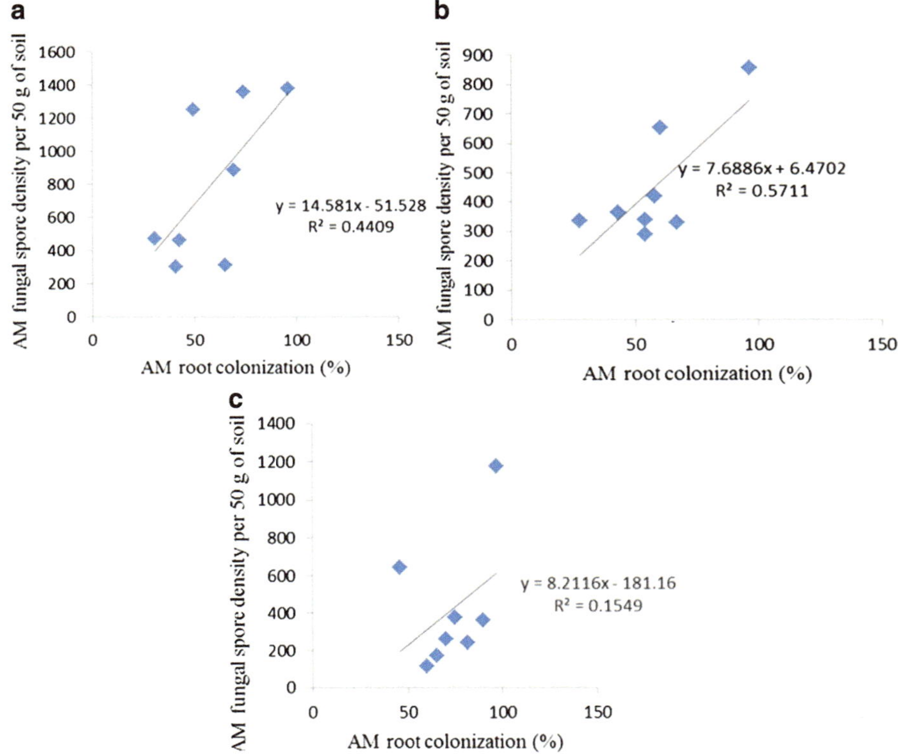

Fig. 17.4 Regression analyses of soil endomycorrhizal spore populations and AM root colonisation during winter (**a**), monsoon (**b**) and summer (**c**) seasons

and different edapho-climatic factors. It is evident from the results that AM root colonisation was affected by Zn ($r=0.814$, $p<0.01$) and Cu contents of soil ($r=0.772$, $p<0.02$) during winter season. However, it was affected by soil pH ($r=0.86$, $p<0.01$) and Ca levels ($r=-0.825$, $p<0.01$) during monsoon season. The N content of soil was negatively correlated with AM root colonisation during summer season ($r=-0.68$, $p<0.05$). Statistically nonsignificant negative correlation was observed between available P and AM root colonisation in all three seasons.

During winter season, AM fungal spore density was negatively correlated with EC ($r=-0.667$, $p<0.05$) and K ($r=-0.669$, $p<0.05$) and positively correlated with Cu ($r=0.787$, $p<0.02$). Similarly, during monsoon season, it was negatively correlated with soil temperature ($r=-0.761$, $p<0.02$) and K ($r=-0.866$, $p<0.01$) levels. During summer season, statistically nonsignificant correlations were observed between soil edaphic factors and spore density.

Relation Between AM Root Colonisation and AM Fungal Spore Density

The regression analysis showed that a strong positive correlation exhibited between AM colonisation and spore density during monsoon season ($R^2=0.5711$, $y=7.6886x-6.4702$) as compared to that of winter ($R^2=0.4409$, $y=14.581x-51.528$) and summer season ($R^2=0.1549$, $y=8.2116x-181.16$) (Fig. 17.4). Discussion-

AM fungi have been described as 'keystone mutualists' in ecosystems due to their unique position at the root-soil interface. In order to more fully understand the basic ecology of AM fungi and their role in natural ecosystems, it is necessary to document seasonal changes of various aspects of these fungi. Despite the importance of these soil fungi in various ecosystems, few studies exist that examine the seasonality of different variables of AM fungi which are directly related to ecosystem functions. The present study was undertaken in order to examine the seasonal dynamics of several soil parameters with specific interest to determine AM root colonisation and AM fungal spore density in the rhizosphere of *Glycine max*, *Saccharum officinarum*, *Zea mays*, *Cajanus cajan*, *Gossypium sativum*, *Triticum aestivum*, *Sorghum vulgare* and *Cicer arietinum*.

It is evident from the data that there is a wide range of seasonal variation in soil edaphic factors in the crop plants studied. During monsoon season, soil parameters such as pH, alkalinity, organic matter, water holding capacity, C, N, P and Ca were found to be elevated. It may be due to the fact that plant receives enough water and moisture and experiences good rainfall and average temperatures during monsoon season [19]. The AM root colonisation rates were more in all crop plants during monsoon season. Similar results were obtained in previous study [20]. The maximum root colonisation in *Withania somnifera* was observed during late rainy season. The high density of AM fungal spores was estimated during winter season in rhizospheric soil samples of crop plants. It is affirmed from the results that AM sporulation and root colonisation have seasonal dynamics, which is directly or indirectly influenced by different intervals of year. The seasonal changes and different range of spore density and root colonisation are due to a wide range of hosts. Sporulation and root colonisation of AM fungi have been found to be host dependent and this is probably one of the reasons that the plants under study have shown different results when compared. de Oliveira and de Oliveira [21] have reported that maximum sporulation and colonisation of AM fungi were registered in rainy season in case of *Theobroma grandiflorum* Schum and *Paullinia cupana* Mart. Similar findings were observed in present study where maximum colonisation and sporulation were reported in *Zea mays* during monsoon and winter seasons, respectively. The maximum sporulation in this season could be correlated with the fact that during this period, most photosynthate is allocated to roots and rhizomes, which help fungal symbionts to produce more spores [22]. The higher root colonisation may be attributed to the capacity of AM fungi to obtain higher profits and growth at abundant rainfall.

The data of Pearson correlation coefficients showed that significant correlation between soil edapho-climatic factors and AM root colonisation sporulation. AM root colonisation was found to be negatively influenced by temperature, EC, Ca, N, K and P content of soil. It is in agreement with the earlier findings [23, 24], which reported the distribution and ecology of AM fungi in *Citrus* nurseries and spatio-temporal variations of AM fungi in papaya plantations. Wu et al. [25] reported significant negative correlation between per cent root colonisation and soil available soil P content indicating that lower available P could accelerate the colonisation on *Citrus* roots. In the present study, AM spore population was found to have seasonal variation with low-moderate levels of phosphorus similar to the results of several workers. The overall growth rate of AM fungi in the present soils correlated with P-deficient soils. When plants have high nutrient availability, a negative response and low AMF spore population should be expected [26].

Regression analysis revealed that AMF colonisation and spore density were found to be positively correlated, which is consistent with the finding of Nadarajah and Nawawi [27] and Muthukumar et al. [28]. Spore production of AM fungi is known to vary greatly in different ecosystems, and it is thus influenced by an array of factors such as environment, host and fungus and spore density, which tends to decrease during root growth but increases during root inactivity or senescence [29]. Zhao et al. [30] suggested that the uneven spatial distribution of AM fungal spores and the complex structure of the underground root component could be major factors affecting spore density of AMF.

Conclusions

The present study reveals the status of AM fungi associated with eight crop plants which are commonly cultivated in Nanded district. It is evident from the data that variations in AM fungal spore density and colonisation may be related with seasonal changes, type of host plant and dynamics of soil edaphic factors. More detailed research is required for a better understanding of seasonal variations in AM fungal species composition and diversity associated with crop plants.

References

1. Harley JL, Smith SE. Mycorrhizal symbiosis. London: Academic; 1983. 483.
2. Brundrett M. Diversity and classification of mycorrhizal fungi. Biol Rev. 2004;79:473–95.
3. Barea JM, Jeffries P. Arbuscular mycorrhizas in sustainable soil plant systems. In: Hock B, Varma A, editors. Mycorrhiza structure, function, molecular biology and biotechnology. Heidelberg: Springer; 1995. p. 521–59.
4. Bagyaraj GJ. Mycorrhizal association in crop plants and their utilization in agriculture. In: Nair MC, Balakrishnan S, editors. Beneficial fungi and their utilization. Jodhpur: Scientific Publication; 1986. p. 59–72.
5. Mosse B. The influence of soil type and Endogone strain on growth of mycorrhizal plants in phosphate deficient soils. Rev Ecol Biol Soc. 1972;9:529.
6. Nemec S. Effects of storage temperatures and moisture on Glomus species and their subsequent effect on citrus rootstock seedling growth and mycorrhizal development. Trans Br Mycol Soc. 1987;89:205–12.
7. Al Karaki GN, Clark RB. Growth, mineral acquisition, water use by mycorrhizal wheat grown under water stress. J Plant Nutr. 1998;21:263–7.
8. Allen EB, Allen MF. Water relations of xeric grasses in the field: interactions of mycorrhizae and competition. New Phytol. 1986;104:559–71.
9. Abbott LK, Robson AD. Factors influencing the occurrence of Vesicular *Arbuscular mycorrhiza*. Agric Ecosyst Environ. 1991;35:121–50.
10. Lugo MA, Gonzalez-Maza ME, Cabello MN. *Arbuscular mycorrhizal* fungi in a mountain grassland: seasonal variation of colonization studies, along with its relation to grazing and metabolic host type. Mycologia. 2003;95:407–15.
11. Mandyam K, Jumpponen A. Seasonal and temporal dynamics of arbuscular mycorrhizal and dark septate endophytic fungi in a tallgrass prairie ecosystem are minimally affected by nitrogen enrichment. Mycorrhiza. 2008;18(3):145–55.
12. Sieverding E. Vesicular arbuscular mycorrhiza management in tropical ecosystem. Germany: GTZ; 1991.
13. Trivedy RK, Goel PK, Trisal CL. Practical methods in ecology and environmental science. 2nd ed. Karad: Enviro Media Publication; 1998.
14. Anderson JM, Ingran JSI. Tropical soil biology and fertility. A handbook of methods. Wallingford: CAB International, Technology and Engineering; 1989. 171pp.
15. Subbiah BV, Asija GL. A rapid procedure for the determination of available nitrogen in soils. Curr Sci. 1956;25:259–60.
16. Phillips JM, Hayman DS. Improved procedure for clearing roots and staining parasitic and vesicular arbuscular mycorrhizal fungi for rapid assessment of infection. Tran Br Mycol Soc. 1970;54:53–63.
17. Nicolson TH. The mycotrophic habit in grass. Thesis paper. Nottingham: University of Nottingham; 1955. pp. 227–86.
18. Gerdmann JW, Nicolson TH. Spores of mycorrhizal *endogone* species extracted from soil by wet sieving and decanting. Trans Br Mycol Soc. 1963;46:235–44.
19. Hindumathi A, Reddy BN. Occurrence and distribution of arbuscular mycorrhizal fungi and microbial flora in the rhizosphere soils of mungbean and soybean from Adilabad, Nizamabad and Karimnagar districts of Andhra Pradesh state, India. Adv Biosci Biotech. 2011;2:275–86.
20. Kumar A, Mangla C, Aggarwal A, Parkash V. Arbuscular mycorrhizal fungal dynamics in the rhizospheric soil of five medicinal plant species. Mid East J Sci Res. 2010;6(3):281–8.
21. de Oliveira AN, de Oliveira LA. Seasonal dynamics of arbuscular mycorrhizal fungi in plants of *Theobroma grandiflorum* Schum. and *Paullinia cupana* Mart. of an agroforestry system in Central Amazonia, Amazonas state, Brazil. Braz J Microbiol. 2005;36:262–70.
22. Gemma JN, Koske RE. Seasonal variation in spore abundance and dormancy of Gigaspora gigantea and in mycorrhizal inoculum potential of dune soil. Mycologia. 1988;80:211–6.
23. Khade SW, Rodrigues BF. Arbuscular mycorrhizal fungi associated with varieties of *Carica papaya* L. in tropical agro-based ecosystem of Goa, India, Tropical and Subtropical. Agroecosystems. 2009;10: 369–81.
24. Khade SW, Rodrigues BF. Spatial variations in arbuscular mycorrhizal (AM) fungi associated with *Carica papaya* L. in a tropical agro-based ecosystem. Biol Agric Hortic. 2008;26:149–74.
25. Wu Q, Xia R, Zou Y. Arbuscular mycorrhizal fungal growth on citrus roots and its correlations with soil available phosphorus content and phosphatase activity. J Appl Ecol. 2006;17:685–90.
26. Dhillon SS, Zak JC. Microbial dynamics in arid ecosystems desertification and the potential role of mycorrhizas. Revista Chilena De Historia Natural. 1993;66:253–70.
27. Nadarajah P, Nawawi A. Zygomycetous endomycorrhizal fungi in some Malaysian soils under rubber (*Hevea brasiliensis*). In: Taylor DA, Kenneth MG, editors. research on multipurpose tree species in Asia. Bangkok: Winrock International Institute for Agricultural Development; 1990.
28. Muthukumar T, Sha LQ, Yang XD, Cao M, Tang JW, Zheng Z. Distribution of roots and Arbuscular mycorrhizal associations in tropical forest types of Xishungbanna, southwest China. Appl Soil Ecol. 2003;22:241–53.
29. Muthukumar T, Sha LQ, Yang XD, Cao M, Tang JW, Zheng Z. Mycorrhiza of plants in different vegetation types in tropical ecosystems of Xishungbanna, southwest China. Mycorrhiza. 2003;13:289–97.
30. Zhao ZW, Xia YM, Qin XZ, Li XW, Cheng LZ, Sha T, Wang GH. Arbuscular mycorrhizal status of plants and the spore density of arbuscular mycorrhizal fungi in the tropical rain forest of Xishungbanna, southwest China. Mycorrhiza. 2001;11:159–62.

Computational Studies on Structural Basis of Isoniazid Resistance in Tuberculosis

18

Vidhyasree Rajasekharan, J. Abhithaj, G. Renganayaki, and Vipin Thomas

Abstract

Tuberculosis (TB) is the biggest killer among infectious diseases worldwide, with nearly two million deaths every year. The excessive use of antibacterial drugs has resulted in the emergence of multiple drug-resistant varieties of Mycobacterium. Resistance to the most widely used drugs, isoniazid and rifampicin, and other second-line drugs has become a major challenge for the treatment and control of TB globally. Resistance to drugs is partly due to single-nucleotide polymorphisms in specific genes of *Mycobacterium tuberculosis*. Mainly, nine genes are known to be associated with resistance to first-line drugs: rpoB for RIF resistance; pncA for pyrazinamide (PZA); katG, inhA, aphC, and kasA for INH; rpsL and rrs for streptomycin (STR); and embB for ethambutol (EMB) resistance. In many cases, mutations in these genes cause structural modifications in the target proteins which in turn modify target–drug interactions. The present study is aimed at investigating the relationship between single-nucleotide polymorphisms (SNPs) which are prevalent in India and consequent structural changes in tuberculosis targets. As the first part, structural basis of resistance to one of the most widely used front-line drugs isoniazid (INH) was studied. The primary target of INH inhibition is the cell wall mycolic acid synthesis pathway in which INH inhibits enoyl-ACP reductase

V. Rajasekharan (✉)
Department of Biosciences, Sree Narayana Guru Institute of Science and Technology, N. Paravoor, Kerala, India
e-mail: vidyasree.rajan@gmail.com

J. Abhithaj
Department of Biochemistry, MHES College of Science and Technology, Cherandathur, Calicut, India

G. Renganayaki • V. Thomas
Department of Computational Biology & Bioinformatics, University of Kerala, Thiruvananthapuram, India

A. Sabu and A. Augustine (eds,), *Prospects in Bioscience: Addressing the Issues*, DOI 10.1007/978-81-322-0810-5_18, © Springer India 2013

enzyme (InhA). Single-nucleotide polymorphisms in InhA at multiple positions confer resistance to INH. To investigate the structural basis of resistance to INH, common mutations in InhA found in India were computationally modeled into the target structure (PDB ID: 2NV6). After *in silico* mutagenesis, the conformations were optimized and were subjected to energy minimization in steepest descent. Analysis of the mutated structures and their binding sites revealed interesting variations in the conformations especially in critical residues involved in ligand binding. Isoniazid was then computationally docked with the mutated target structures to investigate modifications in target–ligand interactions. The changes in binding modes and interactions compared to the native structure were documented. This information will be instrumental in future design of more potent INH derivatives/novel inhibitors with better binding activity on mutant strains.

Keywords
Drug resistance • Single-nucleotide polymorphisms • Computational docking

Introduction

According to the WHO, one-third of the world's population is currently infected with tuberculosis bacteria (nine million in 2010). Persons infected with human immunodeficiency virus (HIV) are more likely to have tuberculosis than HIV-negative individuals. The WHO has reported 26 deaths among 185 incidents per 100,000 populations of HIV negatives in India in the year 2010.

Drug resistance is a major threat in the treatment of tuberculosis. Coinfection with *Mycobacterium tuberculosis* and HIV (TB/HIV), and multidrug-resistant (MDR) and extensively drug-resistant (XDR) tuberculosis, increases the severity of infection. The failure to ensure regular treatment with high-quality existing drugs and the fact that only a few drugs to treat TB are available make the drug resistance dreadful. The last new treatments for TB were developed before 28 years. In the treatment of TB, second-line drugs are more expensive and are administrated intravenously. Chances of side effects and long duration of treatment are the shortcomings in such cases.

Mainly, nine genes are known to be associated with resistance to first-line drugs: rpoB for RIF resistance; pncA for pyrazinamide (PZA); katG, inhA, aphC, and kasA for INH; rpsL and rrs for streptomycin (STR); and embB for ethambutol (EMB) resistance. INH was first reported to be an effective anti-tuberculosis drug in 1952. Increases in INH-resistant and multidrug-resistant tuberculosis jeopardize drug effectiveness [1, 2], and development of INH resistance is often a first step in multidrug resistance [3, 4].

The present study is focusing on resistance due to single-nucleotide polymorphisms in specific genes of *Mycobacterium tuberculosis*. These mutations can confer selective advantages on the bacterial populations, especially in case of survival in the presence of antibiotics. Lots of neutral mutations are seen to be present in *Mycobacterium tuberculosis* strains, but mutations are significant only if they are expressed. Severity of tuberculosis varies in different patients. *Mycobacterium tuberculosis* encountering the human hosts with individual immune systems needs to modulate their virulence associated biological factors accordingly to survive within the host. Drug resistance itself makes the organism in a better position to survive within the hostile intracellular environment.

In many cases, mutations in specific genes cause structural modifications in the target proteins which in turn modify target–drug interactions. The combination of genomics and bioinformatics has the potential to generate the information and knowledge that will enable the conception and development of new therapies and interventions needed to treat this airborne disease.

MDR-TB has resulted in fatal outbreaks in many countries [5]. The condition is more deadly for both HIV-negative and HIV-positive individuals [6]. The primary target of INH inhibition is the cell wall mycolic acid synthesis pathway in which INH inhibits enoyl-ACP reductase enzyme (InhA). Single-nucleotide polymorphisms in InhA at multiple positions confer resistance to INH.

Understanding of the molecular aspects of SNP–drug resistance relationships, structural modifications and protein–drug interactions that are occurring in InhA will be pivotal in developing new resistance-resistant new anti-tubercle drug targets.

Materials and Methods

Isoniazid is the first drug that had been effective against TB. InhA (PDB Id: 2NV6), one of the most important targets of INH, is the protein of present study. To understand the structural basis of isoniazid resistance, SNPs which are prevalent in India were computationally induced to the native structure. Stable conformers of the mutated structure were subjected to interact with the ligand. Analysis of these interactions revealed instrumental details.

In Vitro Mutagenesis

Mutations at target protein were induced computationally on the resident protein structure. Mainly four mutations (S94A, H121V, M147T and M161V) were found to be prevalent in India. All the chosen mutations were SNPs. Swiss PDB Viewer was the tool used to alter the amino acid residues. Energy minimization of the mutated structures was carried out using GROMACS.

Drug–Target Interactions

To understand the target–ligand interactions happening in drug-resistant strains, docking analysis of each mutated structures with known ligand was performed in GOLD. Original ligand (CID_449263) was docked against the active site of each mutated protein.

Results and Discussion

Mutation at S94A

Figures 18.1–18.4 show the best pose of ligand bound to S94A, H121V, M147T and M161V. Analysis of interaction between ligand and S94A clearly revealed the differences in hydrogen bonds and short contacts from that of the native protein. 2NV6 interacts with ligand via three H-bonds – GLY 192, PHE 41 and ILE 194. Mutated structure S94A had shown same number of H-bonds with the ligand but with different residues [(2)ILE 95 and GLY 96]. The change was noticeable even in the case of short contacts. Seven short contacts were present in 2NV6. The number was higher in the mutated structure. Out of 11 short contacts in the mutated residue, GLY14 (length 2.16) was present in common. GOLD score of this mutation (−1843.58) was lower than that with the native protein (Table 18.1).

Mutation at H121V

Interaction of this mutated structure with ligand had shown an RMSD of −1839.28, which was lower than the native protein. Compared to the 3H-bonds in the native protein–ligand interaction, this mutation could generate only 2H-bonds (GLY 96 and PHE 41). H-bond at PHE41 was common in both but with little difference in length – 2.962 in native protein and 3.07 in mutated one. Number of short contacts was 9 in mutated structure. Some of the short contacts were the same as that of the un-mutated structure (MET 199, GLY 14 and GLY 40).

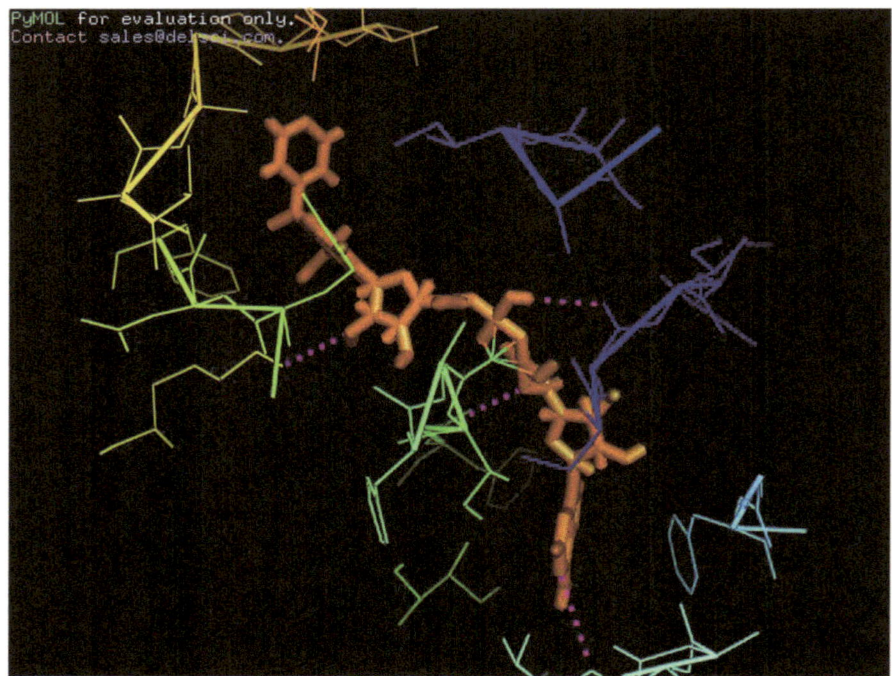

Fig. 18.1 Best pose of ligand bound to S94A

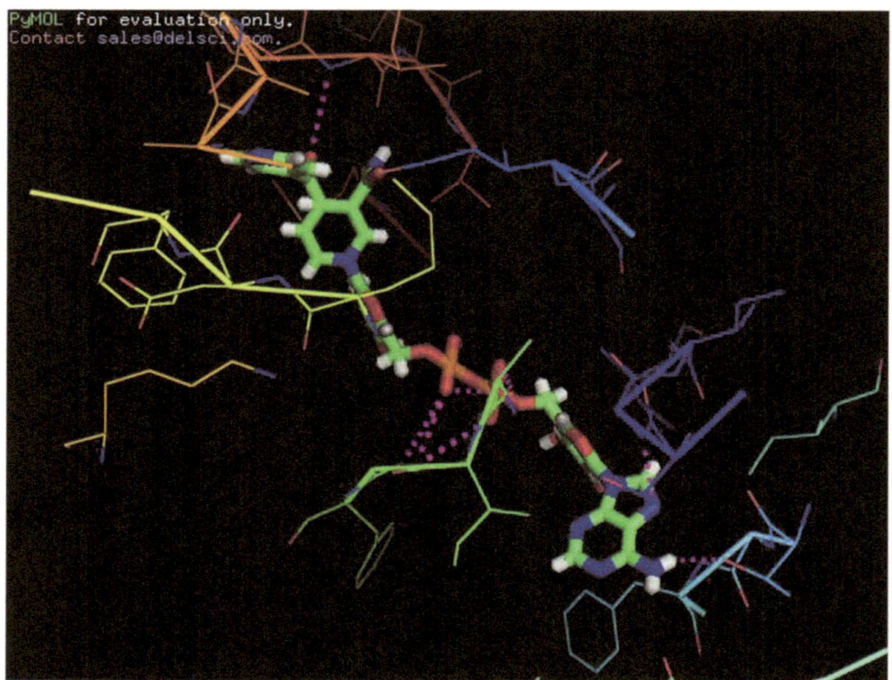

Fig. 18.2 Best pose of ligand bound to H121V

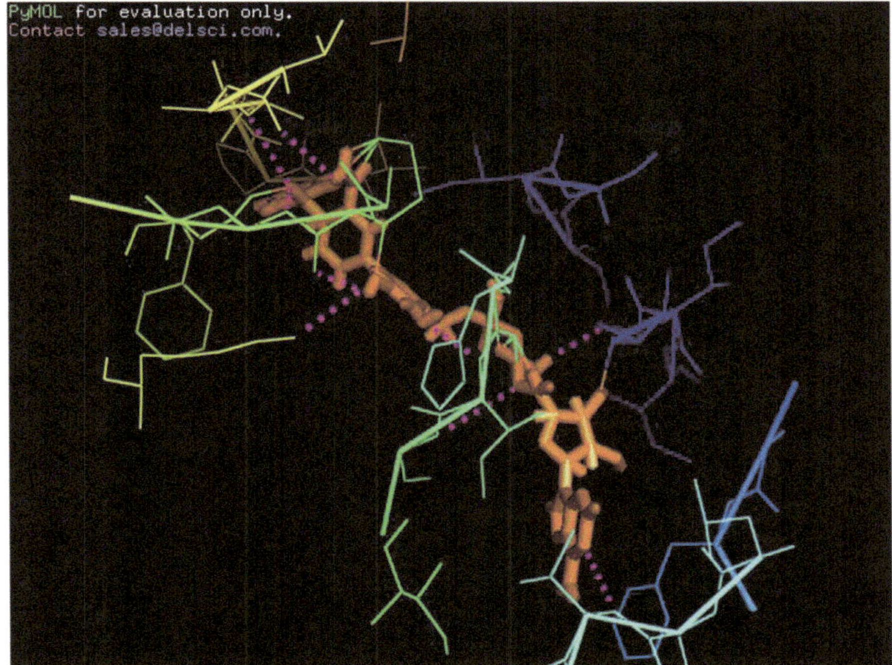

Fig. 18.3 Best pose of ligand bound to M147T

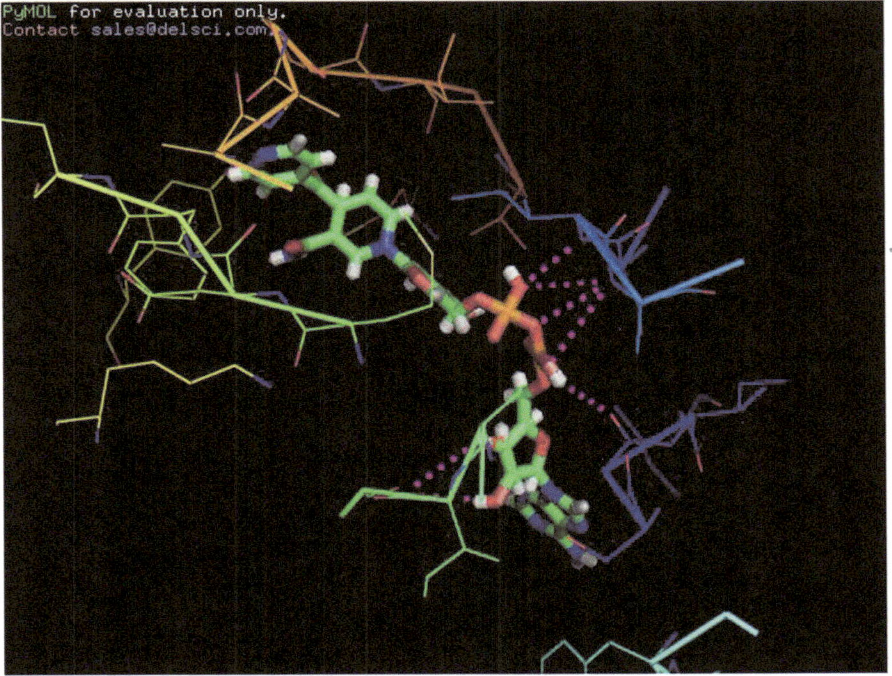

Fig. 18.4 Best pose of ligand bound to M161V

Table 18.1 Hydrogen bonds and short contacts in the best poses of each mutation

Site of mutation	Name of residue	Number of hydrogen bonds	Length of H-bonds	Residue	Number of short contacts	Length of short contacts
M147T				ASP 148	1	2.145
	PHE 95	1	2.957	THR 147	2	2.242, 2.395
	GLY 96	1	2.545	ILE 95	5	2.542, 2.041, 1.822, 2.150, 2.217
	ILE 95	1	3.043	GLY 14	1	1.972
S94A				ALA 94	1	2.217
				ILE 21	3	2.085, 2.150, 2.516
				GLY 192	2	2.150, 2.584
				PHE 194	1	1.908
	ILE 95	2	2.729, 2.642	GLY 212	1	2.447
	GLY 96	1	2.782	SER 94	3	2.717, 2.555, 2.476
				GLY 14	1	2.16
				ILE 95	1	2.426
				PHE 41	2	2.177, 2.700
H121V				ILE 21	2	2.682. 1.823
				MET 199	1	1.768
	GLY 96	1	2.81	GLY 96	1	2.528
	PHE 41	1	3.07	GLY 14	4	2.031, 2.389, 2.244, 2.336
				GLY 40	1	2.141
M161V				PRO 193	1	2.21
	ILE 21	1	2.672	ILE 21	1	1.85
	THR196	1	2.982	ALA 94	2	1.880, 2.480
				ILE 16	2	2.380, 2.209
2NV6 (native protein)				GLY 40	1	2.203
	GLY192	1	3.057	GLY 14	2	2.163, 2.636
	PHE 41	1	2.962	ILE 16	1	2.381
	ILE 194	1	2.356	THR 196	1	2.012
				MET 199	1	1.751
				ILE 194	1	2.269

Mutation at M147T

Analysis of interaction of ligand with this mutated structure also revealed a reliable GOLD score (−1831.94). Even though the number of H-bonds was same as compared to the native structure, none of the residues were the same in the formation of H-bonds. Thirteen short contacts were present between the ligand and the mutated target. H-bond from GLY 14 was seen in both.

Mutation at M161V

Among the four mutations made, only this one had a greater GOLD score than the native structure (−1857.36). The number of H-bonds with the protein and ligand was 2. Short contacts were 6 in number. Only ILE 16 was seen as common in mutated and un-mutated structures.

Conclusions

Study on structural basis of isoniazid resistance had provided a number of instrumental details. Comparison of ligand–target interactions has provided essential data regarding the basis of resistance of isoniazid. Some of the bonds were unique in certain mutations but some were common in both mutated and un-mutated structures. There are changes in binding modes of residues in ligand–target interactions, when compared with the native structure. This information will be instrumental in future design of more potent INH derivatives/novel inhibitors with better binding activity on mutant strains. This study may pave way for treating INH-resistant strains and thereby the complete eradication of TB in future.

References

1. Fang Z, Doig C, Rayner A, Kenna DT, Watt B, Forbes KJ. Molecular evidence for heterogeneity of the multiple-drug-resistant *Mycobacterium tuberculosis* population in Scotland (1990 to 1997). J Clin Microbiol. 1999;37:998–1003.
2. Weis SE, Pogoda JM, Yang Z, Cave MD, Wallace C, Kelley M, Barnes PF. Transmission dynamics of tuberculosis in Tarrant County, Texas. Am J Respir Crit Care Med. 2002;166:36–42.
3. Alland D, Kalkut GE, Moss AR, McAdam RA, Hahn JA, Bosworth W, Drucker E, Bloom BR. Transmission of tuberculosis in New York City. An analysis by DNA fingerprinting and conventional epidemiologic methods. N Engl J Med. 1994;330:1710–6.
4. Filliol I, Motiwala AS, Cavatore M, Qi W, Hazbón MH, Bobadilla del Valle M, Fyfe J, Garcia-Garcia L, Rastogi N, Sola C, Zozio T, Guerrero MI, León CI, Crabtree J, Angiuoli S, Eisenach KD, Durmaz R, Joloba ML, Rendón A, Sifuentes-Osornio J, de Ponce Leon A, Cave MD, Fleischmann R, Whittam TS, Alland D. Global phylogeny of Mycobacterium tuberculosis based on single nucleotide polymorphism (SNP) analysis: insights into tuberculosis evolution, phylogenetic accuracy of other DNA fingerprinting systems, and recommendations for a minimal standard SNP set. J Bacteriol. 2006;188:759–72.
5. Snider DE, Roper WL. The new tuberculosis. Engl J Med. 1992;326:703.
6. Frieden TR, Sterling T, Pablos-Mendez A, Kilburn JO, Cauthen GM, Dooley SW. The emergence of drug-resistant tuberculosis in New York City. N Engl J Med. 1993;328:521–6.
7. Cummings MP, Segal MR. Few amino acid positions in rpoB are associated with most of the rifampicin resistance in Mycobacterium tuberculosis. BMC Bioinformatics. 2004;5:137.

Phylogenetic Analysis of Few Actinobacteria with Potential Antimicrobial Properties, Isolated from the Forest Soils of Western Ghats of Kerala

19

Sugathan Shiburaj and Sreenivasan Preethi

Abstract

The application of molecular techniques to the study of microbial diversity has changed the face of microbial ecology and opened a new era of microbial molecular ecology. Actinomycetes are gram-positive bacteria frequently filamentous and sporulating with rich G+C content varying from 57 to 75%. These soil microorganisms are well known for their ability to produce a wide variety of secondary metabolites, many of which have potent biological activities. Actinomycetes, particularly *Streptomyces* spp., account for approximately half of the known biologically active microbial products, including many commercially important antibiotics, immunosuppressive compounds, animal health products and agrochemicals. We have isolated and screened several actinomycete strains from the forest soils of Neyyar Wildlife Sanctuary of Kerala state, India for their antimicrobial potential. The strains with promising antibacterial/antifungal properties were characterised following a polyphasic approach, employing molecular and conventional taxonomic tools. The phylogeny and extracellular antimicrobial properties of few potential actinomycete isolates, namely, *Streptomyces atroolivaceus* TBG-28S1A46 (MTCC 4148), S. *clavifer* TBG-MNR13 (MTCC 4150), *S. setonii* TBG-19NRA1

S. Shiburaj (✉)
Cell and Molecular Biology Lab, Department of Biology, Dalhousie University, 1355 Oxford Street, Halifax, N.S B3H 4R2, Canada

Jawaharlal Nehru Tropical Botanic Garden and Research Institute, Palode, Thiruvananthapuram, Kerala, 695 562, India
e-mail: drshiburaj@gmail.com

S. Preethi
Department of Fisheries (Government of Kerala), Vikas Bhavan, Thiruvananthapuram, Kerala, 695 033, India
e-mail: preethifisheries@gmail.com

A. Sabu and A. Augustine (eds.), *Prospects in Bioscience: Addressing the Issues*, DOI 10.1007/978-81-322-0810-5_19, © Springer India 2013

(MTCC 3756) and *Streptosporangium nondiastaticum* TBG-75A20 (MTCC 4149), are discussed in this communication.

Keywords
Antimicrobial • Streptomyces • Streptosporangium • 16s rDNA sequence

Introduction

Actinomycetes are gram-positive bacteria forming branching filaments with spore-bearing structures and having high G+C (>55%) content in their DNA. They are most popularly considered as boundary or transition microorganisms having morphological and developmental similarities with certain higher fungi but have more characters of bacteria in cell wall chemistry, susceptibility to bacterial phage and antibacterial compounds, with phylogenetic relationships to gram-positive bacteria [1]. Actinomycete members form a large and important segment of the microflora of most natural environments and have a remarkable place in modern science as the largest producer of bioactive metabolites.

The discovery of actinomycin in 1940 and that of streptomycin in 1943 has been pivotal to the exploration of soil microorganisms as a source of new lead molecules. The microbial products, especially those from actinomycetes, have been a phenomenal success for the past seven decades. During the Golden Age of antibiotic discovery, in the 1950s and 1960s, such well-known antibacterial drugs as tetracycline, erythromycin and kanamycin, antifungal agents like candicidin and nystatin and anticancer drugs such as adriamycin were discovered through the efforts of academic and industrial researchers. After 1970, the rate of discovery of useful compounds declined progressively, although several important agents nevertheless came to light, including the antihelmintic avermectin, the immunosuppressants rapamycin and tacrolimus (FK506) and the natural herbicide bialaphos. Thus, actinomycetes have proved to be a sustained mine of novel antibiotics, which selectively destroys the pathogens without affecting the host tissues. The versatility and immense economic value of anti-infective compounds are extremely noteworthy, as it has a turnover of over US$79 billion in 2009, includes about 166 antibiotics and derivatives, most of which are produced by actinomycetes [2]. Actinomycetes continue to play a highly significant role in drug discovery and approximately 22,500 biologically active compounds that have obtained from microbes, 45% are produced by actinomycetes, 38% by fungi and 17% by unicellular bacteria [3]. Within the actinomycetes, members of the genus *Streptomyces* account for 70–80% of secondary metabolites, with smaller contributions from genera such as *Actinoplanes*, *Amycolatopsis*, *Micromonospora* and *Saccharopolyspora* [4].

Although thousands of antibiotics have been isolated from *Streptomyces* and related organisms, these represent only a small fraction of the repertoire of bioactive compounds produced [5]. Therefore, isolation and screening of actinomycetes from natural resources and characterization of their secondary metabolites is a valuable endeavour. Western Ghats in south Kerala is one of the 12 mega diversity centres in the world and considered as 'hot spots' of biodiversity [6]. Neyyar Wildlife Sanctuary is located in the western slope of the Southern Western Ghats of Kerala, with a vegetation of tropical evergreen and moist deciduous forests. This report describes the isolation and characterization of actinomycete strains producing antibacterial and antifungal secondary metabolites from soil samples collected from different spots of Neyyar WLS.

Materials and Methods

Isolation of Actinomycetes

The soil samples were collected aseptically from the study area at a depth of 6–12 in. The soil samples were air-dried in a hot air oven at 45°C for 1 h to reduce the proportion of bacteria other than actinomycetes [7]. Actinomycete strains were isolated and obtained as pure culture by using standard dilution plate technique using glucose-aspargine-agar plates supplemented with cycloheximide at a concentration of 100 μg/ml [8]. After an incubation of 7–9 days at 27 ± 1 °C, the actinomycete colonies were carefully counted by visual observation, and colony-forming units per gram (cfu/g) of soil were determined. Single spore cultures of actinomycetes were preserved on yeast-extract-glucose-agar slants at 4 °C for 2 months and at −20 °C in the presence of glycerol (15%v/v) for longer periods.

Screening for Antibiotic Activity

The isolates were screened for antibacterial and antifungal properties. Preliminary screening for antibiotic activity of the isolates was done by using streak-plating technique on yeast-extract-glucose-agar medium. Each pure isolates were streaked individually on different agar plates in a single line. The plates were then incubated at 32°C for 5 days to allow the isolates to secrete antibiotics into the medium. After the incubation period, the properly diluted test organisms were cross-streaked along the line of fully grown isolates. Each streaking was started near the edge of the plates and streaked towards the actinobacteria growth line. The plates were then incubated for 18 h at 37°C, and the zone of inhibition was measured.

The isolates showing activity in the preliminary screening were further checked to determine the efficiency of producing antibiotic substances. Five-day-old inoculums were prepared in 20 ml tryptone-yeast extract broth in 100-ml Erlenmeyer flasks and inoculated to 100 ml antibiotic sensitivity broth and in 250-ml Erlenmeyer flasks [9]. The flasks were incubated at room temperature (28 ± 1°C) on a rotary shaker at 120 rpm for 5 days. The culture filtrates were collected after incubation and were analysed separately for antimicrobial activity following Oxford agar cup-plate method using yeast-extract-glucose-agar plates. The diameters of inhibition zone were measured after an incubation period of 24 h for bacterial strains and 3 days for fungal strains at a temperature of 28 ± 1 °C. The experiments were done in triplicates and the antimicrobial activity of the isolates were scored in terms of average diameter (mm) of inhibition zones and categorised according to average inhibition zone as strong (>25 mm), moderate (15–25 mm) and weak (<15 mm).

Test Organisms

Six test organisms were used to test the antibiotic activity of the isolates, including standard bacteria (gram positive and negative) and nonfilamentous fungi. The bacterial strains, *Staphylococcus aureus* subsp. *aureus* MTCC 740, *Bacillus subtilis* MTCC 441, *Pseudomonas fluorescens* MTCC 103, *Escherichia coli* MTCC 443 and the fungal strains *Candida albicans* MTCC 221, *Saccharomyces cerevisiae* MTCC 36 were obtained from Microbial Type Culture Collection and Gene Bank (MTCC), Institute of Microbial Technology (IMTECH), Chandigarh. The bacterial strains were maintained as live pure cultures on nutrient agar slants and nutrient broth at suitable incubation conditions by periodical subculturing. The fungal strains are preserved in Sabouraud's agar slants and Sabouraud's broth under suitable incubation conditions.

Identification of Potential Isolates

Various cultural, physiological and biochemical characteristics of the isolates showing potential antimicrobial activities were studied according to the methods described in International *Streptomyces* Project [10] and Bergey's manual [11]. Micromorphology was determined by phase contrast microscopy and scanning electron microscope (SEM) following cover slip culture method on ISP-2 agar medium [12].

For 16s rDNA homology studies, the producer strain was grown on ISP-2 agar. Total genomic DNA preparation was carried out using the procedure described earlier [13]. The 16s ribosomal DNA was amplified using the primers 8-27 F, 5′-AGAGTTTGATCCTGGCTCAG-3′ and 1495R 5′-CTACGGCTACCTTGTTACGA-3′.

Amplification was performed in automated thermocycler (MJ Research) using 30 ng of genomic DNA as template with 10 pmol of each primers, 1U of Taq DNA polymerase and the recommended buffer system, in a 25 μl reaction volume. The PCR programs were carried out at 95 °C for 3 min, followed by 35 cycles of 95 °C for 1 min, 56 °C for 50 s, 72 °C for 1 min and 72 °C for 10 min for elongation. The PCR products were resolved by electrophoresis on 1% agarose gels. The fragments of interest were excised and then purified by QIAEX II Gel Extraction Kit (Qiagen, USA) and sequenced in an automated sequencer (ABI Prism) with 8-27 F primer. A BLAST (NCBI) homology search was performed with the resultant sequence. The 16S rDNA sequences of the strains were aligned manually using the Clustal W program [14] against corresponding nucleotide sequences of similar strains retrieved from GenBank. Evolutionary distance matrices were generated as described by Jukes and Cantor [15], and a phylogenetic tree was inferred by the neighbour-joining method [16]. Tree topologies were evaluated by bootstrap analysis [17] based on 1,000 resamplings of the neighbour-joining dataset, and a phylogenetic tree was constructed using TREECON® software package [18].

Results and Discussion

Isolation, Screening and Grouping of Antagonistic Actinomycetes

The finding of new bioactive compounds is a never-ending process to meet the everlasting demand for novel biomolecules with antimicrobial properties in order to contest human and plant pathogens. This study was performed with an aim of isolating actinomycete strains with antimicrobial activities using the selective isolation media from virgin forest soils. Seventy-four different actinomycete strains were isolated from nine soil samples collected from different locations of Neyyar Wildlife Sanctuary. All of these strains were isolated by using glucose-asparginate-agar (GAA) media supplemented with cycloheximide (100 μg/ml) to inhibit fungal growth. The GAA media is very specific for the isolation of actinomycetes and most of the *Streptomyces* and related organisms grow well on this medium.

The colony-forming units (cfu/g) were determined by counting the colonies on the dilution plates. Maximum average number of colonies (24.57×10^6 cfu/g soil) was obtained in the soil collected from the evergreen forest habitat. While that in deciduous forest soils is 17.864×10^6 and 15.986×10^6 in grass land samples. The forests of Neyyar Wildlife Sanctuary of Kerala state are rich in biodiversity [6]. The habitat wise correlations of the region show that the evergreen forest soil is rich in actinomycetes flora, evidenced from the number of organisms per gram of soil and the maximum number of isolates as compared to other habitats.

All the isolated actinomycete strains were made into pure cultures and screened for their antibacterial activity on yeast-extract-glucose-agar medium using streak-plating technique. A broad spectrum of antimicrobial activity was observed in 52.7% (39 out of 74) of the total pure isolates. This indicates that actinomycetes are the potential candidates for broad range antimicrobial metabolites, which may be the results from natural selection of the microorganisms in order to survive in a competing environment. Recent exploration studies conducted in various parts of the world considered that virgin forest soils are still continuing as excellent sources for the isolation of actinomycetes particularly *Streptomyces* with diverse potential [19–21]. Remarkably, majority of the isolates obtained from this study were morphologically similar to *Streptomyces* species. Many works of the actinomycetes ecology also showed the widespread occurrence of *Streptomyces* in soil in any region [22, 23].

The isolates showing antimicrobial activity in the preliminary screening were grown in antibiotic sensitivity broth for 5 days at room temperature on a rotary shaker (120 rpm). The culture filtrates (100 μl) were analysed separately for antimicrobial activity following Oxford agar-cup-plate against the test organisms. Among the 39 isolates analysed, 4 isolates designated as TBG-28SA46,

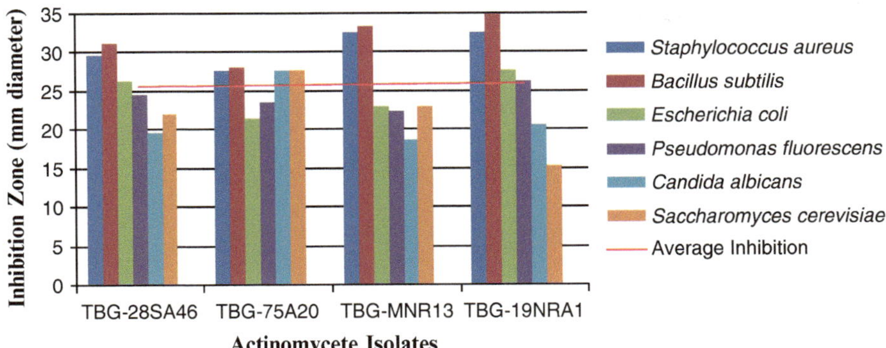

Fig. 19.1 Antimicrobial activity of mycelia free culture filtrate of isolates with strong activity (*average inhibition zone >25 mm diameter) *100µl of cell free culture filtrates were assayed against test organisms using Oxford agar-cup method. The activity was scored in terms of diameter inhibition zones against each test organisms and average value calculated for each of the actinobacteria

TBG-MNR13, TBG-19NRA1 and TBG-75A20 exhibited strong antimicrobial activity with an average inhibition zone >25 mm against gram-positive, gram-negative bacteria and fungi (Fig. 19.1). Thirteen isolates were shown to be moderately active (average inhibition zone of 15–25 mm) and 22 were weak (average inhibition zone of <15 mm) in activity.

Morpho-Physiological Characteristics of the Selected Strains

Culture characteristics of strains were derived on the basis of observations made after 7 and 14 days of incubation on different ISP media. The colony morphology, aerial mass colour, reverse colony colour and pigmentation were noticed after 7–14 days of growth at 28 °C following International *Streptomyces* Project (ISP) method. The morphology of mycelia and spore bearing structures were observed under a phase contrast microscope (Nikon Optiphot-II). Based on morphological and microscopical characters, the isolates were grouped into different genera. The morpho-physiological characters of the strains are summarised in Table 19.1. According to the cultural observations, the isolates TBG-28SA46, TBG-MNR13 and TBG-19NRA1 grew well on ISP-2, ISP-5 and ISP-6 media forming leathery greyish to white, elevated colonies. The microscopic observation shows that these three isolates are forming separate aerial and vegetative mycelium with long spore chains (conidial chains) on aerial mycelia.

The spore chain morphology is considered as one of the important characteristics in the identification of *Streptomyces*, and it greatly varies among the species. On cell wall analysis, LL-diaminopimelic acid (LL-DAP) was observed as major cell wall peptidoglycan in these isolates. Based on these characters, the three isolates were identified as *Streptomyces* spp. The isolate TBG-75A20 showed leathery, elevated, orange-red colonies in all the media. The strain showed the presence of small globose sporangia on aerial mycelia. The cell wall is having meso-diaminopimelic acid (meso-DAP). Based on these characters, the isolate TBG-75A20 is identified as *Streptosporangium* sp.

Comparative 16S rDNA Analysis

The comparison of 16s rRNA sequences is a powerful tool for deducing phylogenetic and evolutionary relationships among bacteria and actinomycetes. The advantages of 16S rRNA sequences for phylogenetic studies have been discussed extensively [24, 25]. PCR-amplified 16S rDNA of TBG-28SA46, TBG-MNR13, TBG-19NRA1 and TBG-75A20 were sequenced

Table 19.1 Morphological and physiological characters of selected actinomycete isolates with potential antimicrobial activity

	TBG-28SA46	TBG-MNR13	TBG-19NRA1	TBG-75A20
Aerial mycelia colour	White to light grey	Grey	White to light grey	Pinkish brown
Vegetative Mycelia colour	Yellow brown	Yellow brown	Yellow brown	Red orange to brown
Soluble pigments	Absent	Absent	Absent	Absent
Spore chain type	*Rectus flexibilis*	*Rectus flexibilis*	*Rectus flexibilis*	Absent
No. of spores/ chain	>20	>20	>20	Nil
Sporangia	Absent	Absent	Absent	Present, globose
Cell wall peptidoglycan	LL-DAP	LL-DAP	LL-DAP	Meso-DAP
Melanin production	Positive	Negative	Negative	Negative
Growth temperature (°C)	10–36	10–36	10–36	10–42
Carbon sources utilised	L-arabinose	L-arabinose	Dextran	L-arabinose
	D-glucose	D-fructose	D-fructose	D-fructose
	L-rhamnose	D-glucose	D-galactose	D-galactose
		Inositol	D-lactose	D-glucose
		D-mannose	D-mannitol	D-mannose
		L-rhamnose	D-mannose	D-mannitol
		Sucrose	L-rhamnose	D-raffinose
		D-xylose	Trehalose	D-xylose

and analysed for the similarities using BLAST (NCBI). Almost complete (around 1,500 bp) 16S rRNA gene sequences determined in this study and retrieved from the GenBank databases were aligned using the CLUSTAL X software. The TREECON® software package for construction and drawing of evolutionary trees was used to construct a phylogenetic tree. Distances were calculated and an evolutionary distance matrix was constructed according to the Jukes and Cantor method with insertions and deletions taken into account, and tree topology was inferred by neighbour-joining method. The NCBI-BLAST search program showed that the sequence data of TBG-28SA46 had high identity (98%) to those of *Streptomyces atroolivaceus*. Strains TBG-MNR13 and TBG-19NRA1 displayed high similarity towards *Streptomyces clavifer* and *Streptomyces setonii*, respectively, while TBG-75A20 showed 98% similarity to *Streptosporangium nondiastaticum*. The evolutionary trees were drawn using the neighbour-joining method (Fig. 19.2).

Based on the 16s rDNA sequence similarity analysis and morphological characters, the isolate TBG-28SA46 is identified as *Streptomyces atroolivaceus* (MTCC 4148), TBG-MNR13 as *S. clavifer* (MTCC 4150), TBG-19NRA1 as *S. setonii* (MTCC 3756) and TBG-75A20 as *Streptosporangium nondiastaticum* (MTCC 4149). The reference cultures of these isolates were deposited to Microbial Type Culture Collection and Gene Bank (MTCC), Institute of Microbial Technology (IMTECH), Chandigarh, India.

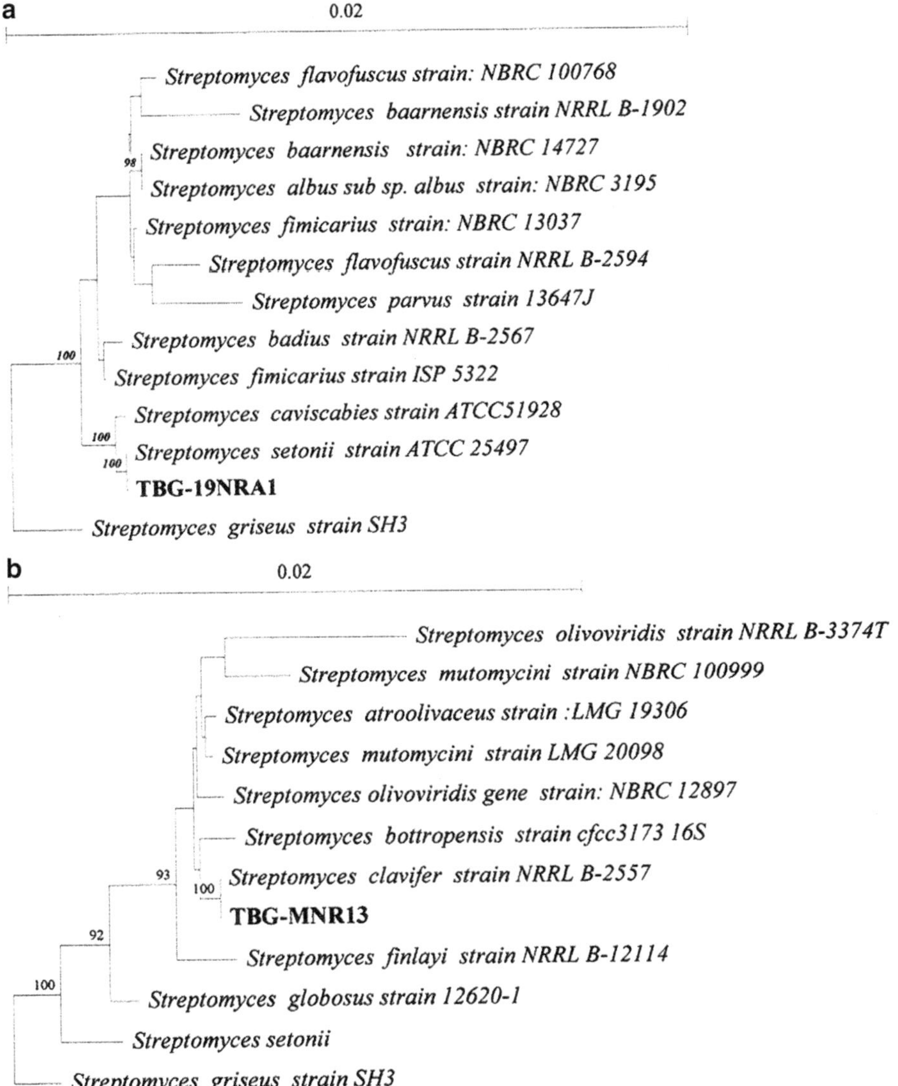

Fig. 19.2 (**a–d**) Neighbour-joining trees constructed using TREECON software, based on 16s rDNA sequences, showing the relationship with selected isolates, retrieved from GenBank databases. The numbers at the nodes indicate the levels of bootstrap support based on 1,000 resampled datasets; only values over 80% are given. Evolutionary distances were calculated to the Jukes and Cantor method with insertions and deletions taken into account. *Bar* denotes 0.2-nt substitutions per nucleotide position

Conclusions

It can be concluded that the forest soils of Neyyar Wildlife Sanctuary are rich in actinomycetes flora and are of the richest sources of potential antimicrobial metabolites. More than 50% of the actinomycetes were isolated with antimicrobial activities against pathogenic bacteria and nonfilamentous fungi. Among the different actinomycete cultures obtained from this sanctuary, the isolates *Streptomyces*

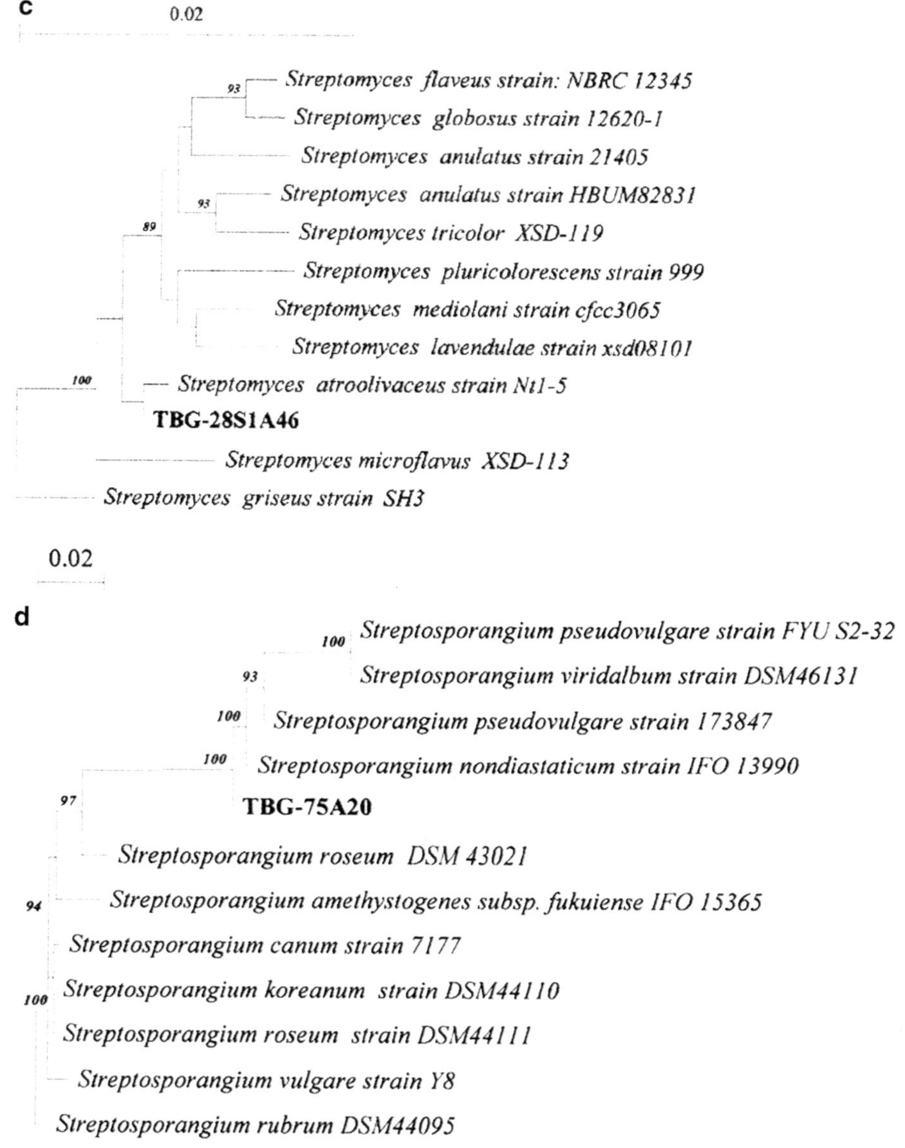

Fig. 19.2 (continued)

atroolivaceus TBG-28S1A46, *Streptomyces clavifer* TBG-MNR13, *Streptomyces setonii* TBG-19NRA1 and *Streptosporangium nondiastaticum* TBG-75A20 showing potential antimicrobial activity in vitro against gram-positive and gram-negative bacteria and nonfilamentous fungi. In view of the decline in the discovery of new lead compounds in recent years, further investigations on these isolates would lead to some useful products.

Acknowledgements Authors acknowledge the Director, JNTBGRI, for providing facilities and Dr. S. Mayilraj, IMTECH, Chandigarh, India, for the confirmation of taxonomic identity of the strains and also acknowledge the Department of Forests and Wildlife, Govt. of Kerala, for financial support.

References

1. Yokota A. Phylogenetic relationship of Actinomycetes. In: Miyadoh S, editor. Atlas of Actinomycetes. Tokyo: Society for Actinomycetes; 1997. p. 194–9.
2. Mahajan GB, Balachandran L. Antibacterial agents from actinomycetes – a review. Front Biosci. 2012;4:240–53.
3. Berdy J. Bioactive microbial metabolites: a personal view. J Antibiot. 2005;58(1):1–26.
4. Challis GL, Hopwood DA. Synergy and contingency as driving forces for the evolution of multiple secondary metabolite production by *Streptomyces* species. Proc Natl Acad Sci. 2003;100 Suppl 2:14555–61.
5. Watve MG, Tickoo R, Jog MM, Bhole BD. How many antibiotics are produced by the genus Streptomyces? Arch Microbiol. 2001;176(5):386–90.
6. Nayar MP. Hot spots of endemic plants of India, Nepal and Bhutan. Thiruvananthapuram: TBGRI; 1996. p. 254.
7. Williams ST, Davies FL, Mayfield CI, Khan MR. Studies on the ecology of the Actinomycetes in soil II: the pH requirements of *Streptomyces* from two acid soils. Soil Biol Biochem. 1971;3:187–95.
8. El-nakeeb MA, Lechevalier HA. Selective isolation of aerobic Actinomycetes. Appl Microbiol. 1963;11:75–7.
9. Atlus RM, Perks LC. Hand book of microbiological media. London: CRC Press; 1993. p. 850.
10. Shirling EB, Gottlieb D. Methods for the characterization of *Streptomyces* species. Int J Syst Bacteriol. 1966;16:313–40.
11. Holt JG, Krieg NR, Sneath PHA, Staley JT, Williams ST. Bergey's manual of determinative bacteriology. 9th ed. Baltimore: William and Wilkins; 1994.
12. Chakrabarti T. Actinomycetes – isolation, screening, identification & gene cloning in streptomyces – laboratory manual. Chandigarh: Institute of Microbial Technology; 1998. p. 94.
13. Murray MG, Thompson WF. Rapid isolation of high molecular weight plant DNA. Nucleic Acids Res. 1980;8:4321–5.
14. Thompson JD, Higgins DG, Gibson TJ. CLUSTAL W: improving the sensitivity of progressive multiple sequence alignment through sequence weighting, position-specific gap penalties and weight matrix choice. Nucleic Acids Res. 1994;22:4673–80.
15. Jukes TH, Cantor CR. In: Munro HN, editor. Mammalian protein metabolism. New York: Academic; 1969. p. 21–123.
16. Saitou N, Nei M. The neighbour-joining method: a new method for reconstructing phylogenetic trees. Mol Biol Evol. 1987;4:406–25.
17. Felsenstein J. Confidence limits on phylogenies: an approach using the bootstrap. Evolution. 1985;39:783–91.
18. Van de Peer Y, De Wachter R. TREECON: a software package for the construction and drawing of evolutionary trees. Comput Appl Biosci. 1993;9:177–82.
19. Ouhdouch Y, Barakate M, Finanse C. Actinomycetes of Moroccan habitats: isolation and screening for antifungal activities. Eur J Soil Biol. 2001;37: 69–74.
20. Lee JY, Hwang BK. Diversity of antifungal Actinomycetes in various vegetative soils of Korea. Can J Microbiol. 2002;48:407–17.
21. Rahman MA, Islam MZ, Ul Islam MA. Antibacterial activities of Actinomycete isolates collected from soils of Rajshahi, Bangladesh. Biotechnol Res Int. 2011;2011:1–6.
22. Williams ST, Wellington EMH. Actinomycetes. In: Page AL, Miller RH, Keeney DR, editors. Methods of soil analysis part 2. Madison: American Society for Agronomy; 1982. p. 969–87.
23. Williams ST, Lanning S, Wellington EMH. Ecology of actinomycetes. In: Goodfellow M, Mordadarski M, Williams ST, editors. The biology of actinomycetes. London: Academic; 1984. p. 481–528.
24. Woese CR. Bacterial evolution. Microbiol Rev. 1987;51:221–71.
25. Embley TM, Stackebrandt E. The molecular phylogeny and systematics of the Actinomycetes. Annu Rev Microbiol. 1994;48:257–89.

Binding of Apolipoprotein A-1 to Jacalin-Sepharose in a Sugar-Dependent Manner

20

P. Priya and V.M. Kannan

Abstract

Lectins are carbohydrate-binding and cell-agglutinating proteins of nonimmune origin. Lectins are used to isolate glycoproteins and follow alterations in glycopart of glycoconjugates in pathological conditions such as cancer. In this study we isolated and characterized proteins from bovine and goat serum that sugar-specifically bound to jackfruit seed agglutinin (Jacalin). Jacalin is an α-galactose-specific lectin from the seeds of jackfruit (*Artocarpus heterophyllus*). Jacalin coupled to Sepharose (Jacalin-Sepharose) was packed to a volume of 0.5 ml in a Pasteur pipette. Serum (3 ml) was loaded on to the column and washed with phosphate buffered saline (PBS), pH 6.5 containing 0.15 M NaCl, (PBS 6.5) and bound proteins eluted with 0.15 M D-galactose in PBS 6.5. Isolated proteins were separated by SDS-PAGE and bands cut out and proteins identified by mass spectrometry. Six protein bands were visualized when the proteins isolated from bovine serum were subjected to SDS-PAGE. Out of these, four major bands were identified by mass spectrometry, and they were found to be inter-alpha-trypsin inhibitor heavy chain H4 precursor (104.7 kDa), factor XIIa inhibitor precursor (79.43 kDa), alpha-2-HS-glycoprotein precursor (Fetuin-A) (66.07 kDa) and apolipoprotein A-I precursor (Apo-AI) (28 kDa). From goat serum by affinity chromatography on Jacalin-Sepharose column, we obtained seven protein bands (125.89, 117.49, 104.71, 97.72, 83.18, 66.07 and 27 kDa). Some of these bands (117.49, 104.71, 66.07 and 27 kDa) are having the similar molecular weight counterparts in bovine serum. It is intriguing that one of the proteins identified (Apo-A1) is a non-glycosylated protein. Apo-A1 is a protein component of high-density lipoprotein (HDL). Since direct recognition of Apo-A1 by Jacalin has been ruled out, it is possible that glycoconjugates associated with Apo-A1 may mediate its binding to the lectin. Such glycoconjugates reported in the literature include lipopolysaccharides and transferrin. Further investigations on this phenomenon are ongoing.

P. Priya • V.M. Kannan (✉)
Department of Zoology, University of Calicut,
Thenjipalam, Kerala, India
e-mail: kannanvadakkadath@gmail.com

A. Sabu and A. Augustine (eds.), *Prospects in Bioscience: Addressing the Issues*,
DOI 10.1007/978-81-322-0810-5_20, © Springer India 2013

Keywords
Apolipoprotein-A1 • Jacalin • Bovine serum • Goat serum

Introduction

Lectins are sugar-binding proteins of nonimmune origin that agglutinates cells or precipitates glycoconjugates [1]. Jackfruit seed agglutinin (Jacalin) is a α- D-galactoside-specific lectin isolated from jackfruit seed (*Artocarpus heterophyllus*) [2]. Jacalin is a tetrameric two-chain lectin (molecular mass 65 kDa) combining a heavy α chain of 133 amino acid residues with a light β-chain of 20–21 amino acid residues. It is highly specific for the α-O-glycoside of the disaccharide Thomsen-Friedenreich antigen (Gal β1-3 Gal NAc), even in its sialylated form. Jacalin has been used as a valuable tool for specific capturing of O-glycoproteins such as mucins and IgA1 [3]. Jacalin showed high affinity for T-antigen (Core-1) and Tn-antigen (alpha N-acetylgalactosamine)-attached peptides. It has wide application in immunological research and used as a tool in cancer research and in purifying α-D-galactoside-containing glycoconjugates.

Generally, glycoproteins bind to lectin column, but some non-glycosylated proteins also bind to lectin columns. Apolipoprotein A-1 (Apo-A1) is a non-glycosylated protein. Apo-A1 is the major protein component of HDL lipoprotein and represents ~1% of the total protein content of plasma. It is a single polypeptide chain with a molecular weight of 28 kDa [4]. Apo-A1 is synthesized in liver and small intestine. Intestinal enterocyte secretes chylomicrons which also contains Apo-A1 but is quickly transferred to HDL in blood stream [5]. Apolipoprotein A1 has specific role in lipid metabolism, where it participates in reverse transport of cholesterol from tissues to the liver for excretion by promoting cholesterol efflux from tissues and by acting as a cofactor for lecithin cholesterol acyltransferase (LCAT) which forms cholesteryl esters in HDL. Apo-A1 Milano is a naturally occurring mutant of Apo-A1, found in a family descended from a single couple of the eighteenth century. Paradoxically, carriers of this mutation have very low HDL cholesterol levels, but no increase in the risk of heart disease. Defects in Apo-A1 cause coronary artery disease (CAD), Tangier disease, amyloidosis, etc. [6].

Materials and Methods

Bovine and Goat blood were collected from local slaughter houses. Blood was allowed to clot. Serum was collected by centrifugation at 2,000 rpm for 10 min in a clinical centrifuge. The collected serum was stored at −20°C until use. Jacalin-Sepharose was a generous gift from the Department of Biochemistry, Sree Chitra Tirunal Institute for Medical Sciences and Technology, Thiruvananthapuram. D-galactose was obtained from Sigma Chemicals, St. Louis, USA. All other reagents used were of analytical grade.

Isolation of Jacalin-Binding Proteins from Serum

Isolation of Jacalin-binding proteins from serum was done by affinity chromatography on Jacalin-Sepharose. Jacalin-Sepharose was packed in a Pasteur pipette to a volume of 0.5 ml. The column was washed (5 column volumes) with phosphate buffered saline, pH 6.5 containing 150 mM NaCl (PBS 6.5). Three-millilitre serum was loaded onto the column slowly, and then washed the column with PBS 6.5 until the flow through becomes protein free. The bound proteins were eluted with PBS 6.5 containing 0.15 M D-galactose. Two hundred and fifty microlitre fractions were collected and checked for the presence of protein using Bradford's reagent. Protein-containing fractions were pooled or individual fractions were used for electrophoresis. The column was regenerated by washing with 5 column volumes of 2 M NaCl followed by 10 column volumes of PBS 6.5.

Separation of Jacalin-Binding Proteins

Protein-containing fractions were pooled and bound proteins separated by 10% sodium dodecyl sulphate-polyacrylamide gel electrophoresis (SDS-PAGE) [7].

Determination of Molecular Weight of Jacalin-Binding Proteins

Molecular weights of the Jacalin-binding protein bands were determined from SDS-PAGE. For this log, molecular weight versus relative mobility of the molecular weight standards were plotted, and from the mobility of Jacalin-binding proteins, their molecular weight was calculated.

Identification of Proteins

After SDS-PAGE, gels were stained with colloidal Coomassie dye without fixing, and protein bands were sliced out and identified by mass spectrometry at University of Munster, Germany.

Results and Discussion

Identification of Jackfruit Seed Agglutinin-Binding Proteins from Bovine Serum

Affinity chromatography of bovine serum on Jacalin-Sepharose column yielded six polypeptides on SDS-PAGE with molecular weights 117.49, 104.71, 79.43, 66.07, 35.48 and 28 kDa (Table 20.1, Fig. 20.1a, b). On mass spectrometric identification, they were found to be inter-alpha-trypsin inhibitor heavy chain H4 precursor (104.7 kDa), factor XIIa inhibitor precursor (79.43 kDa), alpha-2-HS-glycoprotein precursor (Fetuin-A) (66.07 kDa) and apolipoprotein A-I precursor (Apo-AI) (28 kDa). Of these, Fetuin-A was the most abundant protein band.

Interestingly, one of the proteins (Apo-A1) bound to Jacalin-Sepharose is a non-glycosylated protein. Apo-A1 is a protein component present in high-density lipoprotein (HDL). Since direct recognition of Apo-A1 by Jacalin has been ruled out (2011) [8], it is possible that glycoconjugates associated with Apo-A1 may mediate its binding to the lectin. Such glycoconjugates reported in the literature include lipopolysaccharides and transferring [9]. Further investigations on this phenomenon are ongoing.

Jacalin-Binding Proteins from Goat Serum

Seven protein bands were observed on SDS-PAGE of Jacalin-Sepharose-binding proteins from goat serum (125.89, 117.49, 104.71, 97.72, 83.18 and 66.07 kDa and 27 kDa). Of these, four are similar in molecular weight to that of proteins from bovine serum (117.49, 104.71, 66.07 and 28 kDa), and these may represent conserved proteins in two related organisms.

Conclusions

Though Apo-A1 is a non-glycosylated protein, it is possible that glycoconjugates associated with Apo-A1 may mediate its binding to the lectin. Four of the proteins from bovine and goat serum are similar in molecular weight which may

Table 20.1 Four major Jacalin-Sepharose-binding proteins from bovine serum identified by mass spectrometry

Protein	Molecular weight calculated from SDS-PAGE (kDa)	Molecular weight reported from mass spectrometry data (kDa)
(1) Inter-alpha-trypsin inhibitor heavy chain H4 precursor	104.71	102.08
(2) Factor XIIA inhibitor precursor	79.43	51.69
(3) Alpha-2-HS-glycoprotein precursor	66.07	38.39
(4) Apolipoprotein A-1 precursor	28	30.26

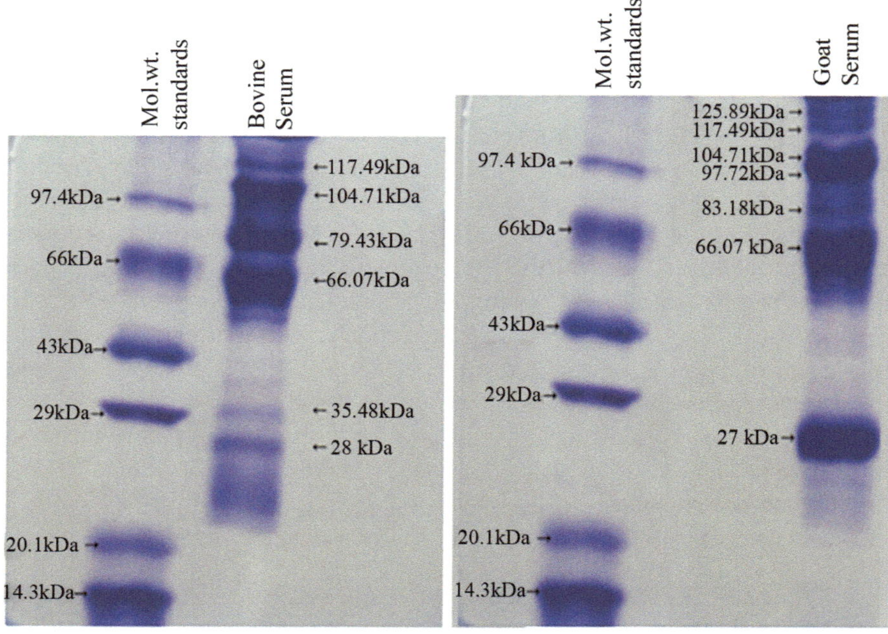

Fig. 20.1 Polyacrylamide gel (10%) electrophoresis of Jacalin-binding proteins from bovine serum (**a**) and goat serum (**b**). Twenty microlitres of eluate loaded in each well and gel-stained with Coomassie brilliant blue stain

represent similar proteins conserved in two related organisms. This protocol making use of lectin affinity chromatography in combination with mass spectrometry will be helpful in identifying proteins indirectly binding to affinity columns. Also, this protocol is useful in identifying altered proteins in pathological conditions like cancer and thereby helping in designing better diagnostic/prognostic or treatment strategies.

Acknowledgements Thanks to Dr. N. Ramani (Head, Dept. of Zoology, University of Calicut), Dr. P. S. Appukuttan (Sree Chitra Tirunal Institute for Medical Sciences and Technology, Trivandrum) for their support and Dr. M. Krishnan (Dept. of Environmental Biotechnology, Bharathidasan University, Trichi) for mass spectrometry analysis.

References

1. Goldstein IJ, Hayes CE. The Lectins: carbohydrate-binding proteins of plants and animals. Adv Carbohydr Chem Biochem. 1978;35:127–340.
2. Kumar SG, Appukuttan PS, Basu D. α-D-galactose-specific lectin from jackfruit (*Artocarpus integrifolia*) Seed. J Biosci. 1982;4:257–61.
3. Tachibana K, Nakamura S, Klang H, Iwasaki H, Tachibanak MK, Cheng L, Hirabayashi J, Narimatsu H. Elucidation of binding specificity of Jacalin towards O-glycosylated peptides: quantitative analysis by frontal affinity chromatography. Glycobiology. 2006;16(1):46–53.
4. Davidson WS, Thompson CB. The structure of apolipoprotein A-I in high density lipoproteins. J Biol Chem. 2007;282:22249–53.
5. Wasan KM, Brocks DR, Lee SD, Sachs-Barrable K, Thornton SJ. Impact of lipoproteins on the biological activity and disposition of hydrophobic drugs: implications for drug discovery. Nat Rev Drug Discov. 2008;7(1):84–99.
6. Baker HN, Gotto Jr AM, Jackson RL. The primary structure of human plasma high density apolipoprotein glutamine(ApoA-1). J Biol Chem. 1975;250:2725–38.
7. Laemmli UK. Cleavage of structural proteins during the assembly of the head of bacteriophage T4. Nature. 1970;227:680–5.
8. Appukuttan PS. Department of Biochemistry, Sree Chitra Tirunal Institute for Medical Sciences and Technology, Trivandrum, India. 2011
9. Carol LN. Lectins: analytical technologies. Oxford: Elsevier Science and Technology Books; 2007. p. 19. Chapter 8.

21 Morphological Basis for Identification of Cup Quality Characteristics in F_1 Hybrids Derived from *Coffea arabica* L. Crosses

Anil Kumar, S. Ganesh, K. Basavraj, M.K. Mishra, and Jayarama

Abstract

The coffee quality, either of arabica (*C. arabica* L. var. arabica) or robusta (*C. canephora* Frohener. var. robusta) is an outcome of genetic behaviours performed under various environmental conditions and post-harvest processing techniques. The combination of these factors in an appropriate proportion produces the coffee quality to the degree of merit. Therefore, to study the genotypic influence on beverage quality and its relationship with morphological traits, 15 F_1 arabica hybrids were experimented at Coffee Research Sub Station, Chettalli, Kodagu District, Karnataka, India. The morphological parameters were recorded and correlated with the coffee quality parameters assessed by the Coffee Board, Quality Control Division, Bangalore. Results indicated the superiority in cup quality of an interspecific hybrid progeny of an arabica variety 'Cauvery' (Catimor) × (CxR) a robusta cultivar among all the hybrids and graded as 'specialty coffee' by the tasters. Sln.5B × Sln.9 scored the second place in quality rating parameters. Though, the other crosses such as Sln.6 × Cauvery, Tafarikela × S.4179 (Catimor progeny) and Sln.5B × S.795 had good liquor aroma, body acidity

A. Kumar (✉)
Faculty of Agriculture & A.H., G.R.U.,
Gandhigram, Dindigul District, Tamil Nadu, India
e-mail: anilsirsi@yahoo.com

Coffee Board, RCRS, Narsipatnam, Visakhapatnam District, Andhra Pradesh 531 116, India

S. Ganesh
Faculty of Agriculture & A.H., G.R.U.,
Gandhigram, Dindigul District, Tamil Nadu, India

K. Basavraj
Coffee Board, No.1, Dr. Ambedkar Veedhi, Bangalore, Karnataka, India

M.K. Mishra • Jayarama
CCRI, Coffee Research Station Post, Chikmagalur District, Karnataka 577117, India

A. Sabu and A. Augustine (eds.), *Prospects in Bioscience: Addressing the Issues*,
DOI 10.1007/978-81-322-0810-5_21,

and other components but indicated low flavour intensity. The progeny of Sln.9×Sln.11 recorded the lowest rank among the F_1 progenies probably due to introgression of diploid genes from liberica coffee being one of the parents in evolution of Sln.11. An interesting finding revealed that the plants with thick stem produced low flavour, and the thick primary induced low acidity in the cup. Besides this, internodal length, leaf length and breadth and size of fruit cluster exhibited positive correlation with other quality parameters like mouthfeel, aftertaste, flavour and overall as well as with 100 bean weight. The F_1 hybrids derived from the parents with robusta genetic background expressed lower acidity in the liquor. The findings of the present study revealed that the coffee plant with its inherent quality characteristics could easily be identified based on its phenotypic parameters such as stem girth, thickness of the primary branches, internodal length, leaf size and tightness of the fruit cluster. Further, the study also widened the scope to use these traits as quality markers to produce coffee to the consumers' choice through genetically blended arabica and robusta varieties rather than physical blends of beans of both the varieties.

Keywords
Coffee quality • Morphological traits • Hybrid progenies • Correlation

Introduction

Coffea arabica and *Coffea canephora* var. robusta are two commercially cultivated species in the several coffee growing countries, and arabica is well known for its superior beverage quality over robusta, while robusta has gained more popularity due to its tolerance to major diseases and insect pest like coffee leaf rust, CBD, white stem borer, and nematodes. This behaviour of robusta prompted the coffee scientists to integrate these robusta genes into arabica cultivars through interspecific hybridization [1]. Undoubtedly, considerable quantum of success was achieved and several resistant arabica cultivars, namely, 'Sln.6', 'Sln.5' (in India), 'Icatu' (in Brazil) and 'Columbia' (in Columbia), and some others were evolved [2, 3]. Simultaneously, deterioration in beverage quality was also realized in arabica-robusta hybrids [4]. The liquor prepared out of arabica beans possesses all the characteristics of a high-quality coffee that is globally acceptable, while the other species like *canephora*, *congensis*, *liberica* and the species of low economic importance could not overtake arabica in its quality performance. Various species contain various levels of chemical constituents in the coffee beans. Among these, caffeine content in the beans plays a crucial role in quality determination [5]. Apart from this, bean density equally influences the cup quality of the particular variety [6]. Since cup quality is dependent on genetic and environmental factors and the interaction of both [7, 8], the environmental factors such as climate, soil type, altitude, agronomic practices, post-harvest processing and roasting temperature are undoubtedly responsible for production of quality coffee [9, 10]. A desirable genotype with inherent quality characteristics and favourable environment [11] and role of certain precursors responsible for formation of aroma compounds upon roasting at ambient temperatures [10] collectively upgrade the quality standards. Several works carried out on quality aspects indicate that the desirable taste of a coffee variety is its inherent trait that forms the quality after passing through the various circumstances commonly known as environmental conditions.

Roche [8] observed that bean size was not the indication of a good cup quality and found no strong relationship between bean size and cup quality. His finding is against the belief that larger the bean size and superior the quality of a cup. Present study was undertaken to evaluate the F_1 progenies of various arabica crosses for beans and cup characteristics and its relationship with morphological characters. The influence of seasonal variation and involvement of different parent cultivars in the crosses on cup uniqueness was also studied.

Materials and Methods

The study was carried out during 2008–2009 and 2009–2010 on 15 F_1 progenies derived from Dwarf × Tall, Tall × Dwarf and Tall × Tall crosses established during 1997–1998 at Coffee Research Sub Station, Chettalli, Kodagu District, Karnataka, India, with an objective to relate the morphological traits with bean density and cup quality parameters. This research station is located at an elevation of about 3,000 ft receiving an average annual rainfall of 60 in., temperature range of 10–30°C and 40–90% relative humidity (RH).

Data on morphological characters such as bush spread, stem girth, primary thickness, internodal length, leaf length and breadth and number of fruits per cluster were recorded in four replications using seven numbers of plants per replication. Similarly, fruit length and breadth were also observed. Three samples were drawn from a lot of each progeny to determine 100 bean weight of 'A' grade size. Experiment was conducted following RBD and analysed statistically. The coffee samples representing all 15 F_1 populations were prepared through wet processing method, and cup quality was evaluated by the Quality Control Division, Coffee Board, Bangalore. The marks awarded for each quality parameters, namely, aroma/fragrance, body, acidity, mouthfeel, aftertaste, flavour and overall, were subjected to correlation analysis. All the F_1 crosses were grouped into three classes, namely, Dwarf × Tall, Tall × Dwarf and Tall × Tall, and each class possesses five cross combinations. The results obtained are presented in tabular form separately for each group of five crosses. Similarly, the results on cup quality and its relation with morphological characters were exhibited graphically.

Results and Discussion

Among the five crosses of Dwarf × Tall cultivars, Cauvery × (CxR) exhibited the highest score of 7 points for aroma/fragrance and 6.5 for other parameters such as body, mouthfeel, aftertaste, flavour and overall. While acidity was highest in Cauvery × Tafarikela cross, this has got a score of 6.5 points followed by 6.0 points in Cauvery × Devamachy progeny. Cauvery × (CxR) progeny scored 5.5 points for acidity that was on par with the points awarded to Cauvery × S.881 (Rume Sudan arabica) and Cauvery × Sln.9 (Table 21.1). Cup quality assessment for two seasons indicated high consistency only in liquor aroma/fragrance in Cauvery × Devamachy, but not in other characters. Besides this, Cauvery × (CxR) progenies also showed consistency in aroma, body, mouthfeel, aftertaste and overall, but acidy and flavour were not found consistent. Though, Cauvery × Tafarikela cross expressed steadiness in aroma, but other characters varied with the seasonal variations. The expressions of stability in body, acidity, mouthfeel, aftertaste and overall were noticed in F_1 generation of Cauvery × S.881cross combination. Similar

Table 21.1 Cup quality ratings in F_1 hybrids of Dwarf × Tall crosses (mean of 2 years)

Sl. no.	Dwarf × Tall crosses	Aroma/fragrance	Body	Acidity	Mouthfeel	Aftertaste	Flavour	Overall
1	Cauvery × Sln.9	5.8	5.8	5.5	5.3	5.3	5.3	5.3
2	Cauvery × S.881	6.3	6.0	5.5	5.0	5.0	5.3	5.0
3	Cauvery × Devamachy	6.5	5.8	6.0	5.3	5.3	5.3	5.3
4	Cauvery × Tafarikela	6.5	6.0	6.5	5.5	5.5	5.3	5.5
5	Cauvery × (CxR)	7.0	6.5	5.5	6.5	6.5	6.0	6.5

Table 21.2 Cup quality ratings in F_1 hybrids of Tall × Dwarf crosses (mean of 2 years)

Sl. no.	Tall × Dwarf crosses	Aroma/ fragrance	Body	Acidity	Mouthfeel	Aftertaste	Flavour	Overall
1	Sln.9 × Cauvery	6.0	6.0	6.0	5.5	5.5	5.0	5.5
2	S.881 × Cauvery	6.3	6.0	5.8	5.3	5.3	5.0	5.3
3	Sln.6 × Cauvery	6.3	6.0	6.0	5.8	5.8	5.5	5.8
4	Sln.5B × Cauvery	6.5	6.0	6.3	6.0	6.0	5.8	5.8
5	Tafarikela × S.4179	6.5	6.0	6.0	6.0	6.0	5.5	6.0

Table 21.3 Cup quality ratings in F_1 hybrids of Tall × Tall crosses (mean of 2 years)

Sl. no.	Tall × Tall crosses	Aroma/ fragrance	Body	Acidity	Mouthfeel	Aftertaste	Flavour	Overall
1	Sln.9 × Sln.11	6.3	5.5	5.8	5.0	5.0	5.0	5.0
2	Sln.5B × Sln.9	6.4	6.3	6.0	5.8	6.0	5.8	6.5
3	Sln5B × S.795	6.5	6.0	6.0	5.9	5.8	5.5	5.9
4	Sln.6 × S.795	6.5	6.0	5.8	5.9	5.8	5.5	5.9
5	Sln.6 × Sln.9	6.3	6.0	6.0	5.8	5.5	5.3	5.5

trend was found in Cauvery × Sln.9 progeny also. Although both the crosses scored 5.5 (above average) points for acidity, they had consistency. This trend of regularity in cup characteristics indicated the genotypic influence, while variability in score pointed out environmental influence on coffee beans.

In the second group of crosses between Tall × Dwarf, Tafarikela × S.4179 and Sln.5B × Cauvery expressed highest score of 6.5 for aroma, while liquor body was same in all the crosses. Sln.5B × Cauvery recorded the higher level of acidity with the score of 6.3 points, whereas mouthfeel and aftertaste was on par with Tafarikela × S.4179 progeny which achieved 6.0 (good) points. Subsequently, superiority of Sln.5B × Cauvery was noticed over other crosses for flavour. The overall rating was highest in Tafarikela × S.4179 cross combination (Table 21.2). As a whole, the performance on cup quality and consistency was superior in Tafarikela × S.4179 over other crosses in this group. However, S.881 × Cauvery, Sln.9 × Cauvery and Sln.5B × Cauvery progenies exhibited balanced cup without much seasonal variations in the cup characteristics except Sln.6 × Cauvery (Table 21.2). Although, flavour component was weak (5.5) but above average in Tafarikela × S.4179 hybrid as compared to the other components of this hybrid, yet it showed higher degree of seasonal stability with higher level of aroma as compared to the remaining crosses of the same group. This behaviour of Tafarikela × S.4179 indicated the intrinsic character of the genotype (Table 21.2). Among all the crosses of Tall × Dwarf, Sln.6 × Cauvery was observed to be highly influenced by the environmental variations.

Tall × Tall cross combinations projected the higher aroma of the liquor in Sln.5B × S.795 and Sln.6 × S.795 along with higher liquor body, acidity, aftertaste and overall scores in Sln.5B × Sln.9 progeny (Table 21.3). Out of 15 genotypes, 8 showed consistency for aroma, while 11 genotypes expressed consistency for body. Similarly, regularity in liquor acidity was observed in seven progenies. Mouthfeel and aftertaste exhibited the stability in four progenies. Steadiness in flavour and overall components was noticed in six and five crosses, respectively. This tendency of evenness in cup quality parameters in more than 50% of the genotypes might be mainly due to the inherent characters of progenies that possess resistance to the environmental changes. In the other cases, where cup components showed stability in few genotypes would be because of their environmental susceptibility. Walyaro [12]

Table 21.4 Correlation of morphological traits with cup characteristics in Dwarf × Tall crosses

Characters	Stem girth (cm)	Primary's thickness (cm)	Internodal length on primary (cm)	Leaf length (cm)	Leaf breadth (cm)	No. of fruit/ cluster	Fruit breadth (cm)	100 fruit wt. (gm)	100 bean wt. (gm)
Aroma/ fragrance	−0.104	−0.152	0.734	0.598	0.568	0.393	0.429	0.727	0.631
Body	−0.234	0.430	0.721	0.71	0.487	0.74	0.172	0.751	0.767
Acidity	−0.394	**−0.920****	0.374	0.297	0.472	−0.220	**0.822***	0.45	0.342
Mouthfeel	−0.549	0.184	**0.838***	**0.925****	**0.817***	**0.943****	0.216	0.534	0.759
Aftertaste	−0.549	0.184	**0.838***	**0.925****	**0.817***	**0.943****	0.216	0.534	0.759
Flavour	−0.248	0.394	**0.818***	**0.834***	0.723	**0.999****	−0.065	0.398	0.614
Overall	−0.549	0.184	**0.838***	**0.925****	**0.817***	**0.943****	0.216	0.534	0.758
100 bean wt. (gm)	−0.783	−0.037	**0.839***	**0.884***	0.707	0.630	0.695	**0.953****	–

also observed considerable (additive) genetic variation in most of the quality components that were influenced by environmental factors also.

It was interesting to note that during the process of cup evaluation, the F_1 progeny of Cauvery × (CxR) drew the special attention of the cup taster due to its consistent striking typical aroma/fragrance in both the seasons and recorded highest score of 7.0 (very good) among all the 15 progenies rated for cup quality. It showed good body (6.5) and slightly lower acidity (5.5) than the other samples of the same group of crosses; however, the acidity level was above average and on par with the acidity level observed in the crosses like Cauvery × Sln.9 and Cauvery × S.881 (Table 21.1). Low acidity in Cauvery × (CxR) was thought to be due to introgression of robusta genes as this variety possesses inherent characteristics of low acidity in the liquor. Moreno et al. [2] reported lowest acidity, body and flavour in *C. canephora* species as compared to the arabica varieties and their progenies. Similar level of acidity was reported by Clifford [13] also.

Correlation Between Cup Characteristics and Morphological Traits

Correlation between the plant components plays a major role in determining the presence or absence of the characters and their behaviours that are related to each other. Some of these are directly related and some indirectly and accordingly one influences the behaviour of other. Hence, an attempt was made to track those quality characters in coffee cultivars that are governed by the morphological components. The present study was aimed at identification of quantitative traits directly related to inherent quality characters. The present study revealed that the phenotypic components such as stem girth, primary thickness, internodal length, leaf length and breadth are highly correlated to cup quality characteristics and influence the beverage performance. Simultaneously, yield components like number of fruits/cluster, fruit length and breadth, 100 fruit weight and 100 bean weight are also an indicator of plants with superior quality.

Correlation between the characters differs from genotype to genotype because of their phenotypic variations. Leaf length and breadth showed positive and strong correlation with 100 bean weight and cup characteristics such as mouthfeel, aftertaste, flavour and overall in Dwarf × Tall crosses. Internodal length of primary shoots also expressed same trend. A significant negative correlation of primary thickness with liquor acidity was noticed. Stem girth also indicated similar tendency of negative correlation with cup quality parameters and bean density, but it was not significant (Table 21.4). This kind of relationship with stem girth and primary thickness indicated that the plants with higher primary thickness and stem girth possessed inherent characteristic of low acidity as this phenomenon has

Table 21.5 Correlation of morphological traits with cup characteristics in Tall × Dwarf crosses

Characters	Stem girth (cm)	Primary's thickness (cm)	Internodal length on Primary (cm)	Leaf length (cm)	Leaf breadth (cm)	Fruit length (cm)	Fruit breadth (cm)	100 fruit wt. (gm)	100 bean wt. (gm)
Aroma/ fragrance	−0.754	0.671	**0.960****	−0.328	−0.256	0.582	0.755	0.562	−0.386
Body	−0.755	−0.166	0.484	−0.171	0.721	0.023	0.167	0.677	−0.345
Acidity	0.605	**−0.915****	−0.668	**0.866***	0.449	−0.663	−0.712	−0.651	0.775
Mouthfeel	−0.723	0.085	0.723	0.120	0.249	−0.123	0.101	0.335	−0.199
Aftertaste	−0.723	0.085	0.723	0.120	0.249	−0.123	0.101	0.335	−0.199
Flavour	**−0.961****	0.620	**0.836***	−0.636	−0.003	0.116	0.295	0.710	**−0.872***
Overall	−0.324	−0.208	0.335	0.508	0.203	−0.600	−0.412	−0.164	0.101
100 bean wt. (gm)	0.779	−0.663	−0.565	**0.893***	0.059	−0.183	−0.264	−0.724	–

been reported in robusta varieties, because robusta is characterized by its robust stem growth and wider bush spread [13]. This finding is in accordance with the results reported by Moreno et al. [2], where the variety Caturra characterized by its thin primary branches had higher acidity, body and flavour as compared to the tall varieties like Typica. Besides this, the Caturra × H-de-T progenies achieved lower score points than pure Caturra. Perhaps, because of improvement in the stem thickness in the progeny caused by the involvement of H-de-T as one of the parents, acidity in the liquor was declined.

One hundred fruit and bean weight were observed to have positive but insignificant influence on all the quality parameters. Whereas, correlation between bean weight and fruit weight was positive and significant. This phenomenon indicated that bean weight was highly influenced by the fruit weight in Dwarf × Tall crosses (Table 21.4). Number of fruits/cluster showed positive and strong relationship with mouthfeel, aftertaste, flavour and overall parameters besides negative but weak relation with acidity. This kind of negative correlation of number of fruits with acidity was in accordance with robusta behaviour which possesses more number of fruits/cluster and low acidity in the liquor. There was an indication of positive correlation between 100 bean weight and quality components in Dwarf × Tall crosses (Table 21.4), but it was not significant.

It showed positive but weak correlation of stem thickness with acidity while negative and strong correlation with flavour besides weak and negative linkage with aroma, body, mouthfeel, aftertaste and overall and 100 bean weight. Primary thickness indicated strong negative linkage with 100 bean weight. Though internodal length of primary shoot exhibited negative and nonsignificant influence on 100 bean weight and acidity, but it had positive and strong effect on aroma and flavour. As the long internodes and thick primaries are the common features of robusta varieties that produce low acidity and thick liquor body [13], a similar trend was apparent in the present study with arabica crosses too. This behaviour of the hybrid progenies indicated the introgression of robusta genes transmitted through the parent cultivars because some of the inbred lines such as Sln.5B, Sln.6, Devamachy and CxR used in hybridization are closely related to *C. canephora* species, while Cauvery and Sln.9 are associated distantly with

H-de-T (a natural hybrid of arabica and robusta) as one of the progenitors. In Tall × Dwarf crosses, the plants with higher leaf length showed higher level of acidity and more bean weight because of their positive and significant association (Table 21.5). Simultaneously, with the increase in the leaf length, a reduction in the flavour was also noticed, though the reduction was not significant. Likewise, leaf breadth had weak and positive correlation with liquor body.

Table 21.6 Correlation of morphological traits with cup characteristics in Tall × Tall crosses

Characters	Stem girth	Primary's thickness	Leaf length (cm)	Leaf breadth (cm)	No. of fruit/ cluster	Fruit breadth (cm)	100 fruit wt. (gm)	100 bean wt. (gm)	Fruit yield
Aroma/ fragrance	0.132	−0.322	−0.287	−0.309	−0.007	**0.851***	0.583	0.612	−0.300
Body	**0.821***	**0.533**	−0.021	0.434	−0.012	0.231	**0.985****	0.472	**−0.80***
Acidity	0.524	0.475	0.772	0.825*	−0.722	0.096	0.729	0.014	−0.912**
Mouthfeel	0.463	0.248	−0.277	−0.008	0.445	0.049	0.738	0.074	−0.581
Aftertaste	0.315	−0.079	−0.319	−0.199	0.218	0.561	0.736	0.423	−0.473
Flavour	0.456	0.036	−0.524	−0.205	0.334	0.544	0.702	0.658	−0.330
Overall	0.373	−0.076	−0.140	−0.019	−0.189	**0.816***	**0.729***	0.738	−0.450
100 bean wt. (gm)	0.449	0.077	−0.454	−0.056	−0.003	0.622	0.376	–	0.030

*Significant at 5% level; ** Significant at 1% level

The weak correlations between the characters might be due to the presence of low dosage of genes pertaining to the characters that caused dilution in the morphological traits and in turn, it affected the related quality characters. Besides this, genetic variation in hybrid population might also be one of the factors to weaken the strength of relationship between morphological and quality characters.

The fruit parameters like fruit length and breadth were positively related to aroma and inversely to acidity and overall ratings. Flavour was inversely and strongly related to bean mass (Table 21.5). Tall × Tall progenies exhibited strong and positive bond of liquor body with stem girth, whereas primary thickness exhibited negative though insignificant relationship with aroma and positive with body and acidity. A nonsignificant correlation of internodal length was observed with aroma and overall scores. Leaf length and breadth, both were observed to have direct positive influence on acidity, but leaf breadth had significant influence. Number of fruits/cluster had relatively insignificant reverse effect on acidity. Aroma and overall beverage components showed positive and significant correlation with fruit breadth, while liquor body with fruit weight (Table 21.6). Variation in the intensity of the correlations in the crosses might be due to variability in the population caused due to frequency of gene association. Correlation between the morphological traits and quality components was observed to have variation from cross to cross. In some of the crosses, characters were closely related and in some distantly. The present findings revealed that the genes for morphological and quality traits of arabica are closer to both the species, namely, *C. canephora and C. eugenioides* as it is confirmed by the findings of Lashermes et al. [14]. As it was reported that arabica exhibited low genetic divergence, but its meiotic behaviour was similar to the diploid species [15–17].

Conclusions

Mean of two seasons of cup quality scores indicated that among the F_1 progenies derived by crossing Dwarf × Tall cultivars, Cauvery × (CxR) produced superior cup quality among all the F_1 hybrids. Cauvery × Tafarikela scored the second position in this group showing somewhat low values for all the cup characteristics and slightly higher acidity than the Cauvery × (CxR) hybrids. Among the Tall × Dwarf crosses, Sln.5B × Cauvery progeny exhibited superior performance for all the quality parameters. The performance of Tafarikela × S.4179 was very similar to the former progeny with slightly low acidity and flavour. Out of five crosses of Tall × Tall cultivars, four expressed almost equal performance for quality behaviours with minimal differences in the numerical scores. The progeny of Sln.9 × Sln.11 had shown comparatively weaker points for the cup characters.

In Tall × Tall crosses, the number of fruits/cluster showed negative correlation with acidity, while stem girth and 100 fruit weight indicated positive correlation with liquor body. Leaf length and breadth and 100 fruit weight showed positive linkage with acidity. Apart from this, there are several other genes responsible for inheritance of other quantitative traits appeared to be independent genes, and their assortment is self-regulatory. The present study indicated significant genetic correlation of morphological traits with the cup quality characteristics that formed the basis for identification of plants of better quality characters and its exploitation on commercial scale.

Acknowledgement Authors are thankful to Dr. Y. Raghuramulu, Coffee Board, Bangalore, and Sri N. Ramamurthy, CRSS, Chettalli, for their valuable suggestions and encouragement. Thanks are also due to Smt. V.S. Amaravenmathy and Dr. T.N. Gopinandan of CRSS, Chettalli, for the help rendered in carrying out the experiments.

References

1. Illy E. How science can help to improve coffee quality. In: ASIC, 17th Colloque, Nairobi; 1997. pp. 29–33.
2. Moreno G, Moreno E, Cadena G. Bean characteristics and cup quality of the Colombia variety (*Coffea arabica*) as judged by International Tasting Panels. In: ASIC, 16th Colloque, Kyoto, Japan; 1995. pp. 574–83.
3. Srinivasan CS, Prakash NS, Padma Jyothi D, Sureshkumar VB, Subbalakshmi V. Genetic improvement of coffee in India. In: Proceedings of the 3rd International Seminar on Biotechnology in the Coffee Agro-industry, Londrina, Brazil; 1999.
4. Van Der Vossen HAM. Agronomy I: coffee breeding practices. In: Clarke RJ, Vitzthum OG, editors. Coffee recent developments. London: Blackwell Science Ltd; 2001. p. 184–201.
5. Sondahl MR, Petracco M, Zambolim L. Breeding for qualitative traits in arabica coffee. In: ASIC, 17th Colloque, Nairobi; 1997. pp. 447–56.
6. Petracco M. Organoleptic properties of Espresso coffee as influenced by coffee botanical varieties. Coffee Biotechnology and Quality. In: Proceedings of International Conference on Coffee Science; 2000. pp. 347–553.
7. Mawardi S, Hulupi R. Genotype-by-environment interaction of bean characteristics in arabica coffee. In: Proceedings of the 15th ASIC Colloquium. Paris: ASIC; 1995. pp. 637–44.
8. Roche D. Coffee genetics and quality. In: ASIC, 16th Colloque, Kyoto, Japan; 1995. pp. 584–8.
9. Wintgens JN. Coffee: growing, processing, sustainable production – a guidebook for growers, processers, traders and researchers. Weinheim: WILEY-VCH Verlag GmbH & Co. KGaA; 2004. p. 3–975.
10. Guilmineau F, Adisson L, Silanes-kenny J, Fernandes J. Study of the thermal generation of aroma compounds in coffee using an in-bean model system. In: 22nd Proceedings of International Scientific Conference on Coffee Science; 2008.
11. Cortez JG, Menezez HC. Recent development in Brazilian coffee quality: new processing system, beverage characteristics and consumers preferences. Coffee Biotechnology and Quality. In: Proceedings of International Conference on Coffee Science; 2000. pp. 339–46.
12. Walyaro DJ. Breeding for disease and pest resistance and improved quality in coffee. In Proceedings of 17th ASIC Colloquium, Nairobi; 1997. pp. 397–405.
13. Clifford MN. Agronomy I: coffee breeding practices. In: Clifford MN, Wilson KC, editors. Coffee botany, biochemistry and production of beans and beverage. Westport (EE.UU.): The AVI Publishing Company, Inc; 1985. p. 184–201.
14. Lashermes P, Combes MC, Robert J, Trouslot P, D'hont A, Anthony F. Molecular characterization and origin of the *Coffea arabica* L. genome. Mol Gen Genet. 1999;261:259–66.
15. Krug CA, Mendes AJT. Cytological observations in *Coffea* – IV. J Genet. 1940;39:189–203.
16. Lashermes P, Paczek V, Trouslot P, Combes MC, Couturon E, Charrier A. Single-locus inheritance in the allotetraploid *Coffea arabica* L. and interspecific hybrid *C. arabica* x *C. canephora*. J Hered. 2000;91:81–5.
17. Lashermes P, Andrzejewski S, Bertrand B, Combes MC, Dussert S, Graziosi G. Molecular analysis of introgressive breeding in coffee (*Coffea arabica* L.). Theor Appl Genet. 2000;100:139–46.

Microbial Transformation of Rifamycin by a Bacterial Strain *Chryseobacterium* Species

22

A.H. Jobanputra and B.A. Karode

Abstract

Chryseobacterium species was used for biotransformation of rifamycin B to rifamycin S by the enzyme rifamycin oxidase (RO), an enzyme known to be involved in this type of biotransformation in fungi. Factors influencing the activity of rifamycin oxidase (RO) enzyme of a bacterial strain (*Chryseobacterium* sp.) were studied. Temperature optima was found to be 50°C, and the enzyme exhibited thermostability as it retained 78% and 43% relative activity at temperatures of 55 and 60°C, respectively. The pH optima of the enzyme was found to be 6.5, and it was stable over the pH range 6.5–7.5. The enzyme obeyed Michaelis-Menten kinetics towards the substrate rifamycin B with apparent Km of 0.71 mmol and Vmax 22.4 μmol/min/mg. 100% transformation was found when 0.3 g/L rifamycin B substrate was provided. Drastic change in enzyme activity was observed under optimized conditions where medium containing malt extract and $CaCl_2$ was used. The activity was found to be 68 IU/mL vis-a-vis control which was 31 IU/mL. Influence of different oils on organism was also investigated which exhibited maximum growth, and 120 and 112% relative activity was retained in the presence of soya oil and mustard oil.

Keywords

Biotransformation • *Chryseobacterium* species • Rifamycin oxidase

A.H. Jobanputra (✉) • B.A. Karode
Department of Microbiology, PSGVPM's S. I. Patil Arts, G.B. Patel Science & S. T. S. K. V. S. Commerce College, Shahada, 425409, Maharashtra State, India
e-mail: arpana_j12@rediffmail.com

A. Sabu and A. Augustine (eds.), *Prospects in Bioscience: Addressing the Issues*,
DOI 10.1007/978-81-322-0810-5_22, © Springer India 2013

Introduction

Rifamycin was developed as one of the antituberculosis drugs in 1966 and has been used since several years [1]. Rifamycin B is the microbial product of *Nocardia mediterranei* [2]. Rifamycin B obtained can be converted to rifamycin S by existing chemical process using strong oxidizing agents like MnO_2, tetrahydrofuran and H_2SO_4. The use of such stringent chemicals results in unsatisfactory yields. Consequently, the process generates large amounts of hazardous acidic products. However, biotransformation is an alternative process over the chemical process because it is one or few step reaction that requires a short duration, is safe, less-energy requiring, eco-friendly and economic. Therefore, it emerges as a more advantageous process over the conventional chemical process.

Rifamycin S obtained from rifamycin B is of considerable economic value because it is the precursor for majority of semi-synthetic rifamycins and is highly active against Gram-positive bacteria like *Mycobacteria*, *Staphylococci*, *Streptococci* and some Gram-negative bacteria, namely, *Brucella*, *Chlamydia*, *Haemophilus*, *Legionella* and *Neisseria* species. Inhibition of DNA-dependent RNA synthesis by rifamycins has been reported [3]. Inhibitory activity against HIV-1 reverse transcriptase has been reported in open ansa chain rifamycin S derivative [4]. This in turn was used to manufacture rifampicin which acts as antituberculotic and antileprotic drug [5, 6].

Biotransformation of rifamycin B to S has also been reported with intracellular rifamycin oxidase from *Humicola* species ATCC 20260 and *Monocillium* species *ATCC* 20621 [7]. Rifamycin oxidase from a bacterial strain *Chryseobacterium* species which brings about the transformation of rifamycin B to rifamycin S has also been reported [8]. Influence of nitrogen sources and heavy metals on the production of rifamycin oxidase by *Chryseobacterium* species has also been reported [8]. This chapter deals with the factors influencing the biotransformation of rifamycin B to S with the help of RO of the organism and influence of temperature and pH on the activity of the enzyme obtained from *Chryseobacterium* species.

Materials and Methods

All the chemicals used were of analytical grade (AR) and guaranteed grade (GR). Casein enzyme hydrolysate was procured from HiMedia, Mumbai. Standard samples of rifamycin B and S were obtained through the courtesy of Mr. Patgaonkar (Lupin Chemicals Ltd, Tarapur, Mumbai). A local bacterial isolate identified by Belgian Co-ordinated Collections of Micro-organims (BCCM) as *Chryseobacterium* species was used for biotransformation. This organism was grown on modified casamino acid medium (CAA medium) containing casein enzyme hydrolysate, 5.0; $K_2HPO_4.3H_2O$, 1.18 and $MgSO_4$ $7H_2O$, 0.25 g/l. Rifamycin oxidase enzyme was obtained from *Chryseobacterium* species. Stock culture was maintained at 4°C on nutrient agar slant.

Enzyme Activity Measurement

Enzyme activity was measured using the method of Seong et al. [9]. Enzyme (0.025 ml) in the form of whole cells of *Chryseobacterium* species was added to 2 mM substrate (0.5 mL) – rifamycin B prepared in phosphate buffer (0.1 M, pH 6.5). The mixture was incubated at 50°C for 60 min, followed by the addition of 4.5 mL of 1:1 (v/v) solution of methanol: phosphate buffer (0.1 M, pH 7.8). Reaction mixture was boiled for 3 min in water bath and centrifuged at 4,000 rpm for 5 min. The absorbance of rifamycin S in supernatant was measured at 525 nm with a double-beam spectrophotometer (Shimadzu model, UV-1601). One unit of enzyme activity is defined as an equivalent of 1 μM rifamycin S produced in an hour under the specified assay conditions. Rifamycin B concentration was measured using the method of Pasqualucci et al. [10]. Two 1 mL portions of the broth in two 50-mL volumetric flasks were made up to the volume, one with solution A (acetate buffer 0.01 M, pH 4.63) and another with solution B (solution A containing 0.1% $NaNO_2$). The absorbance of the solution diluted with the solution A was measured at 425 nm in a 1-cm cell against the solution diluted with solution B. The rifamycin B concentration in the transformed broth was calculated using the following formula:

$$\text{Rifanycin B}\ (\mu\text{g}/\text{ml}) = \frac{A_{425} \times 50{,}000}{21.5}$$

where A_{425} is the absorbance at 425 nm and 21.5 is the absorptivity of rifamycin B. The method described by Chung [11] was used for the estimation of rifamycin S. Two millilitre of the sample was diluted with 4 mL methanol and 4 mL 0.1 M phosphate buffer (pH 6.5), and the solution was hydrolyzed from rifamycin O to rifamycin S by boiling for 3 min. Absorbance was read out at 525 nm with appropriate buffer blank.

Effect of Temperature on Activity and Stability of Enzyme

Biotransformation of rifamycin B to S was studied with the action of rifamycin oxidase. Overnight grown colony of the organism was inoculated in to modified CAA medium. In order to ascertain the effect of temperature on the activity of RO, enzyme in the form of whole cells (0.025 ml) was incubated with 2 mM rifamycin B for an hour at different temperatures ranging from 20 to 60°C. Enzyme activity was measured according to the method of Seong et al. [9]. The temperature value corresponding to highest enzyme activity was taken as temperature optima. The effect of temperature on enzyme stability was determined by incubating enzyme at different temperatures, ranging from 20 to 60°C for 30 min.

Effect of pH on Activity and Stability of Enzyme

As biotransformation is enzyme-mediated process it is sensitive to the pH of the medium which affects the activity of rifamycin oxidase enzyme [12, 13]. To ascertain the pH optima for the maximum enzyme activity, enzyme in the form of whole cells (0.025 mL) was incubated with 2 mM rifamycin B for an hour with the use of the 0.1 M buffers having different pH. Phosphate buffer having pH range 6–8 was used. The pH value corresponding to highest enzyme activity was taken as pH optima. The experiment on effect of pH on enzyme stability was carried out by pre-incubating enzyme at different pH, ranging from 6 to 8 for 30 min, followed by determination of enzyme activity.

Influence of Carbon Sources (Sugars)

Studies on the effect of glucose were carried out with a view to know the effect of metabolites involved in the fermentation for rifamycin oxidase synthesis and thereby explore and understand the control mechanisms (catabolite repression) involved in rifamycin production. Experiments on the effect of glucose on RO production were designed according to the methods followed by Kim et al. and Banerjee and Srivastava [12, 13]. Hundred millilitre CAA medium was separately supplemented with each of glucose, fructose, sucrose and maltose in a concentration of 1 g%. It was inoculated with 1% inoculum and incubated on rotary incubator shaker (Orbital incubator shaker, Remi India Ltd.) at 200 rpm at 30°C. After 12-, 24-, 36- and 48-h intervals, cell mass was harvested from each of the flasks, washed and used for the measurement of enzyme activity. Change of pH in presence of different sugars was critically studied.

Influence of Substrate Concentration with Respect to Biotransformation

As biotransformation is an enzyme-mediated process, it is necessary to check the substrate loading capacity. For this, starvation medium containing different concentrations of rifamycin B was provided in the range of 0.1–1 g/L. The amount of residual rifamycin B was determined according to method of Pasqualucci et al. [10], and transformed rifamycin S was estimated by the method of Chung [11]. Organism was initially subjected to trace levels of substrate rifamycin B before subjecting to the range of concentrations.

Determination of K_M and V_{max}

The enzyme was incubated with various concentrations of rifamycin B substrate ranging from 1 to 5 mM prepared in phosphate buffer (0.1 M, pH 6.5), and the enzyme activity was determined under standard assay conditions. K_M and V_{max} of the enzyme for rifamycin were calculated by using Lineweaver-Burk plot [14].

Influence of Cell Mass

To determine the optimum concentration of cell mass for transformation process, the organism was grown on CAA medium; cells were harvested at late logarithmic phase by centrifuging the broth at 10,000 rpm for 15 min. Biomass obtained was washed twice with sterile distilled water and then used for biotransformation process. Biomass was used in the range of 1–10%. The inoculated flasks containing rifamycin B substrate were incubated at 30°C at 150 rpm, and further estimations were carried out as described earlier.

Effect of Oils on Growth and Enzyme Production

Effect of various oils on growth and RO activity was studied. Oil such as sesame oil, soy oil, groundnut oil and mustard oil were used as carbon source. They were provided in concentration of 1% in CAA medium individually, where an overnight grown colony was inoculated. The medium was incubated for 24 h at room temperature. Growth of the organism and rifamycin activity was checked.

Growth and RO Activity Under Optimized Conditions

In order to increase the growth and RO production/activity, separate set of experiments were run with CAA medium containing extra source, i.e. 1% malt extract as nitrogen source and 2 mM $CaCl_2$ as metal source [8]. pH was adjusted at 6.5, and these sources were used to compute the enzyme activity when provided together.

Results and Discussion

Effect of Temperature on RO Activity and Stability

As biotransformation is an enzyme-mediated biological process, the effect of temperature against RO enzyme was studied for the *Chryseobacterium* species. Measurement of RO activity at different temperatures showed that the enzyme activity increased up to 50°C and then declined progressively. The temperature optima of the enzyme was found to be 50°C as enzyme was most active at this temperature. It has been found that the organism tolerates temperature up to 60°C, while at 55 and 60°C, it retained 78 and 43% of its activity, respectively (Fig. 22.1). On the other hand, RO from *Monocillium* species showed temperature optima of 40°C. The temperature optima of this enzyme was greater than that of reported *Humicola* species [15]. Temperature optima for *Humicola* species was 45°C [9]. The enzyme was found to be stable up to 60°C temperature (Fig. 22.2).

Effect of pH on Enzyme Activity and Stability

From Fig. 22.3, it is clear that enzyme activity changes as the pH changes. However, pH has a critical role to play in biotransformation of rifamycin B to S. pH optima of the enzyme was found to be 6.5 at which maximum enzyme activity was found. At pH 6 and 7, relative activity 71 and 92% was found. The pH optima of *Humicola* species was reported as 7.8 [9], while that for *Monocillium* species pH optima was reported as 7.5–8.5 [7], which was quite above the pH optima of RO enzyme of *Chryseobacterium* species. In case of *C. lunata*, pH optima was reported as 6.5 [12]. Stability of RO was studied where it has been found that at pH over 6.5 and 7, it showed more than 90% of its relative activity, while at pH 8 it was found to be 57% which indicated that the RO enzyme was stable at 6.5–7 pH range (Fig. 22.4).

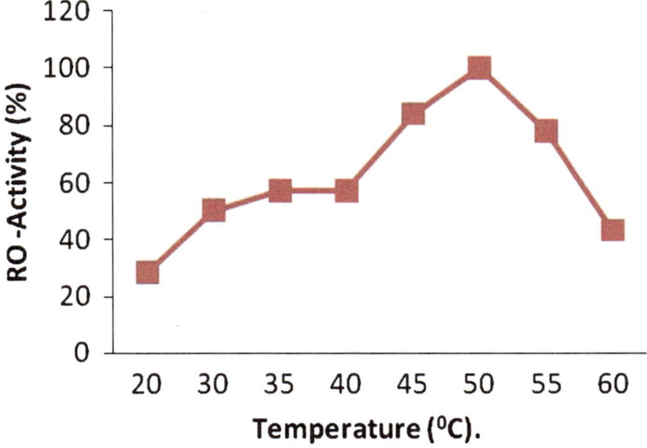

Fig. 22.1 Temperature activity profile of rifamycin oxidase from *Chryseobacterium* species

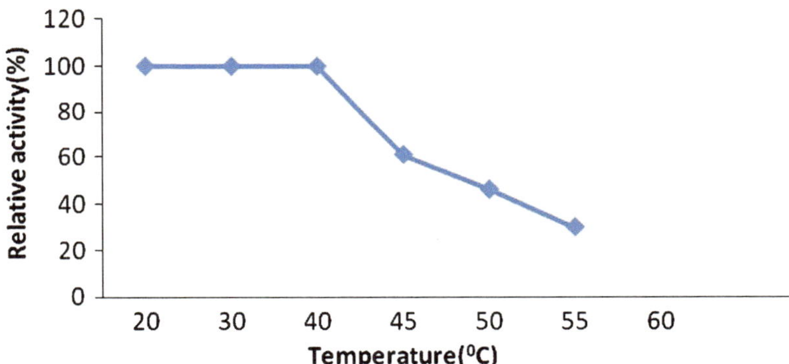

Fig. 22.2 Temperature stability of rifamycin oxidase from *Chryseobacterium* species

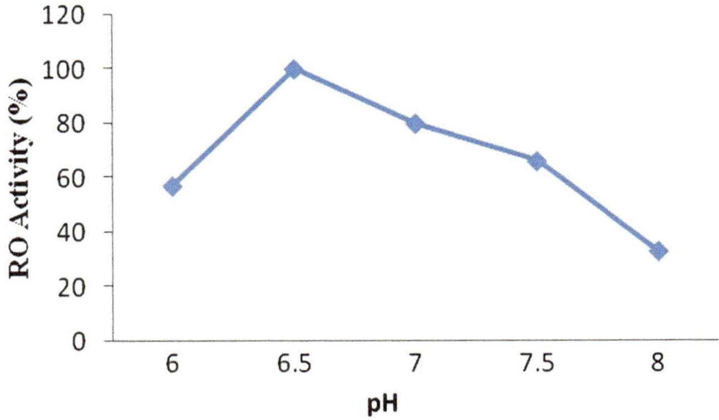

Fig. 22.3 pH activity of rifamycin oxidase from *Chryseobacterium* species at 50°C

Influence of Carbon Sources

Effect of various sugars on growth and RO activity at different time intervals is represented in Fig. 22.5a–d. Growth and RO activity in the absence of any of the sugars, i.e. control, is represented in Fig. 22.5e. Graphical representation of this clearly indicated that growth of *Chryseobacterium* seemed to be suppressed in the presence of glucose, sucrose and maltose.

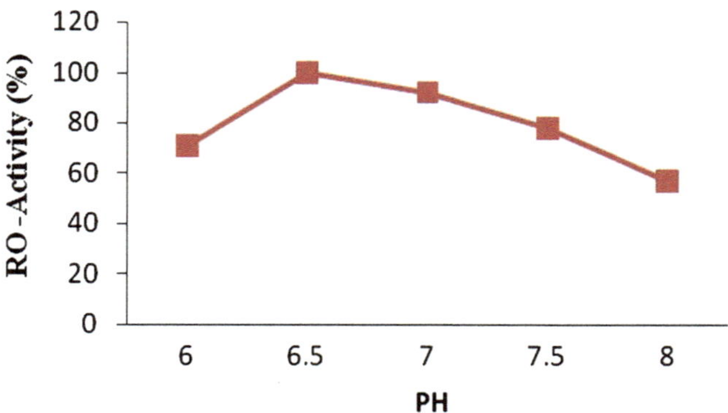

Fig. 22.4 pH stability profile of rifamycin oxidase from *Chryseobacterium* species

In contrary to this, a minor stimulation of growth was observed in the presence of fructose. The change in pH in the presence of all the sugars showed a nearly similar pattern as compared to the control indicating that pH control was essential during biotransformation as observed in the case of *C. lunata* and *Humicola* species [12, 13].

The enzyme activity was found to reach at its maximum (29 $IUml^{-1}$) in 12 h, and thereafter a sharp decline was observed in the control. A drastic fall in RO activity was found in the presence of sucrose and maltose indicating a strong catabolite repression. Reports on the influence of glucose on RO activity for *C. lunata* in YPD medium supplemented with 40 g/L of glucose showed maximum enzyme activity of 10.82 IU/ml. The depletion of glucose was accompanied with reduction of enzyme activity, rise of culture pH and biosynthesis of a black intracellular pigment [12]. Similar characteristics were observed in *Chryseobacterium* species, but there was no deposition of black pigment. For *Humicola* species enzyme, production was not repressed by the high glucose concentration up to 60 g/L^{-1} of the culture broth [13]. However, for *Chryseobacterium* species, only 51.7 and 62% RO activity was retained in the presence of glucose and fructose, respectively, indicating that there was no rise in enzyme activity in the presence of these catabolites. Conclusively, there was repression of RO activity in the presence of all the used sugars under the given experimental conditions.

Effect of Substrate Concentration

Rifamycin B transformation rate was checked with increase in the substrate concentration. It was found that 100% conversion was achieved when substrate concentration was 0.3 gm/l (Fig. 22.6). As the concentration increases the conversion rate decreased, as 75% conversion was achieved at a substrate concentration of 1 gm/l. The activity of RO from *Chryseobacterium* species measured at various concentrations by providing the substrate rifamycin B (1–5 mM) exhibited the rectangular hyperbola which obeyed Michaelis-Menten kinetics. The apparent K_M and V_{max} values obtained from Lineweaver-Burk plot were 0.71 mM and 22.4 μmoel/min/mg, respectively. K_M and V_{max} of rifamycin oxidase enzyme from *Monocillium* species were reported as 0.5 mM and 6.5–7.1 μM/h/mL (Han et al. 1983), whereas in *C. lunata*, K_M and V_{max} were 0.67 mM and 11 μM/h/mL [12]. No inhibition of enzyme was found up to 10 mM rifamycin B (Figs. 22.7 and 22.8).

Effect of Cell Mass

Effect of cell mass on rifamycin biotransformation was studied, and Fig. 22.9 clearly shows that 5% cell mass concentration was effective in biotransformation of about 82%. Beyond this cell mass concentration, there is not much increase in the transformation rate.

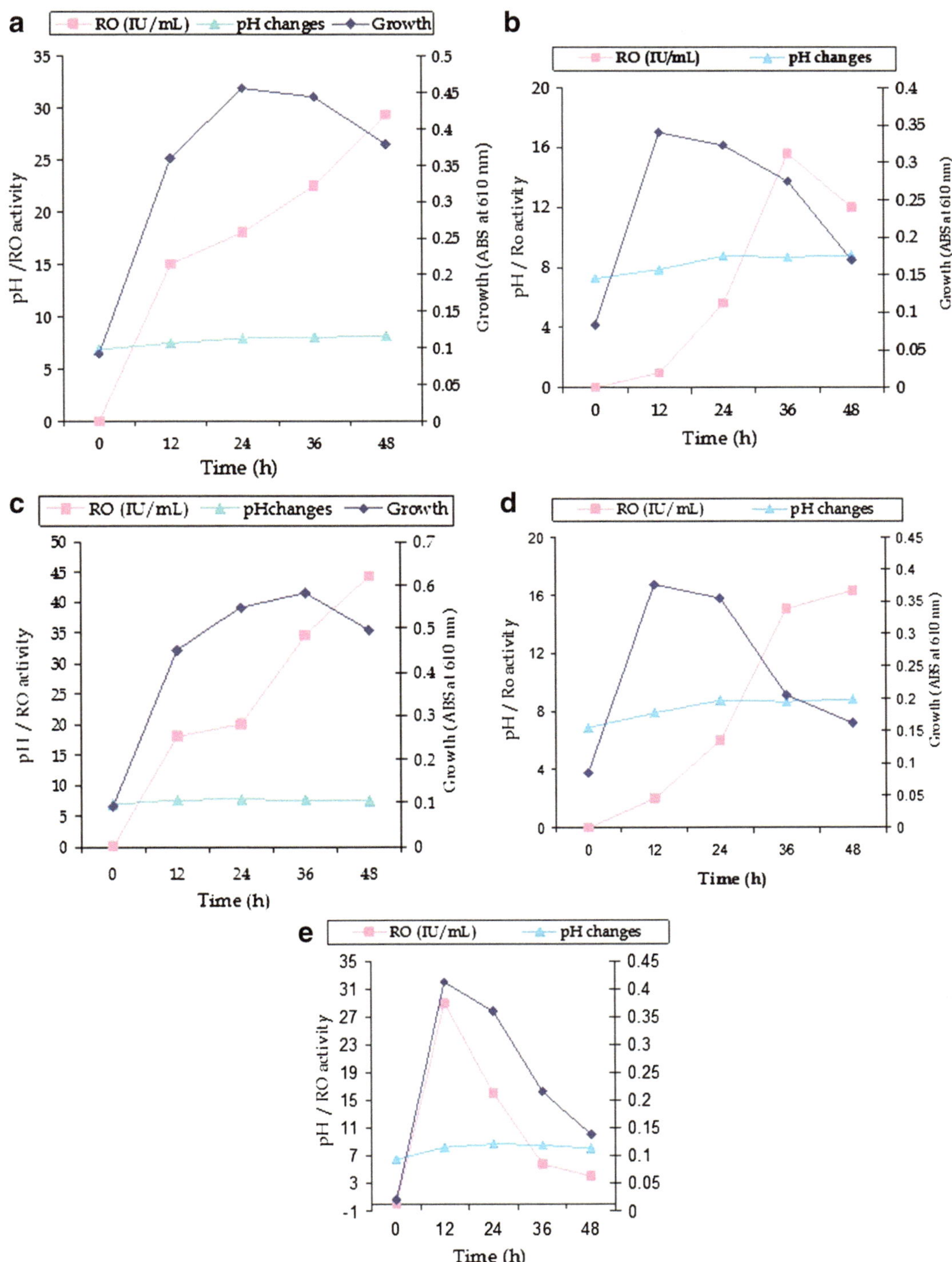

Fig. 22.5 (**a**–**d**): Influence of various sugars on growth and RO activity, (**a**) influence of glucose, (**b**) influence of fructose, (**c**) influence of maltose sugar, (**d**) influence of sucrose sugar and (**e**) growth and RO activity in absence of sugar (control)

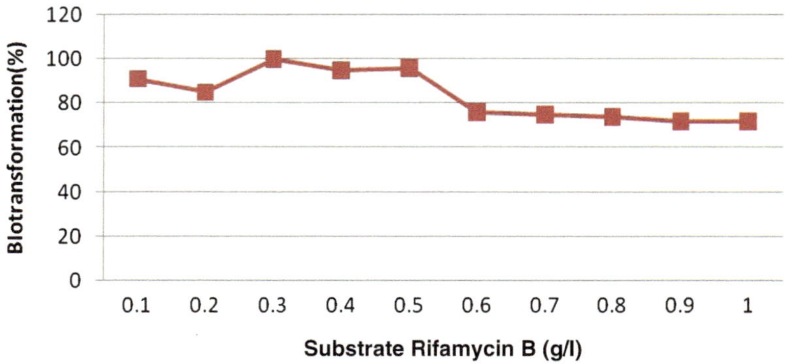

Fig. 22.6 Influence of substrate concentration on biotransformation

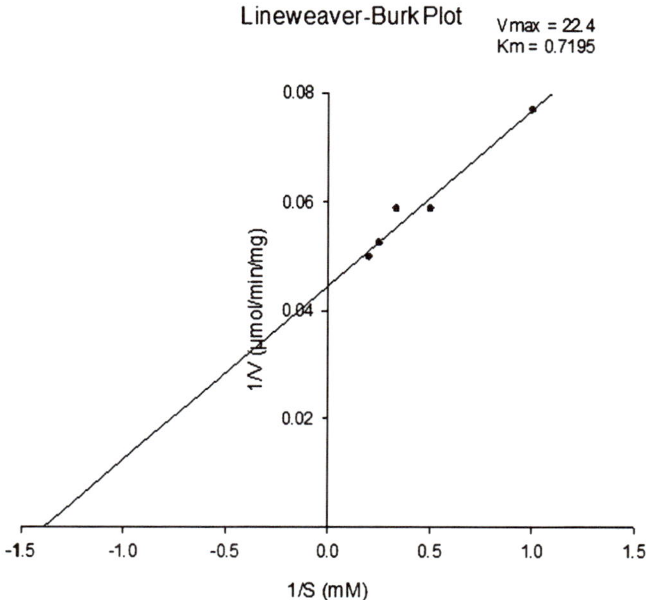

Fig. 22.7 Lineweaver-Burk plot for RO enzyme from *Chryseobacterium* species

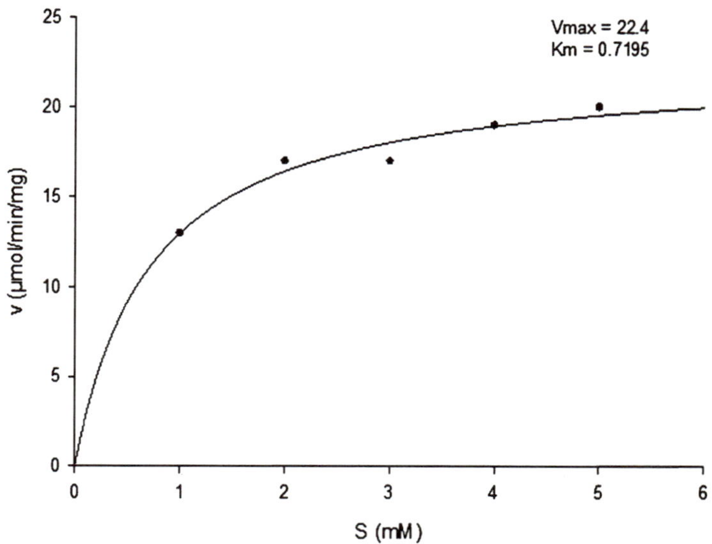

Fig. 22.8 Michaelis-Menten curve for RO enzyme from *Chryseobacterium* species

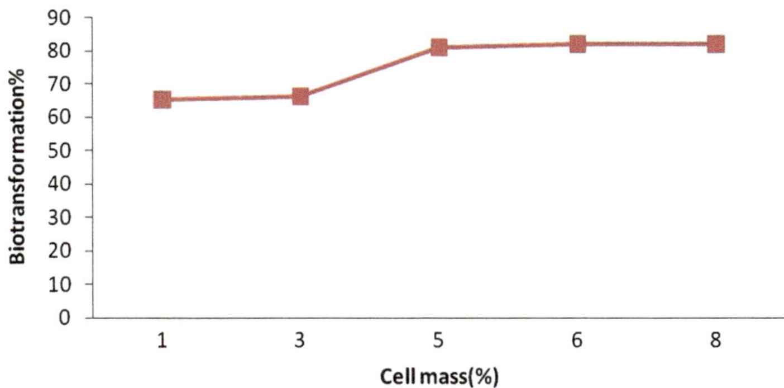

Fig. 22.9 Influence of cell mass on biotransformation

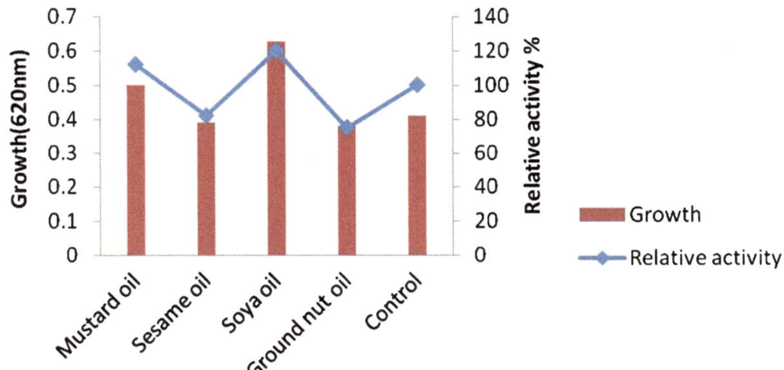

Fig. 22.10 Growth of *Chryseobacterium* sp. and RO activity in presence of respective oils

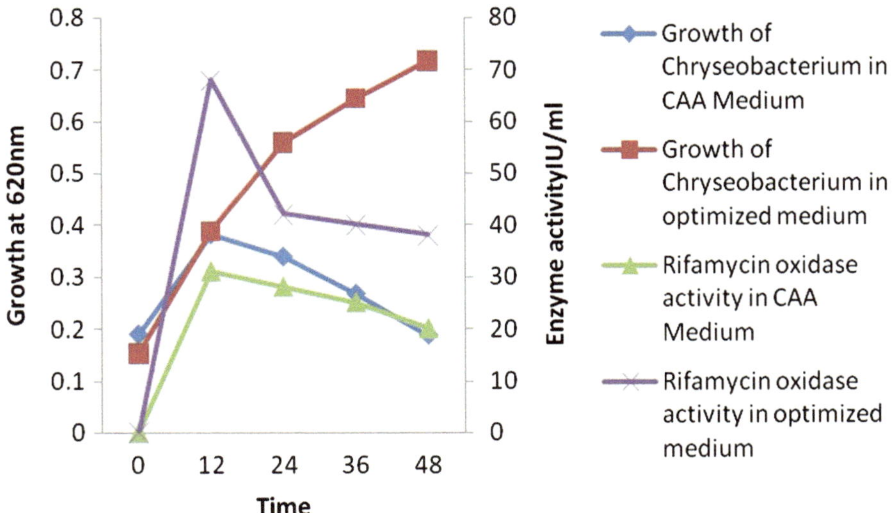

Fig. 22.11 Growth and RO activity of *Chryseobacterium* sp. in CAA and optimized media

Effect of Oils on Growth and Enzyme Production

Effect of different oils on growth of organism and enzyme activity was studied. From Fig. 22.10, it was found that organism exhibited enhanced growth in presence of mustard oil and soya oil and relative activity was found to be 112 and 120%, respectively. An inhibitory effect was found against groundnut oil and sesame oil. It showed that use of soy oil favoured the growth of *Chryseobacterium*, and it boosted up the RO enzyme activity.

Growth and Enzyme Activity in Optimized Condition

Under optimized condition, it was found that enzyme activity was increased (Fig. 22.11), as enzyme activity in unoptimized media was found to be 31 U/mL, while 68 IU/mL was found for the optimized media with increase in growth. From this result, it is clear that malt extract and $CaCl_2$ can boost up the enzyme activity.

Acknowledgements Authors are thankful to the University Grants Commission (UGC), New Delhi, for providing financial assistance for the project and patent office for granting process patent for the work.

References

1. Hidaka T. Current status and perspectives on the development of rifamycin derivative antibiotics. Kekkaku Aidsline Med. 1999;74:53–61.
2. Vohra RM. Novel assay for screening rifamycin producing mutants. Appl Environ Microbiol. 1992;58:403–4.
3. Umezawa H, Mijuno S, Yamazaki H, Nitta K. Inhibition of DNA dependent RNA synthesis by rifamycins. J Antibiot. 1968;21:234–6.
4. Bartolucci C, et al. Synthesis of nucleosidyl rifamycins on inhibitors of human immunodeficiency virus type I. Antivir Chem Chemother. 1997;3:215–21.
5. Shepard CC. A brief review f experiences with short term trials monitored by footpad inoculum. Lepr Rev. 1981;52:299–308.
6. Fox W. Short course chemotherapy for pulmonary tuberculosis and some problems of its program application with particular reference to India. Bull Int Union Tuberc. 1985;60:40–9.
7. Han MH, Seong BL, Son HJ, Mheen TI. Rifamycin B oxidase from *Monocillium* species: a new type of diphenol oxidase. FEBS Lett. 1983;151:36–40.
8. Jobanputra AH, Chincholkar SB. Influence of nitrogen sources and heavy metals on the production of rifamycin oxidase by *chryseobacterium* sp. J Sci Ind Res India. 2007;66:615–7.
9. Seong BL, Son HJ, Mheen FI, Han MH. Microbial transformation of rifamycin B: a new synthetic approach to rifamycin derivatives. J Antibiot. 1983;36:1402–4.
10. Pasqualucci CR, Vigevani A, Radaelli P, Gallo GG. Improved differential spectrophotometric determination of rifamycins. J Pharm Sci. 1970;59:685–7.
11. Chung BH, Chang HN, Han MH. Immobilization and characterization of rifamycin B oxidase in cellulose acetate beads. Korean J Appl Microbiol Bioeng. 1985;13:115–8.
12. Banerjee UC, Srivastava JP. Effect of pH and glucose concentration on the production of rifamycin oxidase by *Curvularia lunata* in batch reactor. J Biotechnol. 1993;28:229–36.
13. Kim EK, et al. Effects of pH and glucose concentration on production of Rifamycin oxidase. J Ferment Technol. 1984;62:117–21.
14. Lineweaver H, Burk D. The determination of enzyme dissociation constants. J Am Chem Soc. 1934;56:658–66.
15. Seong BL, Son HJ, Mheen FI, Park YH, Han MH. Enzymatic oxidation of rifamycin by microorganism of genus Humicola. J Ferment Technol. 1985;63:515–22.

Isolation, Molecular Characterization and Expression of a New *Cry*3a Gene from Native Isolates of *Bacillus thuringiensis* and Its Toxicity Against Asian Grey Weevil *Myllocerus undecimpustulatus undatus* Marshall (Coleoptera: Curculionidae)

23

H.M. Mahadeva Swamy, R. Asokan, Geetha G. Thimmegowda, Riaz Mahmood, and Dilip K. Arora

Abstract

The genetic diversity and distribution of cry genes in *Bacillus thuringiensis* strains vary based on geographical location. Each habitat may contain novel *B. thuringiensis* isolates that have more toxic effects on target spectra of insects. To obtain novel *B. thuringiensis* strains for the production of Cry proteins, isolation of numerous new *B. thuringiensis* strains is becoming a routine activity in many industries. The *B. thuringiensis Cry*3A gene is regulated by a different mechanism from that of most of the other cry genes. 1.935-kb DNA fragment of *Cry*3A gene was PCR amplified using gene-specific primers, cloned in expression vector pQE-80 L and then used for transformation of *E. coli* M15 cells. The optimum expression was obtained with 1 mM IPTG at 37°C for 3 h. Nucleotide sequencing of the *Cry*3A gene revealed an open reading frame of 1,935 bp, encoding a protein of 645 amino acid residues in length, with a predicted molecular mass of 77.4 kDa. These full-length gene sequences were deposited at NCBI GenBank with accession number JQ038134 and JQ038135. Present work is now being directed toward the isolation and expression of the new *Cry*3A gene to access its effect against Asian grey weevil *Myllocerus*

H.M.M. Swamy (✉) • R. Asokan
Bio-Pesticide Laboratory (BPL), Division of Biotechnology, Indian Institute of Horticultural Research (IIHR), Hessaraghatta Lake Post, Bangalore 560089, India
e-mail: clintonbio@gmail.com

G.G. Thimmegowda
Division of Entomology & Nematology, Indian Institute of Horticultural Research (IIHR), Hessaraghatta Lake Post, Bangalore 560089, Karnataka, India

R. Mahmood
Post-Graduate Department of Studies and Research in Biotechnology and Bioinformatics, Kuvempu University, Jnanasahayadri, Shankaraghatta, Shimoga 577451, Karnataka, India

D.K. Arora
National Bureau of Agriculturally Important Micro organisms (NBAIM), Mau Nath Bhanjan 275101, Uttar Pradesh, India

A. Sabu and A. Augustine (eds.), *Prospects in Bioscience: Addressing the Issues*,
DOI 10.1007/978-81-322-0810-5_23, © Springer India 2013

undecimpustulatus undatus (Coleoptera: Curculionidae). The availability of the recombinant protein will also allow the isolation and identification of its receptor in the insect midgut. Additionally, novel activities/specificities based on the sequence of the new *Cry*3A gene can be explored by the use of techniques such as DNA shuffling or point mutations. This system offers an additional method for potentially improving the efficacy of *Bt* insecticidal proteins efficiently, stably, and safely particularly against anti-lepidopteran insect pests. In other words, *Bt* technology will be a key to future growth of agriculture.

Keywords
*Cry*3A • *Bacillus thuringiensis* • Asian Grey Weevil • Coleoptera • Midgut • Insecticidal protein

Introduction

Insect pests, diseases, and weeds inflict enormous losses to the potential agricultural production. Anecdotal evidences also indicate rise in the losses, despite increasing use of chemical pesticides. At the same time, there is a rising public concern about the potential adverse effects of chemical pesticides on the human health, environment, and biodiversity. These negative externalities, though, cannot be eliminated altogether; their intensity can be minimized through development, dissemination, and promotion of alternative technologies such as biopesticides and bioagents as well as good agronomic practices rather relying solely on chemical pesticides. India has a vast flora and fauna that have the potential for developing into commercial technologies [1]. *Bacillus thuringiensis* (*Bt*) is a crystalliferous spore-forming bacterium, which typically contains both a spore and a crystal toxin within its cell wall. *Bt* has a long and safe history of use as a microbial insecticide, and this has prompted research into the nature of the toxin moiety that is central to mortality of target organisms. Recognition that *Bt* has a diverse genetic structure within the range of toxin genes, referred to as *cry* genes, has led to research into the interactions between crystal toxins and potential hosts. Although *Bt* was already the dominant microbial insecticide globally, the ability to identify and incorporate toxin genes into plants has accelerated interest in the agent and has expanded both research and uptake in integrated pest management (IPM). The unique feature of these *cry* genes, the *cry3Aa* and *cry3Bb* genes, is that it is sporulation independent and it may be induced or depressed during stationary phase, probably by transition phase regulators. These sporulation-independent promoters may be useful in improving the production of sporulation-dependent Cry proteins. Heterologous recombination may be used, not only to integrate *cry* genes into a resident plasmid, or into the chromosome, but also to disrupt genes of interest [2]. The *Bt Cry*3A gene is regulated by a different mechanism from that of most of the other *cry* genes. Its expression begins during late exponential growth and not during sporulation as for the other classes of *cry* genes [3] because of its non-sporulation-dependent *Cry*3A promoter. Many studies were conducted and indicated that the *Cry*3A promoter was able to enhance the expression of other *cry* genes and significantly increase the production of those Cry proteins [3–6].

Climate change will have direct and indirect impacts on pests and their natural enemies. Farm management responses to these changes are also starting to affect pest complexes. Transboundary plant pests, animal diseases, and invasive alien aquatic species are a constraint to food security due to their impacts on food availability, food access, food safety, and food stability. The discovery rate of new *Bt* toxin is more because of its genome diversity. Therefore, it is reasonable to search for new *Bt* toxins which are more effective against insect pests. Previous studies on screening of indigenous *Bt* strains for toxicity and molecular characterization of new *Bt* strains revealed difference in nature and composition of *Bt* toxins, between native and standard *Bt* strains. Hence, cloning of toxin genes from new *Bt* strains and their characterization are essential to expand the use of *Bt* and for the management of resistance development in insects. Searching for novel sources of cry genes encoding new Cry toxins active against coleopteran insect pests is important; a *Cry*3A-type gene was isolated from the Andaman and Nicobar Islands and Jammu and Kashmir *Bt* strains. Herein, we report the cloning and characterization of the *Cry*3A gene, its expression in *E. coli* cells, and also its insecticidal specificities and activities toward coleopteran insect pests.

Materials and Methods

Isolation of a *Cry*3A Gene from the *B. thuringiensis* Strain

B. thuringiensis strains were grown in 30 mL Luria-Bertani (LB) broth for 12 h at 200 rpm and 30°C; cells were centrifuged and the pellets were kept at –20°C for DNA extraction. DNA extraction was done using bacterial genomic DNA kit (HiMedia) following the manufacturer's instructions. In order to identify *Cry*3A-type genes among the *cry3*-type genes detected in *B. thuringiensis* total DNA, primers to amplify complete *Cry*3A-type genes (i.e., from the start codon till the stop codon) were designed on the basis of multiple alignment of all previously described *Cry*3A and the consensus sequences of their N- and C-terminal coding regions. PCRs were performed using two sets of primers designed to amplify complete *Cry*3A genes (forward primer: 5′-ATGAATCCGAACAATCGAAGTGAACATGAT-3′, reverse primer: 5′-TTAATTCACTGGAATAAATTCAATTTTGTC-3′), and (forward primer: 5′-GGATCCATGAATCCGAACAATCGAAGTGAA-3′, reverse primer 5′-AAGCTTTTAATTCACTGGAATAAATTCAAT-3′), 30 cycles of amplification at 62°C annealing temperature and high-fidelity LA Taq DNA polymerase (Sigma). The resulting PCR fragment was excised from the gel and purified using the NucleoSpin® Extract II Kit (MN), following the manufacturer's instruction.

Cloning, Sequencing and Sequence Analysis of the *Cry*3A Gene from the *Bt* Strain

The purified PCR fragment putatively corresponding to a *Bt* strain S811 *Cry*3A gene was cloned into the pTZ57T/R (Fermentas, USA) and used to transform *E. coli DH5α* cells. The clone was sequenced using an automated sequence analyzer (MWG). Computer analyses of the cloned DNA sequence were done using the BioEdit software package. Databank comparisons of the cloned DNA sequence with other published cry sequences were made using the BLASTx software from the NCBI databank (http://www.ncbi.nlm.nih.gov). The Conserved Domain Database search (CDD search) from the NCBI site was used to compare motif identity and similarity with the clone DNA sequence and known conserved domains. Sequence alignments and dendrograms were obtained by using CLUSTALW software and were edited with the BioEdit software. Dendrograms were edited using the TreeView software (http://darwin.zoology.gla.ac.uk/~rpage/treeviewx). The molecular mass and predicted pI of the clone deduced protein sequences were determined by the Protein Machine software available at the EXPASY site (http://us.expasy.org/tools/).

Construction of *E. coli* Expression Vector pQE-80 L *Cry*3A

E. coli M15 cells containing the *Cry*3A gene cloned into the pQE80-L expression vector were grown in LB medium in the presence of 100 μg/mL ampicillin, 25 μg/mL kanamycin for 14–16 h at 37°C, and the plasmid DNA was isolated using GeneJET Plasmid Miniprep Kit (Fermentas) following the manufacturer's instruction. PCR was performed using the isolated plasmid as template (forward primer: 5′-GGATCCATGAATCCGAACAATCG

AAGTGAA-3′, reverse primer: 5′-AAGCTTTTA ATTCACTGGAATAAATTCAAT-3′) 30 cycles of amplification at 50°C annealing temperature and high-fidelity LA Taq DNA polymerase (Sigma). The resulting PCR fragment was excised from the gel and purified using the NucleoSpin® Extract II Kit (MN), following the manufacturer's instruction. The purified PCR fragment was sub-cloned into the expression vector pQE80-L (Qiagen), following the manufacturer's instructions. The resulting construction, encoding the recombinant *Cry*3A gene to be expressed fused to a C-terminal His tag, was named pQE806-*Cry*3A.

Recombinant *Cry*3A Toxin Expression and Purification

E. coli M15 cells transformed with pQE-80 L-*Cry*3A were grown at 37°C at 200-rpm agitation, in the presence of 250 µg/mL ampicillin and 75 µg/mL kanamycin until O.D.600 = 0.6. *Cry*3A expression was induced by addition of 1-mM isopropyl-beta-D-thiogalactopyranoside (IPTG) when an O.D.600 = 0.6 was reached. Four hours after induction, the cells were harvested by centrifugation at 4,000 rpm for 20 min. The pellet containing the cells expressing the *Cry*3A His-tagged protein was then resuspended in lysis buffer (50 mM sodium phosphate buffer, 300 mM NaCl, 1% glycerol and 0.5% Triton X-100, pH 7.0). Thaw the cell pellet from an *E. coli* expression culture on ice (if frozen). Resuspend 1 g of pelleted, wet cells in 2–5 mL lysis-equilibration-wash (LEW) buffer (50 mM NaH2PO4, 300 mM NaCl, pH 8.0). Add lysozyme to a final concentration of 1 mg/mL. Stir the solution on ice for 30 min. Sonicate the suspension on ice according to the instructions provided by the manufacturer (e.g., use 10 × 15 s bursts with a 15-s cooling period between each burst). Centrifuge the crude lysate at 10,000× g for 30 min at 4°C to remove cellular debris. Carefully transfer the supernatant to a clean tube without disturbing the pellet. Store supernatant on ice. Purify under native conditions using Protino® Ni-TED (MN) packed columns as per the manufacturer's guidelines. Finally, use protein assay and/or SDS-PAGE analysis to determine which fraction(s) contain(s) the majority of the polyhistidine-tagged protein.

Bioinformatics Analysis

Nucleotide and deduced amino acid sequences were analyzed with the BLAST (www.ncbi.nlm.nih.gov/BLAST). Signal peptide sequence was analyzed using SignalP 3.0 (http://www.cbs.dtu.dk/services/SignalP). Related sequences obtained from databases using the software GENSCAN online tool (www.genes.mit.edu/GENSCAN.html) were used for identification of gene features such as exon and splice sites in genomic DNA. BioEdit (version 7.0.4.1) was used for sequence editing and analysis. Conserved Domains and Protein Classification (http://www.ncbi.nlm.nih.gov/Structure/cdd/cdd.shtml, version CDDv2.32-40526 PSSMs) and the predicted structure were validated using protein structure validation software suite (PSVS) tool. Determination of protein functional analysis obtained from databases using InterProScan (http://www.ebi.ac.uk/Tools/pfa/iprscan/) and ProFunc (http://www.ebi.ac.uk/thornton-srv/databases/profunc/index.html). COBALT is a multiple sequence alignment tool that finds a collection of pair wise constraints derived from Conserved Domain Database, protein motif database, and sequence similarity, using RPS-BLAST, BLASTP, and PHI-BLAST.

Bioassays of the Recombinant *Cry*3A Toxicity Against Insect Larvae

The activity of the Cry3A-type protein obtained from the recombinant *E. coli* strain was tested against coleopteran insect pest *Myllocerus undecimpustulatus undatus* Marshall (Coleoptera: Curculionidae). A preliminary test was performed to determine the specificity of this protein at a relatively high protein concentration (1,200 µg/ml). Bioassays with *Myllocerus undecimpustulatus undatus* Marshall (Coleoptera: Curculionidae) were carried out by dipping leaf disks prepared from mango into the protein solution using second instar larvae (grubs) and adults. Negative controls for all the insects tested were included using the same conditions but without any toxin. Bioassays were repeated at least three times. Bioassays were conducted at 25°C, 60–70% relative humidity (RH), and a 16:8 (light/dark [h]) photoperiod. For all insects bioassayed, mortality was evaluated after 5 days.

Statistical Methods

For all investigated parameters, the analysis of variance (ANOVA) was performed using the GraphPad Prism5 statistical software. The measurements of treatments were compared and grouped using Duncan's multiple range tests at the 0.01 significance level.

Results and Discussion

*Cry*3A Gene

The *Cry*3A gene was amplified, cloned, and sequenced. Total DNA isolated from all native *Bt* isolates from Andaman and Nicobar Islands [6] and Jammu and Kashmir (unpublished data) was amplified (Fig. 23.1a, b) using gene-specific primers. The optimum conditions for amplification of *Cry*3A gene were 250 ng of DNA, 200 lM of dNTPs, 1.5 mM of MgCl2 and 50 pmoles of primers and annealing temperature of 62°C. Gene cloning was confirmed by colony PCR and restriction digestion with *BamHI* and *HindIII* of recombinant plasmid DNA containing *Cry*3A gene (Fig. 23.2). The total length of *Cry*3A gene was 1,935 nucleotides which encode protein of molecular weight 77.4 kDa, approximately. All the sequences of full-length *Cry*3A gene of *Bt* isolates were submitted to GenBank database with accession numbers JQ038134 and JQ038135. The 1.935 kb *Cry*3A gene sequence from SITAB and SLK *Bt* isolates showed 91% homology with *Bacillus thuringiensis* serovar *tenebrionis Cry*3A gene when aligned with already known sequences using BLAST (basic local alignment search tool). The deduced amino acid sequence of toxic region of *Cry*3A proteins of two *Bt* isolates was aligned with Cry3A-like protein of public database. A total of 117 variations were found in toxic domains.

Composition of *Cry*3A of *Bt* Isolates and Phylogenetic Analysis

The deduced amino acid sequence similarity varies from 31 to 82% of *cry*3-like sequences. The amino acid residues ranged between 36.49% (Ala), 28.17% (Thr), 18.19% (Gly), and 17.16% (Cys). A total of 645 amino acids were found in *Cry*3A proteins in which the negatively charged amino acids ranged between 32 and 34 (66), whereas positively charged amino acid ranged between 10 and 37 (70). The phylogenetic relationship of *Cry*3A gene with other *Cry*3A genes reported earlier on from the present study and other laboratories is shown in Fig. 23.2. *Cry*3A reported from the present study are distinctly different from the ones reported elsewhere. The cluster of *Cry*3A genes reported from the present study seems to have evolved differently from the other reported *Cry*3A genes [7, 8].

Recombinant *Cry*3A: Expression in *E. coli*, Purification and Evaluation of Toxicity

The *E. coli* M15 strain harboring the pQE-80 L-*Cry*3A constructs was induced with 1-mM IPTG to express the recombinant His-tagged *Cry*3A protein. SDS-PAGE analysis of *E. coli* extracts after IPTG induction showed a differential protein band corresponding to the expected 77.4-kDa recombinant His-tagged *Cry*3A protein. No additional bands were identified within extracts of non-induced cells with empty pQE-80 L vector (Fig. 23.1c). In these conditions, a large amount of the recombinant His-tagged *Cry*3A protein was produced and, subsequently, purified using Ni-TED affinity chromatography from sonicated *E. coli* crude extract (Fig. 23.1d). The toxicity of purified recombinant His-tagged *Cry*3A was tested in leaf dip bioassays against Asian grey weevil *Myllocerus undecimpustulatus undatus* Marshall (Coleoptera: Curculionidae).

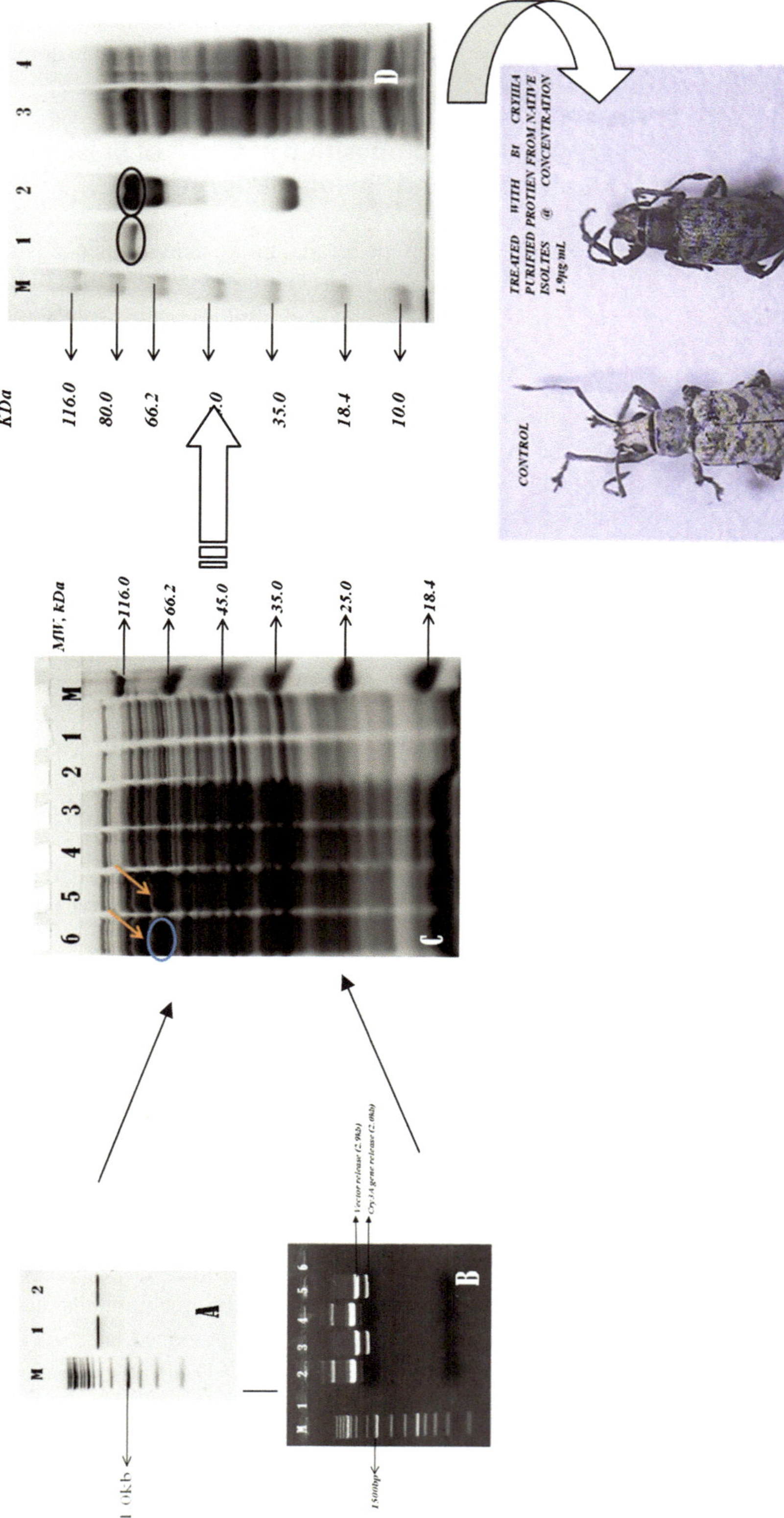

Fig. 23.1 (**a**) Isolation and screening of full-length *Cry*3A gene from the native *B. thuringiensis* strains; (**b**) Restriction digestion analysis of full-length genes; (**c**) Expression of polyhistidine-tagged *Cry*3A gene using IPTG; (**d**) Purification of polyhistidine-tagged *Cry*3A recombinant protein using Protino® Ni-TED; (**e**) Toxicity test analysis on *Myllocerus undecimpustulatus undatus* Marshall (Coleoptera: Curculionidae) adults

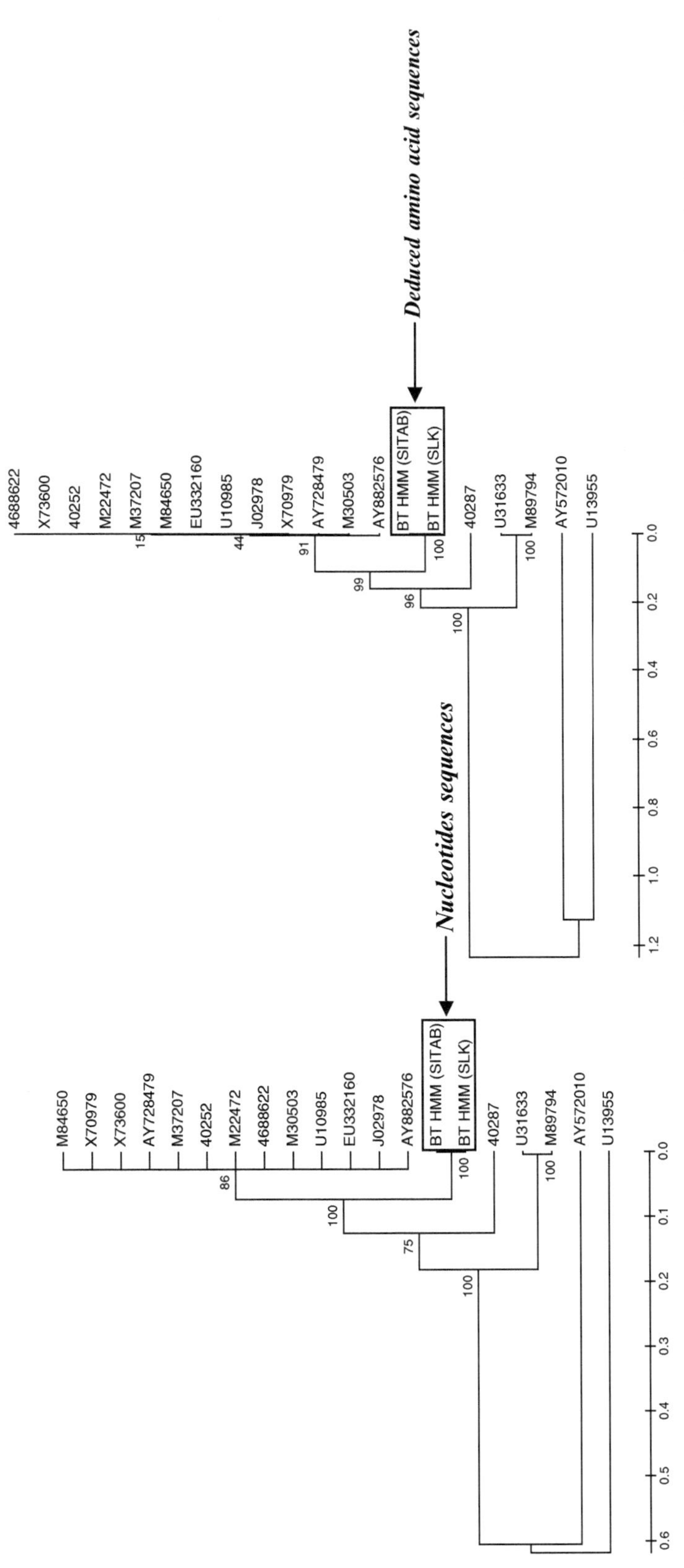

Fig. 23.2 Neighbor-joining tree showing phylogenetic relationship among different *Bacillus thuringiensis cryIIIA* genes based on nucleotide and deduced amino acid sequences using NCBI GenBank accession number U13955 (*cry14* gene) as an out-group

The bioassays revealed that the insect pest adults were susceptible to the purified recombinant His-tagged *Cry*3A (Fig. 23.1e). The purified recombinant His-tagged *Cry*3A was toxic to Asian grey weevil *Myllocerus undecimpustulatus undatus* Marshall (Coleoptera: Curculionidae), being 1,200 and 920 µg/mL the highest concentrations tested to achieve the maximum toxicities.

Conclusions

In conclusion, this study led to the identification of a novel *Cry*3A gene, which is known to be effective against coleopterans. These new genes served as candidate gene against different coleopteran insect pests and also new encounters due to climate change. Present study data is strengthened by an efficient expression, selection, or screening system; hopefully this allows us to reach into the full potential of these environmentally friendly pesticides.

Acknowledgment The authors are grateful to Indian Council of Agricultural Research (ICAR), New Delhi, for funding this study under network project on Application of Microbes in Agriculture and Allied Sectors (AMAAS). Infrastructure facility and encouragement by the Director of the Indian Institute of Horticultural Research (IIHR) are duly acknowledged.

References

1. Birthal PS, Sharma OP. Integrated pest management in Indian agriculture. Proceedings. 2004; pp. 1–279. National Centre for Agricultural Economics and Policy Research (NCAP), New Delhi, India and National Centre for Integrated Pest Management (NCIPM), New Delhi, India.
2. Lenin K, Udayasuriyan V, Kannaiyan S. Diversity in *cry* genes of *Bacillus thuringiensis*. National Biodiversity Authority, India; 2007. pp. 1–30.
3. Salamitou S, Agaisse H, Bravo A, Lereclus D. Genetic analysis of cryIIIA gene expression in *Bacillus thuringiensis*. Microbiology. 1996;142:2049–55.
4. Adams LF, Mathews S, O'Hara P, Petersen A, Gürtler H. Elucidation of the mechanism of CryIIIA overproduction in a mutagenized strain of *Bacillus thuringiensis* var. *tenebrionis*. Mol Microbiol. 1994;14:381–9.
5. Park HW, Bideshi DK, Johnson JJ, Federici BA. Differential enhancement of Cry2A versus Cry11A yields in *Bacillus thuringiensis* by use of the *Cry3A* STAB mRNA sequence. FEMS Microbiol Lett. 1999;181:319–27.
6. Sanchis V, Chaufaux J, Lereclus D. AmeÂlioration biotechnologique de (1996) *Bacillus thuringiensis*: les enjeux et les risques. Annales de l'Institut Pasteur/ActualiteÂs. 1996;7:271–84.
7. Malvar T, Baum I. *Tn5401* disruption of the *spo0F* gene, identified by direct chromosomal sequencing, results in *cryIIIA* overproduction in *Bacillus thuringiensis*. J Bacteriol. 1994;176:4750–3.
8. Mahadeva Swamy HM, Asokan R, Arora DK, Nagesha SN, Ajanta B, Riaz M. Cloning, characterization and diversity of insecticidal crystal protein genes of *Bacillus thuringiensis* native isolates from soils of Andaman and Nicobar islands. Curr Microbiol. 2011;63:420–5.

Characteristics of a Newly Isolated Bacterium for Decolorization of Textile Dyes

24

Joshni T. Chacko, S. Kalidass, and Lubna Lukman

Abstract

In recent years, use of microbial biomass for decolorization of textile effluent is becoming a promising alternative for costlier treatment processes. In the present study, a newly isolated bacterial strain expressing azoreductase activity was utilized to remove the color of textile dyes (Red RR and Black GDN) which come under the group of reactive dyes. The effects of substrate concentrations, medium compositions, and operation parameters (e.g., pH and temperature) on decolorization of the dyes by the strain were systematically investigated to reveal the key factors that govern the performance of dye decolorization. The metabolites resulting from bacterial decolorization were analyzed by high-performance liquid chromatography (HPLC). Azoreductase activity in the crude cell extract was assayed by the method of Zimmermann et al. The results showed that maximum decolorization occurred at 37°C at pH 7–8. The optimum concentration of dye was found to be 20ppm. It was observed that there was an increase in the percentage of decolorization upon supplementation with yeast extract. Batch incubation studies revealed that maximum decolorization occurred within 72 h and it was found to be 98.25 and 78.47% for Red RR and Black GDN, respectively. Azoreductase activity in the crude extract was found to be 44% for Black GDN. We found in the present study that the isolated strain is very efficient in removing color of textile dyes and the key enzyme involved in the process is azoreductase.

Keywords

Azoreductase • Dye decolorization • Textile dyes

J.T. Chacko • S. Kalidass (✉) • L. Lukman
School of Biotechnology & Health Sciences,
Department of Biotechnology, Karunya University,
Coimbatore 641114, India
e-mail: kalidass@karunya.edu

A. Sabu and A. Augustine (eds.), *Prospects in Bioscience: Addressing the Issues*,
DOI 10.1007/978-81-322-0810-5_24, © Springer India 2013

Introduction

Environmental pollution and its consequences have become the hot topic of international society and drawn the attention of the universe. It became worse in the past decade mainly because of the population explosion, increased industrial activity, and increased production of synthetic organic colorants. India is one of the chief producer and consumer of synthetic organic chemicals including synthetic dyes. These dyes are extensively used in textile dyeing, paper printing and color photography, and also as additives in petroleum industry. Two-thirds of the dye industry is dominated by textile industry and discharges large volume of synthetic dyes into the environment [1]. The majority of these dyes are recalcitrant so that they can confer color on various raw materials; moreover, certain dyes and dye precursors and some aromatic amines produced through biotransformation of dye compounds have been shown to be carcinogenic [2]. These toxic effluents discharge into the water bodies are undesirable not only because of their impact on the photosynthesis on the aquatic flora but also due to the carcinogenic and mutagenic nature [3, 4] of many of these dyes and its breakdown products [5].

Removal of dyes, particularly from textile wastewaters, has been a big challenge over the last decades, and up to now there is no single and economically attractive treatment method that can effectively decolorize dyes [6]. Several methods were adapted in the treatment of textile effluents to achieve decolorization. These include physiochemical methods [7] such as filtration, specific coagulation, use of activated carbon, and chemical flocculation [8, 9]. Such methods are often very costly, and although the dyes are removed, accumulation of concentrated sludge creates a disposal problem. Therefore, there is a need to find alternative treatment methods that facilitate a more natural and complete cleanup of the pollutants in a more economical way.

Biotreatment offers a cheaper and environmentally friendlier alternative for color removal in textile effluents. Recently biodegradation by bacteria has been investigated as a good method to transform or mineralize the dyes, since it is less time consuming than fungal systems [10, 11]. Many research reports are available, which explain successful decolorization of dyes by using purified microbial cultures. But these findings do not find much application in practical treatment system due to complexity and heterogeneity of chemical compounds present in textile wastewater. The present work investigated the use of potent microbial populations, which are native part of the textile mill soil samples and can endure the hostile conditions. Moreover, it included decolorization of two commonly used reactive dyes, that is, Red RR and Black GDN in the presence of additional carbon and nitrogen nutrients to study their effect on decolorization. In this study, we also investigated the effect of operation parameters (e.g., medium composition, substrate concentration, temperature, and pH) and the role of azoreductases on dye decolorization. The information obtained from this study is expected to form a solid foundation for the advancement of existing treatment processes.

Materials and Methods

All chemicals used in the study were of analytical reagent grade. Dyes (Red RR and Black GDN) were obtained from Muthukumaran Dyeing Units, Tirupur District, Tamil Nadu, India, and used without further purification. The media used for decolorization study was minimal salt media (MSM) which contains (g/l) K_2HPO_4 1, $MgSO_4.7H_2O$ 0.2, $CaCl_2.2H_2O$ 0.02, KNO_3 0.5, and ferric chloride 0.05, along with supplementation of 0.1% of yeast extract, tryptone, xylose, and ammonium nitrate. Cultivation of bacterial cells for decolorization study was carried out in nutrient broth medium (Himedia, Mumbai, India). The experimental strain was isolated from soil samples of textile industry by enrichment technique.

Inoculum Preparation

Log phase culture of the experimental strain was raised in nutrient broth medium for 48h and harvested by centrifugation at 7,000rpm for 10min at 4°C to pellet form. The pellet was washed with sterile saline thrice and suspended in specific volume of saline for decolorization studies. Inoculum concentration was 2% (v/v) for all the studies.

Optimization of Decolorization Conditions

After aerobic cultivation on growth medium, log phase cultures of the experimental strain were transferred into "decolorization medium" (typically 1g/l yeast extract+MSM) for the removal of dye color. Decolorization experiments were performed under the "static-incubation" condition (i.e., neither aeration nor agitation was employed). For optimization, decolorization conditions like dye concentration (0–50ppm), pH (3, 5, 7, 8, 9, and 11), temperature (25, 37, 45, 55, and 65), and incubation periods were experimented. All the experiments were run by varying a factor at a time and keeping others constant [12]. Decolorization assay was done as mentioned below.

Decolorization Under Different Medium Conditions

To find out the effect of carbon and nitrogen sources on dye decolorization, the decolorization assay of Red RR was carried by changing media composition. The media examined were 0.1% of yeast extract, tryptone, xylose, ammonium nitrate, and combinations of these components along with MSM. The experiments were operated at 37°C, pH 7.0, and with an initial dye concentration of 20ppm.

Decolorization Assay Via UV–Vis Spectroscopy

Samples withdrawn at different time intervals (namely, 24, 48, 72, and 96h) from experimental vials were centrifuged and run through visible spectrophotometer (Systronics Visiscan-167) to check absorbance at λ_{max} of selected dyes (522nm for Red RR and 596nm for Black GDN). The uninoculated dye-free medium was used as blank. All assays of dye decolorization were performed in triplicate, while the decolorization efficiency was expressed as

$$\text{Decolorization efficiency } \% = \frac{I - F}{F} \times 100$$

where I=absorbance of media prior to incubation and F=absorbance of decolorized media

Amount of dyes present in liquid culture media before and after biodecolorization by the tested strain was determined from the respective concentration versus absorbance calibration curves obtained for respective dyes. Standard error and deviation were calculated for obtained data.

Determination of Azoreductase Activity for Crude Cell Extract

Bacterial cells prepared from designated culture conditions were harvested by centrifugation (8,000rpm, 10min) and resuspended in 0.1M sodium phosphate buffer (pH 7.0). The cell solution (approximately 10–15mg cell/ml) was sonicated (60–75W; 5min) with an ultrasonicator (Sigma). The crude cell extract was collected after removal of cell debris from the disrupted cell solution by centrifugation at 12,000×g at 4°C for 20min. Azoreductase activity was determined based on the procedures described by Zimmermann etal. [13]. In general, 1ml of crude extract was added to 0.1M sodium phosphate buffer (pH 7.0) containing 20ppm of Black GDN and 0.35mM of NADH (Sigma); the total volume of the reaction mixture was 12ml. Dye decolorization was followed by monitoring the decrease in color intensity at 596nm at room temperature. One unit (U) of azoreductase activity was defined as the amount of enzyme required to reduce 1μM of dye/min/ml under the assay conditions.

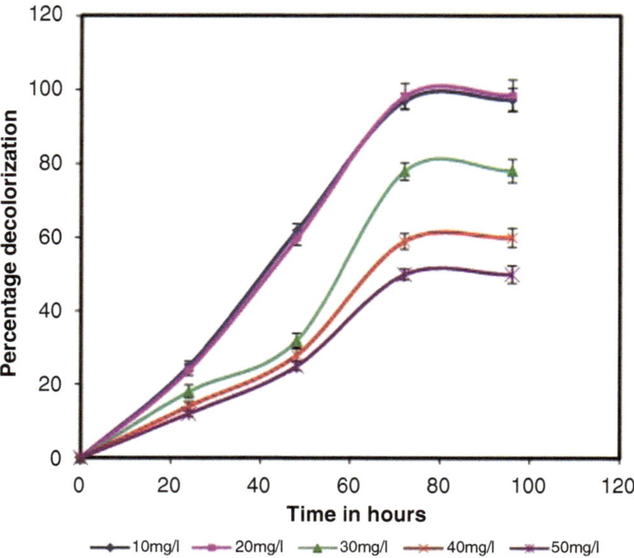

Fig. 24.1 Effect of initial concentration of Red RR on decolorization with time

HPLC Analysis

HPLC analysis of Red RR and its intermediate products on biodegradation were analyzed using Shimadzu HPLC chromatograph consisting of RP-18 (5μM) column, two pumps (model LC-20 AT), manual sample injector (Rheodyne 7725i), and diode array detector (model SPD-$M_2O_{A)}$, with wavelength in the range (190–800) nm. The mobile phase used for the study was acetonitrile: water (60: 40v/v) and was degassed using a sonicator to remove air molecules that might hamper the analysis. Samples were filtered with 0.2mm membrane filter prior to injection into HPLC. Twenty microliters of each sample was injected manually at a flow rate of 0.5ml/min, and the detection was performed at 522 and 254nm for Red RR and the degradation products, respectively. Due to the operation conditions mentioned above, the retention time of residual yeast extract tended to overlap with those of the decolorization intermediates. Thus, the bacterial decolorization intermediates prepared for the HPLC analysis were obtained from yeast-extract-free cultures. The degradation products were identified by comparison of the retention time in the test sample with that of standard solution under the same conditions (mobile phase, flow rate, temperature, and wavelength).

Results and Discussion

Effect of Initial Dye Concentration on Decolorization

Figures 24.1 and 24.2 show the effect of initial concentration (0–50ppm) of Red RR and Black GDN at 37°C and pH 7. It is clear from the figure that percentage removal of dyes increased with an increase in time irrespective of initial dye concentration. Further, percentage removal of dye decreased with an increase in dye concentration. The optimum concentration for both dyes with 2% inoculum was found to be 20ppm at which maximum decolorization was observed. Time period for maximum decolorization was found to be 72h, and it was recorded as 98.25 for Red RR and 78.47 for Black GDN. There was no observable color difference from 72 to 96h. Our study also revealed that increase in inoculum concentration can degrade dyes at higher concentration (data not shown).

Effect of pH on Dye Decolorization

Effect of pH (3–11) on decolorization at 20ppm initial concentration of Red RR and Black

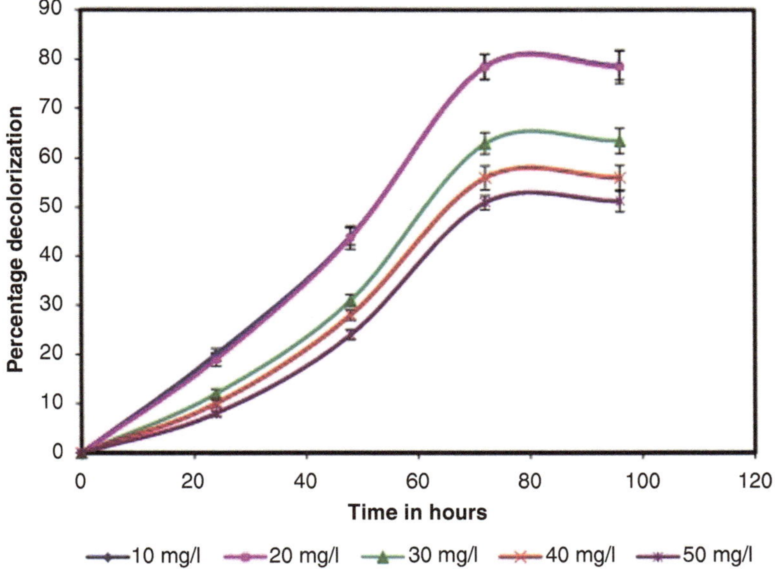

Fig. 24.2 Effect of initial concentration of Black GDN on decolorization with time

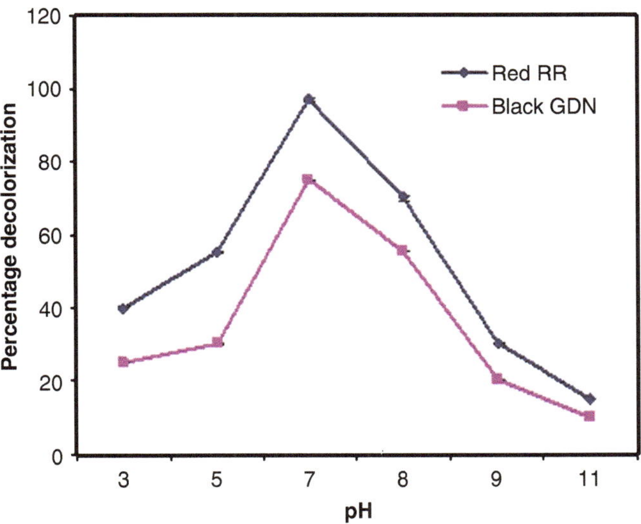

Fig. 24.3 Effect of pH on dye decolorization

GDN at 37°C is shown in Fig.24.3. Maximum color removal of 97 and 75% for Red RR and Black GDN was observed at pH 7 after 72h of incubation period. At pH 8, color removal was moderate for both the dyes. Further increase in pH beyond 8 and decrease in pH below 7 resulted in decreased color removal of dyes. The optimum pH was found to be between 7 and 8 for maximum removal of color. The pH has an important role on the effectiveness of dye decolorization, because pH tolerance of decolorizing bacteria is quite important in treatment methods [14]. The decrease in percentage of color removal may be due to the loss of cell viability or denaturation

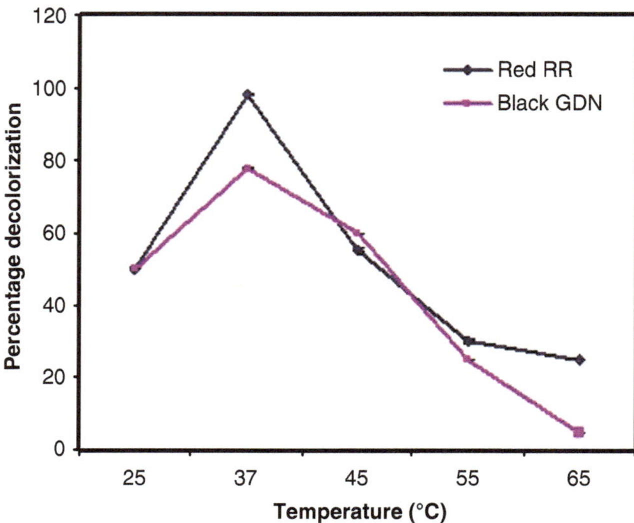

Fig. 24.4 Effect of temperature on dye decolorization

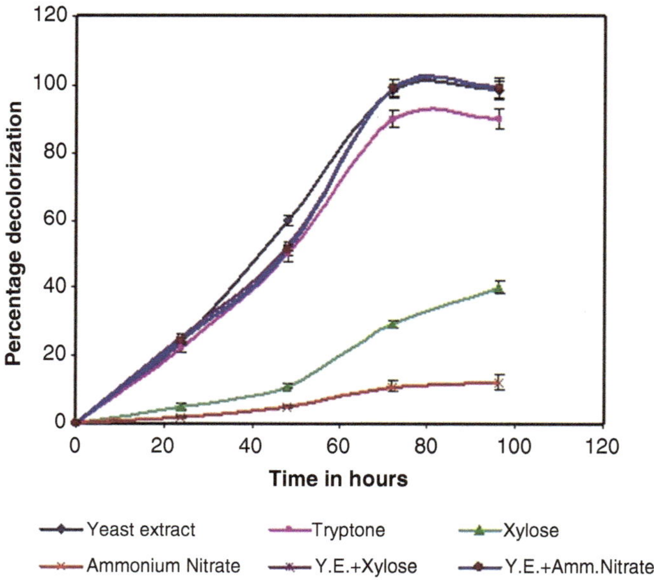

Fig. 24.5 Effect of media composition on decolorization of Red RR

of the key enzyme azoreductase under acidic and highly alkaline conditions [15].

Effect of Temperature on Dye Decolorization

Figure24.4 shows decolorization of dyes at different temperatures (25, 37, 45, 55, and 65°C) after 72h at pH 7.0 and initial dye concentration of 20ppm. It was observed that the percentage of color removal for both dyes increased with an increase in temperature from 25 to 37°C. The percentage removal of dye decreased with further increase in temperature up to 45°C. Decolorizing activity was significantly decreased by raising the temperature from 45 to 65°C. It is reported that at higher temperature, bacterial cell death will be high. The decrease of color removal is normally due to the loss of cell viability [10].

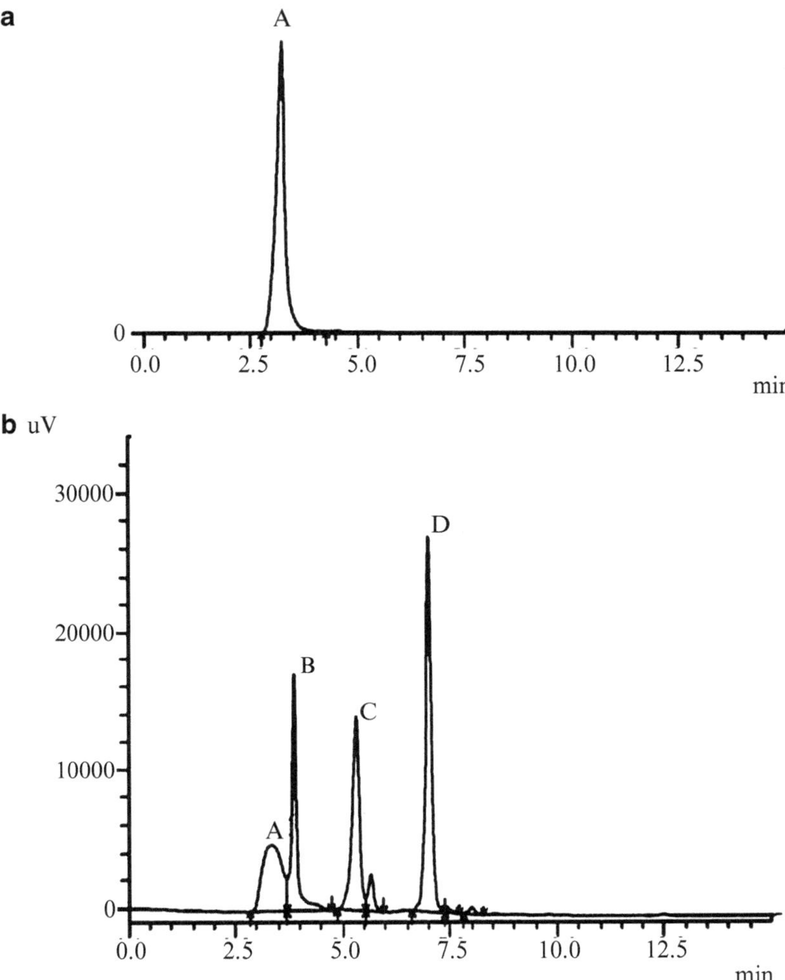

Fig. 24.6 Chromatogram of the HPLC analysis for Red RR. (**a**) At the beginning of static incubation. (**b**) After static incubation for 72h

Effect of Media Composition on Dye Decolorization

Figure 24.5 indicates the color removal of Red RR under different media, namely, yeast extract, tryptone, xylose, ammonium nitrate, and combinations of these media at 37°C and pH 7.0. Results summarize that in presence of 0.1% yeast extract, bacterial cells exhibited a maximum decolorization of 98% within 72h. Presence of same concentration of tryptone also enhanced decolorization to 90% in the same time interval. This may be due to the reductive cleavage of reactive dyes, which is catalyzed by the action of redox mediators that transfer redox equivalents (electron donors), for example, NADH, NADPH, or FADH, to the chromophore group of the dye [10]. But the presence of xylose and ammonium nitrate led to considerable decrease in percentage decolorization. The negative effect of xylose on decolo-rization may be related to a metabolic regulation mechanism, generally recognized as catabolite repression [16], which commonly occurs in bacterial systems. In the present study, yeast extract was selected as the media component since it is cheaper than tryptone.

Azoreductase Activity for Crude Cell Extract

Decolorization study using crude cell extract suggests that the major enzyme involved in the process is azoreductase, since it was initiated by the addition of NADH. NADH is reported as the electron donor for the reduction process by azoreductase [17, 18]. Forty-four percent of the dye was decolorized within 6h, and enzyme activity was found to be 22.4U/ml. This study reveals that oxygen insensitive azoreductases have considerable potential for the purely aerobic treatment of wastewaters, which are polluted by highly recalcitrant dyes. Purification and characterization of azoreductases from the strain contribute to understand the mechanism of degradation of dyes and making it possible for treatment processes.

HPLC analysis for the sample taken at the beginning of static incubation showed a major peak (peak A) appears at a retention time of 3.26min (Fig.24.6a), which represents the retention time of pure reactive Red RR, since decolorization did not occur at the early stage of static incubation. As the decolorization proceeded, the intensity of peak A decreased, along with an increase in intensities of peak B (retention time = 3.873min), peak C (retention time = 5.322min), and peak D (retention time = 7.008min). As the operation lasted for 72h, the intensity of peak A dropped about 80%, and the intensities of peaks B, C, and D were found to be increased. At this point, it is assumed that the bacteria caused the cleavage of reactive Red RR which decomposed to form three intermediates represented by peaks B, C, and D (Fig.24.6b). Elucidation of intermediate compounds has to be done.

Conclusions

The newly isolated bacterial strain has excellent potential for degradation and decolorization of reactive dyes, Red RR and Black GDN. Decolorization efficiency is strongly affected by the medium compositions, in which yeast extract and tryptone are vital to the decolorization triggered by enzymatic reduction of dyes. The strain exhibited good decolorization ability at pH from 7 to 8 and temperature 37°C. Based on the HPLC data, decolorization of reactive red RR by the strain seems to involve breakdown of reactive bonds to form three intermediary compounds. Studies with cell-free extract suggest that azoreductase is the key enzyme for decolorization of dyes. More studies are in progress in our laboratory in order to cram the heterologous expression of this enzyme and to understand its mechanism of action on the selected dyes.

References

1. Siddiqui MS, Andleeb S, Ali N, Ghumro PB, Ahmed S. Up-flow immobilized fungal column reactor for the treatment of Anthraquinone dye Drimarene Blue K2RL. Afr J Biotechnol. 2009;8:5570–7.
2. Olukanni OD, Osuntoki AA, Gbenle GO. Textile effluent biodegradation potentials of textile effluent-adapted and non-adapted bacteria. Afr J Biotechnol. 2006;5:1980–4.
3. Xu M, Guo J, Cen Y, Zhong X, Cao W, Sun G. A dye decolorizing bacterium isolated from activated sludge of a waste-water treatment plant. Int J Syst Evol Microbiol. 2005;55:363–8.
4. Conneely A, Smytha WF, McMullana G. Metabolism of the phthalocyanine textile dye remazol turquoise blue by *Phanerochaete chrysosporium*. FEMS Microbiol Lett. 2006;179:333–7.
5. Sofia N, Rakhshanda N, Muhammad A, Amer J. Accelerated biodecolorization of reactive dyes with added nitrogen and carbon sources. Int J Agric Biol. 2010;12:426–30.
6. Iqbal MK, Nadeem A, Shafiq T. Biological treatment of textile waste water by activated sludge process. J Chem Soc Pak. 2007;29:397–400.
7. Droste RL. Theory and practice of water and wastewater treatment. Singapore: Wiley; 2004.
8. Maier J, Kandelbauer A, Erlacher AC, Paulo A, Gubits GM. A new alkali – thermostable azoreductase from *bacillus* sp. strain SF. Appl Environ Microbiol. 2004;70:837–44.
9. Sapari N. Treatment and reuse of textile wastewater by overland flow. Desalination. 1996;106:179–82.
10. Chang JS, Chou C, Chen SY. Decolorization of azo dye with immobilized cells of *Pseudomonas luteola*. Process Biochem. 2001;36:757–63.
11. Chang JS, Lin YC. Fed-batch bioreactor strategies for microbial decolorization of azo dye using a *Pseudomonas luteola* strain. Biotechnol Prog. 2000;16:979–85.
12. Kalme SD, Parshetti GK, Jadhav SU, Govindwar SP. Biodegradation of benzidine based dye Direct Blue-6

by *P. desmolyticum*. Bioresour Technol. 2007;98: 1405–10.

13. Zimmermann T, Kulla HG, Leisinger T. Properties of purified orange I1 azoreductase, the enzyme initiating azo dye degradation. Electrophoresis. 1982; 203:197–203.
14. Chen KC, Huang WT, Wu JY, Houng JY. Microbial decolorization of azo dyes by *Proteus mirabilis*. J Ind Microbiol Biotechnol. 1999;23:686–90.
15. Aksu Z. Reactive dye bioaccumulation by *Saccharomyces cerevisiae*. Process Biochem. 2003;10: 1437–44.
16. White D. The physiology and biochemistry of prokaryotes. New York: Oxford University Press Inc; 1995.
17. Mazumdar R, Logan JR, Mikell AT, Hooper SW. Characteristics and purification of an oxygen insensitive azoreductase from *Caulobacter subvibrioides* strain C_7D. J Ind Microbiol Biotechnol. 1999;23: 476–83.
18. Ghosh DK, Mandal A, Chaudhuri J. Purification and partial characterization of two azoreductase from *Shigella dysenteriae* type I. FEMS Microbiol Lett. 1992;98:229–34.

Analysis of the Pathway of Phenol Biodegradation by *Alcaligenes* sp. *d2*

25

Merlin Antony, Indu C. Nair, and K. Jayachandran

Abstract

Alcaligenes sp. *d2*, an aerobic bacterium, was proved earlier to be a promising strain for the biodegradation of phenol. The rate-limiting step in the aerobic degradation of phenol by *Alcaligenes* sp. *d2* is the formation of catechol from phenol by the action of the enzyme phenol hydroxylase. The presence of phenol hydroxylase in the culture supernatant strongly supports the fact that *Alcaligenes* sp. *d2* starts the degradation of phenol by converting it to catechol. The optimum condition for the production of phenol hydroxylase by the organism was done by statistical modeling with Design-Expert software 7.1.4 – Plackett-Burman. The optimum conditions for the production of the enzyme were found to be temperature, 40°C; pH, 9.8; incubation period, 16 h; phenol, 200 μl (5%); catechol, 1 mg ml^{-1}; $MgSO_4$, 5 mg ml^{-1}; and KH_2PO_4, 100 mg ml^{-1}. The organism degrades phenol completely to Krebs cycle intermediates through the meta pathway of degradation. This fact is supported by the data obtained from the FT-IR analysis of the samples taken at different time intervals. FT-IR bands obtained from samples taken at 16 h (1,725 cm^{-1}), 24 h (2,960 cm^{-1}, 1,705 cm^{-1}), 32 h (1,703 cm^{-1}), and 40 h (1,729 cm^{-1}) add on to the fact that *Alcaligenes* sp. *d2* follows the meta pathway o f biodegradation.

Keywords

Phenol biodegradation • *Alcaligenes* sp. *d2* • Phenol hydroxylase • Meta pathway

M. Antony
School of Biosciences, Mahatma Gandhi University, Kottayam, Kerala 686560, India

I.C. Nair
Department of Biotechnology, SASSNDP Yogam College, Konni, Pathanamthitta, Kerala, India

K. Jayachandran (✉)
School of Biosciences, Mahatma Gandhi University, Kottayam, Kerala 686560, India,

Department of Biotechnology, SASSNDP Yogam College, Konni, Pathanamthitta, Kerala, India
e-mail: jayansbs@gmail.com

A. Sabu and A. Augustine (eds.), *Prospects in Bioscience: Addressing the Issues*,
DOI 10.1007/978-81-322-0810-5_25, © Springer India 2013

Introduction

Phenol and its derivatives are the basic structural unit in a wide variety of synthetic organic compounds [1]. Phenol is more commonly produced artificially from industrial activities. Currently, the largest use of phenol is as an intermediate in the production of phenolic resins, which are used in the plywood, adhesive, construction, automotive, and appliance industries. Phenol is also used in the production of synthetic fibers such as nylon and for epoxy resin precursors such as bisphenol A. Phenol is water soluble and highly mobile [2], and, hence, waste waters generated from industrial activities contain high concentrations of phenolic compounds [3]. This may eventually reach down to streams, rivers, lakes, and soil representing a serious ecological problem [4]. Phenol is a listed priority pollutant by the US Environmental Protection Agency [5] and is considered to be a toxic compound by the Agency for Toxic Substances and Disease Registry [6].

Pollution due to phenolic compounds has been drawing attention of the research community for the past few years. Many technologies have been investigated for removing phenolic compounds from environment. They include adsorption [7], biodegradation [8, 9], UV/Fe^{+3} [10], extraction by liquid membrane [11], chemical oxidation [12–15], etc. Powdered animal charcoal treatment is one such strategy to eliminate phenol from industrial effluents. But these methods are incompetent as far as the complete removal of the pollutant from environment is concerned. They also have serious drawbacks such as high cost and formation of hazardous by-products.

Because of the widespread occurrence of phenol in the environment, many microorganisms have started utilizing phenol as the sole carbon and energy source. Biodegradation of phenol follows both aerobic and anaerobic pathways. Aerobic biodegradation of phenol has been studied in the early nineteenth century. The common first step in the aerobic degradation of phenols is its hydroxylation to catechol. Phenol hydroxylase (PH) enzyme catalyze the hydroxylation of phenol to catechol. This reaction is considered to be the first and rate-limiting step in the aerobic phenol-degrading pathway [16].

In aerobic degradation of phenol, the enzyme phenol hydroxylase adds a hydroxyl group to phenol utilizing molecular oxygen and a molecule of $NADH_2$ to form catechol (1,2-dihydroxybenzene). Catechol is then degraded via two alternative pathways: ortho pathway or β-ketoadipate pathway and meta pathway. It is further metabolized to Krebs cycle intermediates. In meta pathway, enzyme catechol 2,3-dioxygenase transforms catechol to 2-hydroxymuconic semialdehyde. This compound is metabolized further to intermediates of Krebs cycle.

We have been doing research in phenol biodegradation for the last few years. The organism we used was *Alcaligenes* sp. d_2, isolated from soil through soil enrichment culture technique.

Alcaligenes sp. d_2 is an efficient phenol-degrading microorganism and is capable of withstanding high concentrations of phenol. Its phenol-degrading property and the conditions required for degradation were reported in soil science [17]. We could use this organism for the better treatment of phenolic paper factory effluent [9]. Later we could also establish the fact that the organism while removing phenol was capable of progressively accumulating PHB, the mostly recommended biopolymer [18]. The present work mainly focuses on establishment of the role of phenol hydroxylase and elucidation of the pathway of phenol biodegradation by *Alcaligenes* sp. d_2.

Experimental

Strain and Source

The bacterium *Alcaligenes* sp. d_2 was used in the present study. The strain was collected from the culture collection of School of Biosciences, Mahatma Gandhi University Campus, Kottayam.

Estimation of Phenol

Phenol was estimated as per the protocol by [19].

Biodegradation

Hundred milliliter of mineral salt phenol medium with pH 6.5 and 200 μl of 5% phenol substrate were prepared in different flasks and inoculated with 3% inocula of the strain. The culture was then incubated for different time intervals in a 150 rpm shaker at room temperature. After every 8-h time interval, the media was centrifuged at 10,000 rpm for 15 min, supernatants was collected and assayed for the presence of phenol.

Enzymatic Assay of Phenol Hydroxylase

Phenol hydroxylase was estimated as per the protocol by Gurujeyalakshmi et al. [20]. One unit of phenol hydroxylase activity was defined as 1 μmol of phenol conversion per minute under the assay conditions.

Optimization of the Conditions for Phenol Hydroxylase Production

Hundred milliliter of mineral salt phenol medium at pH 6.5 and 200 μl of 5% phenol as substrate was inoculated with 3% inoculum. The broth was incubated for 24 h at room temperature in a shaker at 150 rpm for enzyme production. After incubation, the broth was centrifuged at 10,000 rpm for 15 min. Supernatant was collected and assayed for the presence of phenol hydroxylase enzyme. The conditions favoring the maximum secretion of the enzyme was found to be the optimum condition. Optimization of the process conditions for the production of phenol hydroxylase was done with Design-Expert software 7.1.4 – Plackett-Burman, downloaded from the website www.statease.com. Eleven parameters with 2 options were given to the software which gave 12 trials to be performed.

Analysis of Degradation Products by GC-MS and FT-IR

The supernatants collected after biodegradation at 8-h time interval were extracted with diethyl ether in a separating funnel. Separating funnel was filled with equal volumes of supernatant and diethyl ether. The mixture was shaken well and kept undisturbed for half an hour. The ether portion was collected in a beaker and kept in water bath at 50 °C for removing ether by evaporation. A control (mineral salt phenol medium without inoculum) was also extracted with diethyl ether. The extracts were collected and subjected to GC-MS and FT-IR analysis. Both control and samples were extracted at different time intervals of 8 h and were subjected to GC-MS and FT-IR analysis. Spectroscopic analysis was done at National Institute of Interdisciplinary Science and Technology (NIIST), Trivandrum.

Results and Discussion

The present research work was carried out to evaluate the pathway of phenol degradation followed by *Alcaligenes* sp. d_2. The medium used was mineral salt phenol medium (MSPM), and the process was carried out under optimized laboratory conditions [17].

Biodegradation of phenol follows both aerobic and anaerobic pathways. *Alcaligenes* sp. d_2 is an aerobic organism. The aerobic biodegradation of phenol occur either through ortho pathway or by meta pathway. In both cases, the first step is the conversion of phenol to catechol by the enzyme phenol hydroxylase.

Phenol hydroxylase is a flavoprotein enzyme, and its action (conversion of phenol to catechol) is considered to be the rate-limiting step in the whole process of phenol biodegradation by aerobic pathway. The most important function of phenol hydroxylase is the hydroxylation of phenol, i.e., addition of one more –OH group to phenol.

This step is very important as it gives room for the subsequent enzymes to act either in ortho

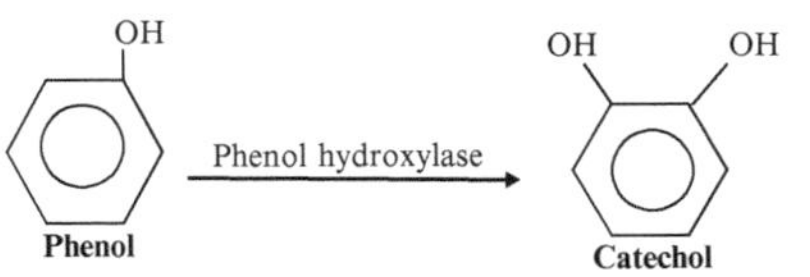

OH
Phenol
Phenol hydroxylase
OH OH
Catechol

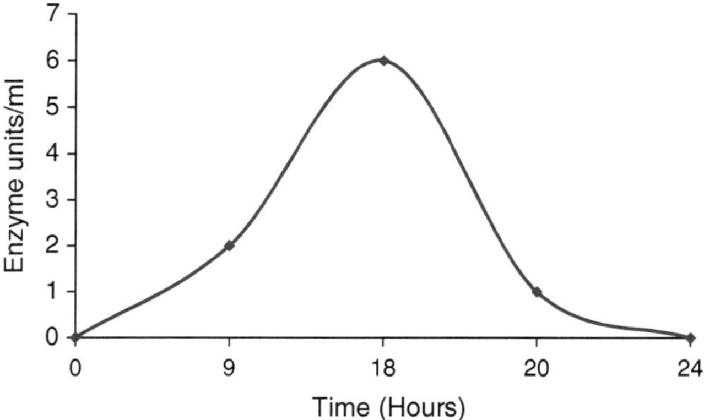

Fig. 25.1 Production of phenol hydroxylase enzyme at different incubation period

pathway or in meta pathway to cleave the benzene ring. Our first objective was to establish the presence of phenol hydroxylase in the culture supernatant. Phenol hydroxylase activity was evaluated in the culture supernatants at various time intervals and was plotted with activity against time (Fig. 25.1).

The strain produced phenol hydroxylase enzyme during the initial stages of phenol degradation, and its rate was found to decrease gradually. Production of enzyme was found at the maximum during 18 h of incubation. The presence of phenol hydroxylase in the culture supernatant indicated that the biodegradation of phenol by *Alcaligenes* sp. d_2 is aerobic in nature.

Phenol hydroxylase is an inducible enzyme and hence may be influenced by the presence of phenol in the medium. In addition to this, several other factors such as temperature, pH, incubation time, salt concentration, buffers, and nitrogen sources can also influence the production of phenol hydroxylase. A high level of phenol hydroxylase production can enhance the rate of phenol degradation, and, hence, it is appropriate to look into the optimum production of phenol hydroxylase enzyme.

Optimization of the production conditions of enzyme can be done by the conventional one at a time approach. This is a time-consuming experiment and requires intensive lab work. More than that, the final optimized conditions may not be contributing to the expected results. This is because in many of the conventional optimization strategies, the individual trials are found to have better yield than the finally optimized conditions. It follows that it is not the individual parameter which is very important, what matters is how the variables interact with each other in combination. A suitable combination of the various factors may be contributing better yield than other conditions.

It is with this objective that scientists all over the world are shifting their experimental strategies of media optimization to statistical modeling. A number of software are currently available for the media optimization. In the present project, we have made an attempt to optimize the conditions of phenol hydroxylase production using one such software. The software used was Design-Expert software 7.1.4 – Plackett-Burman, downloaded from the website www.statease.com.

We have analyzed the optimum production strategies using the first-level application of Design-Expert system. The first-level analysis helped us to find out the most important impact making parameters from the selected variables. The Plackett-Burman statistical model of the Design-Expert software proposes the conduct of "n" trials for $n-1$ variables. The first task before initiating the analysis is to fix up the number and nature of the variables. In this experiment, we have opted 11 variables comprising of temperature, pH, incubation period, phenol concentration, NaCl, KH_2PO_4, $MgSO_4$, $(NH_4)_2SO_4$, $CaCl_2$, and catechol. Temperature, pH, and incubation period have been incorporated as a measure to optimize the external conditions affecting the enzyme production. Phenol has been incorporated to

Table 25.1 Selected parameters and their value applied in the Plackett-Burman model for process optimization

		Values	
Sl. no.	Parameter	Highest	Lowest
1	Temperature	40 °C	30 °C
2	pH	9.8	6.8
3	Incubation time	48 h	16 h
4	5% Phenol	200 μl	0
5	NaCl	5 mg/100 ml	1 mg/100 ml
6	KH_2PO_4	100 mg/100 ml	10 mg/100 ml
7	$MgSO_4$	50 mg/100 ml	5 mg/100 ml
8	$(NH_4)_2SO_4$	100 mg/100 ml	0
9	$CaCl_2$	1 mg/100 ml	0
10	Catechol	1 mg/100 ml	0
11	Distilled water (dummy variable)	B1	B2

evaluate how far it can act as an inducer for the production of phenol hydroxylase enzyme.

In each of these parameters, two values were taken, one lower and the other higher (Table 25.1). Totally 12 trials have been conducted as per the data given by the software (Table 25.2). Finally on analyzing the impact factor contributed by the software analysis, it became evident that the most striking impact making parameters were phenol (1.14) and $(NH_4)_2SO_4$ (1.03) along with KH_2PO_4 (1.31) (Table 25.3).

The results truly agreed with the theoretical expectation that phenol hydroxylase was an inducible enzyme and get induced at a high concentration of phenol (200 μl of 5% phenol). Among the 12 trials conducted, the second trial gave the highest response of 17.70, and the second highest response obtained was 3.18, which is only 18% to the highest response (Table 25.2). Hence, it follows that the parameters fixed for the second trial yielding the highest performance was ideally suited for maximal production of the enzyme. The parameters selected were temperature, 40°C; pH, 9.8; incubation time, 16 h; phenol, 200 μl (5% phenol); NaCl, 5 mg ml^{-1}; KH_2PO_4, 100 mg ml^{-1}; $MgSO_4$, 5 mg ml^{-1}; and catechol, 1 mg ml^{-1}. $(NH_4)_2SO_4$ and $CaCl_2$ were completely absent in the medium.

The presence of phenol hydroxylase enzyme in the initial stages of phenol biodegradation established the fact that phenol biodegradation by *Alcaligenes* sp. d_2 was aerobic and that it followed either ortho pathway or meta pathway. Hence, the second objective was to find whether the organism follows ortho pathway or meta pathway. Ortho pathway has characteristic intermediates like catechol, cis,cis-muconic acid, mycolactone, β-oxoadipate, acetyl CoA, and succinyl CoA, whereas in meta pathway the intermediates are catechol, 2-hydroxymuconic semialdehyde, 2-oxopent-4-enoate, oxaloacetate, acetaldehyde, and pyruvate.

The identification of the nature of degradative pathway followed by *Alcaligenes* sp. d_2 was done by establishing the presence of any one class of the intermediates during the degradation of phenol. The experimental strategy we took for the above objective was GC-MS and FT-IR analysis.

GC analysis was unique in representing the number of structurally different components in the sample. MS was indicative of the mass fractionation of the component peaks obtained from the GC spectrum. GC analysis of the control (non-biodegraded phenol) gave a major peak at 7.008 min, and the MS analysis confirmed that the peak represented phenol with molecular weight 94 (Figs. 25.2 and 25.3).

GC analysis of the samples at varied incubation period gave ample evidence showing the degradation of phenol and generation of new intermediates. The peak of phenol indicated at 7.008 RT in the control gets progressively reduced resulting in the formation of many other peaks in the subsequent samples (Fig. 25.4).

Table 25.2 Yield obtained for phenol hydroxylase as per the trials conducted by the software analysis

Sl. no.	Temperature	pH	Incubation time	Phenol 5% μl	NaCl	KH_2PO_4	$MgSO_4$	$(NH_4)_2SO_4$	$CaCl_2$	Catechol	Distilled water	Response
1	30	6.8	16	200	1	100	50	0	1	1	B2	3.18
2	40	9.8	16	200	5	100	5	0	0	1	B1	17.70
3	30	6.8	16	0	1	10	5	0	0	0	B1	.84
4	40	9.8	16	0	1	100	5	100	1	0	B2	.35
5	30	9.8	48	0	5	100	50	0	0	0	B2	1.20
6	30	9.8	48	200	1	10	5	100	0	1	B2	0
7	40	9.8	48	0	1	10	50	0	1	1	B1	1.13
8	40	6.8	16	0	5	10	50	100	0	1	B2	.212
9	40	6.8	48	200	1	100	50	100	0	0	B1	2.3
10	30	6.8	48	0	5	100	5	100	1	1	B1	1.4
11	30	9.8	16	200	5	10	50	100	1	0	B1	1.4
12	40	6.8	48	200	5	10	5	0	1	0	B2	.84

Table 25.3 Analysis of yields and evaluation of the impact of the variables

Sl. no.	Variables	Temperature	pH	Incubation time	Phenol 5%	NaCl	KH_2PO_4	$MgSO_4$	$(NH_4)_2SO_4$	$CaCl_2$	Catechol	Distilled water
1	ΣH	22.53	21.78	6.87	25.42	22.75	26.13	9.42	5.66	8.3	23.62	24.77
2	ΣL	8.02	8.77	23.68	5.13	7.8	4.42	21.13	24.89	22.25	6.93	5.78
3	Difference ($\Sigma H - \Sigma L$)	14.51	13.01	−16.81	20.29	14.95	21.71	−11.71	−19.23	−13.95	16.69	18.99
4	Effect $\frac{(\Sigma H - \Sigma L)}{6}$	2.42	2.17	−2.80	3.38	2.5	3.62	−1.95	−3.20	−2.32	2.78	3.16
5	Mean square $\frac{(\Sigma H - \Sigma L)^2}{6}$	.488	.392	.653	.952	.520	1.092	.316	.853	.448	.644	–
6	Error Square	–	–	–	–	–	–	–	–	–	–	.832
7	$\frac{\text{Mean Square}}{\text{Error Square}}$ (Impact Factor)	.59	.47	.78	**1.14**	.63	1.31	.38	1.03	.54	.77	–

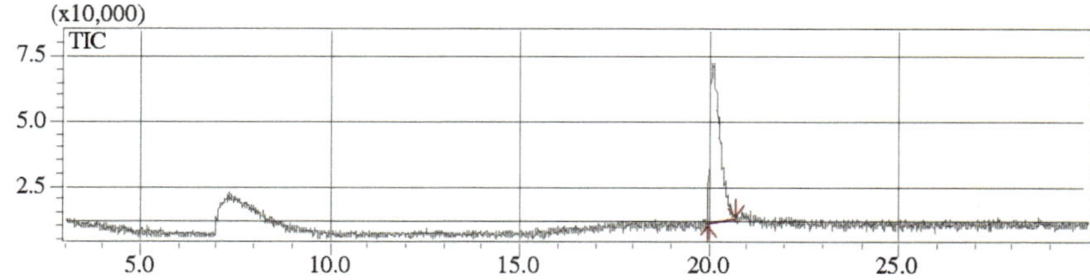

Fig. 25.2 Gas chromatogram of control: S1 (non-biodegraded phenol)

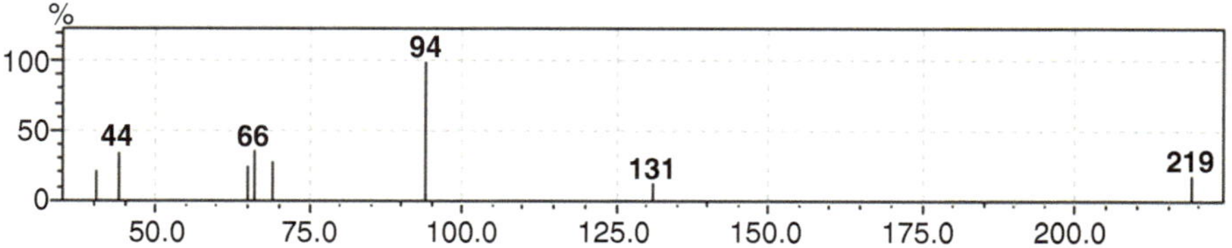

Fig. 25.3 Mass spectrogram of the peaks obtained at 7.008 in the gas chromatogram of the control (non-biodegraded phenol). RT = 7.008

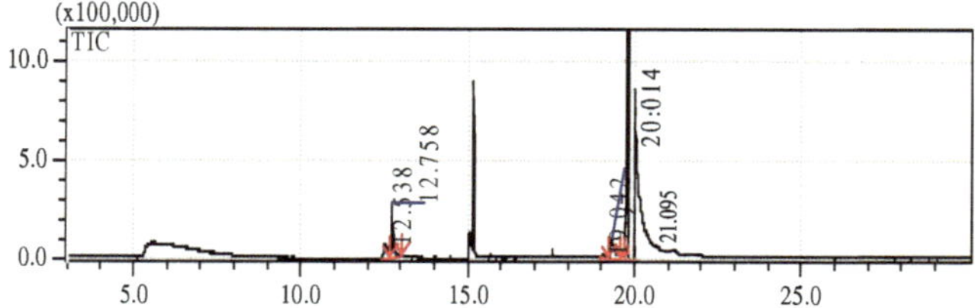

Fig. 25.4 Gas chromatogram of biodegraded phenol after 40 h of incubation: S6

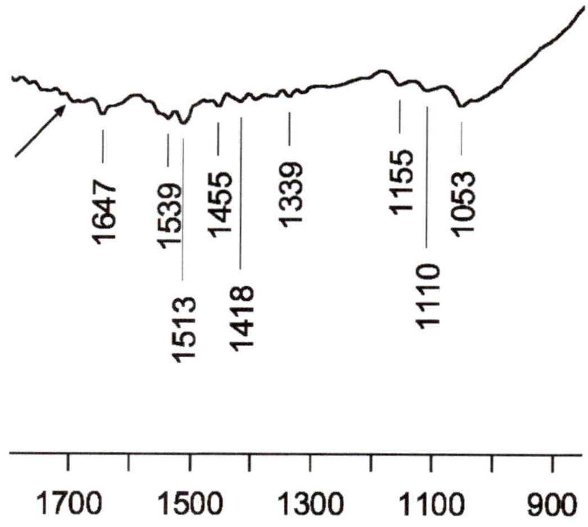

Fig. 25.5 FT-IR of control: S1 (non-biodegraded phenol)

We could analyze the sample extracts at different time intervals using FT-IR. The FT-IR of the control (Fig. 25.5) gave a definite indication of the presence of phenol in the non-biodegraded sample. The characteristic bands were observed at a wave number of 3,731, 1,647, 1,455, and at 1,155 cm^{-1}. There were no bands in the range of 1,700–1,750 cm^{-1}.

The FT-IR analysis of the samples taken at 8 h (Fig. 25.6), 16 h (Fig. 25.7), 24 h (Fig. 25.8),

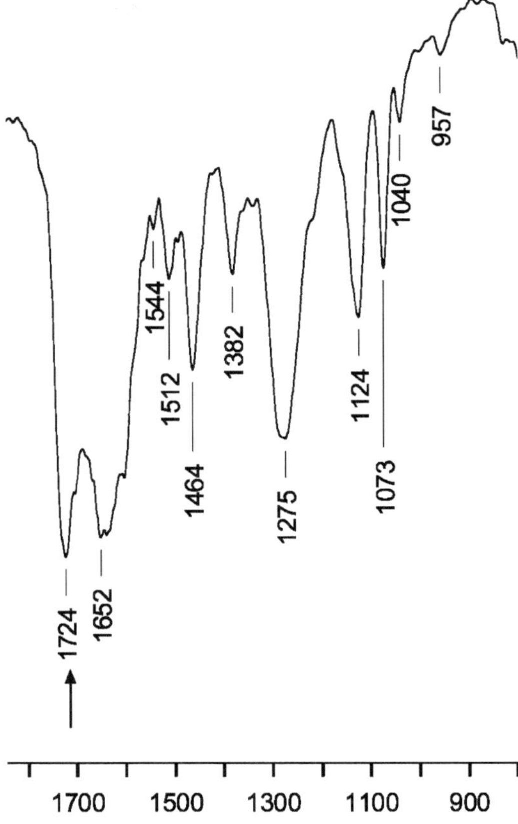

Fig. 25.6 FT-IR of biodegraded phenol after 8 h of incubation

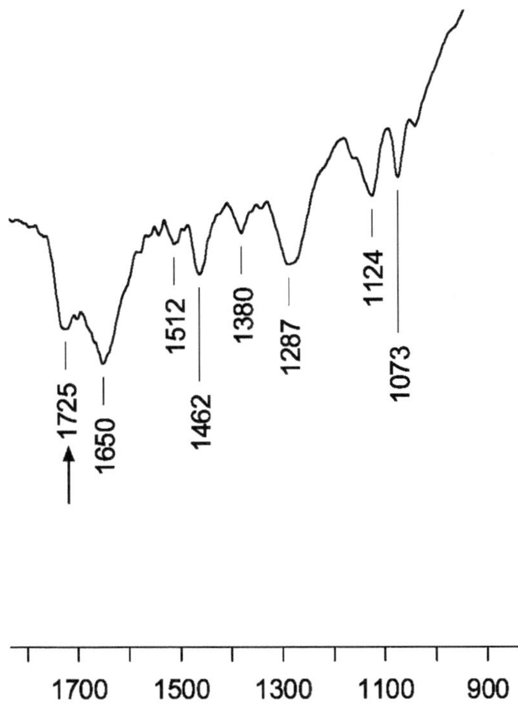

Fig. 25.7 FT-IR of biodegraded phenol after 16 h of incubation

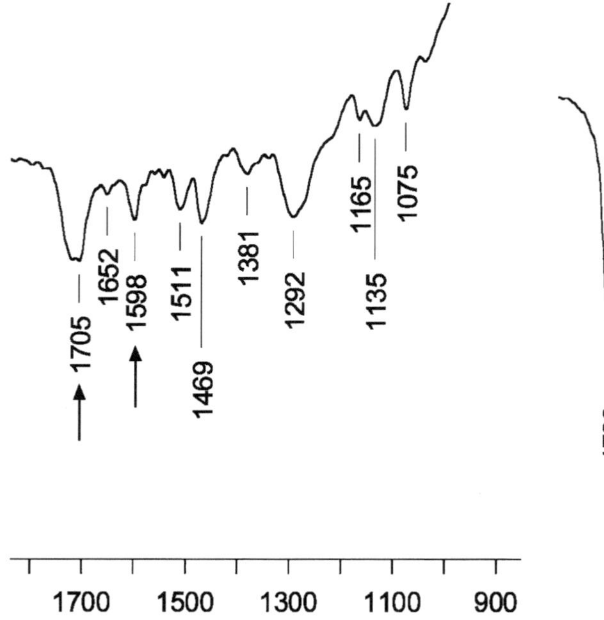

Fig. 25.8 FT-IR of biodegraded phenol after 24 h of incubation

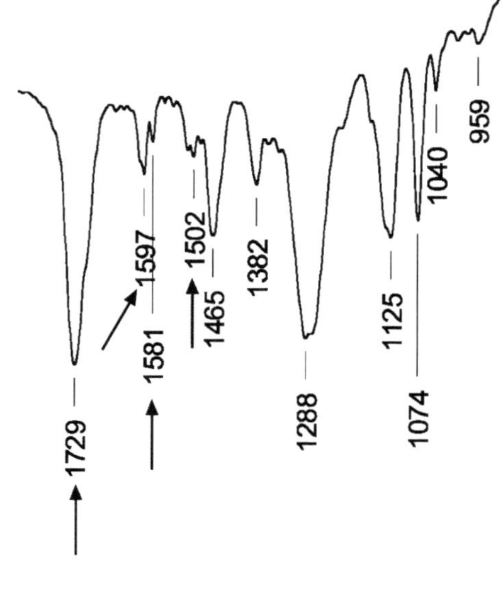

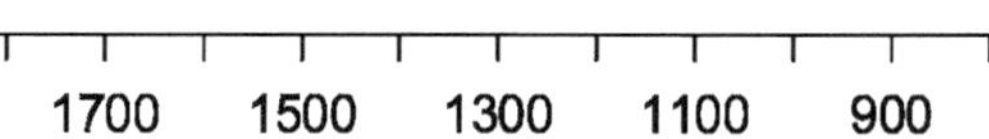

Fig. 25.10 FT-IR of biodegraded phenol after 40 h of incubation

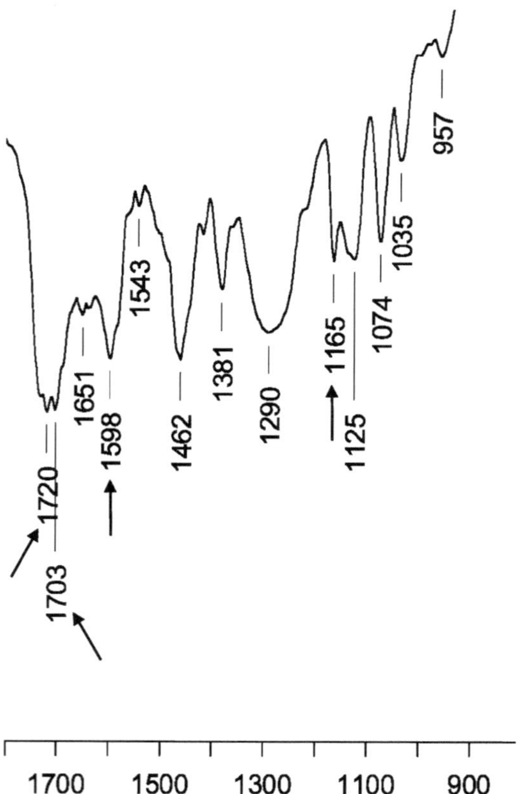

Fig. 25.9 FT-IR of biodegraded phenol after 32 h of incubation

32 h (Fig. 25.9), and 40 h (Fig. 25.10) gave strong characteristic indication of aerobic breakdown of the aromatic ring with new bands appearing in the range 1,750–1,000 cm^{-1}. Many of the bands appeared in this range were characteristic of meta pathway.

The FT-IR analysis of the sample after 8 h (Fig. 25.6) brought new bands at 2,929 and 2,862 cm^{-1}, which were characteristic of acid: hydrogen bond (O-H) stretching. Together with this, there was a characteristic band at 1,724 cm^{-1} which gives the strong indication of the presence of C=O group.

FT-IR analysis of the sample after 16 h (Fig. 25.7) gave strong representation of band at 1,725 cm^{-1}, characteristic of the aldehyde, 2-hydroxymuconic semialdehyde.

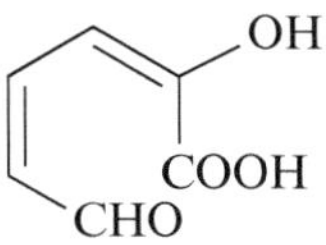

OH
COOH
CHO

2-hydroxymuconic semialdehyde

The presence of aldehyde can be accepted as a strong indication toward meta pathway, as the ortho pathway does not contain any aldehyde intermediates in the beginning.

In the FT-IR spectrum of sample at 24 h (Fig. 25.8), there was a band at 2,960 cm^{-1}, indicating the presence of aliphatic group, which again confirmed the cleavage of aromatic ring inducing aliphatic ring generation. The shifting of the band toward 1,705 cm^{-1} also indicated the oxidation of aldehyde to ketone, which took place in meta pathway during the conversion of 2-hydroxymuconic semialdehyde to 2-oxopent-4-enoate.

The presence of 1,703 cm^{-1} band in the FT-IR of the sample after 32 h (Fig. 25.9) clearly indicated the presence of two carboxylic acids, which took part in the formation of oxaloacetate.

Oxaloacetate

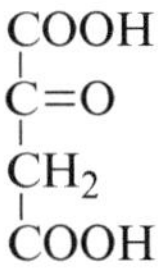

COOH
|
C=O
|
CH_2
|
COOH

Oxaloacetate

Finally, in the FT-IR spectrum of sample taken after 40 h (Fig. 25.10), there was strong indication of CH_3—C—CH_3 and aldehyde. CH_3 was represented by the reappearance of the band at 2,960 cm^{-1}. There was a strong band at 1,729 cm^{-1} indicating the presence of aldehyde. Finally, the spectrum represented the presence of alkyl aldehyde in the sample. This was mostly indicative of acetaldehyde (CH_3—CHO) formed in the last stage of meta pathway. From the above data, we could state that *Alcaligenes* sp. d_2 followed meta pathway for the biodegradation of phenol.

The present study has proved that *Alcaligenes* sp. d_2 is a promising strain for the biodegradation of phenol. It could degrade phenol even at high concentrations. The industrial effluents before being released into the environment can be treated with a suspension of this strain to remove phenol, which reduces its hazardous effect on environment. The strain can also be recombinantly modified to increase its potency in degradation. The intermediary metabolites PHB and catechol, produced by *Alcaligenes* sp. d_2, are important intermediates in the synthesis of biopolymers, pharmaceuticals, agrochemicals, flavors, antioxidants, etc. If utilized effectively, the strain will be a boon to mankind in the fields of food industry, pollution control, pharmaceuticals, biopolymer production, etc.

OH
COOH
CHO
O
COOH

2-hydroxymuconic semialdehyde → **2-oxopent-4-enoate**

References

1. Annadurai G, Balan MS, Murugesan T. Design of experiments in the biodegradation of phenol sing immobilized Pseudomonas pictorum (NICM-2077) on activated carbon. Bioprocess Eng. 2000;22:101.
2. Collins LD, Daugulis AJ. Biodegradation of phenol at high initial concentration in two-phase partitioning batch and fed-batch bioreactors. Biotechnol Bioeng. 1997;55(1):155.
3. Chang YH, Li CT, Chang MC, Sheih WK. Batch phenol degradation by Candida tropicalis and its fusant. Biotechnol Bioeng. 1998;60(3):391.
4. Fava F, Armenante PM, Kafkewitz D, Marchetti L. Influence of organic and inorganic growth supplements on the aerobic biodegradation of chlorobenzoic acid. Appl Microbiol Biotechnol. 1995;43(1):171.
5. Environmental Protection Agency. Phenol ambient water quality criteria. Washington, DC: Office of planning and standards. Environmental Protection Agency; 1979. BB296786.
6. Agency for Toxic Substances and Disease Registry (ATSDR). U.S department of health and human

services, Public health service. Medical management guidelines for phenol. 2003.

7. Rengaraj S, Moon SH, Sivablan R, Arabind B, Murugesan V. Removal of phenol from aqueous solution and resin manufacturing industry wastewater using an agricultural waste: rubber seed coat. Hazard Mater. 2002;89:185.
8. Miland E, Smyth MR, Fagain CO. Phenol removal by modified peroxidases. J Chem Technol Biotechnol. 1996;67:227.
9. Indu CN, Shashidhar S. Treatment of paper factory effluent using a phenol degrading Alcaligenes sp. Under free and immobilized conditions. Bioresour Technol. 2007;98:714.
10. Zhou M, Wu Z, Dahui W. Promoted UV/Fe^{+3} process for p-nitrophenol degradation by electroanalysis. In: Proceedings of The International Water Association Conference on Water and Wastewater management for developing countries, PWTC, Malaysia, vol. 1, p. 222, October 2001.
11. Lin SH, Pan CL, Leu HG. Liquid membrane extraction of 2-chlorophenol from aqueous solutions. Hazard Mater. 1999;65:289.
12. Comninellis C, Pulgarin C. Anoxic oxidation of phenol for wastewater treatment. J Appl Electrochem. 1991;21:703.
13. Tahar NB, Savall A. Mechanistic aspects of phenol electrochemical degradation by oxidation on a Ta/PbO_2 anode. J Electrochem Soc. 1998; 145:3427.
14. Polcaro AM, Palmas S, Renoldi F, Mascia M. On the performance of Ti/SnO_2 anodes in electrochemical degradation of 2-chlorophenol for wastewater treatment. J Appl Electrochem. 1999;29:147.
15. Awad YM, Abuzaid NS. The influence of residence time on the anodic oxidation of phenol. Sep Purif Technol. 2000;18:227.
16. Hinto S, Watanabe K, Takahashi N. Phenol hydroxylase cloned from Ralstonia eutropha strain E2 exhibits novel kinetic properties. Microbiology. 1998; 44:1765.
17. Indu CN, Shashidhar S. Microbial degradation of phenol by a species of Alcaligenes isolated from tropical soil. Soil Sci. 2004;3:47.
18. Indu CN, Shashidhar S. Accumulation of polyhydroxybutyrate in Alcaligenes sp. d_2 under phenol stress. Appl Bxiochem Biotechnol. 2009;159: 545.
19. Mordocco A, Kuck C, Jenkins R. Continuous degradation of phenol at low concentration using immobilized Pseudomonas putida. Enzyme Microb Technol. 1999;25:530.
20. Gurujeyalakshmi G, Oreil P. Isolation of phenol degrading bacillus stearothermophilus and partial characterization of the phenol hydroxylase. Appl Environ Microbiol. 1988;55:500.

26 Improvement of Vermicompost: Influence of Feeding Materials and Inoculation of Nitrogen-Fixing and Phosphate-Solubilising Bacteria

Md. Nasim Ali, S. Chakraborty, Prosanta Saha, and Niladri Lodh

Abstract

Rapid urbanisation, encroachment of fertile area and booming population are leading to generation of massive amount of waste. Vermicompost has been proved as potential technology for converting different organic solid waste including municipal waste and agricultural waste into valuable wealth through a type of biological alchemy of earthworm. The egested cast can be used to improve the fertility and physical characteristics of soil. So definitely the material that ingested by the *epigeic* earthworm could be one of the source of variation of quality of the ultimate product. The enrichment can also be achieved by inoculating microorganisms at initial stage of decomposition. With these two aims, the present work was carried out to study the effect of different organic wastes, namely water hyacinth, cow dung, green gram, vegetable waste and poultry litter, with microbial inoculants on chemical and biochemical properties of vermicompost. Significant variation with respect to microbial count both at initial stage and final product was observed with exception like total bacterial count at initial stage and total fungal count and *Pseudomonas* count in final product. Inoculation of microorganisms increased the P content in vermicompost. During the incubation period, the inoculated bacterial strains proliferated rapidly, solubilised added and native phosphate. It was also observed that the percentage increase in microbial population was found maximum in water hyacinth [327.78%]. It is evident from results that the significant increase of total phosphorus and total potassium and significant decrease in TOC, EC and pH in final vermicompost of different animal/agro and kitchen wastes by the use of earthworm *Eisenia foetida* are due to the vermic activity.

Keywords

Vermicompost • Microbial inoculants • Biochemical properties.

M.N. Ali (✉) • S. Chakraborty • P. Saha • N. Lodh
IRDM Faculty Centre, Ramakrishna Mission Vivekananda University, R.K. Mission Ashrama, Narendrapur, Kolkata 700103, West Bengal, India
e-mail: nasimali2007@gmail.com

A. Sabu and A. Augustine (eds.), *Prospects in Bioscience: Addressing the Issues*,
DOI 10.1007/978-81-322-0810-5_26, © Springer India 2013

Introduction

Vermicomposting has been identified as one of the potential option in managing biodegradable solid waste [1–4] in a cost-effective manner [5, 6]. Different organic waste has been converted into vermicompost by different sp. of earthworm including cattle dung [7], poultry droppings [8], water hyacinth [9] and animal waste [10–12]. The role of *Eisenia foetida* has also been studied by several workers [13, 14]. Reinecke et al. [15] included *Eisenia foetida* as the most eurythermal species of epigeic earthworms. Thermo composting or pre-composting (prior to vermicomposting) of the organic waste for several days is very effective in reduction of mass of the solid waste and pathogen reduction [3]. Because of earthworm activity on pre-composted organic waste, much more fragmented, porous and microbially active manures than parent materials were produced [16] and that could be effectively used as nutrient source of crop plants [17, 18] and also as soil conditioner [4]. Vermicompost has more available nutrient than organic substance from which it is produced [19]. But there are only few studies on the influence of organic waste on physicochemical qualities of the vermicompost and its improvement with respect to plant nutrient content. Nitrogen-fixing bacteria and phosphate-solubilising bacteria could be inoculated for the enrichment of the product [20] because microbial activity has the capacity to increase the availability of nutrient [21]. Nitrogen-fixing bacteria are helpful in fixing N_2 as well in solubilising phosphorus due to production of organic acids and enzymes [22]. Keeping all this in mind, the present study was conducted to assess the potential of *Eisenia foetida* in composting of different types of organic materials and quality of vermicompost thus produced. Attempt was also made to bring changes in N and P components of vermicompost by inoculating of N-fixing and P- solubilising bacteria.

Materials and Methods

Collection of Organic Wastes

Four plant-based (water hyacinth, green gram, paddy straw and vegetable waste from kitchen) and two types of animal waste (cow dung and poultry droppings) were selected for the study. Paddy straw, cow dung and poultry droppings from dairy unit, green gram from agricultural farm, vegetable waste from kitchen of hostels and water hyacinth from different ponds were collected. All the locations were within the premises of Ramakrishna Mission Ashrama, Narendrapur, Kolkata, India.

Collection of Earthworms

Earthworm species *Eisenia foetida* were obtained from vermicompost unit maintained by dairy unit of Ramakrishna Mission Ashrama, Narendrapur, Kolkata, India.

Experimental Setup

The work was carried out following the method proposed by Garg et al. [4] with some modifications. Earthen pots of 45 cm diameter × 23 cm height having capacity of containing 8 kg of partially decomposed waste, with a hole at the bottom, were taken for the present experiment. The composting mixture consisting of different kinds of substrates was mixed with starter material (cow dung, water and soil at a volume of 30, 10 and 1%, respectively) in 5:2 ratio. Total 5 kg of composting mixture was taken in each experimental pot to provide initial favourable environmental conditions for the worms. In the treatment pots, 50 numbers of healthy earthworms of approximately the same size (7–8 cm) and weight (2.5–3 g) were introduced after 15 days of partial decomposition of organic wastes. This was done to avoid exposure of worms to high temperature during the initial thermophilic stage of composting. Water was sprinkled daily on pots using an iron sprayer to maintain the moisture level of 55–60%. The experimental pots were kept under shade and covered with the gunny bags to avoid direct sunlight. There were three replicates for each feed mixture.

Chemical Analysis

Chemical analysis was done at two different sample, i.e. initial product (day of inoculating earthworms) and final product (after getting the mature product). The earthworms were removed manually at the end of the experiment. Determination of pH was done by a digital pH metre (Systronics-335) and electrical conductivity by a conductivity metre (Systronics-306). Total organic carbon was estimated by Walkley and Black [23] and available phosphorus by Olsen et al. [24] and total potassium by flame emission technique were estimated. All the determinations were carried out in triplicate.

Microbial Analysis

Microbial count was conducted in three replications; five plates were taken for each replication.

(a) Testing of Initial and Final Product: Each experimental treatment was analysed at two different samples, i.e. initial product (day of inoculating earthworms) and final product (after getting the mature product). After collecting 1 g of sample vermicompost, standard serial dilution technique was followed. Total numbers of colony-forming units [CFUs] were counted by pouring 1 ml aliquots (10^{-5} dilution for fungal count and 10^{-7} dilution for bacterial count) aseptically in laminar air flow chamber (pour plate technique) in different selective media, i.e. total bacterial count in nutrient agar media, total fungal count in potato dextrose agar media, phosphate-solubilising bacteria in Pikovskaya media, *Rhizobium* in yeast extract mannitol agar media, *Azotobacter* in Jensen's media, and *Azospirillum* in nitrogen-free malate media. It was gently rotated for proper mixing and incubated at 24±2°C and 32±2°C, respectively, for fungal and bacterial colony for 48–72 h in BOD incubator. Then colony-forming units for all the microbes were counted using colony counter (Chemiline-CL920).

(b) Enrichment with Microbial Inoculants: In a separate set of experiment, 500 g of partially decomposed three organic materials, i.e. water hyacinth, green gram and paddy straw, was taken in a plastic jar with perforated lid. Five earthworms were inoculated in each jar. In treatment jar nitrogen-fixing bacteria *Rhizobium* and *Azotobacter*, *phosphate-solubilising bacteria* were mixed with organic waste at 25 ml broth of all three bacterial cultures. In control jars, no external microbial inoculants were mixed. Both treatment and control were conducted in triplicate jar.

Statistical Analysis

All the reported data are the arithmetic means of three replicates. One-way analysis of variance (ANOVA) was done to determine any significant difference among the treatments analysed during vermicomposting at 0.05% level of significance [25].

Result and Discussion

The days required for the vermicomposting ranged from 38 days (cow dung) to 92 days (paddy straw) as shown in Table 26.1. Thermo composting or pre-composting time required for different feeding stock was also different. Differential composition of organic and inorganic components is responsible for duration of vermicomposting process. Besides, the favourable condition required for earthworms' efficiency in decomposition is also varying with chemical nature of organic waste [17]. Because of the presence of different decomposing community like bacteria, fungi, protozoa, nematode and actinomycetes in higher quantity in cow dung, mineralization of cattle dung is done by earthworm more easily than other organic wastes [26].

Table 26.1 Days required for production of vermicompost

Parameter	Water hyacinth (days)	Cow dung (days)	Green gram (days)	Vegetable waste (days)	Poultry litter (days)	Paddy straw (days)
Partial decomposition	12	26	32	24	35	35
Final product	34	12	50	48	39	57

Table 26.2 Chemical analysis of different feeding stock at initial (before earthworm inoculation) and final product (mean ± SD, $n=3$)

Material	Type	pH	EC (mmhos)	Organic C [%]	Available P%	Available K%
Water hyacinth	Initial	7.8 ± 0.02	3.5 ± 0.01	16.5 ± 0.11	0.08 ± 0.003	0.8 ± 0.03
	Final	7.3 ± 0.02	4.7 ± 0.01	12.25 ± 0.23	0.19 ± 0.004	1.1 ± 0.10
Cow dung	Initial	7.2 ± 0.01	2.9 ± 0.01	19.5 ± 0.13	0.16 ± 0.001	1.2 ± 0.01
	Final	6.8 ± 0.02	3.6 ± 0.02	18.2 ± 0.13	0.20 ± 0.007	1.5 ± 0.05
Green gram	Initial	8.1 ± 0.03	3.2 ± 0.03	18.75 ± 0.20	0.20 ± 0.006	1.5 ± 0.03
	Final	7.8 ± 0.01	4.3 ± 0.03	11.5 ± 0.20	0.46 ± 0.001	1.8 ± 0.08
Vegetable waste	Initial	8.0 ± 0.004	3.7 ± 0.02	22.50 ± 0.4	0.24 ± 0.00	2.1 ± 0.02
	Final	7.5 ± 0.03	4.1 ± 0.02	15.2 ± 0.09	0.28 ± 0.008	2.28 ± 0.10
Poultry litter	Initial	8.4 ± 0.03	4.0 ± 0.01	9.7 ± 0.20	0.40 ± 0.013	3.5 ± 0.06
	Final	8.3 ± 0.02	4.5 ± 0.03	7.2 ± 0.03	0.40 ± 0.001	4.2 ± 0.13
Paddy straw	Initial	7.7 ± 0.01	3.0 ± 0.03	18.7 ± 0.15	0.42 ± 0.002	0.9 ± 0.04
	Final	7.6 ± 0.01	3.2 ± 0.01	18.5 ± 0.13	0.61 ± 0.001	1.1 ± 0.04

The chemical parameters were estimated at two different stages (Table 26.2), initial estimation was done prior to earthworm inoculation, i.e. after thermo composting process, and final estimation was made at the end of completion of vermicomposting process. In all types of feeding stock, it was observed that the pH and organic carbon level decreased with comparison to its initial level. The range of pH was 6.8–8.3 where cow dung was having an initial pH of 7.2 and poultry litter having initial pH 8.4. The range of organic carbon was observed 7.2% (for poultry litter) to 18.5% for paddy straw in final product. On the contrary, EC, available P and available K were increased in all treatment. EC ranged between 3.2 (paddy straw) and 4.7 (water hyacinth) at final stage. Available P content was maximum in paddy straw (0.61 ± 0.001%) and that was found minimum in water hyacinth (0.19 ± 0.004%). In water hyacinth available K content was also found in least quantity (1.1 ± 0.01%), whereas this nutrient in available form was obtained in highest quantity in poultry droppings (4.2 ± 0.13%) (Table 26.2). Reduction of pH after decomposition might be reasoned for the production of CO_2, ammonia, NO_3^- and organic acid by microbial decomposition [27]. Rise in EC of vermicomposted materials of all the feeding materials might be due to freely available ions and minerals that are generated during ingestion and excretion by the earthworm [4]. The losses of C in the form of CO_2 as well the conversion of some part of organic fraction into worm biomass are the reason of carbon loss from the substrate [27]. P mineralization depends on quality of bulking materials. When an organic material passes through the gut of earthworm, some amounts of phosphorus are being converted with more availability to plants due to worm's gut phosphatases. The increase of P content could be attributed to the P-solubilising microorganisms present in worm cast [27]. Potassium content increased due to vermicomposted product [28]. The difference in P and K mineralization among different waste materials was found because of variation in chemical structure of the raw feeding materials [29].

Table 26.3 clearly indicates that the number of microbes was decreased in most of the treatments

Table 26.3 Mean microbial count at initial and final product

Feeding	Total fungi [×10^8]		Total bacteria [×10^8]		Azotobacter [×10^8]		Rhizobium [×10^8]		Azospirillum [×10^8]		PSB [×10^8]		Pseudomonas [×10^8]	
Stock	Initial	Final	Initial	Final	Initial	Final	Initial	Final	Initial	Final	Initial	Final	Initial	Final
Water hyacinth	223	192.33	1,065.33	564	593.33	273	86	92.67	398.67	217	322. 67	212	66	58. 67
Cow dung	490	295.67	919.33	338	226	143.33	533.67	297	170.67	119	143.33	104	167.67	187. 67
Green gram	581.33	232	757	353.67	201	255.33	300.33	88.67	68	105	57.667	44	51.67	94. 33
Vegetables	307.67	340	373.67	274.67	229	125.67	114.33	307.33	156.67	116.67	195.67	123.67	240.67	144
Poultry litter	250	213.33	1,021.33	320.67	596.67	154.33	92.33	40	110.67	154.67	138.67	78.33	77	97
Paddy straw	263.67	272.23	787.33	347.23	342.67	423.56	111	173	164.67	231	191.67	175.33	87.33	89.66
Grand mean	150	254.6	352.2	370.2	151.4	190.3	94.7	165.1	81.6	142.4	47.5	112.4	49.6	116.3
SED	28.5	81.7	96.6	64.3	34.1	39.2	22.5	33.2	28.6	33.2	22.1	46.0	15.4	43.1
CD at 5%	63.5	188.6	215.3	148.3	76.1	90.4	50.3	76.7	63.8	76.6	49.3	106.2	34.3	99.4
CV	23.2	39.3	33.6	21.2	27.6	25.2	29.2	24.7	42.9	28.5	36.38	50.1	38.0	45.4

Table 26.4 Effect of microbial inoculants on the physicochemical properties of vermicompost

Plant source	pH		EC		O. C [%]		Available P [%]		Available K [%]		% increasing microbial population
	Before	After	Before	After	Before	After	Before	After	Before	After	
Water hyacinth	7.8±0.02	6.8±0.02 (7.3±0.01)	3.5±0.01	4.8±0.01 (4.7±0.01)	16.5±0.11	11.30±0.12 (12.25±0.23)	0.08±0.003	0.42±0.001 (0.19±0.004)	0.8±0.03	1.4±0.10 (1.1±0.10)	327.78±2.05 (227.78±1.46)
Straw	7.7±0.01	(6.7±0.02) 7.6±0.01	3.0±0.03	4.2±0.02 (3.2±0.01)	18.7±0.15	16.2±0.11 (18.5±0.13)	0.42±0.002	1.41±0.02 (0.61±0.001)	0.9±0.04	2.1±0.02 (1.1±0.04)	67.5±1.52 (59.5±1.01)
Green gram	8.1±0.03	(6.2±0.02) 7.8±0.01	3.2±0.03	4.3±0.03 (4.3±0.01)	18.75±0.20	9.5±0.20 (11.5±0.20)	0.20±0.006	1.2±0.01 (0.46±0.001)	1.5±0.03	2.8±0.01 (1.8±0.08)	78.06±1.2 (62.06±2.1)

Data in parenthesis indicates results found in control (without microbial inoculants) at final product for all the specific microbes and total fungal and bacterial count. But the total count for all the microbes was much higher in all feeding stock. In case of vermicompost produced using water hyacinth, it showed maximum number of bacterial (total) count, 56.4×10^9; *Azospirillum* sp., 21.7×10^9; PSB, 21.2×10^9; and *Pseudomonas* sp., 18.8×10^9 in 1 ml of sample, whereas total fungal colony (19.2×10^9/g of vermicompost) was minimum in this feeding stock. Total fungal colony was found maximum (34×10^9/g of vermicompost) in the product obtained from kitchen waste. Though total bacterial count (27.4×10^9/g of vermicompost) and *azotobacter* count (12.57×10^9/g of vermicompost) were lowest in kitchen waste, individually *Rhizobium* colony-forming units, cfus (30.7×10^9/g of vermicompost), was highest in that. Maximum *Azotobacter* sp. cfus (42.4×10^9/g of vermicompost) and minimum *Pseudomonas* sp. cfus (8.9×10^9/g of vermicompost) were observed in Paddy straw. In poultry droppings, colony of *Rhizobium* sp. (4×10^9/g of vermicompost) was found lowest. Both *Azospirillum* sp. (10.5×10^9/g of vermicompost) and PSB (4.4×10^9/g of vermicompost) were lowest in vermicompost of green gram waste.

In another set of experiment when microbial inoculants (*Rhizobium* sp., *Azotobacter* sp. and PSB) were added with partially decomposed materials of paddy straw, water hyacinth and green gram treatment, the shift of all physico-chemical characters, i.e. pH, EC, organic carbon, available phosphorous and available potassium for all treatment, were found in same direction but in greater extent in comparison to treatment without inoculation of external microbial culture (Table 26.4). In final vermicompost produced using paddy straw, organic carbon (16.2%) and available phosphorous (1.41%) was found maximum, but available potassium was highest in green gram. The shift of organic carbon was higher in green gram-based vermicompost (1.21-folds). Available phosphorous and potassium increased 2.3-fold and 1.6-fold in paddy straw in comparison to control. Similar observation was obtained by many other workers [30, 31]. The initial nitrogen present in the feed materials [32] and also relics on to the extent of nitrogen fixed by free living nitrogen-fixing bacteria [33] could be the possible reasons of getting higher nutrient in the product when feeding materials were treated with external microbial inoculants.

References

1. Adi AJ, Noor ZM. Waste recycling, utilization of coffee grounds and kitchen waste in vermicomposting. Bioresour Technol. 2009;100:1027–30.
2. Garg VK, Kaushik P. Vermistabilization of textile mill sludge spiked with poultry droppings by an epigeic earthworm *Eisenia foetida*. Bioresour Technol. 2005;96:1063–71.
3. Nair J, Sekiozoic V, Anda M. Effect of pre-composting on vermicomposting of kitchen waste. Bioresour Technol. 2006;97:2091–5.
4. Garg P, Gupta A, Satya S. Vermicomposting of different types of waste using *Eisenia foetida*, A comparative study. Bioresour Technol. 2006;97:391–5.
5. Harris RC, Knox K, Walker N. A Strategy for the development of sustainable landfill design and operation. In: IWM proceedings, 1994. 6, pp. 26–9.
6. Logsdson G. Worldwide progress in vermicomposting. Biocycle. 1994;35:63–5.
7. Bansal S, Kapoor KK. Vermicomposting of crop residues and cattle dung with *Eisenia foetida*. Bioresour Technol. 2000;73:95–8.
8. Ghosh M, Chattopadhyay GN, Baral K. Transformation of phosphorus during vermicomposting. Bioresour Technol. 1999;69:149–54.
9. Gajalakshmi S, Ramasamy EV, Abbasi SA. Assessment of sustainable vermiconversion of water hyacinth at different reactor efficiencies employing *Eudrilus engeniae* Kingburg. Bioresour Technol. 2001;80:131–5.
10. Kaplan DL, Hartenstein R, Neuhauser EF, Malecki MR. Physico-chemical requirements in the environment of the earthworm *Eisenia foetida*. Soil Biol Biochem. 1980;12:347–52.
11. Reeh U. Influence of population densities on growth and reproduction of the earthworm *Eisenia Andrei* on pig manure. Soil Biol Biochem. 1992;24:1327–31.
12. Edwards CA, Burrows I, Fletcher KE, Jones BA. The use of earthworms for composting farm wastes. In: Gasser JKR, editor. Composting of agriculture and other wastes. Amsterdam: Elsevier; 1985. p. 229–42.
13. Gunadi B, Blount C, Edward CA. The growth and fecundity of *Eisenia foetida* [savigny] in cattle solids pre-composted for different periods. Pedobiologia. 2002;46:15–23.
14. Chaudhari PS, Bhattacharjee G. Capacity of various experimental diets to support biomass and reproduction of *Perionyx excavatus*. Bioresour Technol. 2002;82:147–50.
15. Reinecke AJ, Viljioen SA, Saayman RJ. The suitability of *Eudrilus eugeniae*, *Perionyx excavatus* and *Eisenia fetida* [Oligochaete] for vermicomposting in

southern Africa in terms of their temperature requirements. Soil Biol Biochem. 1992;24:1295–307.

16. Edwards CA, Bohlen PJ. Biology and ecology of earthworm. 3rd ed. New York/London: Chapman and Hall; 1996.
17. Ndegwa PM, Thompson SA. Integrating composting and vermicomposting the treatment and bioconversion of biosolids. Bioresour Technol. 2001;76:107–12.
18. Ismail SA. Organic waste management. In: Technology appreciation programme on evaluation of biotechnological approaches to waste management held on 26th October 2000. Industrial Association-ship of IIT, Madras, 2000. pp. 28–30.
19. Buchanam MA, Rusell E, Block SD. Chemical characterization and nitrogen mineralization potentials of vermicompost derived from differing organic wastes. In: Edwards CA, Neuhauser EF, editors. Earthworms in environmental and waste management. The Hague: SPB Academic Publishing; 1988. p. 231–40.
20. Kaushik P, Garg VK. Dynamics of biological and chemical parameters during vermicomposting of solid textile mill sludge mixed with cow dung and agricultural residues. Bioresour Technol. 2004;4:203–9.
21. Premono Edi M, Moawad MA, Vlek PLG. Effect of phosphate-solubilizing *Pseudomonas putida* on the growth of maize and its survival in the rhizosphere. Indones J Crop Sci. 1996;1:13–23.
22. Kumar V, Narula N. Solubilization of inorganic phosphates and growth emergence of wheat as affected by *Azotobacter chorococcum*. Biol Fertil Soils. 1999;28:301–5.
23. Walkley A, Black IA. Chromic acid titration for determination of soil organic matter. Soil Sci. 1934;63:251.
24. Olsen SR, Cole CV, Watanabe FS, Dean LA. Estimation of available phosphorus in soils by extraction with sodium bicarbonate. Circular US Department of Agriculture, 1954. p. 939.
25. Gomez KA, Gomez AA. Statistical procedures for agricultural research. New york: A Wiley-Interscience Publication; 1984. p. 130–9.
26. Edwards CA, Fletcher KE. Interaction between earthworms and microorganisms in organic-matter breakdown. Agric Ecosyst Environ. 1988;24:235–47.
27. Suthar S. Vermicomposting of vegetable-market solid waste using *Eisenia fetida*, impact of bulking material on earthworm growth and decomposition rate. Ecol Eng. 2009;35:914–20.
28. Delgado M, Bigeriego M, Walter I, Calbo R. Use of California red worm in sewage sludge transformation. Turrialba. 1995;45:33–41.
29. Suthar S, Singh S. Vermicomposting of domestic waste by using two epigeic earthworms [*Perionyx excavatus* and *Perionyx sansibaricus*]. Int J Environ Sci Technol. 2008;5(1):99–106.
30. Kaviraj SS. Municipal solid waste management through vermicomposting employing exotic and local species of earthworms. Bioresour Technol. 2003;90:169–73.
31. Edwards CA, Lofty JR. Biology of earthworms. London: Chapman and Hall; 1972.
32. Suthar S. Potential utilization of guar gum industrial waste in vermicompost production. Bioresour Technol. 2006;97:2474–7.
33. Kale RD, Bano K, Krishnamurthy RV. Potential of *Perionyx excavatus* for utilizing organic wastes. Pedobiologia. 1982;23:419–25.

Development of SSR Markers and Genetic Variability in Physiological Traits in Bambara Groundnut (*Vigna subterranea* L. Verdc)

27

R. Beena, M.S. Sheshshayee, J.N. Madhura, T.G. Prasad, and M. Udayakumar

Abstract

Bambara groundnut (*Vigna subterranea* L. Verdc) is a widely grown crop in marginal lands mainly in sub-Saharan Africa. Improvement of this crop has tremendous potential for its extension to other places including India. Though bambara groundnut is a drought-tolerant crop, from the agronomic point of view, drought tolerance should be linked with superior growth rates and productivity under water-limited conditions. Before attempting to improve these traits, it would be necessary to assess the genetic variability in the response of these traits to water-limited conditions. We assessed the genetic variability in WUE and the physiological traits associated with these traits. It was observed that on the average, there was a significant increase in WUE when the plants experienced water stress. This increase in WUE was primarily due to a significant reduction in total transpiration. This reduction in total transpiration (CWT) coupled with a reduction in total leaf area leads to a significant reduction in total biomass among these accessions under stress. There was significant reduction in oxygen evolution, potential photosynthesis and the chlorophyll content, all of which decreased significantly under stress. A strong correlation was noticed between oxygen evolution and potential photosynthesis, indicating that the chloroplast mechanism is severely affected by stress among bambara accessions. Marker-assisted breeding is expected to enhance the speed and accuracy of achieving such improvements. However, lack of locus-specific codominant marker systems in this crop is the

R. Beena (✉) • M.S. Sheshshayee • J.N. Madhura • T.G. Prasad
Department of Plant Breeding and Genetics, R.A.R.S., Kerala Agricultural University, Pattambi, Palakkad, 679 306, India
e-mail: beenaajithkumar@gmail.com

M. Udayakumar
Department of Crop Physiology, University of Agricultural Sciences, GKVK, Bangalore, 560 065, India

A. Sabu and A. Augustine (eds.), *Prospects in Bioscience: Addressing the Issues*, DOI 10.1007/978-81-322-0810-5_27, © Springer India 2013

major constraint for molecular breeding. Therefore, development of SSRs (simple sequence repeats) markers in bambara groundnut that can be subsequently employed in identification of QTL (quantitative trait locus) for specific physiological traits relevant for drought tolerance is the need of hour. A pre-cloning enrichment (selective hybridisation) strategy was adopted to fish out the SSRs from bambara groundnut genome to develop a microsatellite-enriched library. This strategy involves the hybridisation of digested genomic DNA with repeat oligomers, leading to selective enrichment. Based on this technique, we obtained 2,239 microsatellite-enriched clones, of which 734 clones were sequenced and analysed for the presence of microsatellite repeat regions. Among these, clones revealed microsatellite regions with more than 15 bp. Initially, primers were designed and synthesised for 143 SSR regions. These primers have been validated for their amplification using the genomic DNA of different land races. Of these, locus-specific amplification was observed in 99 primer pairs. These validated SSR markers are being used for molecular characterisation of different bambara groundnut landraces.

Keywords
Bambara groundnut • SSR markers • Drought

Abbreviations

A	Assimilation rate
Ci	Intercellular CO_2 concentration
CWT	Cumulative water transpired
DMSO	Dimethyl sulfoxide
DNA	Deoxyribonucleic acid
FC	Field capacity
FYM	Farm yard manure
Gs	Stomatal conductance
HI	Harvest index
LAD	Leaf area duration
MSI	Membrane stability index
MTR	Mean transpiration rate
NAR	Net assimilation rate
PP	Potential photosynthetic ability
QTL	Quantitative trait locus
ROS	Rainout shelter
RWC	Relative water content
SCMR	SPAD chlorophyll metre reading
SLA	Specific leaf area
SSR	Simple sequence repeat
T	Transpiration rate
TDM	Total dry matter
TLA	Total leaf area
WUE	Water use efficiency

Introduction

While the exact effects of climate change combined with population growth are not known, it seems highly likely that agriculture in large regions of the world may need to undergo significant adjustment over the next four decades for us to have any chance of meeting needs [1]. Water availability is perhaps the most important constraint that would affect productivity of most crop species. Besides improving yield under drought conditions, enhancing the traits that contribute to productivity appears to be the most relevant strategy in achieving food security. International consensus that has emerged clearly emphasises that further crop improvement in any species is possible only through pyramiding relevant traits onto a single genetic background. Research over the past couple of decades has enumerated a number of physio-morphological traits such as rapid phenological development, paraheliotropic movements, rapid stomatal response, development of pubescence, deposition of waxes on cuticles, etc. However, though most of these traits are highly relevant in avoiding stress injury, these traits are often counterproductive.

Global efforts have been initiated to evolve water saving strategies in agriculture through breeding for more drought-tolerant crop cultivars. The other noteworthy strategy has also been to identify underutilised crop species to either complement or even to replace several conventionally cultivated crops. Bambara groundnut (*Vigna subterranea*) [$2n=22$] [2–5] is an underutilised African legume grown primarily in sub-Saharan Africa with a centre of origin in the region around Cameroon. Prospects in this area have identified the most likely ancestor as *V. subterranea* var. *spontanea*, and this relationship has been confirmed through isozyme and other molecular analysis [6–11]. The cultivated crop exists as landraces and is grown extensively in sub-Saharan Africa.

This crop holds great potential as a food crop in the arid and semiarid regions of the world due to good drought tolerance and a remarkable ability to maintain cell turgor even under prolonged drought conditions [12]. It is also able to produce a yield of pods in dry environments when peanut (*Arachis hypogaea*), the widely introduced and morphological similar South American species, fails. It is an attractive crop because of the high edible protein content (16–25%) rich in lysine and methionine. Therefore, it has a beneficial complementary effect when consumed with cereals that are low in lysine. Furthermore, due to a very low to negligible levels of lipids, bambara has the potential to substitute conventional legumes like peanut [13].

Any trait is relevant only when it is associated with superior crop growth rate. Water harvesting from deeper profiles of soil associated with the root system, efficiency of water use for biomass production, water use efficiency (WUE) and water conservation strategies have been demonstrated to have greatest significance in trait-based crop improvement. Furthermore, to have a holistic improvement in crop performance under water-limited condition, it is equally important that a genotype possesses adequate intrinsic tolerance to stress at cellular level.

Thus, breeding for improved drought tolerance in crop plants is clearly a formidable challenge. A major factor that reduces bambara groundnut yields in farmers' field is its inherent low yield due to the lack of improved varieties. Breeders must have access to diverse landraces with a broad range of agronomic traits that can be matched with improved agronomic practices, e.g. optimal date of sowing. The ability to use molecular technology to exploit information at the gene level, based on genome mapping, marker-trait analysis and marker-assisted breeding, offers the opportunity to apply these approaches to further our understanding of bambara groundnut genetics and spearhead improvement efforts.

With the recent progress achieved in advanced molecular biological tools, DNA marker-assisted QTL introgression has emerged as the most appropriate strategy for pyramiding complex physiological traits onto a single genetic background. However, identification of DNA markers and QTLs almost entirely depends on the abilities for accurate phenotyping of these traits in large sets of germplasm accessions and/or mapping populations. Before attempting to improve these traits, it would be necessary to assess the genetic variability of these traits under water-limited conditions.

Materials and Methods

Five bambara accessions SB42, DIPC, S193, S165A and Uniswa Red were raised in carbonised rubber containers filled with 15 kg of sandy loam soil mixed with FYM in 3:1 proportion (*w*/*w*). Two water regimes were maintained by gravimetric approaches. While one set of plants was maintained at 100% FC, the other set was maintained at 60% FC. Two plants were raised in each of these containers, and a set of ten pots was maintained for each genotype. All cultural practices such as fertilisation were followed as recommended, and appropriate prophylactic measures were taken to

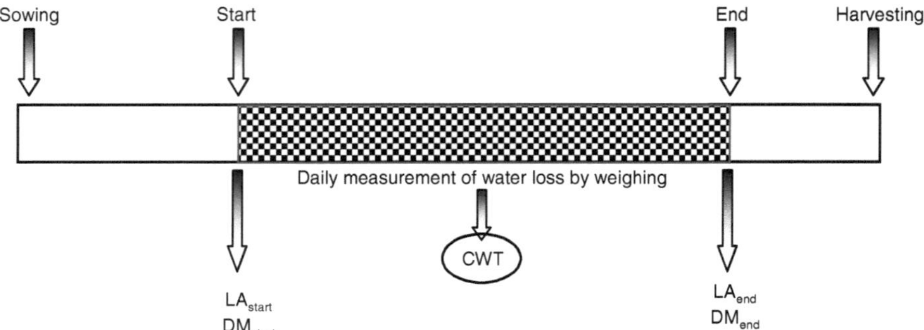

Fig. 27.1 Flow chart indicating the sequence of events for the determination of WUE and other associated parameters by gravimetric approach

raise healthy plants in each container. WUE and other physiological traits were determined gravimetrically. WUE was also determined at single leaf level by adopting gas exchange approaches.

Methodology

There were seven replications per genotype in each of the water regime treatment. Water stress treatment was imposed and maintained by gravimetric approach. The following is the description for the determination of physiological traits by gravimetric technique.

This approach relies on the accurate determination of water content in the soil by weighing the containers at regular intervals, at least once every day. Since transpiration is being monitored, it is extremely essential to ensure that there is no addition of water to the containers by any external source such as rain. Thus, it is preferable that the experiment is laid out under a mobile rainout shelter (ROS). This structure can be moved over the area where containers are placed during night periods and/or during any rain episodes [14]. The protocol is diagrammatically described in Fig. 27.1. The novelty of the gravimetric approach is that, besides determining water use efficiency, a few important physiological traits such as net assimilation rate (NAR), mean transpiration rate (MTR) and leaf area duration (LAD) can also be computed.

Initially, it is necessary to assess the weight of empty container (W_E) and then its weight after filling dry soil (or a specific rooting mixture) (W_{ES}). That is, the amount of dry soil in each container would be $W_S = W_{ES} - W_E$. A hanging load cell balance mounted on a mobile gantry with a provision for movement on rails on either side of the ROS is used to access each container for weighing.

At the "start" of the experiment, the soil should be brought to 100 % field capacity by adding the "appropriate" volume of water, which can be determined by considering the bulk density of the soil and its water-holding capacity. At this stage, the drainage holes of the containers are closed so that water added does not drain out of containers. The soil surface should be covered with plastic pieces or any other suitable mulching material to minimise water loss due to surface evaporation. These arrangements ensure that almost all the water added is available for transpiration only. However, a set of containers without plants (bare containers) should be maintained with soil, water and mulching exactly as is in the case of containers with plants. Weighing these bare containers will provide an accurate measure of surface evaporation.

The weight of individual containers with soil at 100% FC, plastic pieces and the plant is recorded at least once daily to measure evapotranspiration. A set of four to five containers without plants ("bare" containers) needs to be maintained to assess direct water loss despite all care taken to minimise this water loss. The difference in pot weight on the subsequent weighing was corrected by adding water, to bring back to 100% FC [14]. While 80 % of water transpired was replenished for the "control" treatment, only 50% of the daily water loss was replenished for the "stress" treatment.

The daily water added over the entire experimental period is summed to arrive at the cumulative water added (CWA_p) to the pots with plants. Despite having taken sufficient care, there would still be some amount of direct evaporational losses, and thus, the CWAp would represent evapotranspiration. Total transpiration can be determined by subtracting cumulative water added to the "bare" containers (CWA_b) from CWA_p, i.e.

$$\text{Cumulative water transpired } (CWT) = \sum(CWA_p) - \sum(CWA_b)$$

At the "start" of the experiment, total biomass and leaf area are determined in a set of three containers. The soil is washed with a jet of water carefully to remove the roots, and the plant parts (leaves, stem and roots) are separately oven dried at 70 °C for 3 days. Biomass and leaf area are once again recorded in plants from the remaining containers of all the genotypes at the "end" of the experiment. Assuming a linear growth during the experimental period, WUE and other physiological parameters are computed as follows:

$$WUE(g \cdot Kg^{-1}) = \frac{(BM_{end} - BM_{start})}{CWT}$$

where BM_{end} and BM_{start} are the total dry matter ($g \cdot pot^{-1}$) measured at the "end" and "start" of the experiment, respectively,

Leaf area duration (LAD) (cm^2 days)= $[(LA_{end}+LA_{start})/2] * 35$

where LA_{end} and LA_{start} are the leaf area (in $cm^2 \cdot plant^{-1}$) measured at the "end" and "start" of the experiment, respectively.

The net assimilation rate (NAR) and the mean transpiration rate (MTR) are time-averaged measures of photosynthetic rate and transpiration rate and are computed as follows:

$$NAR(g \cdot cm^{-2} day^{-1}) = \frac{BM_{end} - BM_{start}}{LAD}$$

$$MTR\ (ml \cdot cm^{-2} day^{-1}) = \frac{CWT}{LAD}$$

Gas Exchange Measurements

A portable photosynthesis system (Li6400, LICOR Inc., Nebraska, USA) was used for the determination of gas exchange parameters such as assimilation rate (A), stomatal conductance (gs), intercellular CO_2 concentration (Ci), and transpiration rate (E and T) were recorded on the top fully expanded leaves of each of the genotypes of bambara accessions under full sunlight. The automated leaf chamber was used to maintain the light intensity at 1,700 $\mu mol \cdot m^{-2}\ s^{-1}$ and the chamber temperature at 28 °C. All measurements were made between 9 a.m. and 12 noon (local time).

Oxygen Evolution by Oxygen Electrode

Oxygen electrode is a basic system for studies of photosynthesis and respiration by oxygen evolution and uptake in samples held in gas phase. Samples are typically leaf discs, which are sealed into the leaf chamber of the LD1/2 unit. Changes of oxygen partial pressure are determined by the integral oxygen electrode mounted in the base of the unit and give the result in terms of rate of oxygen evolution measured as μmoles $m^{-2}\ min^{-1}$.

Potential Photosynthesising Ability

Relative differences in between the control and stressed plants in potential photosynthesis ability of the leaves are measured by studying $^{14}CO_2$ fixation at high ambient concentrations of CO_2. Leaf punch of known weight is placed on the leaf chamber, and to maintain the required humidity, filter paper soaked in water is placed below the leaf chamber. Then, a bottle containing 5 ml of 100-Mm sodium bicarbonate solution mixed with 50 µl of 0.1 µCi ^{14}C sodium bicarbonate is kept inside the chamber of the potential photosynthesis equipment, and then, the chamber is closed air tight. A source of red light of light intensity of 400µEinstein $m^{-2}\ min^{-1}$ for 30 min at 30 °C is provided inside the equipment. To liberate CO_2 inside the leaf chamber, a set up is made where

5 N HCl is made to fall drop by drop into the bottle containing 100-Mm sodium bicarbonate+ radioactive 14C sodium bicarbonate which would release radio-labelled CO_2 which is taken up by leaf punches for photosynthesis. After 30 min, the leaf samples were removed and ground in pestle and mortar with 1 ml of distilled water, and it is transferred into centrifuge tubes; these are centrifuged at 6,000 rpm for about 5 min, and then, 1 μl of the supernatant is added to scintillation bottles containing 1 ml of scintillation cocktail. The pellets were also taken separately in scintillation bottles containing 1 ml of scintillation cocktail. The radioactivity is measured in terms of count per minute (cpm) using liquid scintillation counter (Model: WALLAC 1409).

Relative Water Content (RWC)

Samples from control and stressed leaves were used for RWC determination. The youngest fully expanded leaf punches were cut into a clean, dry, preweighed and numbered glass vial and capped airtight. The leaf samples were brought to the laboratory, and the fresh weight (FW) of the leaf was recorded along with the capped vial using an electronic analytical balance having a readability of 0.0001 g. The leaf punches were allowed to rehydrate for 4 h by adding 2 ml of distilled water along the sides of the vial using pipette. The rehydrated leaf punches were taken out of the vial using clean forceps, gently blotted to remove external moisture using Kimwipe paper and weighed for turgid weight (TW). Since the detached leaf punches lose water rapidly, care was taken to avoid delay in measuring TW. The water in the vial was poured off, and the leaf punches were transferred to the vial and left uncapped. The leaf punches were then oven dried at 80 °C for 24 h along with the uncapped vial. After oven drying, the vial was allowed to stand for 10 min at room temperature, and the dry leaf weight (DW) was recorded. Leaf RWC was calculated as

$$\mathrm{RWC} = \frac{(\mathrm{FW} - \mathrm{DW})}{(\mathrm{TW} - \mathrm{DW})} \times 100$$

where FW, TW and DW are fresh, turgid and oven dry weights of the leaf sample, respectively, and are expressed in percentage [15] .

SPAD Chlorophyll Metre Reading (SCMR)

Chlorophyll content of leaf was measured using SPAD metre. A chlorophyll metre (SPAD 502), designed by the soil plant analysis development (SPAD) section, Minolta Camera Co., Ltd., USA, was used to record SPAD readings. SPAD 502 is a dual wavelength metre for non-destructive measurements. The measuring head of the SPAD 502 includes a red light-emitting diode (LED) and infrared LED, which emits light in a sequence through the leaf.

The SPAD 502 is self-calibrated for variability in the output of LEDs and has built-in error codes that help to prevent irregular measurement. Leaf chlorophyll absorbance is measured at a wavelength of 650 nm, and non-chlorophyll absorbance is measured at a wavelength of above 700 nm. A microprocessor calculates the SPAD value, which is proportional to the relative optical density, based on the ratio between the two wavelengths [16, 17]. Measurements were taken from topmost fully expanded leaf. Five readings were taken from each plant [18].

Membrane Stability Index

Samples collected from both control and stress subjected plants were washed three times in deionised water to remove electrolytes adhered on the surface. The samples were kept in a capped vial (20 ml) containing 10 ml of deionised water and incubated in the dark for 24 h at room temperature. The conductance was measured with a conductivity metre. After the first measurement, the vials were autoclaved for 15 min to kill the leaf tissue and release the electrolytes. After cooling, the second conductivity reading was taken. These two measurements were carried out individually for both control and stress treatment plants.

MSI (%) = $[1-(T_1/T_2)/(1-C_1/C_2)]\times 100$ where T and C refer to the stress and control samples, respectively. The subscripts 1 and 2 refer to the initial and final conductance readings, respectively [19].

Chlorophyll Content

The content of chlorophyll was estimated by dimethyl sulfoxide (DMSO) method. A known weight of leaf was taken and incubated in test tube containing a mixture of 10 ml of DMSO and 80 % acetone (1:1) overnight at room temperature in dark. After the tissues were completely bleached, absorbance was recorded at 645 and 663 nm using UV-visible spectrophotometer (UV-2450, Shimadzu, Japan). Chlorophylls a and b and total chlorophyll were calculated by using the following equations:

$$\text{Chl a} = (12.7\text{A}_{663} - 2.69\text{A}_{645})(\text{V}/1{,}000)(1/\text{wt})$$

$$\text{Chl b} = (22.9\text{A}_{645} - 4.68\text{A}_{663})(\text{V}/1{,}000)(1/\text{wt})$$

$$\text{Total Chl} = (8.02\text{A}_{663} + 20.20\text{A}_{645})(\text{V}/1{,}000)(1/\text{wt})$$

where A_{663} and A_{645} are the absorbance at 663 and 645 nm, respectively.

Isolation of Microsatellites/Simple Sequence Repeat (SSR) Markers: Pre-cloning Enrichment Strategy

A genomic DNA library with the microsatellite-enriched DNA fragments was followed to develop microsatellite markers from bambara groundnut genome. A pre-cloning enrichment (selective hybridisation) strategy was adopted for this purpose. Genomic DNA was extracted from fresh leaves of bambara accession and digested using suitable restriction enzymes. M13 adapters were ligated to the DNA fragments and were hybridised with repeat oligomers. This strategy leads to selective enrichment.

Enrichment by Selective Hybridisation: The selective hybridisation protocol primarily implies hybridisation of size-fractionated adaptor-ligated genomic DNA with complementary repeat oligomers followed by elution of the captured DNA fragments and subsequent amplification using adaptor primers and cloning in a competent vector.

Results and Discussion

Presently, groundnut (*Arachis hypogaea*) is essentially cultivated globally for oil as well as for table purposes. Yields of groundnut are threatened by the looming water crisis, especially in the tropical regions of the world. Thus, in order to achieve food and nutritional security, scientists have constantly been searching for attractive alternate crops. Bambara groundnut (*Vigna subterranea*) is one such crop that has the potential to be exploited. It is an indigenous African legume cultivated mainly by subsistence farmers in Africa under traditional low-input agricultural systems.

Bambara groundnut yields are low because the production environments are characterised by various abiotic and biotic stresses. However, under optimal condition also, yields are variable and unpredictable, and this is partly due to variability in growth and development of individual plants within landraces [20]. Bambara groundnut is known to be drought-hardy crop, and recent studies have documented some relevant traits such as maintenance of leaf turgor as possible reasons for drought tolerance [12]. However, systematic investigations addressing the reasons for the high drought tolerance have not been initiated.

From the agricultural point of view, drought tolerance should be linked with improved growth rates [21]. Thus, assessing the traits that are associated with growth rates under water-limited conditions is essential before attempting to improve

Table 27.1 Genetic variability in WUE and associated parameters among bambara accessions under control (80 % FC) and stress (50 % FC)

Genotype	Treatment	SLA ($cm^2 \cdot g^{-1}$)	TLA (cm^2)	LAD (m^2 days)	dTDM ($g \cdot pl^{-1}$)	CWT (l)	WUE (g Kg^{-1})	MTR ($ml \cdot m^{-2} day^{-1}$)	NAR ($g \cdot m^{-2} day^{-1}$)
DIPC	Control	182.82	0.09	2.14	9.17	6.81	1.35	3.18	4.29
	Stress	124.45	0.06	1.52	6.90	2.77	2.49	1.82	4.54
SB42	Control	162.61	0.12	2.85	13.22	8.15	1.62	2.86	4.64
	Stress	130.46	0.07	2.12	10.60	7.71	1.37	3.64	5.00
S193	Control	153.25	0.07	1.85	10.30	5.55	1.86	3.00	5.57
	Stress	125.00	0.04	1.29	6.81	4.68	1.46	3.64	5.30
UR	Control	171.57	0.07	2.07	7.51	4.85	1.55	2.35	3.64
	Stress	129.67	0.04	1.45	4.70	0.66	7.12	0.46	3.24
S165	Control	175.41	0.10	2.45	10.97	5.43	2.02	2.22	4.48
	Stress	179.14	0.08	1.91	6.90	1.74	3.97	0.91	3.61
CD(0.05)		23.6	0.02	0.56	2.24	1.72	1.2	1.69	1.34

drought tolerance in crop plants. Among several such traits, total amount of water transpired, the efficiency of water use for biomass production (WUE) and the partitioning of biomass to harvestable plant parts (HI) are known to be most important components of yield [22].

Significant genetic variability in WUE and associated physiological parameters were noticed among the bambara accessions (Table 27.1). It was observed that on an average, there was a significant increase in WUE when the plants experienced water stress. However, there was a significant genetic variability for this response. Uniswa Red recorded the highest increase in WUE when stressed while SB42 showed a marginal reduction in WUE under stress. This increase in WUE was primarily due to a significant reduction in total transpiration (Table 27.1). This reduction in total transpiration (CWT) coupled with a reduction in total leaf area leads to a significant reduction in total biomass among these accessions under stress. A strong positive relationship between CWT and total biomass (TDM) accumulated over a specific growth period was noticed among the bambara accessions. Such relationship between TDM and CWT has been reported for many crop species as well as perennial tree species. Higher transpiration is possible when roots supply sufficient amount of water demand from the canopy. Further, the number and extent of opening of stomata are the predominant factor that determines transpiration. Stomata besides controlling transpiration also determine the entry of CO_2 for photosynthesis. Hence, a positive relationship between biomass production and total transpiration is quite discernible. Under conditions of water limitations, increasing water loss though is associated with growth is not a useful approach. However, deeper root system could be very useful to harness water from deeper soil profiles. A simultaneous improvement in WUE would be a plausible approach to achieve improved growth rates.

Total biomass production has been known to be a function of light intercepted by the leaf area cover and the total transpiration. Accordingly, total biomass was significantly positively correlated with both leaf area duration (Fig. 27.2) and total transpiration (Fig. 27.3).

Because of this strong relationship between total transpiration and biomass production, any increase in WUE is often associated with a reduced biomass production. In such types, stomata seem to control the differences in WUE. Understanding this physiological control is extremely essential before attempting to improve productivity through selection for WUE. A significant inverse relationship between mean transpiration rate and WUE were observed (Fig. 27.4). This indicates that the increase in WUE is associated with reduction in transpiration rate. Because of the association of transpiration with

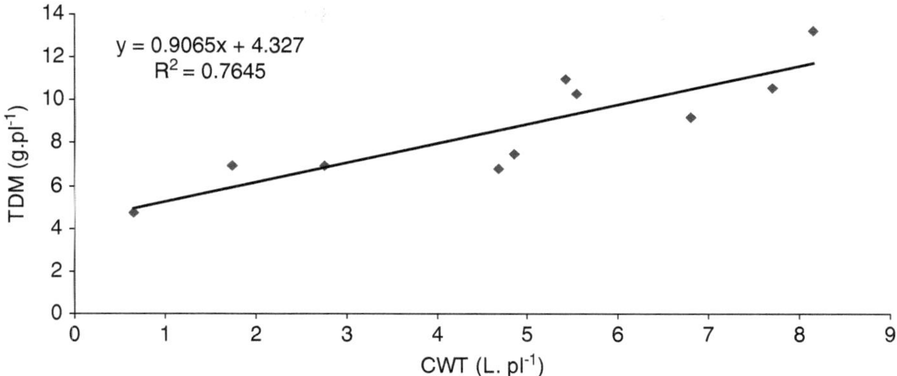

Fig. 27.2 Relationship between total transpiration and biomass among bambara accessions

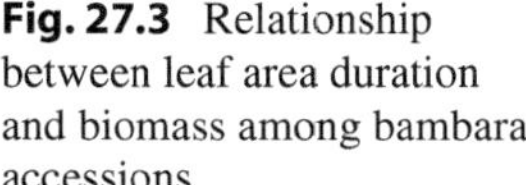

Fig. 27.3 Relationship between leaf area duration and biomass among bambara accessions

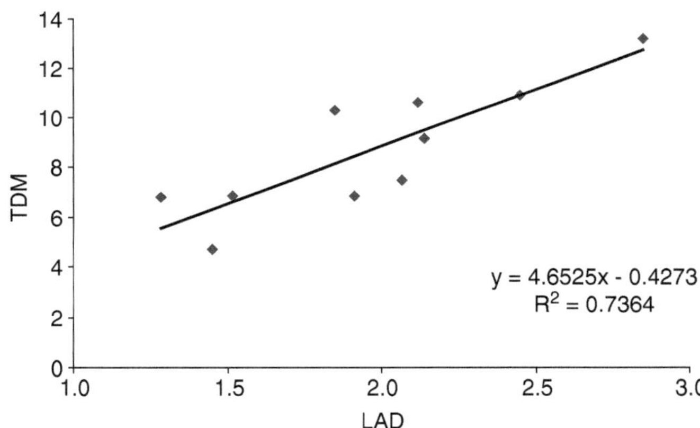

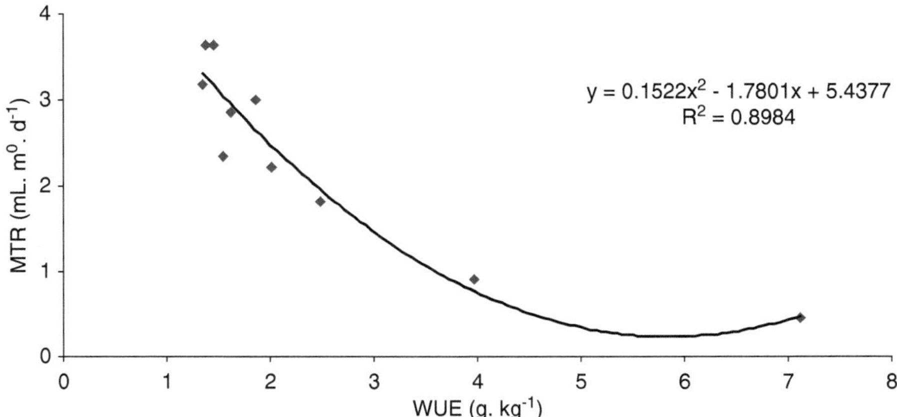

Fig. 27.4 Relationship between WUE and mean transpiration rate among bambara accessions

Table 27.2 Genetic variability in gas exchange parameters among bambara accessions under control and water-stressed conditions

Genotype	Treatment	A (μmol·m^{-2} s^{-1})	Gs (mol·m^{-2} s^{-1})	Ci (ppm)	T (mol·m^{-2} s^{-1})	A/gs	A/T
DIPC	Control	29.93	0.58	269.27	7.34	51.48	4.08
	Stress	5.96	0.10	482.36	2.67	62.17	2.23
SB42	Control	22.81	0.43	271.40	5.46	53.56	4.18
	Stress	4.77	0.10	169.00	0.72	47.74	6.63
S193	Control	29.29	0.36	201.84	5.80	81.68	5.05
	Stress	26.90	0.30	213.00	5.55	89.67	4.85
UR	Control	23.84	0.29	208.37	4.70	82.98	5.07
	Stress	22.30	0.19	181.00	3.38	116.75	6.60
S165	Control	22.80	0.22	192.00	3.94	103.64	5.79
	Stress	19.83	0.19	186.16	3.55	105.25	5.58
DODR	Control	21.60	0.25	197.06	3.58	85.29	6.03
	Stress	12.98	0.14	325.10	3.15	94.88	4.13
CD(0.05)		2.76	0.04	24.06	1.32	15.28	0.578

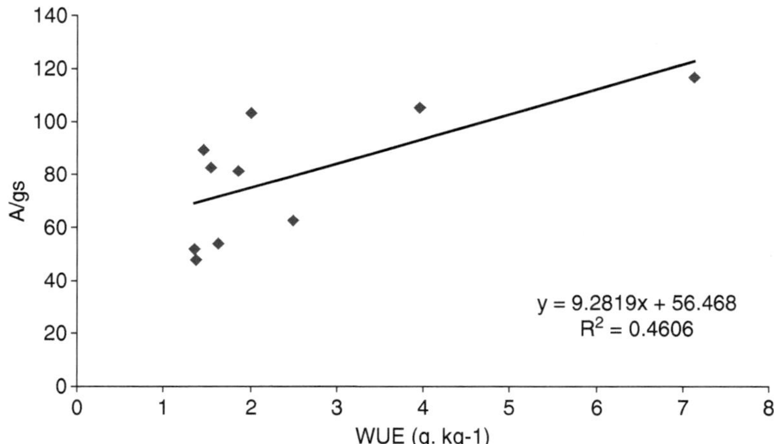

Fig. 27.5 Relationship between the WUE values measured by gravimetry and gas exchange approaches

biomass accumulation, selection for WUE may not be useful. Water use efficiency was also measured at the single leaf level using a portable photosynthesis system. The results revealed a significant reduction in gas exchange parameters in the stressed plants (Table 27.2).

However, accessions like Uniswa Red showed significant tolerance to water stress. This accession showed a mere 6.5 % reduction in photosynthetic rate. The highest reduction of 80% was noticed in DIPC and SB42. The gas exchange measurements also revealed an increase in WUE under stress. A positive correlation between A/gs and WUE, although weak, was a demonstration of the fact that gas exchange can also be used as a rapid measure of WUE at the single leaf level (Fig. 27.5).

To further understand the physiological mechanisms of stress induced changes in photosynthesis, we determined the oxygen evolution using a gas-phase oxygen electrode (Hansatech). There was significant reduction in oxygen evolution, potential photosynthesis and the chlorophyll content; all these parameters decreased significantly under stress, except for Uniswa Red (Table 27.3). A strong correlation was

Table 27.3 Genetic variability in a few traits associated with photosynthesis and membrane characteristics among control and stressed bambara accessions

Genotype	Treatment	Oxygen evolution (μmoles $m^{-2}min^{-1}$)	Chl a (mg g^{-1})	Chl b (mg g^{-1})	Total chl (mg g^{-1})	SCMR	PP[a]	MSI (%)	RWC (%)
DIPC	Control	1.446	1.25	0.27	1.52	40.33	12483.52	91.77	72.98
	Stress	0.469	1.11	0.27	1.37	39.73	3456.86		59.18
SB42	Control	1.091	1.37	0.34	1.71	40.00	8464.63	96.35	73.90
	Stress	0.725	0.83	0.17	1.00	38.00	7034.75		54.26
S193	Control	1.906	1.30	0.29	1.59	43.13	15145.40	90.82	80.07
	Stress	1.000	1.05	0.20	1.25	40.83	4583.16		51.12
UR	Control	0.858	1.40	0.33	1.73	44.70	6540.21	90.49	81.29
	Stress	1.154	1.11	0.24	1.35	39.13	6494.12		79.63
S165	Control	1.336	1.51	0.37	1.88	42.00	10491.29	77.56	80.67
	Stress	1.020	0.86	0.17	1.03	36.10	12273.00		78.56
DODR	Control	1.012	1.35	0.32	1.67	40.87	3543.00	77.95	83.87
	Stress	0.339	0.80	0.17	0.97	37.20	2513.74		80.79
CD		0.634	0.67	0.06	0.27	3.91	1580.8	8.0	11.06

[a]*PP* potential photosynthetic ability (cpm per gram leaf weight)

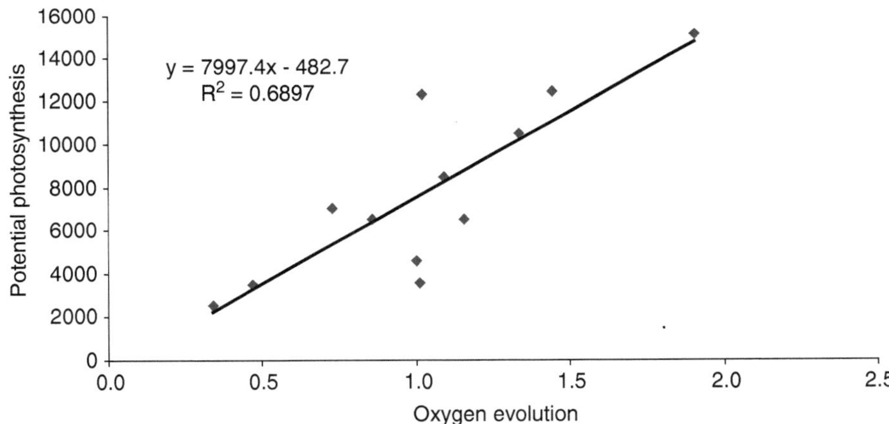

Fig. 27.6 Relationship between oxygen evolution and potential photosynthesis among bambara accessions

noticed between oxygen evolution and potential photosynthesis (Fig. 27.6), indicating that the chloroplast mechanism is severely affected by stress among bambara accessions. Uniswa Red that maintained a superior chloroplast mechanism recorded the highest increase in WUE measured by gravimetric methods (see Table 27.1). This suggests that though bambara is a conductance type, accessions like Uniswa Red have the potential to serve as trait donor parents for further improvement or can also be developed into a good cultivars after looking into the agronomic requirements for this landrace.

The reduction in gas exchange and photosynthetic characteristics matched well with the changes in membrane permeability index as well as relative water content of the leaf tissue (Table 27.3).

Development of SSR Markers

Development of microsatellite markers from the pre-cloning enrichment strategy around 2,239 colonies was screened by colony PCR using the universal M13 primers. A representative gel of the colony PCR analysis is depicted

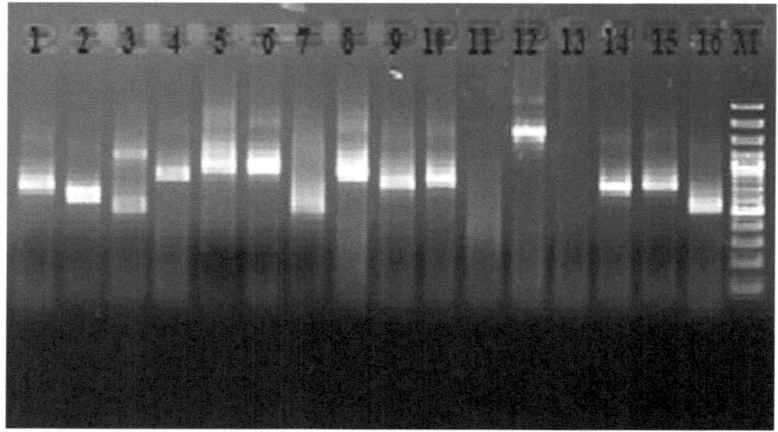

Fig. 27.7 PCR amplification of recombinant *E. coli* colonies using M13 primers ligated to the SSR-enriched DNA fragments. Lanes 1–16, M13 primer amplified *E. coli* colonies, lane M – 100-base-pair ladder

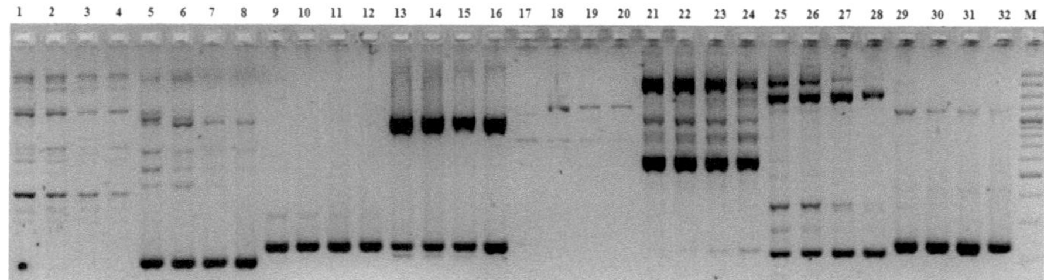

Fig. 27.8 Representative gel showing the standardisation of annealing temperature for primers. Lanes 1–4 (BG7), lanes 5–8 (BG4), lanes 9–12 (BAM37), lanes 13–16 (BAM14), lanes 17–20 (BAM66), lanes 21–24 (BAM66), lanes 25–28 (BAM6), lanes 29–32 (BAM14), lane M – 100-base-pair ladder

in Fig. 27.7. So far, around 638 clones have been sequenced and annotated for the presence of microsatellite regions using the tandem repeat finder software. Of these 638 clones, only 241 clones had microsatellite repeats of >15 bp, and 143 SSR primers were designed and synthesised for the microsatellite regions present in the recombinant clones.

Validation of SSR Primers

The SSR primer pairs designed for the microsatellite regions were initially standardised with the genomic DNA of bambara to arrive at the optimum annealing temperature conditions. A few representative gels showing the standardisation of annealing temperature, yielding the expected product size, are given in Fig. 27.8. Out of 143 primers, 99 primers showed locus-specific amplification. Figure 27.9 depicts the locus-specific amplification of a few standardised SSR primers.

The fact that the markers are robust, codominant and transferable between labs enables datasets to be generated which can be added to, if the same SSR markers are used in the subsequent analysis. This is critical for linking loci between different genetic maps, generating substantial diversity datasets and devising protocols which can be transferred to research and breeding stations alike. The development of microsatellite markers in this programme underpins the development of molecular resources for breeding and genetic analysis in bambara groundnut. With these markers, precise selection becomes feasible, together with critical quality control measures for

Fig. 27.9 Gel showing the locus-specific amplification of SSR primer BG27 among 24 bambara groundnut accessions. Lane 1–12 and 14–25 (24 bambara groundnut accessions), lane M – 100-base-pair ladder

breeding programmes and research. For bambara groundnut, there is the added complication of genetic structure within landraces. Microsatellite markers allow this to be addressed in a coherent fashion. All of these measures are important to be able to make genetic improvement and to gain a genetic understanding for bambara groundnut which will underpin the physiological, nutritional and end-user analysis. This firm foundation is important to be able to fully use bambara groundnut for food security in the future.

Acknowledgement The authors are thankful to European Union for the financial support.

References

1. Mayes S, Massawe FJ, Alderson PG, Roberts JA, Azam-Ali SN, Hermann M. The potential for underutilized crops to improve security of food production. J Exp Bot. 2012;63(3):1075–9.
2. Frahm-Leliveld JA. Some chromosome numbers in tropical leguminous plants. Euphytica. 1953;2:46–8.
3. Doku EV, Karikari SK. Operational selection in wild Bambara groundnut. Ghana J Sci. 1971;11:45–56.
4. Linnemann AR, Azam-Ali SN. Bambara groundnut (*Vigna subterranea* (L.) Verdc.). In: Williams JT, editor. Underutilized crops: pulses and vegetables. London: Chapman Hall; 1993. p. 13–58.
5. Basu S, Mayes S, Davey M, Roberts JA, Azam Ali SN, Mithen R, Pasque RS. Inheritance of domestication traits in Bambara groundnut. Euphytica. 2007; 157:59–68.
6. Pasquet RS, Schwedes S, Gepts P. Isozyme diversity in Bambara groundnut. Crop Sci. 1999;39:1228–36.
7. Amadou HI, Bebeli PJ, Kaltsikes PJ. Genetic diversity in Bambara groundnut (*Vigna subterranea* (L.) Verdc) germplasm revealed by RAPD markers. Genome. 2001;44:995–9.
8. Massawe FJ, Dickinson M, Roberts JA, Azam-Ali SN. Genetic diversity in Bambara groundnut (*Vigna subterranea* (L.) Verdc) landraces revealed by AFLP markers. Genome. 2002;45:1175–80.
9. Massawe FJ, Schenkel W, Basu S, Temba EM. Artificial hybridization in bambara groundnut (*Vigna subterranea* (L.) Verdc.). In: Massawe F, editor. Conference proceedings of the International bambara groundnut symposium, Gaborone, Botswana, 5–12 Sept 2004; 2004. pp. 193–209.
10. Ntundu WH, Bach IC, Christiasen JC, Andresen SB. Analysis of genetic diversity in Bambara groundnut (*Vigna subterranea* (L.) Verdc.) landraces using amplified fragment length polymorphism (AFLP) markers. Afr J Biotechnol. 2004;3:220–5.
11. Singrün C, Schenkel W. Fingerprinting of Bambara groundnut germplasm with molecular markers. In: Massawe F, editor. Proceedings of the international symposium on Bambara groundnut, Botswana College of Agriculture, 8–12 Sept 2003; 2004.
12. Collinson ST, Azam-Ali SN, Chavula KM, Hodson DA. Growth, development and yield of bambara groundnut (*Vigna subterranea*) in response to soil moisture. J Agric Sci. 1996;126:307–18.
13. Brough SH, Taylo AJ, Azam-Ali SN. The potential of Bambara groundnut (*Vigna subterranea*) in vegetable milk production and basic protein functionality systems. Food Chem. 1993;47:277–83.
14. Udayakumar M, Rao RCN, Wright GC, Ramaswamy GC, Ashok B, Stephan R, Gangadhar GC, Aftab Hussain IS. Measurement of transpiration efficiency in field condition. J Plant Physiol Biochem. 1998;1:69–75.
15. Barrs HD, Weatherley PE. A reexamination of the relative turgidity technique for estimating water deficits in leaves. Aust J Biol Sci. 1962;15:413–28.
16. Minolta. SPAD 505 owner's industrial meter division. Minolta Crop. Ramseer, N.I. 1989.
17. Monje OA, Bugbee B. Inherent limitation of non-destructive chlorophyll meters: A comparison of two types of meters. Hort Sci. 1992;27:69–71.
18. Peng S, Felipe S, Garcia V, Rebacca C, Laza A, Cassman KG. Adjustment for specific leaf weight improves chlorophyll meters estimate of rice leaf nitrogen concentration. Agro J. 1993;85:987–90.

19. Blum A, Ebercon A. Cell membrane stability as a measure of drought and heat tolerance in wheat. Crop Sci. 1981;21:43–7.
20. Squire GR, Connolly H, Crawford J, Collinson ST, Sesay A. Linking vegetative and reproductive variability in landraces of bambara groundnut (*Vigna subterranea* L.). In: Proceeding of the International symposium on Bambara groundnut, University of Nottingham, UK, 23–25 July 1996; 1997.
21. Udayakumar M, Prasad TG. 13C isotope discrimination in plants – a potential technique to determine WUE. In: Rao RCN, Wright GC, editors. Selection for WUE in grain legumes, a report of workshop held at ICRISAT center, Andhra Pradesh, India; 1994. pp. 42–5.
22. Passioura JB. Resistance to drought and salinity: avenues for improvement. Aust J Plant Physiol. 1986;13:191–201.

28 Biochemical Characterization of Oxidative Burst During Interaction Between Sesame (*Sesamum indicum* L.) in Response to *Alternaria sesami*

A.S. Lubaina and K. Murugan

Abstract

Induction of plant defence against pathogen attack is regulated by a complex network of different signals. The oxidative burst or rapid and transient production of large amount of reactive oxygen species (ROS) belongs to the fastest and earliest active defence responses to microbial infection known in plants. The aim of this study was to investigate the intensity and timing of the ROS formation, lipid peroxidation and expression of antioxidant enzymes as initial response of sesame (*Sesamum indicum* L.) against the invading pathogen *Alternaria sesami*. The concentration of hydrogen peroxide (H_2O_2) was 384 times higher at 72 h postinoculation, and lipid peroxidation was 5.5 times higher at 72 h postinoculation in the extracts of inoculated leaves than in the control. An increase in total phenolic content was also detected in inoculated leaves. The activities of the antioxidative enzymes, viz. superoxide dismutase (SOD, EC 1.15.1.1), catalase (CAT, EC 1.11.1.6), guaiacol peroxidase (GPX, EC 1.11.1.7) and ascorbate peroxidase (APX, EC 1.11.1.11), increased in response to pathogen inoculation. SOD activity at 72 h postinoculation in leaves was 92 times than the control. CAT activity also showed a decrease after 24 h postinoculation, and the increase in activities of GPX and APX was insignificant after 24 h post inoculation in the inoculated leaves. The oxidative burst generated in the interaction between sesame and *Alternaria sesami* may be an early first line of defence mounted against the invading pathogen. However, seemingly less efficient antioxidative system (particularly the decrease of CAT activity after 24 h postinoculation) leading to sustained accumulation of ROS and the observed higher rate of lipid peroxidation indicate that the biochemical events are largely in

A.S. Lubaina • K. Murugan (✉)
Plant Biochemistry and Molecular Biology Lab, Department of Botany, University College, Thiruvananthapuram, Kerala, India
e-mail: harimurukan@gmail.com

A. Sabu and A. Augustine (eds.), *Prospects in Bioscience: Addressing the Issues*, DOI 10.1007/978-81-322-0810-5_28, © Springer India 2013

favour of the pathogen, thus making this host–pathogen interaction a compatible combination. It is discussed that the oxidative burst served as a weapon for the pathogen because the antioxidative system was not significant enough to impede the pathogen ingress in the host.

Keywords

Antioxidative enzymes • Postinoculation • Lipid peroxidation • *Sesamum indicum* • Oxidative burst • Reactive oxygen species

Introduction

Resistance in many plant–pathogen interactions is associated with multifaceted defence system. The individual components of such systems include ion fluxes across plant membranes, the generation of reactive oxygen species (ROS), phosphorylation of specific proteins, activation of cell wall strengthening enzymes, structural barriers like lignin, hydroxyl proline-rich cell wall proteins, transcriptional activation of several defence-related genes, induction of phytoalexins, localized cell death at infection sites – hypersensitive response (HR) – and induction of systemic acquired resistance in distal plant organs [1]. Proper recognition and judicious regulation of defence responses are essential for host plants, as these responses often have measurable deleterious effects on plant growth and metabolism [2]. Fungal pathogens deploy different strategies to escape host surveillance and establish themselves within the host depending on their nutritional requirements.

The transient production of ROS, in an oxidative burst, is frequently an early plant response to pathogen attack [3]. ROS have been suggested to be involved in plant defence responses in several ways. H_2O_2 is the major ROS of the oxidative burst in plants, since it is the most long lived and is able to cross plant cell membranes and thereby acts as a diffusible and relatively lasting signal. ROS production including H_2O_2 has been especially well established in several plant tissue [4] and suspension-cultured cell systems associated with the expression of an HR and SAR. In contrast, very little is known about oxidative metabolism in plant resistance reactions to pathogens that do not induce HR, such as the necrotrophic fungi that invade the plant vascular system. Sesame (*Sesamum indicum* L.), a member of Pedaliaceae, is perhaps the oldest oilseed known and used by human beings. It is an important annual crop in the tropics and warm subtropics, known as the "queen of oilseeds" because of its nutrient qualities [5]. Sesame seed oil has long shelf life and has remarkable antioxidant function. India is among the largest vegetable oil economies in the world [6]. India ranks first in area and production among the sesame-growing countries. In this work, the aim is to investigate biochemical responses in leaves of sesamum cultivar (Tilarani) after *Alternaria sesami* inoculation in order to find reliable biochemical parameters that could be correlated with pathogen resistance mechanisms.

Materials and Methods

Sesamum indium cultivar was raised from seeds in healthy conditions in a glasshouse. For *in vitro* fungal inoculation studies, mature plants were inoculated with 20 μl of *Alternaria sesami* conidial suspension (1×10^3 conidia ml^{-1}) or 20 μl of water (mock inoculation). The inoculated plants, along with their respective healthy controls, were then maintained at 30°C in a temperature-controlled glasshouse under a photoperiod of 12/12 h (light/dark) and 60% RH. After the development of symptoms in infected plants, the experiment was terminated, and the leaves harvested regularly from 24 to 120 h for analysis.

Estimation of O_2.– and H_2O_2

H_2O_2 concentration was quantified using the protocol of Bellincampi et al. [7]. Superoxide anion was quantified following the method of Doke [8].

Quantification of Lipid Peroxidation (LPX)

The level of lipid peroxidation in the cells was measured as malondialdehyde (MDA) content by the thiobarbituric acid (TBA) reaction [9].

Superoxide Dismutase (SOD)

The SOD enzymatic samples were prepared from biological tissues and assayed as per the method of Elstner et al. [10].

Ascorbate Peroxidase (APX)

APX was isolated from the experimental samples, and the activity determination was carried out according to the modified method of Yanagida et al. [11].

Catalase (CAT)

In CAT assay, the reaction mixture consisted of 50 mM K-phosphate buffer (pH 7.0) containing 10 mM H_2O_2 (0.95 ml) and 0.05 ml enzyme extract. Immediately after adding the enzyme to the buffer, the initial rate of absorbance at 240 nm was determined. The molar absorption coefficient of H_2O_2 (0.04 mM cm^{-1}) was used to calculate the enzyme activity [12].

Peroxidase

Peroxidase was isolated following the method of Goliber [13]. POX activity was assayed using the method of Ingham et al. [14].

Phenylalanine Ammonia-Lyase (PAL)

Isolation of PAL was made following the method of Morrison et al. [15], and the activity was estimated by the method of Whetten and Sederoff [16].

Quantification of Total Phenol

Total phenol was isolated and quantified by the method of Mayr et al. [17].

Reverse-Phase High-Performance Liquid Chromatography (RP-HPLC) of Phenols

Quantitative fractionation of various phenolic acids in the samples was studied by RP-HPLC analysis. Phenolic acids extracted from fresh tissues in aqueous methanol were used for the study [18].

Results and Discussion

Changes in Hydrogen Peroxide (H_2O_2) and Superoxide Anion ($O_2^{\cdot-}$)

The production of reactive oxygen species (ROS) is one of the earliest cellular responses following successful pathogen recognition [19]. As shown in Fig. 28.1, infection by *A .sesami* significantly increased the levels of H_2O_2 and $O_2^{\cdot-}$ production in sesamum leaves, and its levels increased with increasing the time of infection (H_2O_2 and $O_2^{\cdot-}$ levels increased by 384 and 86-fold, respectively, after 72 h from infection). ROS was considered as the first defence line against pathogen and may act as direct antimicrobial agent against phytopathogen attack [20]. Therefore, high production of H_2O_2 and $O_2^{\cdot-}$ in sesamum leaves is an important element of disease resistance mechanism which are involved directly or indirectly in restricting pathogen growth and giving the time for plants to mobilize further defence reactions. Accumulation of H_2O_2 levels may lead to the accelerated senes-

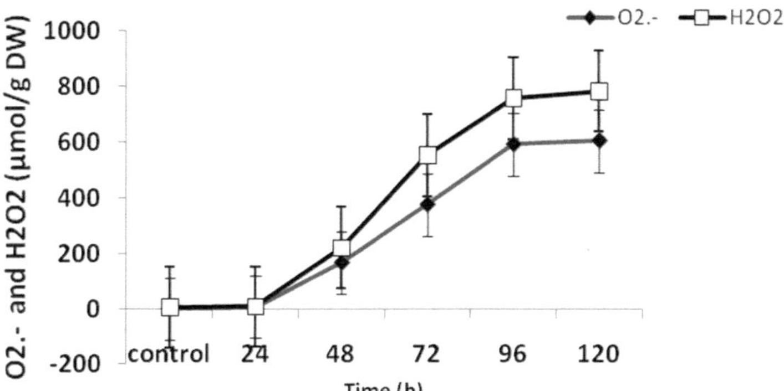

Fig. 28.1 Reactive oxygen species such as superoxide anion ($O_2^{\cdot-}$) and hydrogen peroxide (H_2O_2) content in control and infected sesamum leaves from 24to 120 h. Values are means of three individual experiments with duplicates

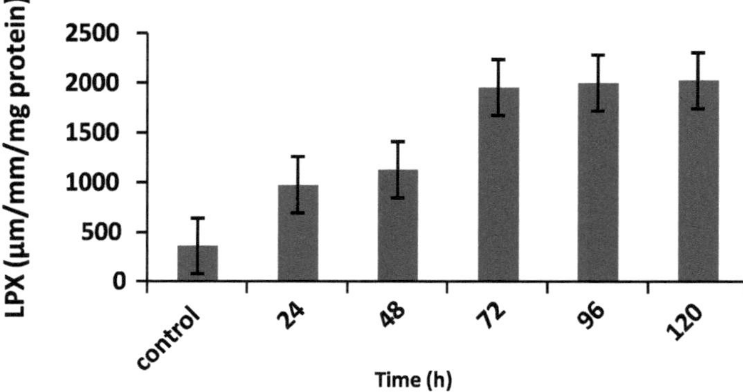

Fig. 28.2 Changes in lipid peroxidation in control and infected sesamum leaves from 24 to 120 h. Values are means of three individual experiments with duplicates

cence and decreased photosynthesis in infected tomato leaves as observed by El-Khallal [21]. Induction or suppression of ROS generation in leaves of these treatments could be related to the activity of antioxidant enzymes which decrease or increase the levels of H_2O_2 (either by direct decomposition or oxidation or by its dismutation). ROS was proposed to act synergistically in a signal amplification loop with shikimic acid-dependent pathways to drive the HR and the establishment of systemic defence. Similarly, jasmonic acid induced H_2O_2 production in tomato plants [22] and pepper [23] which triggers expression of plant resistance and HR – related genes also support our findings. Thus, present results suggest that the high production of ROS (H_2O_2 and $O_2^{\cdot-}$) and the capacity of plants to control its concentrations might contribute to increased resistance against *A. sesami* attack and giving the time for plants to mobilize further defence reactions. ROS have been involved in plant defence responses in multifold ways: (a) reinforcing plant cell wall through cross-linking reactions of lignin and proteins, (b) acting as toxic agents either against the host plant cells, with development of HR and SAR, or against the pathogen through killing them or stopping their growth and (c) participating as second messengers in signalling routes leading to the activation of plant defence-related genes [24].

Changes in Lipid Peroxidation

ROS (H_2O_2 and $O_2^{\cdot-}$) activity frequently causes membrane damage through peroxidation of fatty acids, which may be coincided with the activity of lipoxygenase in plants as a consequence of infection (Fig. 28.2). Levels of MDA gradually increased in leaves of infected sesamum plants with increasing the time of infection. When plants are subjected to pathogen attack, the equilibrium between production and scavenging of ROS is

Table 28.1 Activities of antioxidant enzymes in control and infected sesamum leaves. Results are means of three individual experiments with duplicates

	Control	24	48	72	96	120
SOD (U/mg protein)	6.48±0.2	17.7±0.3	354±0.2	597.6±0.1	713.4±0.5	788.5±0.8
APX (U/mg protein)	22.4±0.5	52.6±0.5	214±0.3	319.4±0.3	343.6±0.5	369.8±0.5
CAT (U/mg protein)	22.8±0.1	18.4±0.6	39.6±0.2	42.3±0.8	43.4±0.1	42.8±0.3
POX (U/mg protein)	4.5±0.3	11.9±0.5	84.5±0.1	112.7±0.7	126.6±0.2	128.4±0.3
PAL (U/mg protein)	2.4±0.4	5.4±0.1	42.3±0.1	89.8±0.6	96.5±0.7	99.4±0.5

broken, resulting in oxidative damage of proteins, DNA and lipids. MDA is the marker for lipid peroxidation released from cellular membranes of tissues and are formed by the reaction of ROS (H_2O_2 or/and $O_2^{\cdot-}$) with lipid molecules [24]. Thus, lipid peroxidation in sesamum plants has been proved to be induced by pathogens, and the subsequent products have been shown to posses antimicrobial properties and signalling function [25].

Changes in the Activity of Antioxidant Enzymes

Organisms protect themselves against oxidative stress by the synthesis of various antioxidant enzymes. Results in Table 28.1 showed that infection by *A. sesami* significantly increased SOD, APX and CAT activities in leaves of sesamum plants at different stages of infection as compared with noninfected control. In banana plants, Subramaniam et al. [26] reported that levels of H_2O_2 and other enzyme activities were increased to the levels of tolerance or susceptibility to Fusarium wilt disease. On the other hand, accumulation of H_2O_2 (Fig. 28.1) in plants of infected compared to control plants could be related to the increase in SOD activity and not coordinated with the same increase in H_2O_2-scavenging enzyme (CAT and APX). Accordingly that increase in CAT and APX activity in pathogen-infected plants did not seem to be high and/or enough to avoid the development of disease symptoms. Durner and Klessing [27] have presented evidence suggesting that salicylic acid-mediated inhibition of CAT and APX probably results from peroxidative reactions. It appears that shikimic acid-dependent pathways require H_2O_2 to potentate lipid peroxidation as translocation signal that invokes PR genes and establishes SAR [28]. Finally, results in Figs. 28.1 and 28.2 suggested that increase in levels of ROS, built up by either enhanced production and decreased scavenging potential, may contribute to the resistance reaction in sesamum against *Alternaria* infection. Finally, imbalance between H_2O_2 generation and scavenging enzymes in leaves may reflect a defence mechanism in sesamum or a pathogenicity strategy of the fungus.

Changes in Phenolic Compounds and Its Related Enzymes

Many plant phenolic compounds are known to be antimicrobial, function as precursors to structural polymers such as lignin or serve as signal molecules [29]. Results in Fig. 28.3 showed that total phenolic compounds markedly increased in leaves of all treated sesamum plants at 72 h from pathogen inoculation. Similarly, *Hypericum perforatum* showed a six fold increase in phenolic compounds in cell suspension after jasmonic acid elicitation. [30]. However, levels of certain phenolic acids greatly changed in leaves of sesamum plants in response to pathogen inoculation as compared with control (Fig. 28.4a, b). Sinapic, coumaric, vanillic, chlorogenic, ferulic and hydroxyl benzoate contents markedly increased in all treatments under pathogen infection. However, contents of gallic acids in sesamum plants decreased marginally as compared with noninfected control. Accordingly, the induction of total phenol and changes in the contents of various phenolic acids in leaves of treatments play an important

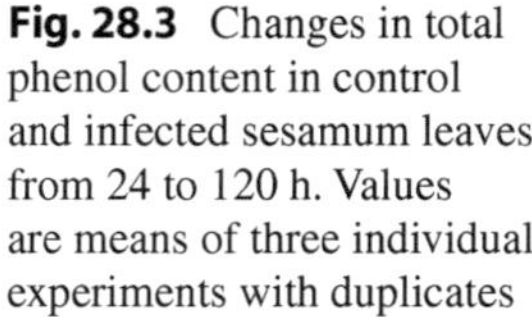

Fig. 28.3 Changes in total phenol content in control and infected sesamum leaves from 24 to 120 h. Values are means of three individual experiments with duplicates

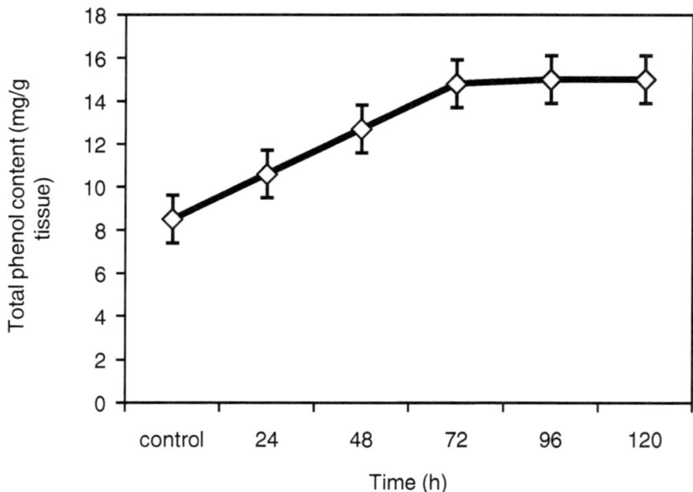

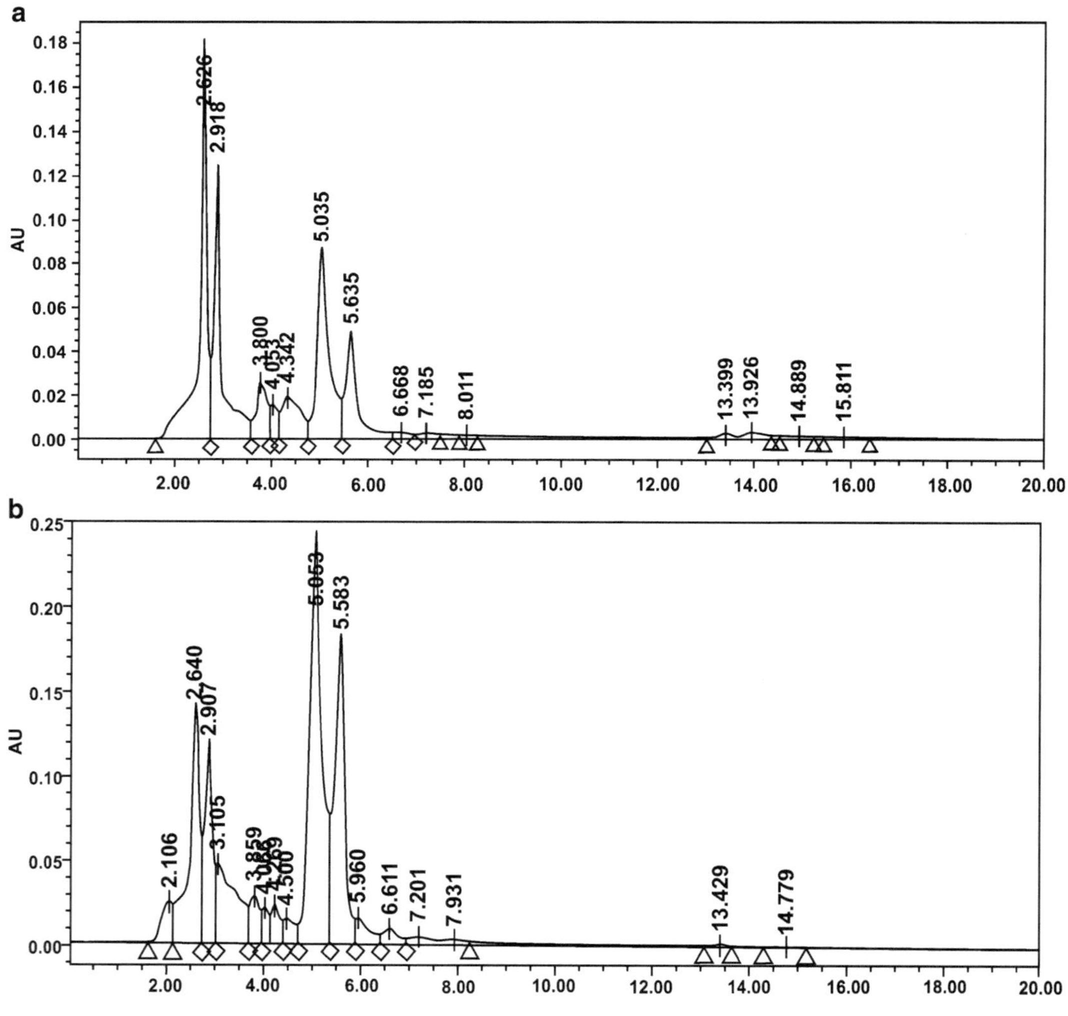

Fig. 28.4 RP-HPLC chromatogram showing the peaks of phenolic acids in control (**a**) and infected (**b**) sesamum leaves

Table 28.2 Amount of phenolic acid by RP-HPLC from control and *A. sesami* infected sesamum leaves (μg/g tissue)

	Sinapic acid	Coumaric acid	Cinnamic acid	Vanilate	Gallate	Chloro-genate	Ferulate	Phlorog-lucinol	Catechol	HBA
Control	1,765.01	2302.6	–	240.4	222.3	–	–	–	85.5	43.9
Infected	2,039.89	2661.2	–	245.07	220.7	235.6	737.6	67.08	91.04	164.7

role in plant resistance and defence against *A. sesami* infection, which are intimately connected with ROS (Fig. 28.1). These results agree with the general speculation that when plant cells are recruited into infection, switch from normal primary metabolism to a multitude of secondary metabolism defence pathway and activation of novel defence enzymes and genes take place [31]. Also, rapid esterification of phenolic compounds such as cinnamic and ferulic acids into plant cell walls is a common and early response to fungal attack, resulting in cell wall strengthening and then enhancing resistance to pathogen penetration [32]. High accumulation of coumaric acid (Table 28.2) might be related to the activation of phenylpropanoid pathway through increased PAL activity (Table 28.1) which reflects a component of a defence response of the plant against *A. sesami* penetration. However, induction in ferulic and chlorogenic acids in infected leaves indicated that both acids are highly antifungal [33] and supported its role in reducing disease through the formation of defence barriers and activation of defence responses. Although hydroxybenzoic acid was more effective and greatly phytotoxic against soilborne root-infecting fungi, its content increased in leaves of infection treatments of sesamum plants, and this might be because HBA converted to both bound salicylic acid and gentisic acid [34]. Parallely, in addition to the induction of various phenolic acids in sesamum, leaves also showed significant activities of POX and PAL in response to *A. sesami* (Table 28.1).

Further investigations are needed to determine whether our results support the induced susceptibility theory of host–biotrophic pathogen interactions. In sum, results from this present work showed the induction of the antioxidant enzyme system and other oxidative stress markers during *A. sesami* in sesamum, suggesting that changes in oxidative metabolism may be a quite general plant defence response not only restricted to foliar pathogens causing resistant reaction through HR and necrotic processes. Our results also suggest that increased levels of ROS, built up by either enhanced production and decreased scavenging potential, may contribute to the resistance reaction. Studies now in progress about direct ROS estimation and induction of ROS-forming enzymes, non-enzyme antioxidants and compartmentation of antioxidant responses at the apoplast level will probably provide more conclusive insights about the production of an oxidative burst and related responses and their role in the pathogenesis of *A. sesami* on sesamum.

References

1. Garcia-Limones C, Hervas A, Navas-Cortes J, Jimenez-Diaz RM, Tena M. Induction of an antioxidant enzyme system and other oxidative stress markers associated with compatible and incompatible interactions between chickpea (*Cicer arietinum* L.) and *Fusarium oxysporum* f. sp. ciceris. Physiol Mol Plant Pathol. 2002;61:325–37.
2. Gupta S, Chakraborti D, Sengupta A, Basu D, Das S. Primary metabolism of chickpea is the initial target of wound inducing early sensed *Fusarium oxysporum* f. sp. ciceri race I. PLoS One. 2010;5(2):e 9030–41.
3. Glazebrook J. Contrasting mechanism of defense against biotrophic and necrotrophic pathogens. Annu Rev Phytopathol. 2005;4:205–27.
4. Apel K, Hirt H. Reactive oxygen species metabolism, oxidative stress and signal transduction. Annu Rev Plant Biol. 2004;55:373–401.
5. Prasad R. Text book of field crops production. New Delhi: Indian Council of Agricultural Research; 2002. p. 821.
6. Ashri A. Sesame breeding. Plant Breed Rev. 1998;-16:179–228.
7. Bellincampi D, Dipierro N, Salvi G, Cervone F, De Lorenzo G. Extracellular H_2O_2 induced by oligogalacturonides is not involved in the inhibition of the auxin regulated rolB gene expression in tobacco leaf plants. Plant Physiol. 2000;122:1379–85.
8. Doke N. Involvement of super oxide anion generation in the hypersensitive response of potato tuber tissue to

infection with an incompatible race of *Phytophthora infestans* and to the hyphal wall components. Physiol Plant Pathol. 1983;23:345–57.

9. Zhang JX, Kirkham MB. Enzymatic responses of the ascorbate-glutathione cycle to drought in sorghum and sunflower plants. Plant Sci. 1996;113:139–47.
10. Elstner EF, Youngman R, Obwald W. Superoxide dismutase. In: Bergmeyer J, Grabl BM, editors. Methods of enzymatic analysis, vol. III. Enzymes oxidoreductases. 3rd ed. Weinheim: Verlag-Chemie; 1995. p. 293–302.
11. Yanagida M, Mino M, Iwabuchi M, Ogawa KI. Expression of spinach ascorbate peroxidase isoenzymes in response to oxidative stresses. Plant Physiol. 1999;119:743–54.
12. Shanker AK, Djanaguiraman M, Sudhagar R. Differential antioxidative response of ascorbate glutathione pathway enzymes and metabolites to chromium speciation stress in green gram (*Vigna radiata* (L) R. Wilczek. Cv CO4) roots. Plant Sci. 2004;166:1035–43.
13. Goliber TE. Gravitational stress and lignification in aerial vs submerged shoots of *Hippuris vulgaris*. Physiol Plant. 1989;75:355–61.
14. Ingham L, Lindsa M, Parker ML, Waldtom K, Keith W. Peroxidase changes in soluble and bound forms during maturation and ripening of apples. Physiol Plant. 1998;102:93–100.
15. Morrison TA, Kessler JR, Hatfield RD, Buxton DR. Activity of two lignin biosynthesis enzymes during development of a maize internode. J Sci Food Agric. 1994;65:133–9.
16. Whetten RW, Sederoff RR. Phenylalanine ammonia lyase from loblolly pine. Purification of the enzyme and isolation of complementary DNA clones. Plant Physiol. 1992;98:380–6.
17. Mayr V, Treutter D, Santos-Buelga C, Bauer H, Feucht W. Developmental changes in the phenol concentration of golden delicious apple fruits and leaves. Phytochemistry. 1995;38:1151–5.
18. Beta T, Rooney LW, Marovatsanga LT, Taylor JRN. Phenolic compounds and kernel characteristics of Zimbabwean sorghums. J Sci Food Agric. 1999;-79:1003–10.
19. Grant MI, Brown S, Adams M, Kinght A, Ainslie A, Mansfield J. The RPMI plant disease resistance increase in cytosolic calcium that is necessary for the oxidative burst and hypersensitive cell death. Plant J. 2000;23:441–50.
20. Shetty NP, Mehrabi R, Lutken H, Haldrup A, Kema GH, Collenge DP, Jorgenson HJ. Role of hydrogen peroxide during the interaction between the hemibiotrophic fungal pathogen *Septoria tritici* and wheat. New Phytol. 2007;174(3):637–47.
21. El-Khallal SM. Induction and Modulation of resistance in tomato plants against *Fusarium* wilt disease by bioagent fungi (arbuscular mycorrhiza) and hormonal elicitors (jasmonic acid and salicylic acid):1 Changes in growth, some metabolic activities and endogenous hormones related to defense mechanism. Aust J Basic Appl Sci. 2007;1(4):717–32.
22. Orozcon-Cardenas ML, Narvaez-Vasquez J, Ryan CA. Hydrogen peroxide acts as a second messenger for the induction of defense genes in tomato plants in response to wounding, systemin and methyl jasmonate. Plant Cell. 2001;13:179–91.
23. Ueeda M, Masaharu K, Nishi K. Contribution of jasmonic acid to resistance against *phytophthora* blight in *Capsicum annuum* cv. SCM 334. Physiol Mol Plant Path. 2006;67:149–54.
24. Shimzu N, Hosogi N, Hyon GS, Jiang S, Inoue K, Park P. Reactive oxygen species (ROS) generation and ROS induced lipid peroxidation are associated with plant membrane modifications in host cells in response to AK-toxin from *Alternaria alternata* Japanese pear pathotype. J Gen Plant Path. 2006;72:6–15.
25. Melan MA, Dong X, Endara ME KR, Ausubel FM, Peterman TK. An *Arabidopsis thaliana* lipoxygenase gene can be induced by pathogen, abscisic acid, and methyl jasmonate. Plant Physiol. 1993;101:441–50.
26. Subramaniam S, Moziah M, Sariah M, Puad MP, Xavier R. Bioassay method for testing *Fusarium* wilt disease tolerance in transgenic banana. Sci Hortic. 2006;108:378–89.
27. Durner J, Klessig DF. Salicylic acid is a modulator of tobacco and mammalian catalase. J Biol Chem. 1996;271:28492–501.
28. Rao MV, Gopinadhan P, Ormrod DP, Murr DP, Watkins CB. Influence of salicylic acid on H_2O_2 production, oxidative stress, and H O metabolizing enzymes. Plant Physiol. 1997;115:137–49.
29. Hammerschmidt R. Phenols and plant-pathogen interactions: the saga continues. Physiol Mol Plant Path. 2005;66:77–8.
30. Gadzouska S, Delaunnay A, Spasenoski M, Maury S, Joseph C, Haðege D. Jasmonic acid elicitation of *Hypericum perforatum* L. cell suspensions and effects on the production of phenylpropanoids and naphtodianthrones. Plant Cell Tissue Org. 2007;89(1):1–13.
31. Tan JB, Schneider A, Svatos P, Bendnarek J, Liu HK. Universally occurring phenyl propanoid and specific indolic metabolites in infected and uninfected *Arabidopsis thaliana* roots and leaves. Phytochemistry. 2004;65:691–9.
32. Stadnik MJ, Buchenauer H. Inhibition of phenylalanine ammonia-lyase suppresses the resistance induced by benzothiadiazole in wheat to *Blumeria graminis* f. sp. tritici. Physiol Mol Plant Path. 2000;57:25–34.
33. Agrios GN. Plant pathology. 4th ed. San Diego: Academic; 1997. p. 93–114.
34. Belles JMR, Carro J, Fayos P, Navarro J, Prino CV. Gentisic acid as a pathogen inducible signal, additional to salicylic acid for activation of plant defenses in tomato. Mol Plant Microb Interact. 1999;12(3):227–35.

Mass Spectrometric Characterization of a Novel Antimicrobial Peptide Isolated from *Clitoria ternatea*

29

Neethu Ajayakumar, Walter Schrenk, Bipin G. Nair, and Sudarslal Sadasivan Nair

Abstract

Increasing microbial resistance to common antibiotics has become a serious threat in maintaining public health. Due to their distinctive features, antimicrobial peptides have become attractive molecules as novel antibiotics. Plants are constantly exposed to attack from a large range of pathogens, and under such conditions, they synthesize antimicrobial peptides as part of their innate defense mechanism. In an effort to identify potential antimicrobial peptides, we have isolated and characterized a novel eight-residue linear peptide from *Clitoria ternatea*, a perennial plant which belongs to Fabaceae family. Liquid chromatography-electrospray ionization mass spectrometry (LC-ESI-MS) shows a UV-absorbing fraction at 230 nm with a measured molecular mass of 787.4 Da. The sequence of the peptide, QAANSVAK, was derived through *de novo* approach using tandem mass spectrometric (MS/MS) data obtained after collision-induced dissociation (CID). The sequence was further confirmed through chemical derivatization and proteolysis followed by data-dependent MS/MS analysis. Antimicrobial studies of the nearly homogenous peptide fraction after solid-phase extraction (SPE) show activity against *Bacillus subtilis*. The peptide bears no sequence similarity with any of the linear peptides isolated thus far from other organisms.

Keywords

Mass spectrometry • Antimicrobial peptide • *De novo* sequencing • *Clitoria ternatea*

N. Ajayakumar • W. Schrenk • B.G. Nair • S.S. Nair (✉)
Amrita School of Biotechnology, Amrita Vishwa Vidyapeetham, Amritapuri,
Clappana P.O., Kollam, Kerala 690 525, India
e-mail: sudarslal@am.amrita.edu

A. Sabu and A. Augustine (eds.), *Prospects in Bioscience: Addressing the Issues*,
DOI 10.1007/978-81-322-0810-5_29, © Springer India 2013

Introduction

Antimicrobial peptides (AMPs) have gotten remarkable attention as alternative to classical antibiotics to combat drug-resistant bacteria [1]. Most of the pathogenic bacteria become multidrug resistant, and as a solution to this crisis, AMPs have been isolated from different sources such as plants, animals, and microorganisms [2, 3]. Unlike animals, plants lack typical immune system, and they use AMPs as part of their immune defense mechanism. Here we report mass spectrometry-based *de novo* sequencing of a linear antimicrobial peptide isolated from *Clitoria ternatea*, a herbaceous perennial plant belonging to the Fabaceae family. *C. ternatea* is widely distributed in the southern part of India. The medicinal properties of this plant are highly impressive and have been extensively used for the treatment of various ailments in folklore medicine [4]. Many bioactive compounds including cyclic peptides were isolated from *C. ternatea* that show potent antimicrobial and anti-insecticidal properties [5]. However, characterization of a linear peptide with specific antimicrobial property is not yet reported from *C. ternatea.*

Materials and Methods

Peptide Extraction and Purification

Leaves of *Clitoria ternatea* were collected from the surrounding areas of Amritapuri, Kollam, Kerala. Two hundred and fifty gram wet weight of leaves were ground and extracted with methanol: dichloromethane in 1:1 ratio (v/v) and stirred overnight at room temperature. The extract was partitioned with dichloromethane and water followed by the separation of the aqueous fraction. The aqueous fraction was further concentrated on a rotary evaporator and subjected to solid-phase extraction (SPE) using C18 disk cartridges (Empore, Supelco). 1 ml of crude sample was loaded on to the SPE cartridge pre-equilibrated with methanol and water. After washing with water, the sample was fractionated by increasing the concentration of eluent (acetonitrile). The eluted fractions after vacuum concentration were subjected to functional assays and mass spectrometric characterization.

Acetylation I

The peptide fraction after SPE purification was dried and resuspended in 1:1 ratio of 100 mM ammonium bicarbonate buffer, pH 8.0, and acetic anhydride and incubated at room temperature for 15 min. The resultant solution was vacuum dried and reconstituted in 9:1 (v/v) water and acetonitrile containing 0.1 % formic acid and analyzed using LC-ESI-MS.

Acetylation II

In the second acetylation protocol, the vacuum-dried peptide sample was redissolved in 1:1 (v/v) acetic anhydride and pyridine and incubated for 5 min at room temperature. The resultant mixture after vacuum drying was resuspended in 9:1 (v/v) water and acetonitrile containing 0.1% formic acid and analyzed using LC-ESI-MS.

Trypsin Digestion

The SPE-purified peptide fraction was digested with proteomics grade trypsin (Sigma) at a ratio of 50:1 (w/w) and incubated at 37°C, overnight. The final solution was analyzed by LC-ESI-MS.

Mass Spectrometry

Electrospray ionization mass spectra (ESI-MS) were acquired on an Agilent ion trap mass spectrometer (6340 Series) coupled to an Agilent 1200 series HPLC. The samples were infused to the mass spectrometer through a reversed-phase column (Zorbax SB-C18, 2.1×35 cm) with solvent A (0.1% formic acid in water) and solvent B (0.1% formic acid in acetonitrile). The flow rate was maintained at 0.2 ml min^{-1}. Collision-induced dissociation (CID) data were collected

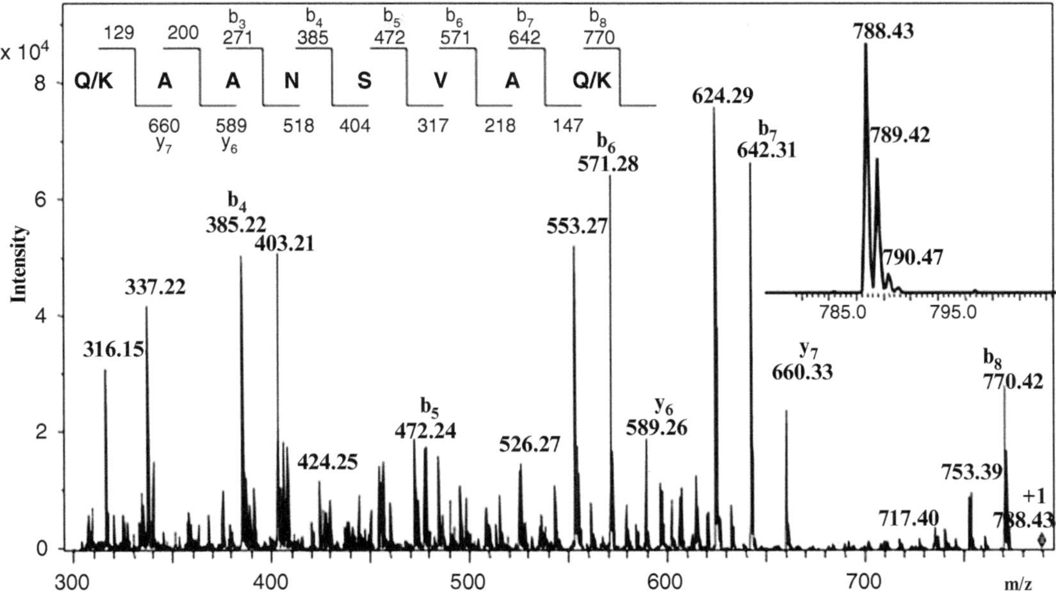

Fig. 29.1 ESI-MS/MS spectrum from the singly protonated *Clitoria ternatea* peptide. Monoisotopic m/z values of b- and y-ion series as proposed by Biemann [7] are marked. The *inset* shows electrospray spectrum of the singly protonated peptide

in a data-dependent manner with helium as the collision gas. The MS data were acquired over a range of 200–1,500 m/z, and the MS/MS data were collected over the m/z range 100–1,000. All data were acquired in positive ionization mode and were further analyzed using Bruker Data Analysis software.

Antimicrobial Activity

The antimicrobial activity of the SPE-purified fractions was checked through standard well diffusion method [6]. Appropriate positive (streptomycin) and negative (distilled water) controls were included for comparison.

Results and Discussion

The crude leaf extract was fractionated and purified to near homogeneity using solid-phase extraction. Each peptide fraction thus obtained was analyzed by mass spectrometry and screened for its antimicrobial properties. LC-ESI-MS of the peptide fraction showed a prominent singly protonated species, $[M+H]^{+1}$, at 788.43 m/z (Fig. 29.1, inset), suggesting neutral monoisotopic molecular mass as 787.42 Da. To derive primary amino acid sequence of this peptide, tandem mass spectrometric data was collected through data-dependent MS/MS; the product ions thus obtained were predominantly b-ion in nature. However, MS/MS of the doubly charged (394.73 m/z) species of the peptide show more representative data with equally distributed b- and y-ions (Fig. 29.2). By following Biemann's nomenclature [7], a tentative sequence Q/K-A-A-N-S-V-A-Q/K was derived through manual *de novo* sequencing approach. Since the data was collected from ion trap mass spectrometers, it was not easy to differentiate the isobaric amino acids, Q and K (the nominal mass is 128 Da), and this ambiguity was attended by acetylating the molecule using acetic anhydride under different alkaline conditions (acetylation of peptides modifies its free N-terminus and lysine residues and increases the masses of peptides by multiples of 42 Da). Acetylation of the *C. ternatea* peptide with acetic anhydride in ammonium bicarbonate buffer generated a prominent doubly charged molecular species at 415.7 m/z, yielding neutral

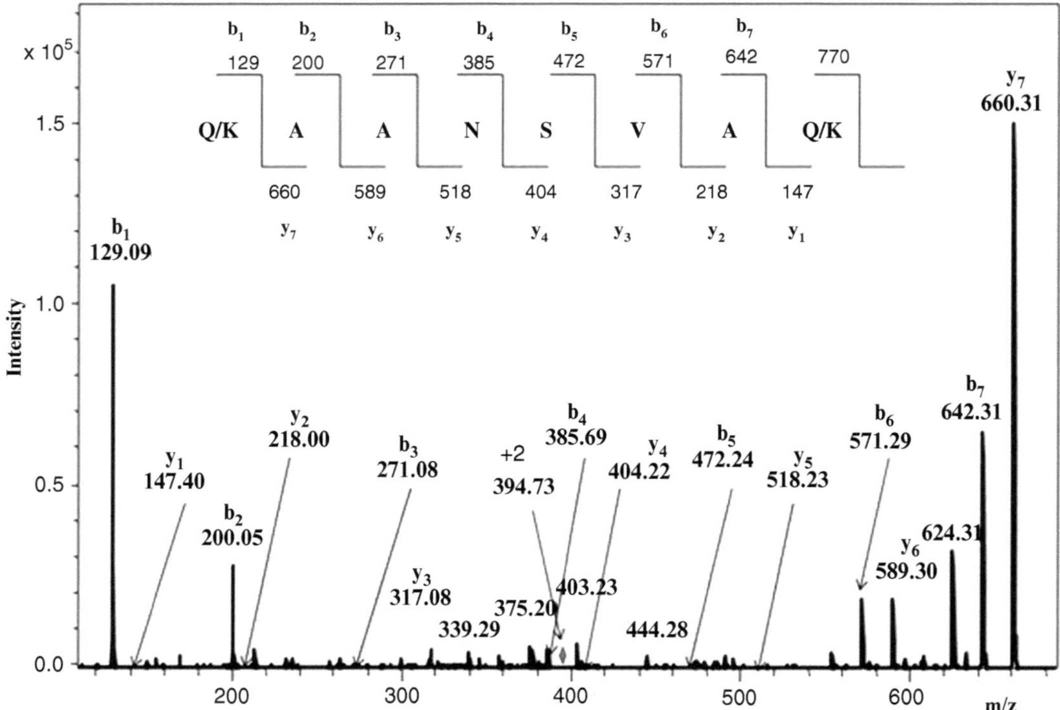

Fig. 29.2 ESI-MS/MS spectrum from the doubly charged *Clitoria ternatea* peptide. Observation of more b- and y-ions facilitates confident sequence assignment

mass of the modified peptide as 829.4 Da ($\Delta m = +42$ Da) indicative of the presence of one primary amine. The primary amine can be either the peptide's free N-terminus or ε-amino group of lysine. Subsequent MS/MS data indicates that the amino acid residue at the C-terminus side of the peptide got acetylated (Fig. 29.3) and confirms the presence of a lysine residue. However, acetylation in the presence of pyridine, where pyridine functions as a catalyst facilitating acetylation reaction, yielded a doubly charged species at 457.73 m/z which in turn corresponds to a neutral molecular mass of 913.46 Da. The increase in molecular mass of 126 Da ($\Delta m = 3 \times 42$) revealed the presence of three acetylation sites. Figure 29.4 shows MS/MS of the triply acetylated peptide which on analysis disclosed presence of one acetylation site at the N-terminus and two acetylation sites at the C-terminal lysine residue (multiple acetylations at lysine residues are not uncommon and have been reported in many biological events [8]). With the above-mentioned acetylation reactions, the following conclusions could be made: (1) the C-terminal residue of the *C. ternatea* peptide is a lysine, (2) the peptide has free N-terminus, and (3) the amino acid residue at the N-terminus side is a glutamine. Consequently, the peptide sequence was assigned as Q-A-A-N-S-V-A-K.

To confirm the presence of an N-terminal glutamine residue, the *C. ternatea* peptide was digested with trypsin followed by data-dependent MS/MS. The resultant spectra did not show a peptide mass corresponding to amino acid sequence, A-A-N-S-V-A-K, which would have generated if the N-terminal amino acid was a lysine. Hence, the final sequence of the peptide was validated as Q-A-A-N-S-V-A-K.

Preliminary investigations on the antimicrobial activity of the SPE-purified *C. ternatea* peptide fraction were tested by standard agar well diffusion method. Among several gram-positive

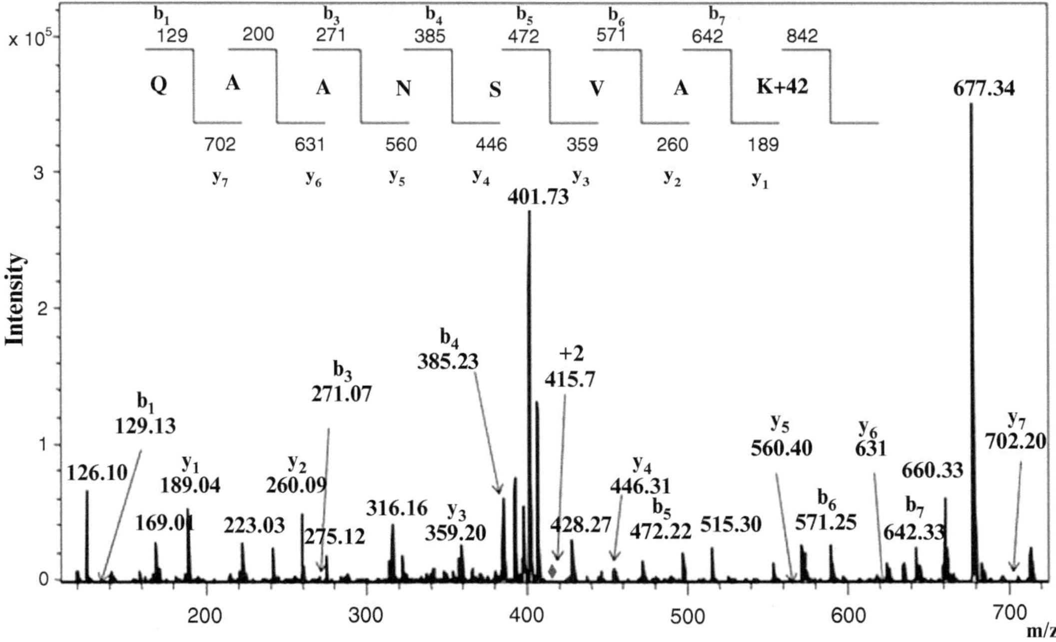

Fig. 29.3 ESI-MS/MS spectrum of the doubly charged peptide after acetylation with acetic anhydride in ammonium bicarbonate buffer

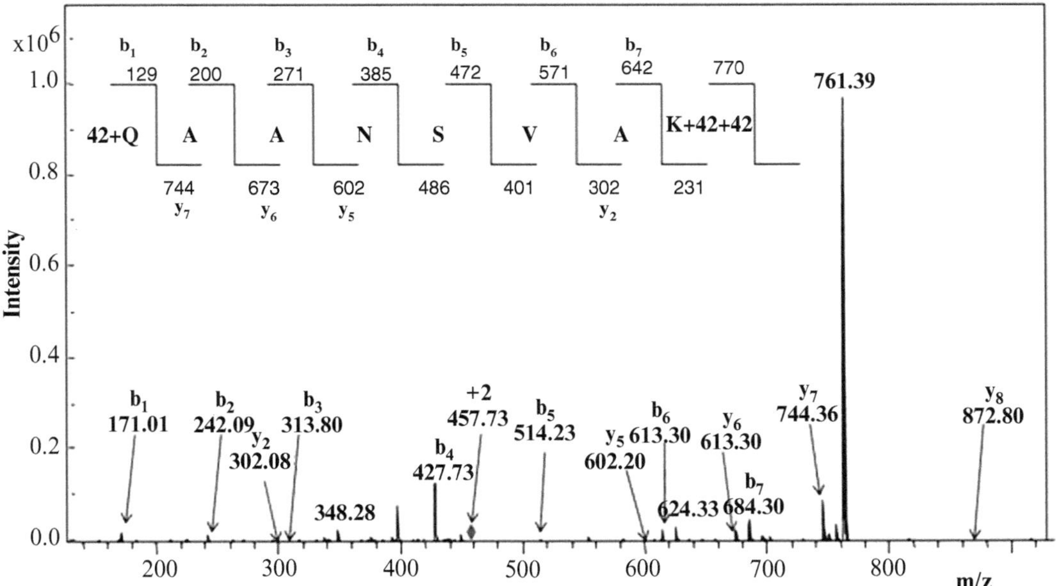

Fig. 29.4 ESI-MS/MS spectrum of the doubly charged peptide after acetylation with acetic anhydride in pyridine

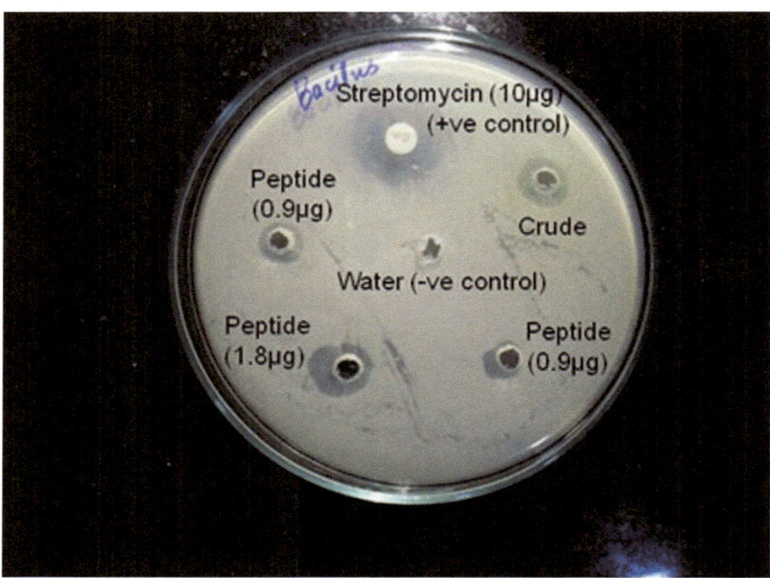

Fig. 29.5 Plate showing antimicrobial activity of the peptide against *Bacillus subtilis*, where streptomycin is used as a positive control

and gram-negative bacteria screened, *Bacillus subtilis*, a gram-positive bacterium, showed significant zone of inhibition (Fig. 29.5). Considerable antimicrobial activity was observed with 0.9 μg of peptide.

Conclusions

In this study, we describe the identification and mass spectrometry-based characterization of a novel linear peptide from the leaf extract of *Clitoria ternatea*. The primary amino acid sequence of the peptide was determined through mass spectral data-based manual *de novo* sequencing. The sequence of the peptide was confirmed through chemical derivatization as well as proteolytic digestion protocols. The peptide shows significant antimicrobial activity against *Bacillus subtilis*, a gram-positive bacterium. Further investigations to unravel the mechanistic aspects of the antimicrobial specificity of the peptide are in progress.

Acknowledgements We would like to acknowledge Agilent Technologies India Pvt. Ltd. for providing the LC-MS system and for sponsorship of consumables. The authors are grateful to Prof. Ashoke Banerji and his lab members for their kind support.

References

1. Oyston PC, Fox MA, Richards SJ, Clark GC. Novel peptide therapeutics for treatment of infections. J Med Microbiol. 2009;58:977–87.
2. Brown KL, Hancock RE. Cationic host defense (antimicrobial) peptides. Curr Opin Immunol. 2006; 18:24–30.
3. Jenssen H, Hamill P, Hancock RE. Peptide antimicrobial agents. Clin Microbiol Rev. 2006;19:491–511.
4. Trease GE, Evans WC. Pharmacognosy. 15th ed. Edinburgh: W.B. Saunders Publishers Ltd; 2002.
5. Poth AG, Colgrave ML, Lyons RE, Daly NL, Craik DJ. Discovery of an unusual biosynthetic origin for circular proteins in legumes. Proc Natl Acad Sci USA. 2011;108:10127–32.
6. Jorgensen JH, Turnidge JD, Washington JA. Antibacterial susceptibility tests: dilution and disk diffusion methods. In: Manual of clinical microbiology. 7th ed. Washington, DC: ASM Press; 1999. p. 1526–43.
7. Biemann K. Sequencing of peptides by tandem mass spectrometry and high-energy collision-induced dissociation. Methods Enzymol. 1990;193:455–79.
8. Yang F, Zhang L, Li J, Huang J, Wen R, Ma L, Zhou D, Li L. Trichostatin A and 5-azacytidine both cause an increase in global histone H4 acetylation and a decrease in global DNA and H3K9 methylation during mitosis in maize. BMC Plant Biol. 2010;10:178–88.

A Novel Plate Assay System for Screening of Antimicrobial Agents and β-Lactamase Inhibitors from Natural Sources

30

S. Prasanth, Shahana Valsan, Ambika Devi, D.S. Preethidan, M. Haridas, and Abdulhameed Sabu

Abstract

β-Lactam class of antibiotics like amoxicillin, ticarcillin, and piperacillin is still used widely especially in developing countries against gram-positive as well as gram-negative organisms. It is used mainly for the treatment of respiratory infections and urinary infections. Resistance to β-lactam antibiotics is often attained by synthesizing β-lactamase enzymes, which act by attacking the β-lactam ring. To overcome this resistance, β-lactam classes of antibiotics are generally prescribed in combination with β-lactamase inhibitors like tazobactam, clavulanic acid, or sulbactam. There are reports of new strains with β-lactamase enzymes resistant to these inhibitors. One of the major sources of novel therapeutic agents is from the natural sources like plants, algae and other microorganisms. The existing methods of β-lactamase assays are based on calorimetric method and may not be always suitable for coloured organic solvent extracts of plants. Also, the varying pH of many of the extracts often becomes a problem in conducting these assays. Here, we report a novel agar plate assay method by which it is possible for screening antimicrobial agents as well as β-lactamase inhibitors simultaneously with a single assay. The method is based on Kirby-Bauer disc diffusion assay using Mueller-Hinton agar. The β-lactamase-producing organism is swabbed on two plates, one with a β-lactam antibiotic and the other without the antibiotic. Paper discs are impregnated with the extract in investigation and applied on the two plates. If it is showing zone of inhibition in both the plates, it can be due to a possible antimicrobial agent, but if it is showing zone of inhibition in only the plate with β-lactam antibiotic, then this may be due

S. Prasanth • S. Valsan • A. Devi • D.S. Preethidan
• M. Haridas • A. Sabu (✉)
Department of Biotechnology & Microbiology and
Inter-University Centre for Biosciences,
Kannur University,
Thalassery Campus, Palayad P.O., Kannur,
Kerala 670 661, India
e-mail: drsabu@gmail.com

A. Sabu and A. Augustine (eds.), *Prospects in Bioscience: Addressing the Issues*,
DOI 10.1007/978-81-322-0810-5_30, © Springer India 2013

to the presence of a possible β-lactam inhibitor. This system was tested good using five different MRSA strains, three different β-lactamase inhibitors and diethyl ether extract of *Holoptelea integrifolia*.

Keywords
β-Lactamase inhibitor assay • Antibiotic resistance • MRSA

Introduction

Globally the antibiotic sales fetch about US$ 42 billion in 2009 [1], and β-lactam antibiotics sale contributing to half of it is generating a revenue of about US$ 20 million. The major β-lactam antibiotics sold are penicillins, monobactams, cephalosporins, penems and carbapenems. Antibiotics like amoxicillin, ticarcillin and piperacillin are still used widely especially in developing countries against both gram-positive and gram-negative organisms, mainly for the treatment of respiratory infections and urinary infections. Due to the heavy and uncontrolled use of antibiotics, the number of strains developing resistance is increasing day by day. The β-lactamase enzymes are responsible for developing resistance to β-lactam class of antibiotics. Penicillin class of antibiotics has a long history of use and so the rise of resistant strains to penicillin. Abraham and Chain in 1940 [2] reported the first β-lactamase which was a penicillinase from gram-negative *E. coli.* The success of penicillin with the rise of resistant strains encouraged for the development of new penicillin derivatives which gave rise to the β-lactam antibiotics in clinical use today (penicillins, narrow- and extended-spectrum carbapenems, monobactams and cephalosporins). The core part of these antibiotics is the β-lactam ring which is a four-member cyclic amide [2]. Bacteria develop resistance to β-lactam antibiotics by two strategies; one way is by producing β-lactamase enzymes which are able to cleave the β-lactam ring and deactivate the antibiotic. The second way is by point mutations at the active site of penicillin-binding proteins, the target of β-lactam antibiotics [3].

β-Lactamases are generally classified based on amino acid sequence homology, but they are also classified based on substrate and inhibitor specificity [4, 5]. Based on substrate specificity, β-lactamases are classified into four Classes, from A to D. Classes A, C and D are serine β-lactamases and Class B is metallo-β-lactamase and has a single Zn^{2+} ion or a pair of Zn^{2+} aligned to histidine, cystine and aspartate residues in the active site. A successful strategy for overcoming β-lactamase-mediated resistance is the use of compounds which are usually structurally similar lactams which can bind to the active site of the enzyme. This method can be employed either by using substrate analogs which can bind to the active site of the enzyme with enhanced affinity but form unfavourable steric interactions. Another method is by the use of "suicide inhibitors"; here a molecule must bind initially at the enzyme-active site, but it is converted into an inactivator through catalytic action of the enzyme itself [6]. Examples of the former are extended-spectrum cephalosporins, monobactams or carbapenems which form acyl enzymes and adopt catalytically incompetent conformations that are poorly hydrolyzed. Other class of compounds includes clavulanic acid which was the first suicide inactivator of β-lactamases to be described [7], tazobactam and sulbactam. The real problem with currently marketed β-lactam antibiotics is that they are not active against all the β-lactamases, for example, clavulanic acid does not inhibit class B metallo-β-lactamases, class C cephalosporinases and some of class A serine β-lactamase [5]

In this scenario, there is a need for developing novel β-lactamase inhibitors having a wide activity range. Natural sources had always been a source of novel therapeutic agents, and man has exploited this ever since science has existed. Around 70% of today's drugs are from natural

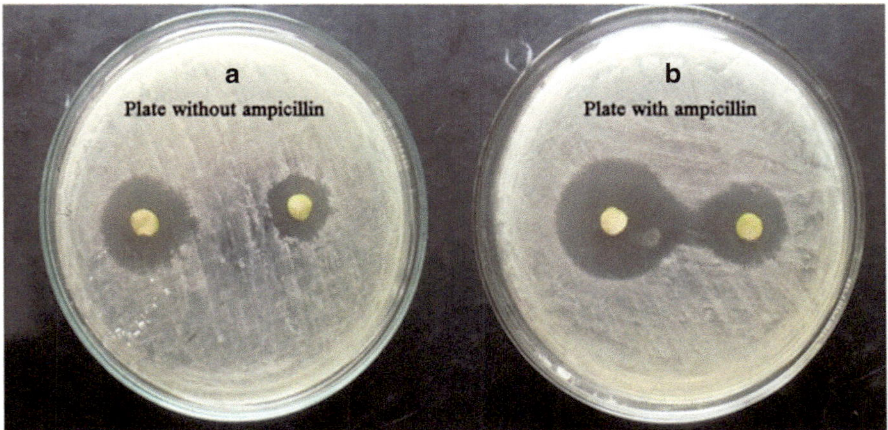

Fig. 30.1 Discs with methanol extract of *Tamarindus indica*, *plate A* without ampicillin and *plate B* with ampicillin

sources [8]. Many β-lactamase assays are employed for the screening of β-lactamase, but they all are based on calorimetric assay [9–11]. In iodometric assay, the starch-iodine complex is broken by the penicillanic acid produced by β-lactamase, and based on the extent of acid produced, the decolorization of starch-iodine complex occurs and is read by a spectrophotometer. The main disadvantage of these assays is that they are based on change of colour due to acid production. These assays are often useless in screening crude organic extracts of natural origin, like medicinal plant extracts, because the pH of the crude extracts is often varying, and also these extracts are coloured and can mask the violet or blue colour of the starch-iodine complex. Here, we report a novel agar plate assay method by which it is possible for screening antimicrobial agents as well as β-lactamase inhibitors simultaneously in a single assay.

Materials and Methods

Microbial Strains

Bacteria used for this study were pure cultures obtained from Microbial Type Culture Collection (MTCC) of Institute of Microbial Technology, Chandigarh, India. Following organisms were procured: *Escherichia coli* MTCC No. 1687 and *Klebsiella pneumoniae* MTCC No. 109. Three methicillin-resistant *Staphylococcus aureus* strains (MRSA) were obtained from Government Hospital, Thalassery, Kerala, India.

Media

For culturing the strains, nutrient agar was used with a composition of peptone, 5 g/L; beef extract, 1 g/L; yeast extract, 2 g/L; NaCl, 5 g/L and agar, 15 g/L, pH, 7.0 ± 0.2.

Antibiotics

Ampicillin (HiMedia, India), potassium clavulanate, tazobactam sodium and sulbactam (Sigma-Aldrich, USA) were used.

Plant Extract

Dried leaves of *Holoptelea integrifolia* samples were successively extracted with hexane, diethyl ether, acetone and water using a Soxhlet apparatus. The solvents were evaporated to dryness and used for the β-lactamase activity [12].

Dried *Tamarindus indica* leaves were successively extracted with chloroform followed by methanol, and the solvents were removed using a vacuum flash evaporator [13], and the methanolic extract was used for antimicrobial activity.

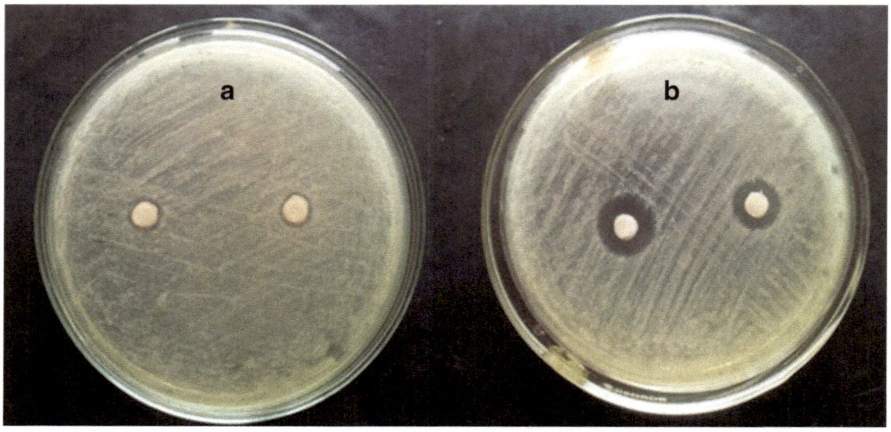

Fig. 30.2 Discs with ether extract of *Holoptelea integrifolia*, *plate A* without ampicillin and *plate B* with ampicillin

Preparation of Plate

A swab culture with test strains was made on 90 mm × 15 mm sterile Petri dishes with the culture media. To every set of the experiment, a replica plate containing nutrient agar and the replica having same media composition but with ampicillin (300 ppm) were also made.

Screening Test Method

For each β-lactam inhibitor or plant extract which has to be screened, the β-lactamase-producing organism was swabbed on two plates, one with a β-lactam antibiotic (ampicillin, 300 ppm) and the other without the antibiotic. Paper discs were impregnated with the plant extract, or other organic solvent extract which has to be screened for β-lactamase inhibitors was applied on the two plates. The plates are then incubated for 12 h at 37 °C. The experiments were carried out in duplicate. Inhibition was recorded by measuring the zone of inhibition.

Results and Discussion

Four sets of experiment were done; in each pair one plate was with nutrient agar media and the other with the same media but with 300 ppm ampicillin added to it. In the first set, which was taken as a control, plate containing amoxicillin, impregnated with discs having β-lactam inhibitors (tazobactam, clavulanic acid and sulbactam), gave clear zone, while the other plate in the pair, which does not have the antibiotic, does not show inhibition. This clearly implies that in the plate having amoxicillin, due to inhibition of β-lactamase enzyme by the β-lactamase inhibitors, bacteria could not grow (Fig. 30.3).

The second set of experiment was done to test whether plant extracts with β-lactamase activity could be screened with this assay. Diethyl ether extract of *Holoptelea integrifolia* [12] was used to impregnate the disc and then applied to both the plates (Fig. 30.2). After incubation, plate with antibiotic showed zone of inhibition, while the plain nutrient agar plate showed none. This infers that, as reported [12], *Holoptelea integrifolia* extract has β-lactamase inhibitory activity, and this method can be used for screening such organic solvent extracts from natural sources.

The third set was designed to test the antimicrobial activity screening with the same assay method. Each plate was impregnated with disc containing methanolic extract of *Tamarindus indica*. Here, both the plates showed zone of inhibition implying that the extract has antimicrobial activity but no β-lactamase inhibitory action. The plate with ampicillin showed higher zone diameter due to the antibiotic pressure (Fig. 30.1.)

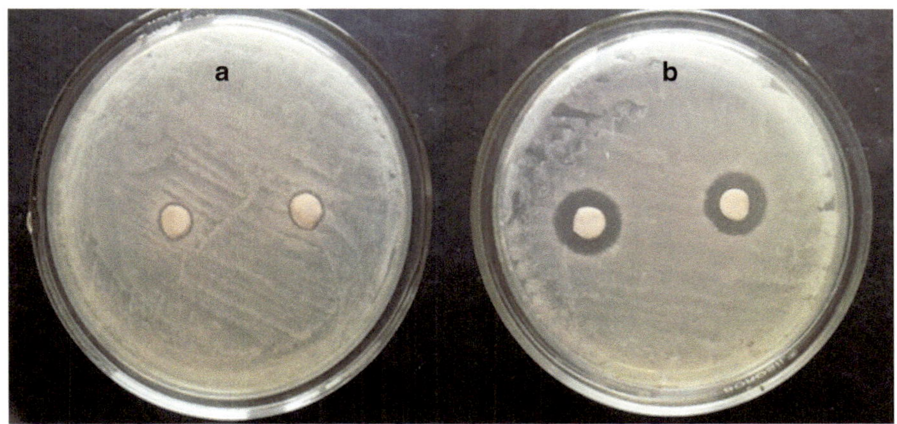

Fig. 30.3 Discs with potassium clavulanate, *plate A* without ampicillin and *plate B* with ampicillin

Results obtained from this study clearly shows that the assay system can be used for screening β-lactamase inhibitors as well as antimicrobial compounds in a single experiment. As seen from the data, it is clear that neither the colour nor the pH of different plant extracts is a problem, and hence, this method can be successfully employed in screening of different organic extracts from natural sources.

References

1. Hamad B. The antibiotics market. Nat Rev Drug Discov. 2010;9(9):675–6.
2. Abraham EP, Chain E. An enzyme from bacteria able to destroy penicillin. 1940. Rev Infect Dis. 1988;10(4):-677–8.
3. Doumith M, Ellington MJ, Livermore DM, Woodford N. Molecular mechanisms disrupting porin expression in ertapenem-resistant *Klebsiella* and *Enterobacter* spp. clinical isolates from the UK. Antimicrob Chemother. 2009;63(4):659–67.
4. Ambler RP, Coulson AF, Frère JM, Ghuysen JM, Joris B, Forsman M, Levesque RC, Tiraby G, Waley SG. A standard numbering scheme for the class A beta-lactamases. Biochem J. 1991;276(Pt 1):269–70.
5. Bush K, Jacoby GA, Medeiros AA. A functional classification scheme for beta-lactamases and its correlation with molecular structure. Antimicrob Agents Chemother. 1995;39(6):1211–33.
6. Bush K. Beta-lactamase inhibitors from laboratory to clinic. Clin Microbiol Rev. 1988;1(1):109–23.
7. Brown AG, Butterworth D, Cole M, Hanscomb G, Hood JD, Reading C, Rolinson GN. Naturally-occurring beta-lactamase inhibitors with antibacterial activity. J Antibiot. 1976;29(6):668–9.
8. Newman DJ, Cragg GM. Natural products as sources of new drugs over the last 25 years. J Nat Prod. 2007;70(3):461–77.
9. Zaichang Y, Xiaosheng Y, Yule N. A novel and rapid method to screen for β-lactamase inhibitors from extracts of herbs. Int J Integ Biol. 2008;6(2):62–4.
10. Rubin FA, Smith DH. Characterization of R factor β-lactamases by the acidimetric method. Antimicrob Agents Chemother. 1973;3(1):68–73.
11. Catlin BW. Iodometric detection of *Haemophilus influenzae* beta-lactamase: rapid presumptive test for ampicillin resistance. Antimicrob Agents Chemother. 1975;7(3):265–70.
12. Vinod NV, Shijina R, Dileep KV, Sadasivan C. Inhibition of beta-lactamase by 1,4-naphthalenedione from the plant *Holoptelea integrifolia*. Appl Biochem Biotechnol. 2010;160(6):1752–9.
13. Escalona-Arranz JC, Péres-Roses R, Urdaneta-Laffita I, Camacho-Pozo MI, Rodríguez-Amado J, Licea-Jiménez I. Antimicrobial activity of extracts from *Tamarindus indica* L. leaves. Pharmacogn Mag. 2010; 6(23):242–7.

31 Diurnal Phototactic Migration of Phytoplankton in Cochin Waters and Its Correlation with Limiting Nutrient Flux

M. Ratheesh Kumar, V.S. Sudhanandh, A.K. Faisal, S. Raj Vishnu, R. Shibu, M.I. Sreejith, and K. Anoop Krishnan

Abstract

A comprehensive, 24-h sampling with an interval of 3 h has been conducted at Cochin shore waters in order to investigate the variation in marine phytoplankton density and its relation with phototactic movements, either positive (diel vertical migration) or negative (reverse diel vertical migrations) displacement towards light gradient. Chlorophyll *a* showed an exponential increase during day cycle, ranging from 6.54 to 26.17 mg m^{-3}, compared to night cycle (1.76–2.92 mg m^{-3}). The total phytoplankton counts in night hours were lowered by one third that of day hours. Phytoplankton community was observed to be dominated by dinoflagellate species, especially *Ceratium furca* which showed a mean count of 1,180 cells/l during night hours and 3,248 cells/l during day hours. The positive phototactic displacement among phytoplankton can be observed during day hours. N/P ratio and Si/P ratio were found to be positively correlated with phytoplankton distribution. The mean N/P ratio during night hours was low (5.42 μmol L^{-1}) compared to day hours (9.40 μmol L^{-1}), and Si/P ratio showed similar trend (<5 μmol L^{-1}) during night hours. Study suggests that optimisation of nutrients coupled with light uptake in the vertical gradient contributes migrating behaviour of phytoplankton, especially dinoflagellates.

Keywords

Phototaxis • Nutrients • Phytoplankton

Introduction

Phytoplankton plays vital importance in the functioning of food web. It forms the basis of food chain and contributes largely to the functioning of marine ecosystem. Being photosynthetic

M.R. Kumar (✉) • V.S. Sudhanandh • A.K. Faisal • S.R. Vishnu • R. Shibu • M.I. Sreejith • K. Anoop Krishnan
Chemical Science Division, Centre for Earth Science Studies, Akkulam, Thiruvananthapuram, Kerala, India
e-mail: ratheesh9767@gmail.com

A. Sabu and A. Augustine (eds.), *Prospects in Bioscience: Addressing the Issues*, DOI 10.1007/978-81-322-0810-5_31, © Springer India 2013

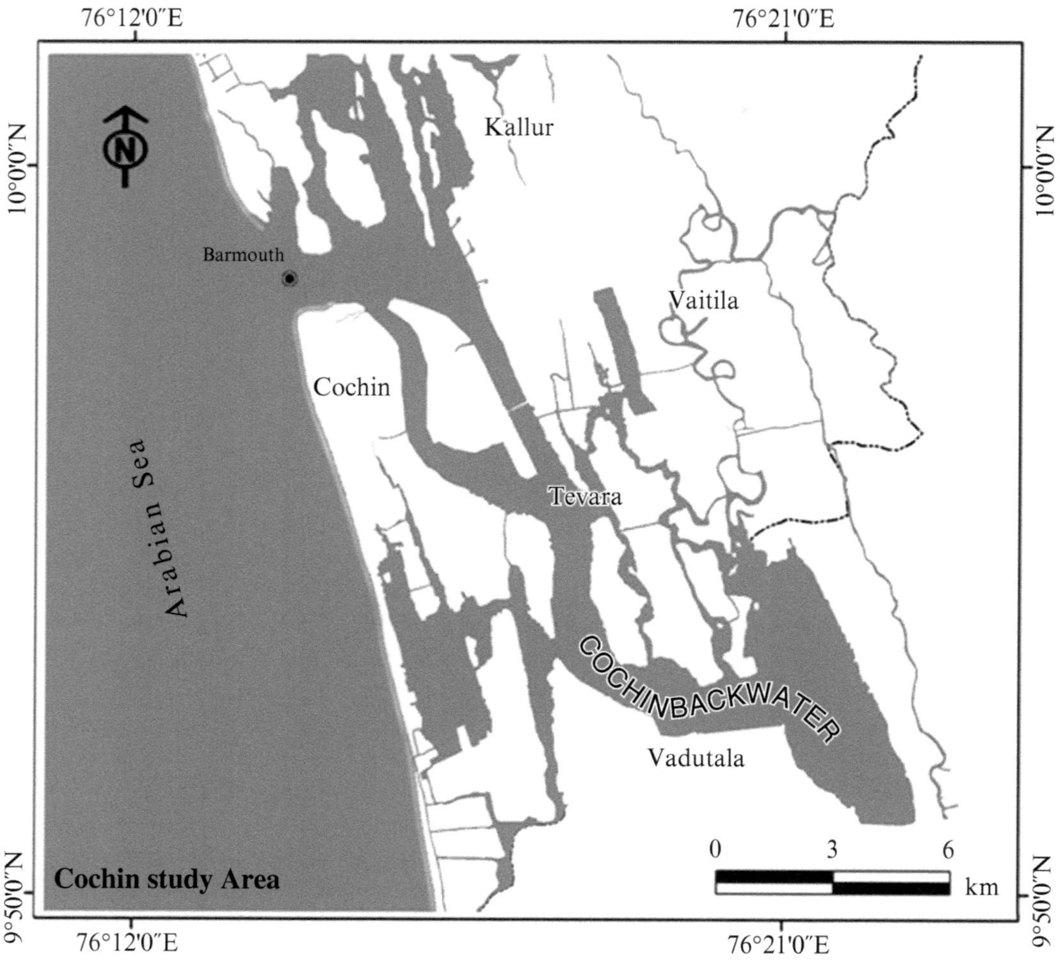

Fig. 31.1 Map showing sampling position (Cochin barmouth)

organisms, phytoplanktons are forced to move to and to stay in the top layer of water column illuminated by solar radiation, the "photic zone" [1]. Phototaxis refers to the positive or negative displacement of an object towards light gradient. Many marine phytoplanktons perform both diel vertical migration (DVM) and reverse diel vertical migrations induced prominently by phototaxis and driving forces behind it. In response to the gradients of essential resources such as light and nutrients, phytoplankton groups are predicted to show differential vertical distributions as they optimise growth conditions according to their specific physiologies and motilities [2]. A 24-h water sampling study was conducted on May 2011 (pre-monsoon season) at Cochin estuary (barmouth) in order to investigate the variation of phytoplankton density and the factors inducing the phototactic movements responsible for fluctuation in plankton density.

Materials and Methods

The sampling station of Cochin barmouth lies between 09°58′16″N latitude and 76°14′43″E longitude (Fig. 31.1). The sampling station recorded a depth ranging from 11.60 to 14 m. Eight sets of surface water samples were taken from intertidal zone of Cochin estuary at an interval of 3 h namely, 06:00, 09:00, 12:00, 15:00 and 18:00 h (day cycle) and 21:00, 00:00 and 03:00 h (night cycle). Standard mercuric calibrated thermometer was used for measuring both atmospheric and water temperatures. The salinity, pH and conductivity were measured using thermo-scientific Eutech probe (Waterproof CyberScan PCD 650). DO values were determined by Winkler's method [3]. Standard chemical analysis was carried out for nutrients such as phosphorous, silicate, nitrite, nitrate, total nitrogen and total phosphorous. Spectrophotometric method (UV-visible, model Shimadzu UV-1800, Japan) was applied for obtaining nutrient values [4]. Pearson correlations were done with the help of SYSTAT 8.0 software packages. For phytoplankton analysis, 100 L of water sample was filtered through 53-μ plankton sieve. The sample thus collected was resuspended in known volume of filtered sea water, and 4% neutralised formaldehyde was added for preservation. Species level analysis of phytoplankton samples was carried out by spreading 1 ml of sample on Sedgwick rafter counting chamber using binocular phase contrast microscope. Chlorophyll values were estimated by filtering the water sample free of zooplankton through GF/C glass filter paper, and the pigments, chlorophyll a, b and c, were extracted from phytoplankton by using 90% acetone. The solution was centrifuged for about 20 min under 5,000 rpm, and the supernatant solution was considered for determination of optical density with the aid of a spectrophotometer [5]. Standard publications were used for taxonomic identification of phytoplankton. The data obtained was treated with simple correlation to study the variation of phytoplankton density in relation with limiting nutrients and light intensity triggering phototactic movements, (Fig. 31.2).

Results and Discussion

Hydrochemical Parameters

Atmospheric temperature throughout the study ranged between 30 and 33.50°C (mean value, 31.03C), and the mean water temperature was found to be slightly lower than atmospheric temperature (30.31C). pH value of respective hours indicated the alkaline nature of surface water samples (mean value is 7.69) (Table 31.2). Forenoon hours (6:00, 9:00 and 12:00) registered a comparatively low salinity (corresponding values of 7.49, 7.65 and 7.59 psu) than evening and night hours. Significant fluctuation in salinity values could be observed after 12:00 h (7.90 and 8.00 psu during 15:00 and 18:00 h), and a mean value of 7.65 psu was noted during night hours. Even at a single place in the estuary, salinity would fluctuate with movement of tides, mixing of nutrients and water by wind. Seasonally, the salinity values were low. Freshwater discharge from various rivers located along the west coast of India and rainfall over the Western Ghats form a layer of saline waters over eastern Arabian Sea, having short life cycle [6]. DO values were ranged between 5.14 mg L^{-1} (18:00 h) and 5.60 mg L^{-1} (15:00 h). Oxygen has limited solubility in water, ranging from 5 to 14 mg L^{-1}[7]. DO values indicated comparatively low range as eutrophic condition prevails in Cochin estuary [8]. Total nitrogen value was found to be highest at 9:00 h (28.03 μmol L^{-1}) and lowest at morning 03:00 h (10.35 μmol L^{-1}). Silicate values showed somewhat similar trend with highest value at morning 6:00 h (18.76 μmol L^{-1}) and lowest value at morning 3:00 h (8.22 μmol L^{-1}). Nitrogen/phosphorous ratio and silicate/ phosphorous ratio indicated a clear distinction between day cycle (6:00, 9:00, 12:00, 15:00, 18:00 h) and night cycle (21:00, 00:000, 03:00 h) (Fig. 31.2). During 15:00 h, both N/P ratio (12.40 μmol L^{-1}) and SiO_4/P ratio (8.26 μmol L^{-1}) reported maximum value, and at morning 00:00 hours, the lowest values of both N/P ratio (6.02 μmol L^{-1}) and SiO_4/P ratio (3.65 μmol L^{-1}) were observed (Fig. 31.3). Pearson correlation values are shown in Fig. 31.2. N/P ratio and SiO_4/P ratio show positive correlation with dominating dinoflagellate species (0.804 and 0.916 in case of *Ceratium furca*).

Chlorophyll Values

Variation in chlorophyll concentration at Cochin barmouth was used to indicate the plankton community structure and biological productivity. Significantly, higher values of chlorophyll a were observed during day cycle (highest value, 26.17 mg m^{-3} at 15:00 h), while night hours exhibited very low values ranging from 1.76 (00:00 h)

Table 31.1 Total phytoplankton distribution of Cochin barmouth station

Species (cells·L^{-1})	6:00	9:00	12:00	15:00	18:00	21:00	0:00	3:00
Amphora lineolata	0	0	0	240	0	20	40	20
Asterionella japonica	200	80		160	100	40	60	50
Aulacodiscus sp.	0	0	80	40	0	0	40	120
Bacillaria paradoxa	0	40	0	0	0	0	0	0
Biddulphia reticulate	320	240	260	160	80	80	360	0
Biddulphia sinensis	150	300	80	400	260	300	120	60
Ceratium furca	1,060	1,680	480	160	980	1,100	950	1,490
Ceratium fusus	0	40	0	0	0	80	140	240
Ceratium macroceros	80	120	40	40	120	0	0	200
Ceratium pentagonium	0	0	0	160	0	60	0	100
Chaetoceros indicus	300	80	80	60	120	210	100	40
Corethron hystrix	80	0	0	0	0	0	0	0
Cosinodiscus excentricus	0	0	240	0	0	0	0	0
Cosinodiscus gigas	140	200	300	120	120	160	380	100
Cosinodiscus radiatus	150	140	60	40	60	80	40	40
Cyclotella sp.	200	40	320	400	280	0	0	160
Dinophysis caudata	60	80	240	320	320	120	160	480
Diploneis sp.	0	0	0	120	0	0	40	40
Ditylum brightwellii	0	0	80		300	100	40	80
Fragilaria oceanica	40	0	450	360	120	80	240	200
Gyrosigma sp.	80	0	0	0		0	0	140
Nitzschia longissima	120	200	40		40	320	60	0
Peridinium sp.	3,700	4,200	2,740	3,120	3,100	1,280	800	1,460
Peridinium steinii	80	140	120	0	0	0	150	70
Phaeocystis sp.	0	100	40	0	0	0	0	0
Pinnularia alpina	0	0	0	160	0	0	0	
Planktoniella sol	300	80	80	120	80	0	60	120
Pleurosigma sp.	160	160	120	80	400	80	0	0
Prorocentrum sp.	40	100	100	160	120	40	120	80
Protoperidinium depressum	0	40	0	0	0	240	80	0
Pyrophacus steinii	1,240	1,600	2,000	1,120	700	700	460	1,200
Rhabdosphaera sp.	0	0	120	0	0	40	0	0
Rhizosolenia curvata	60	0	0	80	0	0	60	0
Rhizosolenia stolterfothii	80	0	160	0	0	0	0	0
Skeletonema costatum	840	580	700	480	1,260	320	200	140
Triceratium favus	0	0	0	40	0	0	40	0

to 2.92 mg m^{-3} (3:00 h). Similar trend was observed with chlorophyll b and c and phaeophytin (Fig. 31.4). The mean chlorophyll, a value during day cycle, was found to be 15.06 mg m^{-3} (which strictly pronounces eutrophic nature), and during night cycle it was 2.25 mg m^{-3}.

Qualitative Distribution

Phytoplankton density depicted a picture of domination of dinoflagellates. Plankton density was observed to be controlled by certain species of dinoflagellates namely, *Ceratium furca* and

Table 31.2 Results of hydrochemical parameters of Cochin barmouth station

Time	Atmospheric temp.	Water temp.	Susp. solids	pH	Salinity	DO	NH_4	Total N	Total P	SIO_4	N/P	SiO4/P
Hours	°C	°C	mg/l		psu	mg/l	µmol/l	µmol/l	µmol/l	µmol/l	µmol/l	µmol/l
6:00	30.500	31.000	11	7.49	23.4	5.28	0.34	24.31	3.78	18.76	6.42	4.97
9:00	30.800	31.000	9.7	7.65	20.6	5.53	1.20	28.03	2.43	11.00	11.49	4.55
12:00	32.000	29.800	11.8	7.59	21.4	5.18	0.75	18.71	1.9	13.07	9.84	6.88
15:00	33.500	31.300	12.8	7.9	28.6	5.60	1.09	13.19	1.06	8.76	12.40	8.26
18:00	30.700	30.300	18.45	8.00	32	5.140	1.24	16.15	2.35	12.80	6.85	5.45
21:00	30.000	29.700	14.8	7.67	26.6	5.370	0.60	15.46	3.93	11.27	3.93	2.87
0:00	30.300	29.700	14.3	7.62	25.4	5.150	0.76	15.05	2.5	9.12	6.02	3.65
3:00	30.300	29.700	14.3	7.62	25.4	5.150	0.76	10.35	1.66	8.22	6.24	4.96

Correlations											
Parameter	Sl. No	1	2	3	4	7	8	9	10	11	12
Atmospheric temperature	1	1									
Water tempertaure	2	0.523	1								
Total suspended solids	3	-0.24	-0.459	1							
pH	4	0.4	0.23	0.643	1						
Salinity	5	0.09	-0.016	.873**	.827*						
Dissolved oxygen	6	0.522	.734*	-0.511	0.183						
Ammonia	7	0.369	0.291	0.237	.774*	1					
Total nitrogen	8	-0.123	0.503	-0.656	-0.362	-0.007	1				
Total phosphorous	9	-0.693	-0.135	-0.027	-0.435	-0.606	0.415	1			
Silicate	10	-0.18	0.289	-0.237	-0.337	-0.486	0.615	0.621	1		
Nitrogen/phosphorous ratio	11	.817*	0.663	-0.523	0.255	0.584	0.3	-0.675	-0.17	1	
Silicate/phosphorous ratio	12	.930**	0.494	-0.159	0.39	0.346	-0.137	-.744*	-0.04	.763*	1
Ceratium furca	13	.795*	0.675	-0.326	0.435	0.626	0.329	-0.561	0.041	.916**	.804*
Ceratium fusus	14	.808*	0.519	-0.079	0.285	-0.015	-0.223	-0.453	0.173	0.464	.870**
Ceratium macroceros	15	-0.239	0.401	-0.295	-0.4	-0.463	0.408	0.352	.726*	-0.144	-0.01
Ceratium pentagonium	16	0.505	0.307	-0.014	0.234	-0.026	-0.539	-0.39	-0.38	0.208	0.445
Dinophysis caudata	17	.828*	0.541	-0.181	0.337	0.274	0.085	-0.535	0.258	0.699	.944**
Prorocentrum sp.	18	0.696	0.276	0.16	0.658	.753*	-0.28	-.835**	-0.55	0.667	0.641
Protoperidinium sp.	19	-0.515	-0.066	-0.117	-0.461	-.710*	0.253	.926**	0.481	-0.617	-0.65
Pyrophacus steinii	20	0.378	0.247	-0.402	-0.049	0.24	0.431	-0.322	0.312	0.582	0.549
Peridinium sp.	21	0.597	0.622	-0.191	0.524	.744*	0.392	-0.456	0.049	.808*	0.631
Peridinium steinii	22	-0.27	-0.165	-0.362	-0.465	0.035	0.472	-0.043	0.082	0.13	-0.2

*. Correlation is significant at the 0.05 level (2-tailed). **. Correlation is significant at the 0.01 level (2-tailed).

Fig. 31.2 Pearson correlation values

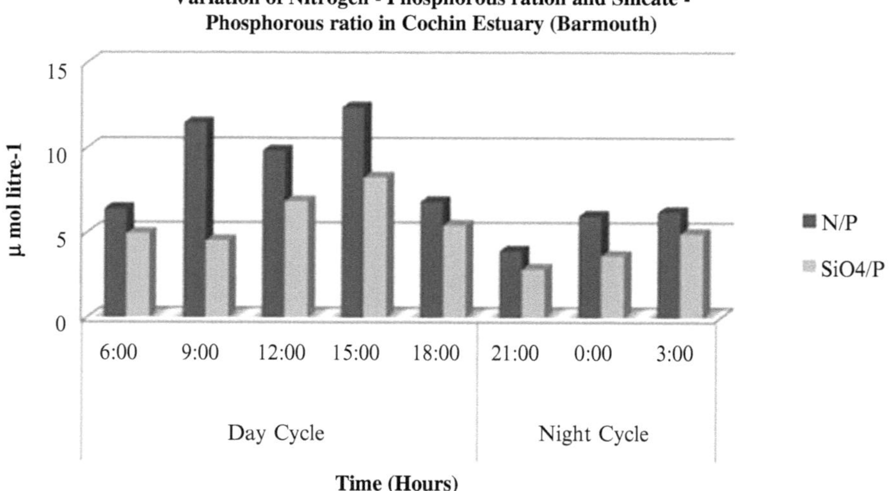

Fig. 31.3 Graph showing the variation of N/P ratio and SiO_4/P

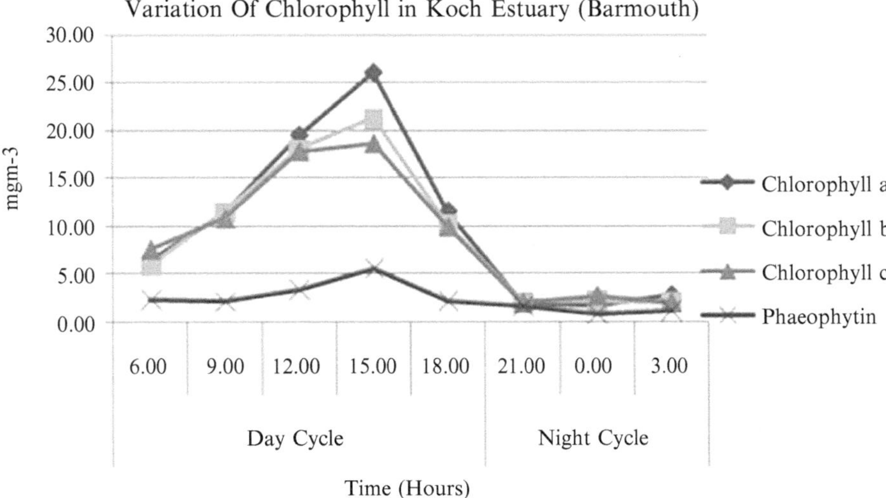

Fig. 31.4 Graph showing variation of chlorophyll pigments

Peridinium sp. A total of thirty nine species were identified, namely, *Amphora lineolata, Asterionella japonica, Aulacodiscus* sp., *Bacillaria paradoxa, Biddulphia reticulate, Biddulphia sinensis, Ceratium furca, Ceratium fusus, Ceratium macroceros, Ceratium pentagonium, Chaetoceros indicus, Corethron hystrix, Cosinodiscus eccentricus, Cosinodiscus gigas, Cosinodiscus radiatus, Cyclotella* sp., *Dinophysis caudata, Dinophysis punctata, Diploneis* sp., *Ditylum brightwellii, Fragilaria oceanica, Gyrosigma* sp., *Nitzschia longissima, Nitzschia closterium, Peridinium* sp., *Peridinium steinii, Phaeocystis* sp., *Pinnularia alpina, Planktoniella sol, Pleurosigma* sp., *Prorocentrum* sp., *Protoperidinium depressum, Pyrophacus steinii, Rhabdosphaera* sp., *Rhizosolenia curvata, Rhizosolenia stolterfothii, Skeletonema costatum, Triceratium favus and Triceratium reticulata.* Among dinoflagellates (Fig. 31.5), *Ceratium furca* dominated throughout day cycle and night cycle (Table 31.3). The highest count was observed at 15:00 h (4,140 cells L^{-1}) during day cycle and at 3:00 hours (1,490 cells L^{-1}) during night cycle, though it was too low as compared to day cycle. *Ceratium furca* showed a mean count of 1,180 cells L^{-1} during night hours and 3,248 cells L^{-1} during day hours. The qualitative distributions of phytoplankton were shown in Table 31.1.

The data elucidates timescale variation between hydrochemical parameters and phytoplankton community. A strict positive correlation could be observed among chlorophyll values, hydrochemical parameters and phytoplankton cell density. The chlorophyll concentration indicates the sampling station Cochin as biologically productive in terms of plankton community structure. The chlorophyll *a* values ranged from 1.76 to 2.92 mg m^{-3} during night hours, and an exponential increase was noted during day hours with a value ranged from 6.54 to 26.17 mg m^{-3}. The major nutrients required by phytoplankton for growth and reproduction are nitrogen and phosphorous [9]. In addition to these nutrients, other inorganic nutrients in small amounts act as limiting factor for phytoplankton productivity. Graphical data along with correlation results accentuate the decisive importance of major nutrients in phytoplankton cell growth. N/P ratio and SiO_4/P ratio could be numerically correlated with plankton density, especially dinoflagellates. Each phytoplankton species has a different set of favourable conditions that promote growth and reproduction [10]. The total phytoplankton counts in night hours were lowered by one third than that of day hours. A significant variation in N/P and SiO_4 ratio could be seen during night cycle. Phytoplankton community was observed to be

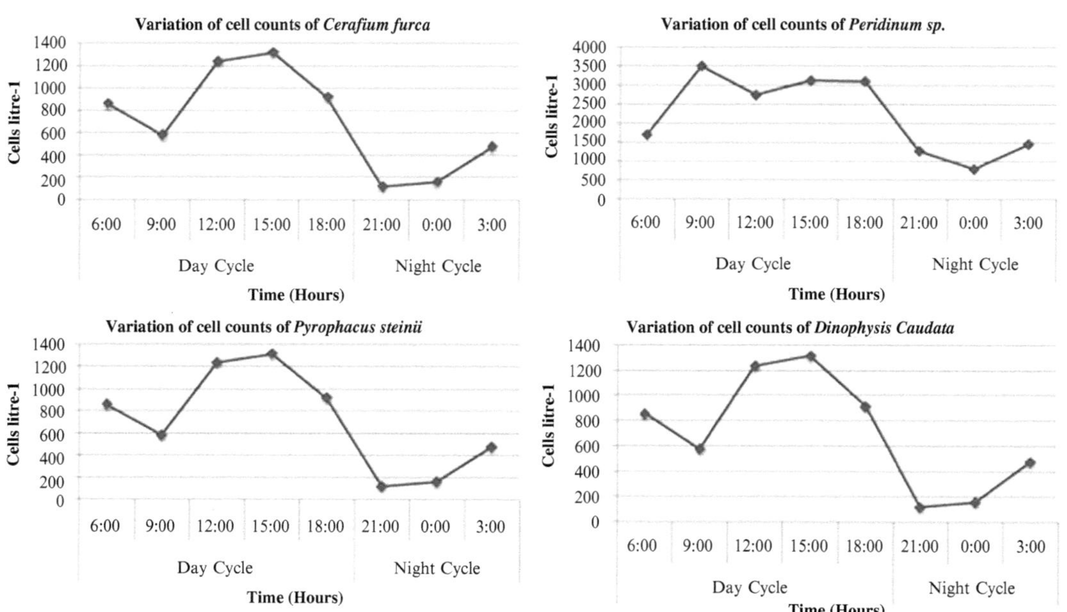

Fig. 31.5 Graph showing variation in cell density of dominating dinoflagellates

Table 31.3 Distribution of dinoflagellates (cells L^{-1}) in Cochin barmouth station

	Day cycle						Night cycle			
	6:00	9:00	12:00	15:00	18:00	Mean	21:00	0:00	3:00	Mean
Ceratium furca	2,060	3,680	3,480	4,140	2,980	3,268	1,100	950	1,490	1,180
Ceratium fusus	460	40	400	720	300	384	80	140	240	153
Ceratium macroceros	480	120	40	40	120	160	0	0	200	66
Ceratium pentagonium	40	0	0	160	0	40	60	0	100	53
Dinophysis caudata	860	580	1,240	1,320	920	984	120	160	480	253
Prorocentrum sp.	40	100	100	160	120	104	40	120	80	80
Protoperidinium sp.	200	40	0	0	0	48	240	80	0	107
Pyrophacus steinii	1,240	1,600	2,000	1,120	1,300	1,452	700	460	1,200	787
Peridinium sp.	1,700	3,500	2,740	3,120	3,100	2,832	1,280	800	1,460	1,180
Peridinium steinii	80	140	120	0	60	80	0	150	70	73

controlled by certain dinoflagellate species such as *Ceratium furca and Peridinium* sp. which are significantly contributing to total density. Plankton biomass is generally highest in areas where physical process supplies adequate nutrients [11]. More than 16 major and several minor industries situated on the banks of Periyar river discharge partially treated or untreated wastes in to estuary during day time. In addition to it, drastic increase in dinoflagellate density was attributed to the light intensity of particular area. Diel cycling of light, as well as its modulation over the course of day, leads to a synchronisation of growth and division of phytoplankton population [12]. Positive phototactic behaviour of phytoplankton in general and dinoflagellates in particular was noted throughout the study, and the variation key factors like nutrient ratios (N/P, SiO_4/P) corresponding to the variation in phytoplankton cell density were correctly established. The distribution of *Ceratium* sp. is mostly explained as partly being the result of phototactic movements. Experiments with *Ceratium* sp. have demonstrated that this species is positively phototactic [13]. Ability of dinoflagellate to migrate up and down through the water column means they can take the advantage of increased nutrient levels at greater depth during night and return closer to surface in order to photosynthesise during day [14]. They may also be better adapted to take the advantage of nutrients at frontal system due to their mobility. The most extensive migration takes place at dawn, and dusk and concurrent changes in light intensity have long been considered an important casual factor [15]. It is established that at least for dinoflagellates this strategy could be energetically advantageous and is often proved to be primary adaptive reason for phototactic vertical migration. Optimisation of nutrients coupled with light uptake in the vertical gradient also contributed to the migrating behaviour of algae [16].

Acknowledgment The authors are thankful to Dr. N. P. Kurian, Director, CESS, Dr. K. V. Thomas, Head, MSD, Centre for Earth Science Studies (CESS) and Coastal Ocean Monitoring and Prediction System (COMAPS) and Ministry of Earth Science for their valuable suggestions, comments and assistance for the study.

References

1. Hader DP. Vertical migration and distribution of primary producers in aquatic ecosystems – the effects of enhanced solar UVB. Photochem Photobiol. 1997;65:263–4.
2. Klausmeier CA, Litchman E. Algal games – the vertical distribution of phytoplankton in poorly mixed water columns. American Society of Limnology and Oceanography 2001;46(8):1998–2007.
3. Winkler LW. Die Bestimmung in Wassarg elosten sauerstoffes. Dusche Chem Qes Ber. 1883;21:2843–55.
4. APHA. Standard methods for the examination of water and wastewater. 20th ed. Washington, DC: American Public Health Association; 1998. p. 1268.
5. Santhanam R, Ramanathan N, Venkataramanaujam K, Jagatheeshan G. Phytoplanktons of Indian Sea. New Delhi: Daya Publishing House; 1987. p. 121–34.
6. Rao AD. Variability of coastal ocean processing along west coast of India. Indian J Mar Sci. 2010;39(4):75–84.
7. Connell DW, Miller GJ. Chemistry and ecotoxicology of pollution. New York: Wiley; 1984.
8. Martin GD, Nisha PA, Balachandran KK, Madhu NV, Nair M, Shaiju P, Joseph T, Srinivas K, Gupta GVM. Eutrophication induced changes in benthic community structure of flow restricted tropical estuary (Cochin Backwaters). Environ Monit Assess. 2011;176:427–38.
9. Tomas CR. Identifying marine phytoplankton. San Diego: Academic; 1997. p. 120–720.
10. Diersing N. Phytoplankton blooms – the basics. Key West: Florida Keys Natural Marine Sanctuary; 1999.
11. Falkowski PG, Greene RM, Geider RJ. Physiological limitations on Phytoplankton productivity in the Ocean. Oceanography. 1992;5(2):84–90.
12. Vaulot D. Phytoplankton, encyclopaedia of life sciences. Macmillan publishers Ltd/Nature Publishing Group: London; 2001.
13. Hasle RG. Phototactic vertical migration in marine dinoflagellates. Oikos. 1950;2:160–5.
14. Olney M. Dinoflagellates, Microfossil Image Recovery And Circulation for Learning and Education (MIRACLE). University College London:London; 2002.
15. Ringelberg J. The photobehaviour of Daphnia spp. as a model to explain diel vertical migration in zooplankton. Biol Rev. 1999;74:397–423.
16. Haupt F. Plankton vertical migrations; Implications for the pelagic ecosystem, der -Ludwig-Maxmilians-Universitat Munchen: Munich; 2011. pp. 12–16.

Cryo-preservation of Pollen for Hybrid Seed Production in Hot Pepper

32

Rakesh C. Mathad, S.N. Vasudevan, and S.B. Patil

Abstract

Hot pepper is one of the important vegetable crops of Asia, and India is fast developing as hub for its seed production. In hot pepper, seed production is mainly done using cytoplasmic male sterile lines, where sterile female plants were grown in fields of seed growers. The male line is grown in a centralised facility, and only male flower buds were distributed to growers for pollen extraction and crossing of female buds. Sometimes it is difficult to supply male flower buds to growers because of distance and possibility of crop failure. Further, hot pepper seed production can be done in 9–10 months in a year in various sowing windows and in different areas. An effort to preserve pollen from these male buds was made under refrigeration and cryo-preservation which can be used to supply pollen directly instead of buds and to longer distances. The pollen from these male lines was harvested and stored at normal room temperature (control), in deep refrigeration (−20 °C) and in liquid nitrogen (−196 °C). The pollen extracted was preserved in capped vials for refrigeration, and semen straws were used for storing in liquid nitrogen. The semen straws containing pollen are kept in IBP cryo-containers which are normally used for artificial insemination. The pollen had lost viability and failed to produce seeds in female buds within 3 days when stored under room temperature. The pollen stored in deep refrigeration produced seeds up to 6 days. In case of cryo-preservation, the pollen can be stored up to 47 days and produced seeds when used for pollination up to 48 days. The pollen viability was tested using 2% acetocarmine staining, and pollen germination by hanging drop technique which showed 85–2% pollen was viable in this period. The molecular genetic purity test using protein banding pattern (to know the genetic purity) of these hot pepper hybrid seeds shows no variations for the characters of F1 as

R.C. Mathad (✉) • S.N. Vasudevan • S.B. Patil
Department of Seed Science and Technology,
University of Agricultural Sciences,
Raichur 584102, India
e-mail: rakeshcmrnr@gmail.com

A. Sabu and A. Augustine (eds.), *Prospects in Bioscience: Addressing the Issues*,
DOI 10.1007/978-81-322-0810-5_32, © Springer India 2013

described by the breeder. Also there is no effect on storability and viability of seeds in modified atmospheric storage. The use of cryo-preservation for storing pollen can be effectively used for hybrid seed production which helps in reducing cost of production, genetic integrity and germ plasm security.

Keywords
Pollen cryo-preservation • Male sterility • Pollen germination

Introduction

India has the unique distinction of being a major seed production hub among the Asian countries due to its diversified agroclimatic regions. Hence, most of tropical vegetable seeds are produced in India with the exception of some *Brassica* seeds which need temperate conditions. Among the important vegetables the Asian type hot pepper is one of the important vegetables in which the commercial hybrid seed production is going on large scale. The seed production is carried out by farmers, which gives them high returns per unit area. With a land area as low as half acre, one can earn profits more than the commercial crops. The seed production of hot pepper is a skilled activity which involves emasculation and pollination. The seed production is carried out using cytoplasmic male sterile lines. The usual practice is to grow male sterile lines and fertile male plants separately, and pollen from fertile male plants is used to pollinate the male sterile lines. To grow fertile lines to supply pollen is a costly affair, and crop failure may jeopardise the entire seed production programme. In an experiment conducted in UAS, Raichur, an effort was made to store pollen of hot pepper in liquid nitrogen which can be used for pollination at various sowing windows without growing fertile lines separately, which is economically viable. Earlier studies reported that hot pepper pollen can be stored in liquid nitrogen [1, 2]. Another study reported that pollen of capsicum can be stored for more than 42 months without any effect on the pollen viability and fertility [3].

Materials and Methods

Collection of Flower Buds, Pollen Extraction and Storage

To conduct the study, the flower buds from fertile plants were collected from the crop grown in the previous season. The flower buds were plucked from the plants and stored in wet cotton cloth. Flower buds were then exposed to sunlight to induce anther dehiscence. The pollen was collected in a container made of china clay and then transferred to various containers based on the type of storage. The pollen is stored in capped vials for normal storage (under cool conditions) and for deep refrigeration. The semen straws are used for storing pollen under liquid nitrogen. These straws were normally used for artificial insemination in Department of Animal Husbandry. The IBP cryo-cans were used for cryo-preservation. The commercial grade liquid nitrogen was obtained from Karnataka Milk Federation. The cryo-cans were monitored regularly as per the earlier methods [4] and filled with fresh liquid nitrogen after every fortnight.

Study of Pollen Viability and Germination

For pollen viability study, 2% acetocarmine was prepared by dissolving 2 g carmine powder in 55 ml of distilled water and 45 ml of acetic acid. This mixture was boiled gently for 5 min and filtered with No. 75 Whatman filter paper after cooling. For studying pollen viability, the pollens were placed on a cavity slide containing 2 drops of 2% acetocarmine. These were mixed using a clean needle and examined under microscope for staining pattern. The staining patterns were classified as dark, light and unstained. The pollen grains which were stained dark and light were taken as viable. Pollen grains were observed randomly from average of five microscopic fields, and numbers of viable and nonviable pollens were recorded. The pollen germination was studied using hanging drop technique [5].

Study of Fruit and Seed Quality Parameters

The pollen after storage in various parameters was used to pollinate the flower buds of male sterile plants. Each day, ten previously tagged flower buds were pollinated. For ease in identification, various colour tags were used. The seed quality parameters such as fruit set (%), average seed set per fruit (No.) and seed germination (%) were recorded.

Collection of Data and Analysis

The data was collected every day for the first 7 days, later once in 5 days up to 47th day. The data recorded for three replicates and the experimental results were analysed statistically with randomised block design with factorial concept adopting analysis of variance technique, and critical differences were calculated whenever 'F' test was significant. Wherever necessary, the percentage data was transformed into arcsin root transformation [6].

Results and Discussion

The results of the experiment were summarised in Table 32.1. The results indicate that the pollen germination was better when pollen was stored for 3 days under ambient room temperature, for 6 days when stored under deep refrigeration and for 47 days when stored in liquid nitrogen. Though the pollen germination was observed in all storage methods, a decreasing trend was noticed. The pollen germination proves the earlier findings of pollen that can tolerate extreme ultra low temperature and effect normal fertilisation due to very low biological activity at these temperatures [7].

The fruit set % was better when pollen is stored for 3 days (99, 88 and 56%) under ambient room temperature. In case of deep refrigeration, fruit set can be obtained up to 7 days (99–24%) and better fruit set % was obtained up to 47 days, when pollen is used which was stored in liquid nitrogen (99–4%) with decreasing trend. This finding proves earlier finding of higher fruit and seed set % by using fresh pollen [8]. This also proves the fact that the normal fertilisation and fruit set is possible when pollen survival in liquid nitrogen is at least 20% [9].

The results also showed that higher seed setting per fruit was obtained when pollen is used which was stored in liquid nitrogen up to 47 days (5–168 days). Very low seed setting was observed for 3–6 days when pollen is stored under ambient room temperature and deep refrigeration. The seed germination showed no significant reduction with respect to storage methods. The seed germination ranged between 64 and 84%. This finding showed us the fact that there will not be any effect of storage method on germination which will help in storability studies as future line of work.

Conclusions

The current study on pollen cryo-preservation showed that pollen can be stored for very long period of time in liquid nitrogen and can be used for hybrid seed production. Also the pollen can tolerate extreme ultra low temperatures and can affect normal fertilisation. The pollen expresses some kind of dormant behaviour under these low temperatures and biological activity. This method can be used for short-term pollen storage

Table 32.1 Influence of cryo-preservation on pollen and seed quality parameters

	Pollen germination %				Fruit set %				Average seed set no.				Seed germination %			
Storage days (T)	C1	C2	C3	Mean	C1	C2	C3	Mean	C1	C2	C3	Mean	C1	C2	C3	Mean
T1–1 DAS	85	85	85	85	99	99	98	99	161	166	168	165	84	84	83	84
T2–2 DAS	63	78	86	76	88	92	94	91	132	146	155	144	83	84	83	83
T3–3 DAS	17	77	75	56	56	86	91	78	32	125	153	103	76	82	82	80
T4–4 DAS	0	51	72	41	20	76	90	62	0	117	145	87	76	81	82	80
T5–5 DAS	0	28	68	32	3	66	89	53	0	73	139	71	74	77	82	78
T6–6 DAS	0	11	63	25	0	54	88	47	0	23	133	52	0	77	82	53
T7–7 DAS	0	1	62	21	0	24	86	37	0	0	128	43	0	76	80	52
T8–12 DAS	0	0	53	18	0	0	68	23	0	0	107	36	0	0	75	25
T9–17 DAS	0	0	51	17	0	0	62	21	0	0	92	31	0	0	73	24
T10–22 DAS	0	0	46	15	0	0	53	18	0	0	72	24	0	0	69	23
T11–27 DAS	0	0	38	13	0	0	48	16	0	0	62	21	0	0	69	23
T12–32 DAS	0	0	34	11	0	0	42	14	0	0	47	16	0	0	66	22
T13–37 DAS	0	0	31	10	0	0	43	14	0	0	21	7	0	0	68	23
T14–42 DAS	0	0	23	8	0	0	33	11	0	0	10	3	0	0	65	22
T15–47 DAS	0	0	7	2	0	0	13	4	0	0	5	2	0	0	63	21
Mean	11	22	53	29	18	33	67	39	22	43	96	54	26	37	75	46
	S.Em ±		CD at 5%		S.Em ±		CD at 5%		S.Em ±		CD at 5%		S.Em ±		CD at 5%	
T	0.74		1.81		0.56		1.81		0.65		1.81		0.33		1.81	
C	1.65		3.10		1.25		3.10		1.46		3.10		0.73		3.10	
T×C	2.86		1.60		2.17		1.60		2.53		1.60		1.27		1.60	

Legends
C1 Pollen stored in ambient room temperature
C2 Pollen stored in deep freezer (–20° C)
C3 Pollen stored in liquid nitrogen (–196° C)
DAS Days after storage

(less than 3 months) and can be used for hybrid seed production which also helps in reducing cost of production, to ensure genetic integrity and germ plasm security. The cryo-preservation also has application in breeding programmes, distributing and exchanging germ plasm among locations as well as for studies in physiology, biotechnology and *in vitro* fertilisation [10].

References

1. Kristof EB, Barnabes B. Deep freezing storage of paprika pollen. Capsicum Newsl. 1986;5:27–8.
2. Benzoickova A. Viability of sweet pepper pollen stored at cryogenic temperature. Capsicum Newsl. 1988;7:30–1.
3. Alexander MP, Ganeshan S, Rajasekharan PE. Freeze preservation of capsicum pollen in liquid nitrogen (−196 °C) for 42 months-effect on pollen viability and fertility. Plant Cell Incompat Newsl. 1991;23:1–4.
4. Rajasekharan PE. Seed cryopreservation: problems and prospects. Paper presented at the ICAR Short Course on In Vitro Conservation and Cryopreservation-New Options to Conserve Horticultural Genetic Resources, Bangalore, India, 21–30 Sept 2006; 2006. pp. 51–65.
5. Van Tiegham PM. Ann. Sci. Natbotan, S., Ser, T. Cited in Stanley, R. G. and H. F. Linskens. Pollen Biol Biochem Manag. 1869;XII:314 (69, 195, 2.1).
6. Panse VG, Sukhatme PV. Statistical methods for agricultural workers. Indian Council of Agricultural Research, New Delhi, India; 1978. pp. 112–119.
7. Sacks EJ, St. Clair DA. Cryo-preservation of tomato pollen. Hort Sci. 1995;30(4):797.
8. Rajashekhar PE. Studies on seed production and pollen viability on KBSH-1 hybrid sunflower. MSc-Agri thesis, University of Agricultural Sciences, Bangalore (India); 2000.
9. Reed BM. Implementing cryogenic storage of clonally propagated plants. Cryo Letters. 2001;22:97–104.
10. Towill LE, Walters C. Cryopreservation of pollen. In: Engelmann F, Takagi H, editors. Cryopreservation of tropical plant germ plasm. Current Research Progress and Application, Barak, Japan, JIRCAS/IPGRI; 2000. pp. 115–129.

Evaluation of *Pseudomonas* sp. VSMKU-4046 for Suppression of Sheath Blight (ShB) of Rice in Detached Leaf Assay

33

C. Sathish Kumar, H. Harikrishnan, R. Charulatha, and V. Shanmugaiah

Abstract

The study of root-associated bacteria and their antagonistic potential is important not only for understanding their ecological role in the rhizosphere and the interaction with plants but also for the suppression of soilborne plant pathogens. In this context, a total of 206 fluorescent pseudomonads (FPs) were isolated from rice rhizosphere from southern part of Tamil Nadu. Initially 113 isolates were found which were active against *Rhizoctonia solani*, of which 89 were strongly active against *R. solani* with zone of inhibition range from 10 to 32 mm. Production of lytic enzymes such as chitinase, cellulase, protease, amylase and pectinase was reported along with other characteristics including phosphate solubilisation and siderophore formation in 89 antagonistic fluorescent pseudomonads. Antagonistic fluorescent pseudomonads were also tested towards other fungal pathogens such as *Macrophomina phaseolina, Fusarium oxysporum, Fusarium udum* and *Alternaria alternata*. Among 89 the strain VSMKU-4046 has high antagonistic potential against *R. solani*. Strain VSMKU-4046 significantly controls sheath blight of rice compared to control using sclerotia of *R. solani* through detached leaf assay. The lesion length around the sclerotium was measured, and ShB severity was rated by the relative lesion height (RLH) method with the following formula:% RLH = 100 x Total height of lesions / Total leaf height.

Keywords

Pseudomonas sp. • *R. solani* • Biocontrol

Introduction

Rice (*Oryza sativa* L.) is an important crop worldwide, and over half of the world population is dependent on it for food [1]. However, the net production of rice is greatly reduced due to losses caused by a number of diseases, and sheath blight caused by fungal pathogen

C. Sathish Kumar • H. Harikrishnan • R. Charulatha
V. Shanmugaiah (✉)
Department of Microbial Technology,
School of Biological Sciences, Madurai Kamaraj University, Madurai 625 021, Tamil Nadu, India
e-mail: vshanmugaiah@yahoo.co.in

A. Sabu and A. Augustine (eds.), *Prospects in Bioscience: Addressing the Issues*,
DOI 10.1007/978-81-322-0810-5_33,

Rhizoctonia solani is a devastating disease accounting for 24–50% economic losses across the rice cultivation zones of the world [2, 3]. Due to public concerns on pesticide use in crops, exploration of alternative control methods has been a global effort for food security. Indeed, a number of rhizobacteria, in particular fluorescent pseudomonads, have already been reported to inhibit the growth of fungal pathogens [4, 5]. Fluorescent pseudomonads are ubiquitous soil microorganisms and common inhabitants in the rhizosphere of crop plants. It protects the plants from infection caused by fungal and bacterial plant pathogens [6]. Plant growth stimulating efficiency of bacterial inoculants is also affected by certain soil nutritional condition. Bacterial inoculation yields much better results in nutrient-deficient soil than in the nutrient-rich soil [7]. Plant plays an important role in selecting and enriching the type of bacteria by their root exudates. Bacteria which is present in the rhizosphere develop depending on nature and concentrations of organic constituents of exudates; certain bacteria have the capability to utilise these as a source [8]. Rhizospheric bacteria can be used as source for plant growth-promoting agent in agriculture [9]. Plant growth-promoting bacteria isolated from rhizosphere soils stimulate growth directly by nitrogen fixation [10]. Fluorescent pseudomonads (FPs) are a group of PGPR that can promote growth and suppress plant pathogens. Their applicability as biocontrol agents has drawn wide attention because of the production of secondary metabolites such as siderophore, antibiotics, volatile compounds, HCN, lytic enzymes and phytohormones [11]. Till now no genetic resistance has been reported for sheath disease, and all the rice cultivars and most of the crops are susceptible to the pathogen *R. solani*. The use of biological control agents as an alternative to fungicides is increasing rapidly in the present day agriculture. Plant growth-promoting rhizobacteria (PGPR) are a group of bacteria that are used for enhancing crop growth and suppressing plant diseases. Therefore, the present study was aimed at isolation, characterisation and *in vitro* screening of several rhizosphere isolates of fluorescent pseudomonads from rice rhizosphere against *R. solani*. Also studies were conducted to determine the control of sheath blight disease through detached leaf assay using *Pseudomonas* sp. VSMKU-4046, culture filtrate of VSMKU-4046 and commercial fungicide carbendazim in order to devise strategies for rice sheath blight disease management.

Materials and Methods

Isolation of *Fluorescent Pseudomonads* from Rice Rhizosphere

About 10 g of soil collected from each rhizosphere was taken in a conical flask. To which 90 ML of sterile distilled water was added, and the flasks were placed on a rotary shaker at 120 rpm for 30 min. The resulted suspension was serially diluted and spread plated on to King's B agar (KBA) [12]. The plates were incubated at room temperature (28 ± 2°C) for 48 h and observed under UV light. Distinct single colonies of fluorescent bacteria were picked, subcultured to purity and maintained in KBA slants and also in 30% glycerol vials.

Screening of *Fluorescent Pseudomonads*

All the 206 isolated fluorescent pseudomonads (FPs) were subjected to primary screening against *Rhizoctonia solani* on PDA by dual culture assay [13]. The mycelial disc 9 mm was cut out from 3-day old culture of *R. solani* on PDA and inoculated at the centre of a fresh PDA Petri plate. Four/two FPs were streaked in the periphery of each Petri plate and incubated at room temperature. The growth of *R. solani* and FPs was observed periodically and the inhibition zone was measured. Among the 206 FPs, a strain designated as VSMKU-4046 isolate from the rice rhizosphere was selected for further studies as it effectively inhibited *R. solani* than other isolates. 113 antagonistic strains were subjected to secondary screening against other fungal pathogens such as *Macrophomina phaseolina, Fusarium udum, Alternaria alternata and Fusarium oxysporum* by the abovementioned method. Among 113 strains, VSMKU-4046 had shown prominent antagonistic activity compared to other FPs and control.

Identification of Selected Strain

The protocol of King's B agar [12] was used for fluorescence production. Bacteria were streaked on King's B agar and incubated at 28 ± 2°C for 48 h. At the end of the incubation, the plates were observed under UV light for production of fluorescence. Physiological and biochemical tests were carried out according to Bergey's Manual of Determinative Bacteriology [14].

Lytic Enzyme Production

Fluorescent pseudomonads were streaked on the respective substrate (0.5%) amended agar medium, and then the plates were incubated at room temperature. After 24 h, chitinase and cellulase production was observed by flooding 1% Congo red over the colloidal chitin and cellulose agar plates. Similarly Gram's iodine was flooded over the starch agar plate and tested for amylase activity. Mercuric chloride was flooded over the gelatin and casein agar plates to check the zone of clearance for protease.

Screening of Phosphate Solubilisation

To detect the phosphate, solubilising FPs were streaked onto Pikovskaya's agar medium, which contains (per litre) 0.5 g yeast extract, 10 g dextrose, 5 g Ca3(PO4)2, 0.5 g (NH4)2SO4, 0.2 g KCl, 0.1 g MgSO4·7H2O, 0.0001 g MnSO4H2O, 0.0001 g FeSO4·7H2O and 15 g agar. After 3 days of incubation at 28°C, strains that induced clear zone around the colonies were considered as positive [15].

HCN Production

Production of hydrocyanic acid (HCN) was determined by method of Lorck, 1948 [16]. Bacterial strains (24-h-old) were separately streaked on nutrient sucrose agar with 4.4 g glycine with simultaneous addition of a filter paper soaked in 0.5% picric acid in 1% Na_2CO_3 in the upper lid of Petri plate. The plates were sealed with parafilm. The control plates did not receive inoculum. After incubation at 28 ± 1°C, changes in colour from yellow to reddish brown were examined.

Siderophore Production

Siderophore production by different strains FPs *was* tested by chrome azurol S (CAS) assay [17]. The strains were streaked on cetrimide agar and incubated for 48 h at 30°C. After incubation, a thin layer of CAS reagent in 0.7% agar was spread on the bacterial growth, and plates were again incubated for 24 h at 30°C; formation of yellow orange zone around the colonies indicates siderophore production.

IAA Production

It was done as per the protocols of Patten and Glick (1996) [18]. FPs were grown in Luria broth supplemented with L-tryptophan (1 μg ml^{-1}) for 72 h. At the end of the incubation, cultures were centrifuged at 10,000 *g* for 10 min and the supernatants were collected. One ml of this culture filtrate was allowed to react with 2 ml of Salkowsky reagent (1 ml of 0.5 M $FeCl_3$ in 50 ml of 35% HCIO4) at 28 ± 2°C for 30 min. At the end of the incubation, pink colour is developed which indicates the presence of IAA. Quantification of IAA was done by measuring the absorbance in a spectrophotometer at 530 nm.

Detached Leaf Assay

Healthy rice leaves were cut into 10-cm length segments, surface sterilised with 2% sodium hypochlorite for 3 min and washed with sterile distilled water. Rice leaves were dipped in culture, culture filtrate of *Pseudomonas* sp. VSMKU-4046, carbendazim and water for 3 min and then

Table 33.1 Antagonistic activity of rice rhizosphere FPs against fungal pathogen

Fungal pathogens	No. of antagonistic FPs	Zone of inhibition cm
Rhizoctonia solani	113	1.0–3.2 cm
Macrophomina phaseolina	101	1.0–2.5 cm
Fusarium oxysporum	105	0.7–2.7 cm
Fusarium udum	93	0.5–2.5 cm
Alternaria alternata	89	0.4–1.1 cm

allowed for air-dried. After air dry, the uniform size of sclerotium was placed on the above treated rice leaves. Then leaves were placed in the moist chamber for 7 days. After 7 days sheath blight lesion development in rice leaf segments was observed. The lesion length around the sclerotium was measured, and ShB severity was rated by relative lesion height (RLH) method with the following formula:

$$\% \text{ RLH} = 100 \times \text{total height of lesions} / \text{total leaf height}.$$

Results and Discussion

Isolation and Screening of Antagonistic Fluorescent Pseudomonads

A total of 206 fluorescent pseudomonads (FPs) were isolated from rice rhizosphere soils and screened for antifungal activity against *R. solani.* Among them, 113 fluorescent pseudomonads were showed inhibitory activity against *R. solani.* The zone of inhibition of mycelia growth of the above fungal pathogens ranged from 1.0 to 3.2 cm. Further secondary screening of 113 isolates towards other fungal pathogens such as *Macrophomina phaseolina, Fusarium oxysporum, Fusarium udum and Alternaria alternata* were performed. The zone of inhibition of mycelia growth of the above fungal pathogens ranged from 1.0 to 2.5 cm, 0.7 to 2.7 cm, 0.5 to 2.1 cm and 0.4 to 1.1 cm (Table 33.1).

Table 33.2 Lytic enzymes, secondary metabolites and IAA production by rice rhizosphere FPs

Lytic enzymes, secondary metabolites and IAA	No. of strains (positive)
Amylase	4
Cellulase	1
Pectinase	4
Chitinase	8
Protease	61
Phosphate solubilisation	50
HCN	38
Siderophore	52
IAA	52

Identification of VSMKU-4046

The selected strain VSMKU-4046 was identified as *Pseudomonas* sp. based on the physiological tests.

Production of Lytic Enzymes, Phosphate Solubilisation, HCN, Siderophore and IAA

Among 89 fluorescent pseudomonads, more protease and chitinase enzyme production was observed than the other lytic enzymes such as amylase, cellulase and pectinase (Table 33.2). Significant production of phosphate solubilisation, hydrogen cyanide, siderophore and IAA was found among 89 FPs.

Detached Leaf Assay

Detached leaf assay was confirmed that treatment with culture and culture filtrate of *Pseudomonas* sp. VSMKU-4046 remarkably inhibited the sclerotial viability. Moreover, the sclerotia failed to establish sheath blight symptoms on detached leaves, whereas other treatments showed different levels of lesion development (Table 33.3).

The aim of current investigation was the screening and selection of significant fluorescent pseudomonads from rice rhizospheric soil with antagonistic activities against phytopathogenic fungi and prospective plant growth stimulating characteristics with purpose for further field

Table 33.3 Evaluation of *Pseudomonas* sp.VSMKU-4046 against rice sheath blight

Treatments	Sheath blight severity (%)
R. solani (sclerotium)	85
R. solani+carbendazim	20
Pseudomonas sp.VSMKU-4046	15
Culture filtrate of VSMKU-4046	28

application. Isolation of bacteria from innate of rhizosphere of the target crop is essential for successful identification of potential biocontrol and biofertilising agents [19, 20]. The present result demonstrates that a total of 206 fluorescent pseudomonads were isolated from rice rhizosphere. Of which 89 fluorescent pseudomonads were found to be effectively against *R. solani.* Among 89 antagonistics, a strain designated as VSMKU-4046 was selected because of its broad spectrum antagonistic activity against various phytopathogens [5, 21].

The selected strain VSMKU-4046 was identified as *Pseudomonas* sp. based on the physiological tests. Further, 16SrRNA sequencing analysis is in progress to identify the species. Production of lytic enzymes such as chitinase, cellulase, protease, amylase, pectinase, phosphate solubilisation, hydrogen cyanide, siderophore and IAA was recorded with 89 antagonistic fluorescent pseudomonads. Antagonistic FPs were also tested towards other fungal pathogens such as *Macrophomina phaseolina, Fusarium oxysporum, Fusarium udum* and *Alternaria alternata* with zone of inhibition ranging from 1.0 to 2.5 cm, 0.7 to 2.7 cm, 0.5 to 2.1 cm and 0.4 to 1.1 cm.

The strain VSMKU-4046 has significantly reduced the ShB severity when compared to the control. The disease severity in this significant strain VSMKU-4046 ranged from 15 to 85%. Among the four different tests, maximum inhibition of lesion development was obtained with *Pseudomonas* sp.VSMKU-4046 with 15% of disease severity (Table 33.3). The next best reduction of ShB was noticed with culture filtrate of VSMKU-4046 and carbendazim with 28 and 20% compared to water shocked sclerotium control. This effective strain VSMKU-4046 is now being screened for their effects on development of sheath blight disease in the greenhouse. The detached leaf inoculation technique was earlier attempted for determining the morphological and pathological variability in rice isolates of *R. solani* [22]. The assay was found to be useful in the host-specific toxin production by *R. solani* in rice [23]. The detached leaf assay for control of sheath blight is less time-consuming compared to rice plant assays under greenhouse conditions.

Acknowledgement The authors greatly thank University Grant Commission, New Delhi (F.No.39-214/2010) (SR), for financial assistance. We also would like to acknowledge the Head, Department of Microbial Technology, the Co-ordinator, NRCBS, School of Biological Sciences, Madurai Kamaraj University, Madurai, Tamil Nadu, India.

References

1. Yang Q, Lin F, Feng S, Wang L, Pan Q. Recent progress on molecular mapping and cloning of Blast resistance genes in rice (*Oryza sativa L.*). Scia Agric Sin. 2009;42:1601–15.
2. Ou SH. Rice diseases. Kew survey: Commonwealth Mycological Institute; 1985. p. 256–368.
3. Padaria JC, Singh A. Molecular characterization of soil bacteria antagonistic to *Rhizoctonia solani*, sheath blight of rice. J Environ Sci Health B. 2009; 44:397–402.
4. Shanmugaiah V. Biocontrol potential of phenazine-1-carboxamide producing plant growth promoting rhizobacterium, *Pseudomonas aeruginosa* MML2212 against sheath blight disease of rice. Ph.D. thesis, University of Madras, Chennai, India; 2007.
5. Shanmugaiah V, Mathivanan N, Varghese B. Purification, crystal structure and antimicrobial activity of phenazine-1-carboxamide produced by a growth-promoting biocontrol bacterium, *Pseudomonas aeruginosa* MML2212. J Appl Microbiol. 2010; 108(703–711):703.
6. Defago G, Hass D. *Pseudomonads* as antagonists of soil borne plant pathogens: modes of action and genetic analysis. Soil Biol Biochem. 1990;6:249–92.
7. Egamberdiyeva D. The effect of plant growth promoting bacteria on growth and nutrient uptake of maize in two different soils. Appl Soil Ecol. 2007;36(Suppl 2–3):184–9.
8. Curl EA, Truelove B. The rhizosphere. Berlin/ Heidelberg: Springer; 1986. p. 288.
9. Chaiharn M, Chunhaleuchanon S, Kozo A, Lumyong S. Screening of rhizobacteria for their plant growth promoting activities. KMITL Sci Technol J. 2008;8 suppl 1:18–23.

10. Han J, Sun L, Dong X, Cai Z, Sun X, Yang H, Wang Y, Song W. Characterization of a novel plant growth promoting bacteria strain *Delftia tsuruhatensis* HR4 both as a diazotroph and a potential biocontrol agent against various plant pathogens. Syst Appl Microbiol. 2005;28(1):66–76.
11. Gupta CP, Dubey RC, Kang SC, Maheshwari DK. Antibiosis mediated necrotrophic effect of *Pseudomonas* GRC2 against two fungal plant pathogens. Curr Sci India. 2001;81:91–4.
12. King EO, Ward MK, Raney DE. Two simple media for the demonstration of pyocyanin and fluorescein. J Lab Clin Med. 1954;44:301–7.
13. Haung MC, Hoes JA. Penetration and infection of *sclerotinia sclerotiorum* by *coniothyrium minitans*. Can J Bot. 1976;54:406–10.
14. Williams ST, et al., editors. Bergey's manual of determinative bacteriology. 9th ed. Baltimore: Williams and Wilkins Co.; 1994.
15. Pikovskaya RI. Mobilization of phosphorus in soil connection with the vital activity of some microbial species. Microbiologia. 1948;17:362–70.
16. Lorck H. Production of hydrocyanic acid by bacteria. Physiol Plant. 1948;1:142–6.
17. Schwyn B, Neilands JB. Universal chemical assay for the detection and determination of siderophores. Anal Biochem. 1987;160(1):47–56.
18. Patten C, Glick CR. Bacterial biosynthesis of indole-3-acetic acid. Can J Microbiol. 1996;42:207–20.
19. Shanmugaiah V, Ramesh S, Jayaprakashvel M, Mathivanan N. Biocontrol and plant growth promoting potential of a *Pseudomonas* sp. MML2212 from the rice rhizosphere. In: Zeller W, Ullrich C, editors. Proceedings for the 1st international symposium on Biological Control of Bacterial Plant Diseases. Federal Institute of Biological Control, Dormstadt, Germany; 2006. pp. 320–324.
20. Nagarjkumar M, Bhaskaran R, Velazhahan R. Involvement of secondary metabolites and extracellular lytic enzymes produced by *Pseudomonas fluorescens* in inhibition of *Rhizoctonia solani*, the rice sheath blight pathogen. Microbiol Res. 2004;159:73–81.
21. Mathivanan N, Shanmugaiah V. Management of sheath blight disease in rice by *Pseudomonas aeruginosa* MML2212. In: Reddy MS, Wang Qi, editors. Proceedings for the 2nd Asian PGPR Conference, 21–24 Aug 2011, Beijing, PR China; 2011.
22. Devi VT, Malarvizhi R, Sakthivel N, Gnanamanickam SS. Biological control of sheath blight of rice in India with antagonistic bacteria. Plant Soil. 1989; 119:325–30.
23. Pan XB, Rush MC, Sha XY, Xie QJ, Stetina SD, Oard JH. Major gene, nonallelic sheath blight resistance from the rice cultivars "Jamine 85" and "Tequing". Crop Sci. 1999;39:338–46.

Antimicrobial and Immunomodulatory Activities of Jellyfish (*Chrysaora quinquecirrha*) Venom

34

K. Suganthi and S. Bragadeeswaran

Abstract

Marine invertebrates, especially jellyfish (Cnidarians), are of particular concern in the production of natural toxins. In Cnidarians, the toxin is present in specialised structures called the nematocysts, which inject the venom after mechanical or chemical stimulation and so protect themselves from the attack of prey. The present study deals with the antimicrobial and immunomodulatory activity of fractionated venom of the jellyfish *Chrysaora quinquecirrha.* In this study, the fractionated extracts were tested for antibacterial and antifungal activity. It showed moderate antibacterial activity against 10 pathogens, and *Salmonella paratyphi* was the most sensitive against n-butanol extracts (8.0±0.81 mm). The fungi *Aspergillus niger* was the most sensitive against n-butanol extracts (9.0±1.63 mm). Immunomodulatory activity of the fractionated extract has been showed as significant activity at higher concentrations. The extracts of *C. quinquecirrha* exerted an immunostimulating effect of 40–25% magnitudes at lower concentrations and a suppressive effect of 33% at higher concentration. The fractionated extract exhibited immunostimulation ranging from 25 to 40% at concentration up to 1,000 μg but showed immunosuppressive effects at 20 μg. From this study, we conclude that the venom of *C. quinquecirrha* has potent antimicrobial and immunomodulatory activity. Molecular weight of *C. quinquecirrha* extract fraction *ranging* from 15 to 105 kDa revealed the presence of medium-sized proteins and categorised as (1) medium-sized cytolytic actinoporins (~20 kDa), (2) cardiostimulatory proteins (~28 kDa) and (3) cytolysin with or without phospholipase (~40 kDa) in both crude and fractionated proteins. The FTIR spectra of jellyfish sample experimentally observed

K. Suganthi • S. Bragadeeswaran (✉)
Marine Biotoxinology Lab, Centre for Advanced Study in Marine Biology, Annamalai University, Parangipettai 608502, Tamil Nadu, India
e-mail: drpragathi@gmail.com

A. Sabu and A. Augustine (eds.), *Prospects in Bioscience: Addressing the Issues*,
DOI 10.1007/978-81-322-0810-5_34, © Springer India 2013

that frequencies of crude and fractionated venoms displayed in the O–H stretching frequency appeared at 3,567 cm^{-1} and the N–H stretch in secondary amide (polypeptides) appeared at 3,348 cm^{-1}. The overall frequencies showed an increase in amide I, amide II and amide III regions.

Keywords
Cnidarians • Toxin • Nematocysts • Immunomodulatory

Introduction

Marine toxins have drawn worldwide attention because of their involvement in human intoxication and the socioeconomic impacts brought by some incidents [1–3]. Marine toxins are notoriously more potent substances as useful drugs. Coelenterates or Cnidarians are simple metazoans having primary radial, biradial or radiobilateral symmetry. There are 9,000 species of Coelenterates, of which approximately 100 species are toxic to humans. The presence of tentacles equipped with nematocyst is a unique character of this group. The toxin is present in the nematocysts, which inject the venom after mechanical or chemical stimulation and so themselves from the attack of prey. Among various constituents of the Coelenterates (Cnidarians), the cytolytic properties have been demonstrated [4]. The sea nettle (*Chrysaora quinquecirrha*) is a species of jellyfish occurring all over the world. Toxins derived from some of the jellyfishes as models for the development of new drug also have promising applications in cardiovascular medicine and target medicine of nerve molecular biology. The present study was undertaken to investigate the biological properties of crude and fractionated extracts of *C. quinquecirrha* venom.

Materials and Methods

Sample Collection and Extraction of Venom

The jellyfish *C. quinquecirrha* was collected from the Vellar estuary, during post-monsoon season. The crude venom was extracted following the methodology of Yanagihara [5]. The live animals were kept in distilled water during stress condition, and the nematocysts were extracted from the tentacles, and the procedure was repeated thrice. The nematocysts containing toxins were collected with 0.5 mm mesh sieve and filtered by Whatman No. 1 filter paper. The supernatant was kept for lyophilisation which was later transformed into crystalline powder and stored at 4 °C for further use.

Partial Purification Using Column Chromatography

The crude extract was partially purified using DEAE cellulose [1, 7]. The crude toxin was dissolved in phosphate buffer (pH, 7.4) at 5 mg/ml and eluted through a DEAE-cellulose column (height of 30 cm and diameter of 2 cm). One unabsorbed fraction of crude toxin was eluted with phosphate buffer (pH, 7.4), and then 10 absorbed fractions were eluted with a linear gradient of 15 ml stored at −20° C.

Dialysis Dehydration

Dialysis dehydration was followed by the method of Li et al. [8] with slight modification. In brief, the collected fraction samples were dialysed separately by using Sigma (USA) dialysis membrane 500 (average flat width, 24.26 mm; average diameter, 14.3 mm; approximate capacity, 1.61 ml/cm) against D-glucose to remove excess water. The dialysed sample was washed with the phosphate buffer (pH 7.4, 10 mmol/l) and lyophilised (FreeZone® Freeze Dry Systems, Labconco, USA).

Antimicrobial Activity

Antibacterial Activity

Antibacterial activity was determined against *E. coli*, *V. cholera*, *S. paratyphi*, *K. pneumonia*, *P. aeruginosa*, *P. mirabilis*, *L. vulgaris* and *K. oxytoca* by using the standard disc diffusion method [9].

Antifungal Activity

Antifungal activity was determined against *A. alternaria*, *A. flavus*, *A. niger*, *C. albicans*, *E. floccosum*, *Mucor* sp., *Penicillium* sp. *Rhizopus* sp., *T. mentagarophytes* and *T. rubrum* using the standard disc diffusion method [10].

Immunomodulatory Activity

Preparation of Test sample

Samples for *in vitro* studies were prepared by dissolving 5 mg/ml of crude and fractionated extracts in 5 ml of distilled water to obtain concentration ranging from 10, 20, 50, 100 to 1,000 μg/ml.

Neutrophil Locomotion and Chemotaxis Test

Neutrophil cell suspension was prepared in phosphate buffer saline solution (PBS) at about 10^6 cells/ml. The lower compartment of chemotaxis chamber was adjusted to a pH of 7.2, e.g. chamber 1-PBS solution (control), chamber 2- casein 1 mg/l (standard), and chamber 3, 4, 5, 6, 7 with different concentrations (10, 25, 50, 100 and 1,000 μg/ml) of test sample. The upper compartment (1-ml syringe) was placed with suspended neutrophil, and wet filter (Millipore) 3-mm pore size was fixed at the bottom of the upper compartment. The same was incubated at 37 °C for 3 h.

The upper compartment was removed and inverted to empty the fluid. The lower surface of the filter was fixed with 70% ethanol for 2 min and then stained with haematoxylin dye for 5 min. The fixed filters were observed under microscope using 100x lenses, and the number of neutrophil cells reached to the lower surface was counted.

Phagocytosis of Killed *Candida albicans*

Preparation of *Candida albicans* suspension was followed by the method of Ponkshe [11]. The *Candida albicans* culture was incubated in Sabouraud broth overnight and centrifuged, and supernatant was discarded. The cell button was washed with sterile Hank's balanced salt solution (HBSS) and centrifuged again. This was done 3–4 times. The final cell button was mixed with a mixture of sterile HBSS and human serum in proportion of 4:1. The final cell suspension of concentration 1×10^8 was used for the experiment.

Phagocytosis Evaluation

The mean number of Candida cells phagocytosed by PMNs on the slide was determined microscopically for 100 granulocytes using morphological criteria. This number was taken as phagocytic index (PI) and was compared with basal PI of control. This procedure was repeated for different concentrations (10, 25, 50, 100 and 1,000 μg/ml) of test sample. The percentage of immunostimulation was calculated by using the following equation:

$$\text{Stimulation}(\%) = \text{PI}(\text{test}) - \text{PI}(\text{control}) \times 100 / \text{PI}(\text{control}).$$

Characterisation of *C. Quinquecirrha* Venom

Molecular Weight Determination in SDS-PAGE

The proteinaceous venom *of C. quinquecirrha* was subjected to electrophoresis following the method of Laemmli [12] and confirmed by running the crude sample in 12% SDS-PAGE. SDS was being used to make uniform charge all over the protein sample, and β-mercaptoethanol was used to break the disulphide linkage, which made all protein in the same shape, so migration of protein in the gel was only according to their molecular weight with the molecular marker [13].

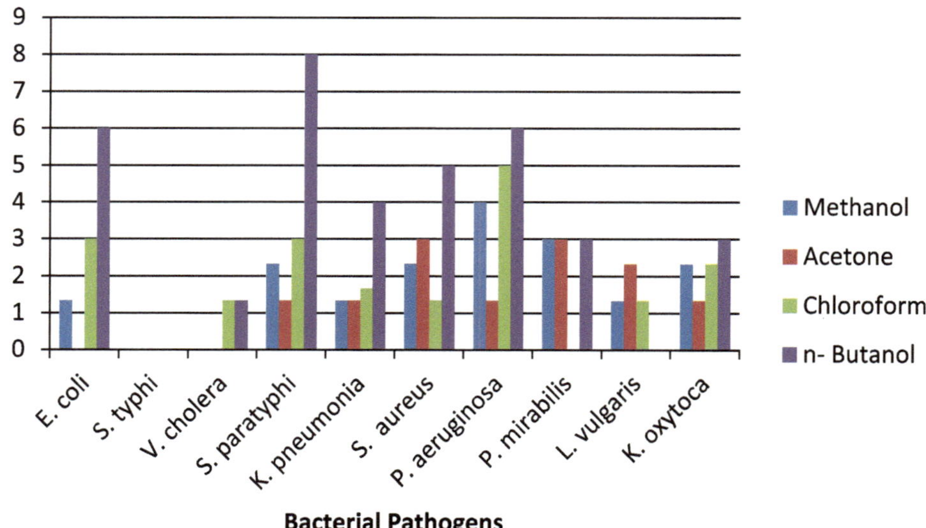

Fig. 34.1 Antibacterial activities of *C. quinquecirrha* against bacterial pathogens

FTIR: Sample Preparation

The powdered sample and KBr were again lyophilised to remove moisture which may interfere with the measurement of the amide I band. Approximately 5 mg of sample was mixed with 1,000 mg of dried KBr and was subjected to a pressure of 5×10^6 Pa and made into clear pellet of 3 mm diameter and 1 mm thickness. Absorbance spectra were recorded using Nicolet Avatar-360 FTIR spectrometer equipped with a KBr beam splitter and an air-cooled DTGS detector. The absorption of light intensity of the peak was calculated using the base line method. The frequencies for all sharp bands were accurate to 0.01 cm^{-1}.

Results and Discussion

Antibacterial Activity

The crude, methanol, acetone, chloroform and n-butanol showed moderate antibacterial activity (Fig. 34.1). Among the test bacteria, *S. paratyphi* was the most sensitive against n-butanol extracts (8.0 ± 0.81 mm).

Antifungal Activity

The extracts showed moderate antifungal activities against 10 pathogens assayed (Fig. 34.2). Among them n-butanol extracts were the most sensitive fractions (9.0 ± 1.63 mm).

Immunomodulatory Activity

Crude and fractionated extracts showed significant activity at higher concentrations. The neutrophil locomotion and chemotaxis showed significant activity in all concentrations; fractionated extract showed significant activity at 20 μg/ml concentration (Fig. 34.3). In case of phagocytosis of killed *Candida albicans*, the extract showed significant activity even at low concentration of 10 μg/ml. The extract showed predominantly very good activity at a concentration of 20 μg/ml (Fig. 34.4).

Phagocytosis

The extracts of *Chrysaora quinquecirrha* exerted an immunostimulating effect of 40–25%

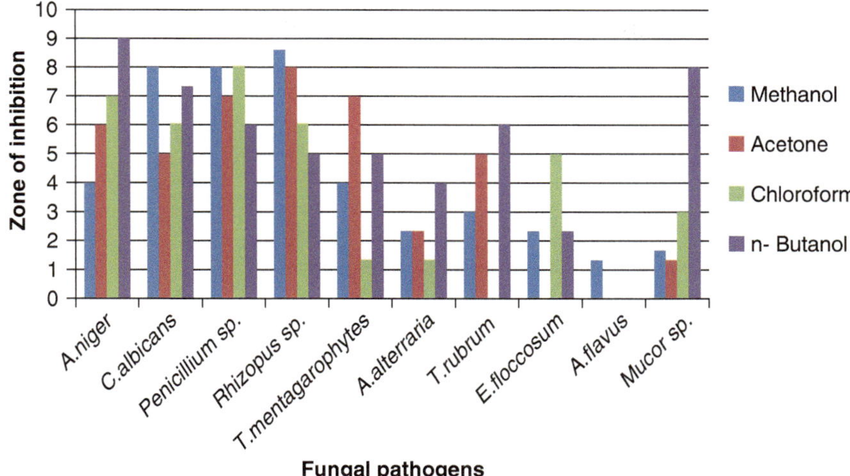

Fig. 34.2 Antifungal activities of *C. quinquecirrha* against fungal pathogens

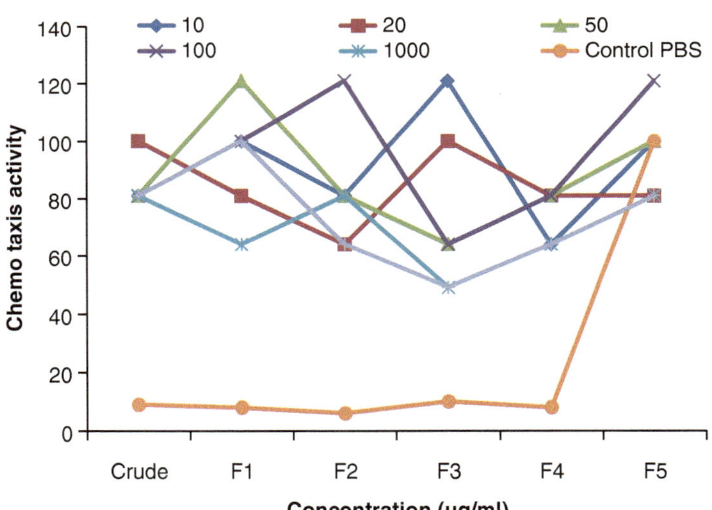

Fig. 34.3 Immunomodulation produced by *C. quinquecirrha* extracts (**Good activity at 20 µg/ml*)

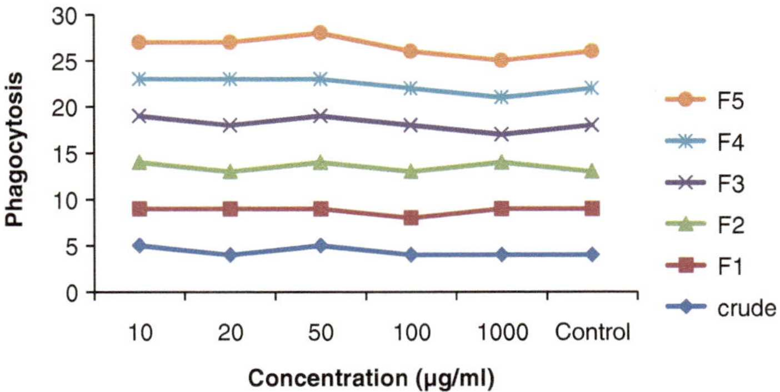

Fig. 34.4 Phagocytosis by *C. quinquecirrha* extracts

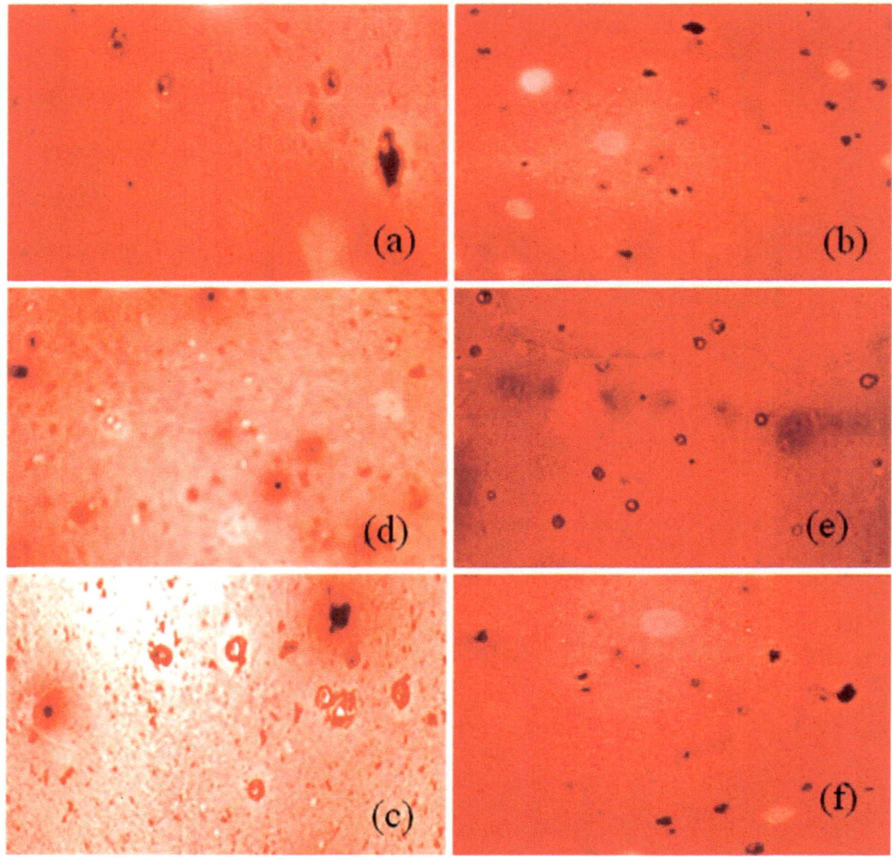

Fig. 34.5 Phagocytosis of *Candida albicans* by PMN when treated with *C. quinquecirrha* extract at different concentrations: (**a**) 10 μg/ml, (**b**) 20 μg/ml, (**c**) –50 μg/ml, (**d**) 100 μg/ml, (**e**) 1,000 μg/ml and (**f**) control

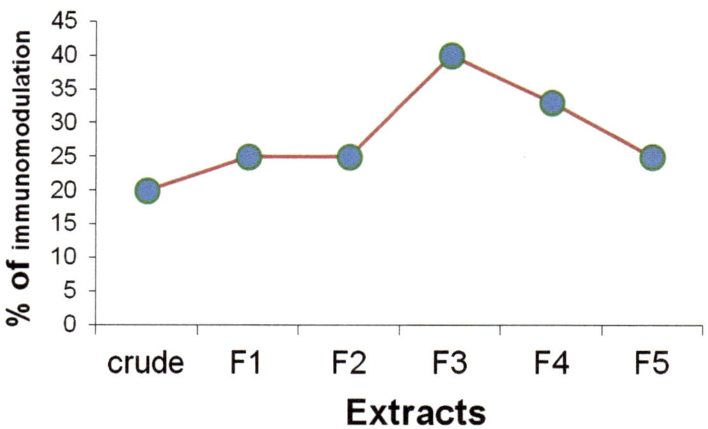

Fig. 34.6 Immunomodulation produced by C. *quinquecirrha* extracts

magnitudes at lower concentrations and a suppressive effect of 33% at higher concentration (Fig. 34.5). The fractionated extract exhibited immunostimulation ranging from 25 to 40% at concentrations up to 1,000 μg but showed immunosuppressive effects at 20 μg (Fig. 34.6).

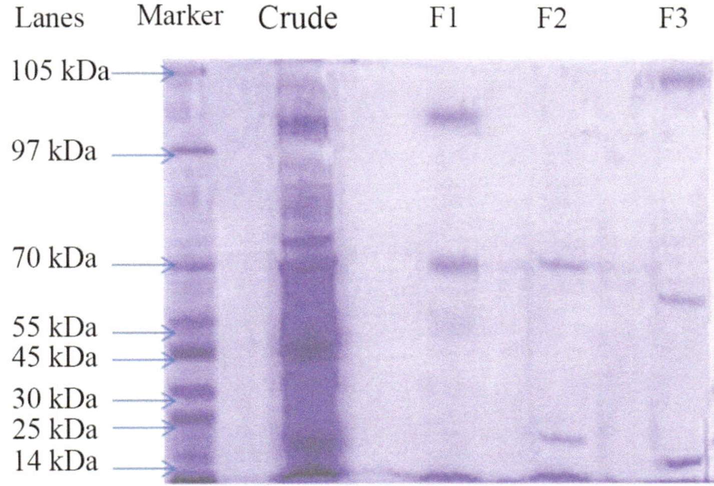

Fig. 34.7 SDS-PAGE analysis of the crude and fractionated proteins of *C. quinquecirrha* venom

SDS Gel Electrophoresis

Separation of native *C. quinquecirrha* venom protein was achieved by ion-exchange chromatography across the molecular mass separation range, 14–105 kDa. Reducing SDS-PAGE analysis of the crude venom and purified fractions revealed that all the three major fractions were significantly apparent and the molecular weights were almost the same. The four distinct clear bands showed molecular weights of 17, 35, 50 and 70 kDa in the crude and 105, 65 and 9 kDa in fractionated samples (Fig. 34.7).

FTIR

The FTIR spectra of *C. quinquecirrha* fractionated venom samples were shown in Fig. 34.8. The O–H stretching frequency appeared at 3,567 cm^{-1} and the N–H stretch in secondary amide (polypeptides) appeared at 3,348 cm^{-1} in C–H stretching. C–CH_3 group appeared at 2,918 cm^{-1} and C–H stretching frequency in O–CH_3 group appeared at 2,851 cm^{-1}. Substituted benzene ring stretches appeared at 2,057 cm^{-1} and the C–O stretch on tertiary amide (amide I) appeared at 1,634 cm^{-1}. The N–H deformation frequency of secondary amide (II) appeared at 165 cm^{-1} and the amide (III) group C–N stretch at 419 cm^{-1}.

The CH_3 group anti-symmetry frequency appeared at 1,465 cm^{-1} and CH_3 symmetry deformation at 1,216 cm^{-1}. The C–O–C stretching frequency appeared at 216 cm^{-1} and the alcohol C–OH stretching frequency at 1,123 cm^{-1}. The cyclic compound carbon ring frequency appeared at 1,042 cm^{-1} and the phenol O–H out-of-plane deformation at 7,178 cm^{-1}. While using fractionated extract (F3) venom, almost the same results were found. Hence, both the results showed the presence of polypeptide chains.

The antimicrobial activity was performed to detect the defensive action from *C. quinquecirrha* extract. Of the pathogens tested, *S. paratyphi* showed moderate antibacterial activity in n-butanol (8.0±0.81 mm) and *A. niger* was the most sensitive against n-butanol extracts (9.0±1.63 mm). The present study supported by the earlier studies from the sea anemone *Actinia equina* showed considerable inhibitory activity against Gram-negative bacteria [14]. The present results on antimicrobial

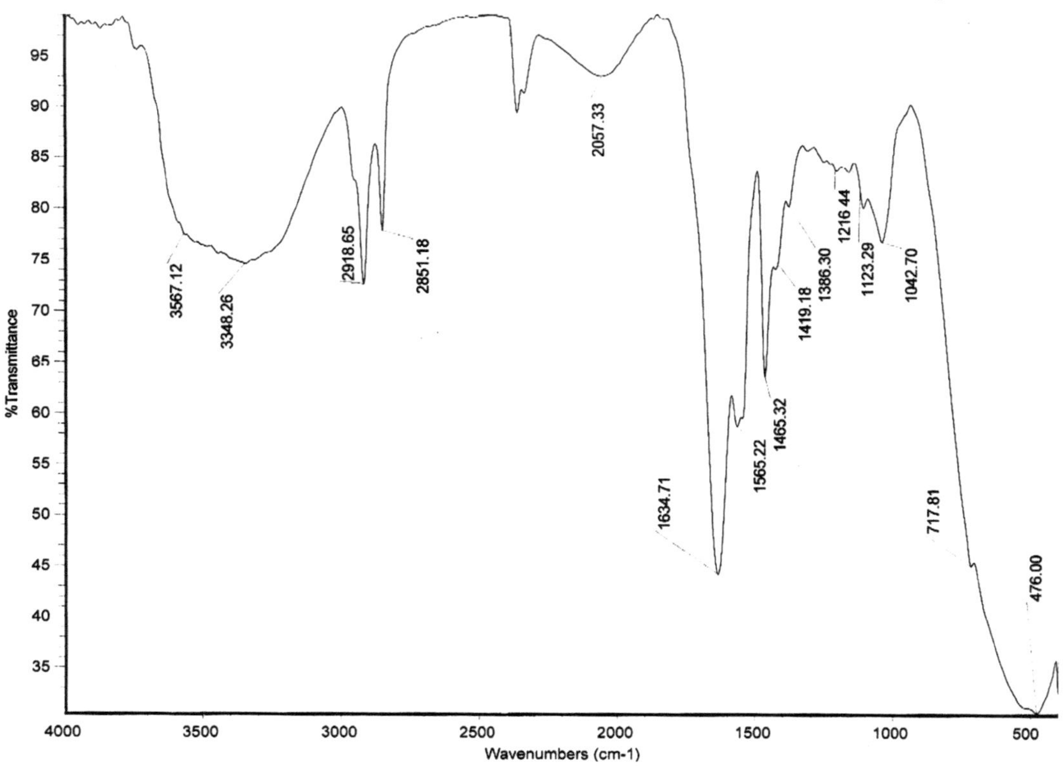

Fig. 34.8 FTIR analysis of the crude and proteins of *C. quinquecirrha* venom

activity of *C. quinquecirrha* extracts indirectly correspond to earlier findings of Thakur et al. [15]. The use of immune stimulants for prevention of diseases in fish considered an alternative and promising area [16]. The present study demonstrates the immunostimulant potential of the jellyfish venom extract for the first time. Immunomodulation was found to decrease with the increasing concentration of the crude and fractionated extracts.

Immunomodulation of marine toxin is poorly studied [17]. The wound healing activity exhibited by the epidermal secretion of the Gulf catfish was associated with immunomodulation as well as prostaglandin pathway [18]. The results of the *in vitro* PMN function test showed a significant increase in the percentage of phagocytosis and phagocytic index for jellyfish venom. It indicates that these extracts enhanced the phagocytic efficacy of the PMN cells by causing more engulfment of the Candida cells control, thereby stimulating a nonspecific immune response. As the *C. quinquecirrha* extracts showed promising immunostimulant activity in the *in vitro* test, the fractions eluted from the extract seemed to be the most active. On the contrary, the fractions eluted from the crude folklore preparation resulted as more active. The extract of *C. quinquecirrha* was reported to have a significant immunostimulant activity [19]. This resistance may be due to the increased activity of phagocytic cells [20] with subsequent increase in lysozyme activity. These outcomes are encouraging enough to pursue structure elucidation of the active components.

The molecular weight of jellyfish venom was evaluated [21, 22]. Separation of native *C. quinquecirrha* venom protein was achieved by ion-exchange chromatography across the molecular mass separation range between 15 and 100 kDa. Reducing SDS-PAGE analysis of the crude venom and purified fractions revealed that all the three major fractions are significantly purified, but apparent molecular weight remains the same. Subsequent results were found from the Cnidarian

L. danae with molecular weight of 58 and 62.5 kDa [23]. Jellyfish venoms from different species vary in activity and composition. It was reported that a major component in *C. quinquecirrha* venom had 19-kDa molecular weight, whereas major protein component in *Chrysaora achlyos* venom was 55 kDa [24]. Thus, it could be concluded from the present study that toxic proteins obtained from the jellyfish *C. quinquecirrha* could be categorised as (1) medium-sized cytolytic actinoporins (~20 kDa), (2) cardiostimulatory proteins (~28 kDa) and (3) cytolysin with or without phospholipase in both crude and fractionated proteins (~40 kDa).

The FTIR spectra of *C. quinquecirrha* venom showed increase in amide I, amide II and amide III regions. Similarly, Wang et al. [25] have used pressure tuning FTIR spectra to find the structural differences between the connective epithelial and malignant epithelial cervical tissues. Wang et al. [26] have investigated normal and malignant tissues by pressure turning of the FTIR spectra in terms of structural changes at the molecular level. From the present study, it can be concluded that the venom *C. quinquecirrha* may have many biologically active principles, which need further study in future. Thus, the importance of biologically active compounds present in the venom of jellyfish has become evident and can be useful tools for probing biological or pharmacological activity.

References

1. Yasumoto T. The chemistry and biological function of natural marine toxins. Chem Rec. 2001;1:228–42.
2. Diaz J, Morera V, Delfin J, Huerta V, Lima G, Rodriguez de la Veega M, Garcia B, Padron G, Assfalgmachleidt L, Chavez M. Purification and partial characterization of a novel proteinase inhibitor from the sea anemone *Stichodactyla helianthus*. Toxicon. 1998;36:1275–6.
3. Li RA, Tomaselli GF. Using the deadly μ-conotoxins as probes of voltage-gated sodium channels. Toxicon. 2004;44(2):117–22.
4. Halstead BW. In: Halstead BW, editor. Poisonous and venomous marine animals of the world, vol. 1. Washington, DC: U.S. Government Printing Office; 1965. p. 297–536.
5. Yanagihara A, Kuroiwa J, Oliver L, Chung J, Kunkel DD. Ultra structure of a novel eurytele nematocyst of *Carybdea alata* (Cubozoa: Cnidaria). Cell Tissue Res. 2002;308:307–18.
6. Yasumoto T. The chemistry and biological function of natural marine toxins. Chem Rec. 2001;1:228–42.
7. Pani Prasad V, Venkateshwaran K. Microhaemolytic assay, international training manual on advance techniques in marine biotoxinology. Mumbai: CIFE; 1997. 41 pp.
8. Li CP, Yu HH, Liu S, Xing RE, Guo ZY, Li PC. Factors affecting the protease activity of venom from jellyfish *Rhopilema esculentum* Kishinouye. Bioorg Med Chem Lett. 2005;15(24):5370–4.
9. Murugan A, Santhana Ramasamy M. Biofouling deterrent natural product from the ascidian *Distaplia nathensis*. Indian J Mar Sci. 2003;32:162–4.
10. National Committee for Clinical Laboratory Standards. Reference method for broth dilution antifungal susceptibility testing of Yeast. NCCLS document M 27-A2, National Committee for Clinical Laboratory Standards, Wayne; 2002. pp. 45–65.
11. Ponkshe CA, Indap MM in vivo and in vitro evaluation for immunomodulatory activity of three marine animal extracts with reference to phagocytosis. Indian J Exp Biol. 2002;40:1399–402.
12. Laemmli UK. Cleavage of structural proteins during the assembly of the head of bacteriophage T4. Nature. 1970;227:680–5.
13. Sambrook J, Russell DW. Molecular cloning-Laboratory manuals 3. Cold Spring Harbor/New York: Cold Spring Harbor laboratory Press; 2001. A. 8.49.
14. Hutton DCM, Smith VJ. Antibacterial defenses in anthozoans with special reference to the sea anemone *Actinia equina*. In: Stolen JS, Fletcher TC, Bayne CJ, Secombes CJ, Zelikoff JT, Werdok LT, Anderson DP, editors. Modulators of immune responses, the evolutionary trail. Fair Haven: SOS Publications; 1996. 15–30 pp.
15. Thakur NL, Hentschel U, Krasko A, Pabel CT, Anil AC, Muller WEG. Antibacterial activity of the sponge *Suberites domuncula* and its primmorphs: potential basis for epibacterial chemical defense. Aquat Microb Ecol. 2003;31(1):77–83.
16. Sakai M. Current research status of fish immunostimulants. Aquaculture. 1999;172:63–92.
17. Hudson L, Hay F. Practical immunology. 3rd ed. England: Blackwell; 1991. 138 pp.
18. Al-Hassan JM, Thomson M, Summers B, Criddle RS. Purification and properties of a hemagglutination factor from the Arabian Gulf catfish *Arius thalassinus* epidermal secretion. Comp Biochem Physiol. 1986; 85:31–9.
19. Purushottama GB, Venkateshvaran K, Pani Prasad K, Nalini P. Bioactivities of extracts from the marine sponge *Halichondria panicea*. J Venom Anim Toxins Incl Trop Dis. 2009;15(3):444–5.
20. Macek P. Polypeptide cytolytic toxins from sea anemones (*Actiniaria*). FEMS Microbiol Immunol. 1992; 105:121–30.
21. Jiang Y, Lee A, Chen J, Ruta V, Cardene M, Chalt BT, Mackinnon R. X-ray structure of the voltage-dependent K^+ channel. Nature. 2003;423:33–41.
22. Robertsen B, Engstad RE, Jorensen JB. Beta glucan as an immunostimulant in fish. In: Stolen JS, Fletcher TC, editors. Modulators of fish immune responses:

models for environmental toxicology, biomarkers and immunostimulators. Fair Haven: NJ, SOS Publications; 1994. p. 83–99.

23. Rodriguez SJ, Cruz-Vazquez K. Isolation and biological characterization of neurotoxic compounds from the sea anemone *Lebrunia danae*. Arch Toxicol. 2006;80(7):436–41.
24. Radwan FFY, Gershwin L, Burnett JW. Toxinological studies on the nematocyst venom of *Chrysaora achlyos*. Toxicon. 2000;38:1581–91.
25. Wang S, Huang MX, Sun Z. Artificial breeding of edible medusa. J Fish China. 1991;15(4):35–40.
26. Wang YS, Huang MX, Zhang QS. Release of tagged jellyfish. J Zhejiang Coll Fish. 1993;13(3):201–4.

Diplazium esculentum: A Wild Nutrient-Rich Leafy Vegetable from Western Ghats

35

G. Nair Archana, S. Pradeesh, M. Devi Chinmayee, I. Mini, and T.S. Swapna

Abstract

Diplazium esculentum (Reytz.) belongs to the family Dryopteridaceae and is an edible fern commonly used by Paniya and Chetti tribes of Western Ghats. Young fronds are stir-fried as a vegetable or used in salads. Wild leafy vegetables remain underutilized due to lack of awareness and promotion of appropriate technologies for their effective utilization. Evaluation of nutritional and anti-nutritional factors of this plant along with multiplication will help to identify a potential source to solve nutrition deficiency problem and also help to maintain biodiversity. Different biochemical parameters like crude fibre, reducing sugar, total carbohydrate, free fatty acids, and proteins were quantitatively estimated using standard estimation methods. Various micronutrients and anti-nutritional factors like phytic acid, trypsins and tannins were also tested. Results showed that this wild leafy vegetable has important nutritional factors. It is rich in micronutrients, beta-carotene, folic acid and minerals such as Ca, Fe and P. Anti-nutritional factors like phytic acids, trypsin and tannins were present but in quite safe quantities. So *Diplazium esculentum* with rich nutrient factors opens the possibility of the potential utilization of this plant in food or in nutritional/food supplement programmes.

Keywords

Diplazium esculentum • Nutritional factors • Antinutritional factors • Leafy vegetables

Introduction

Leafy vegetables are also referred to as green vegetables. They are plant leaves consumed as vegetables and sometimes accompanied by the tender petioles and shoots [1]. There is considerable information in the literature on the nutritive and anti-nutritive composition of most cultivated

G.N. Archana • S. Pradeesh • M. Devi Chinmayee • I. Mini • T.S. Swapna (✉)
Department of Botany, University College, Thiruvananthapuram-34, Kerala, India
e-mail: swapnats@yahoo.com

A. Sabu and A. Augustine (eds.), *Prospects in Bioscience: Addressing the Issues*,
DOI 10.1007/978-81-322-0810-5_35, © Springer India 2013

leaves [2–4]; however, little information is available on the proximate composition of wild leafy vegetables. The few documented information revealed that these vegetables contain appreciable amounts of bioactive compounds and thus may serve as important sources of vitamins, minerals and certain hormone precursors in addition to protein and energy [5]. In times of scarcity when the staple food is in short of supply, tribal people collect many types of wild edible ferns for their meagre food available at home [6].

Diplazium esculentum is found in the rural areas and is amongst the leafy vegetables consumed by Paniya and Chetti tribes of Western Ghats. The plant is a fern that grows wild on the bank of streams and rivers. The edible parts are the young fronds and crosiers that are used as ingredients for preparations of soup and salads. Various parts of the leaves of *Diplazium esculentum* are consumed by various people throughout the world [7]. Despite the great nutritive value and physiological importance of *Diplazium esculentum* leaves, very little information has been documented. Therefore, in the present investigation, an attempt has been made to understand the chemical composition and anti-nutritional factors of this wild edible plant.

Materials and Methods

Collection and Preparation of Samples

Fresh leaves of *Diplazium esculentum* were collected from Western Ghats. The leaves were shade-dried and were ground well using mechanical blender into fine powder and transferred into airtight container for future studies. The fine powder is then subjected to extraction in a Soxhlet apparatus using different solvents like methanol, acetone, petroleum ether and chloroform. Standard methods were used for the estimation of total protein, total carbohydrates, reducing sugar, total phenols, pigments, vitamins, amino acids etc., using fresh leaves. Dried powder was used for the analysis of phytic acid, free fatty acids, tannic acid and different secondary metabolites. A Pye Unicam atomic absorption spectrophotometer with acetylene flame was used to analyse calcium, iron, magnesium, manganese, copper, zinc and aluminium [8].

Nutritional analysis was carried out in young fronds, mature fronds, young petiole and mature petiole and experiments were done in triplicate. Total moisture, free fatty acid, crude fibre, lipids, reducing sugar, total carbohydrate and proteins were determined by standard procedure [9]. Moisture content was estimated following the method of IUPAC in which fresh plant sample was weighed and placed in a preweighed Petri dish, closed with a lid and kept in an oven at 110°C for about 3 h, cooled and then weighed. Moisture content was calculated from the difference in weight and expressed in percentage. For the estimation of reducing sugar, fresh samples were extracted with distilled water. Total reducing sugar present in the sample was estimated by dinitrosalicylic acid method by Miller [10] and the absorbency was read at 540 nm against a blank. The amount of total carbohydrate present in the sample was estimated by Anthrone method [11]. Total protein was estimated by Lowry's method [12].

For chlorophyll estimation, fresh tissue was homogenized in 80% acetone and read the absorbance at 645, 663, 652 nm and the chlorophyll present was calculated by Arnon's formula. The amount of amino acids present in the samples was estimated as per the method of Moore and Stein [13]. Folin-Ciocalteu method was used for total phenol estimation [14]; trypsin inhibitor was determined as per Kakade et al. [15]. Phytic acid was extracted using 37% trichloroacetic acid and the absorbance was read at 480 nm for estimation of phytic acid [16]. Tannic acid was estimated as per the procedure of Schanderl [17].

Phytochemical analysis of extract was done as described by Harbone [18]. The phytochemicals like reducing sugar, glycosides, flavonoids, alkaloids, tannins, steroids, terpenoids, coumarins, saponins, anthraquinones, phlobatannins and iridoids were tested.

Results and Discussion

This study was carried out to justify the potentials of the leaves of *Diplazium esculentum* for its nutritional and probable medicinal relevance.

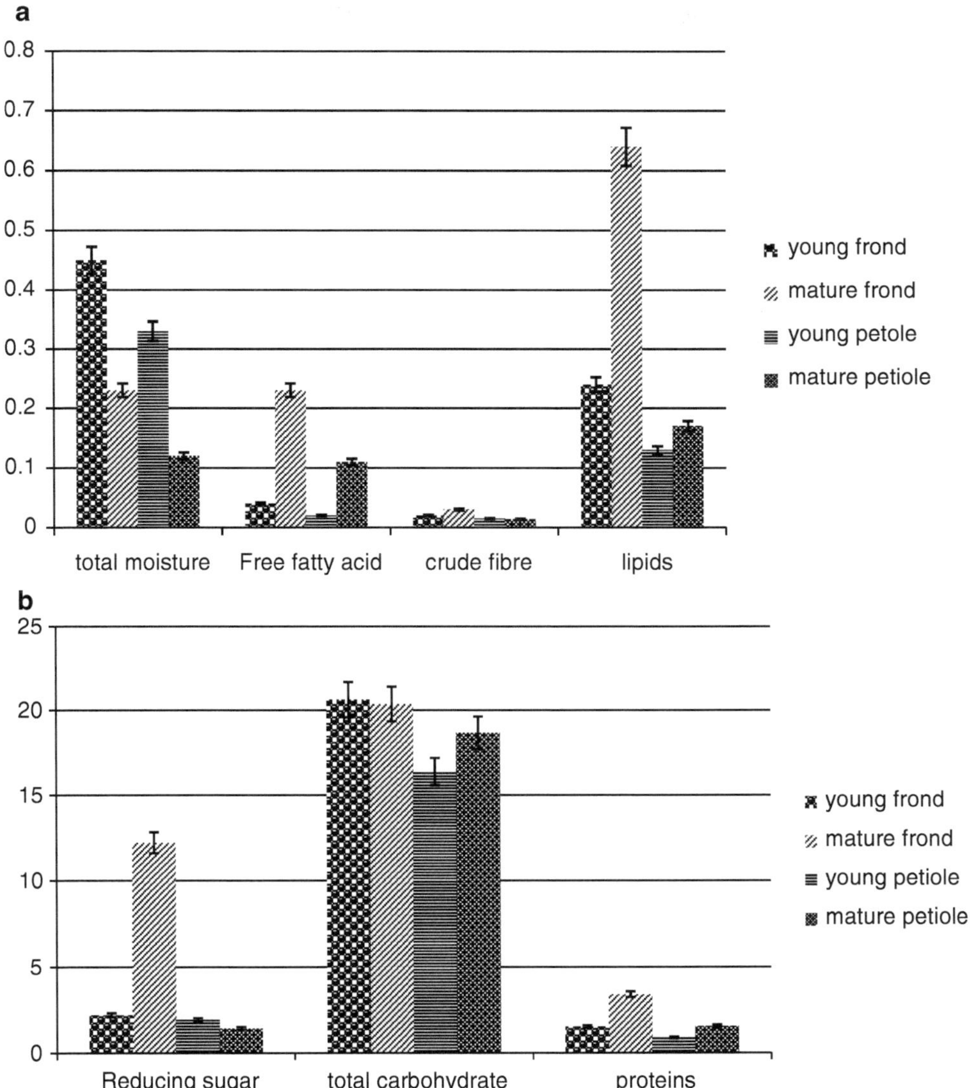

Fig. 35.1 (**a**) Nutritional factors in *Diplazium esculentum* (in mg g^{-1}). (**b**) Nutritional factors in *Diplazium esculentum* (in mg g^{-1}). (**c**) Nutritional factors in *Diplazium esculentum* (in mg g^{-1}). (**d**) Nutritional factors in *Diplazium esculentum* (in mg g^{-1})

The moisture content of food is used as a measure of stability and susceptibility to microbial contamination [19]. The percentage of moisture was high in young frond (0.45%) compared to other parts as shown in (Fig. 35.1a). The result showed high free fatty acid and lipid content in the mature frond of the plant compared to other parts. In *Diplazium esculentum,* free fatty acid content was higher in mature frond (0.23 mg g^{-1}) (Fig. 35.1a). The crude fibre content is commonly used as a measure of the nutritive value of poultry and livestock feeds and also in the analysis of various foods and food products to detect adulteration, quality and quantity [20]. Analysis showed that higher amount of crude fiber was present in mature frond (0.03%) compared to other plant parts (Fig. 35.1a).

Reducing sugar from mature leaf, young leaf, stem and root were extracted in distilled water. Results showed high sugar content (12.24 mg g^{-1}) in the mature frond of the plant (Fig. 35.1b) compared to other parts. Carbohydrates are the

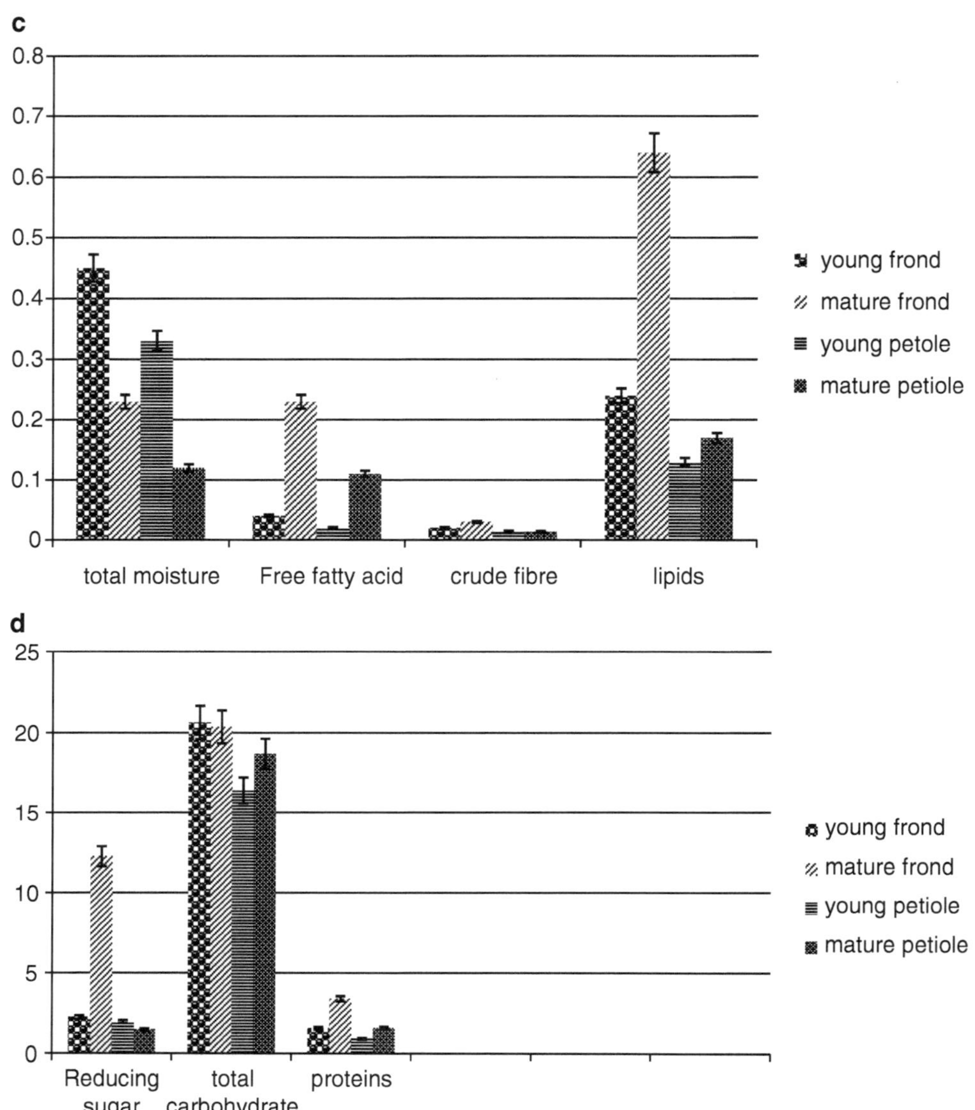

Fig. 35.1 (continued)

important components of storage and structural materials in plants. They exist as free sugars and polysaccharides. Total carbohydrates present in young petiole were less compared to the other parts (Fig. 35.1b). Protein is essential in daily diet and daily protein requirement for children and adults are 23–36 g and 44–56 g, respectively [21]. The analysis of plant parts in *Diplazium* showed least amount of protein in mature frond (3.4 mg g^{-1}) compared to young frond (1.54 mg g^{-1}), young petiole (0.9 mg g^{-1}) and mature petiole (1.56 mg g^{-1}) (Fig. 35.1b). This shows the photosynthetic efficiency of young leaves.

Amount of chlorophyll in a green plant or any crop plant is an index of photosynthetic ability, which in turn is an index of its productivity level, primarily of primary metabolites [22]. The amount of chlorophyll was measured and compared in

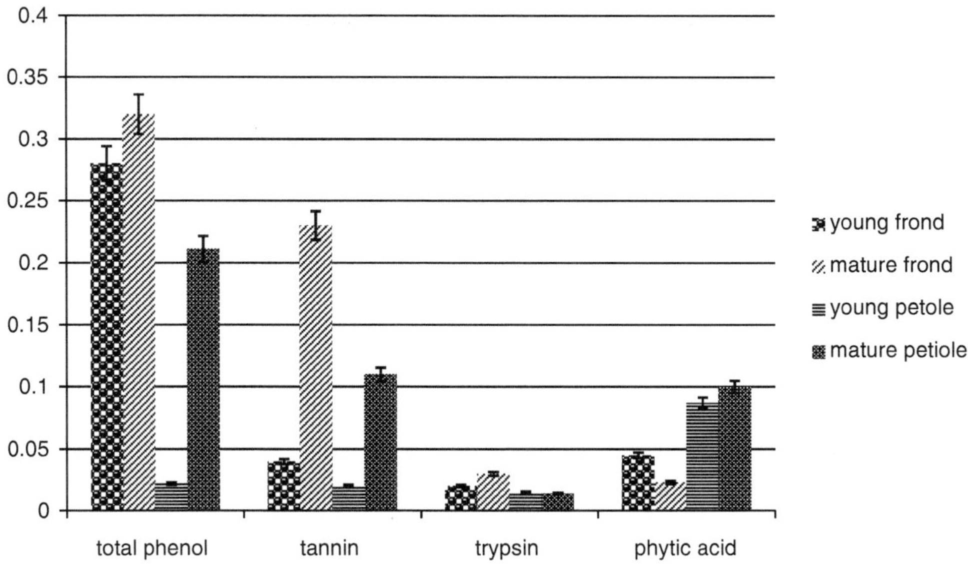

Fig. 35.2 Anti-nutritional factors in *Diplazium esculentum* (in mg g^{-1})

mature frond, young frond, mature petiole and young petiole of *Diplazium esculentum*, and it was found that contents of chlorophyll a, chlorophyll b, total chlorophyll and carotenoids present in mature frond were higher than those in young frond (Fig. 35.1c).

Vitamin A is an essential nutrient required for maintaining immune function, playing an important role in the regulation of cell-mediated immunity and in humoral antibody responses [23]. In *Diplazium esculentum*, β-carotene content was higher in young petiole (3.8 mg g^{-1}) (Fig. 35.1d). Ascorbic acid is an important dietary antioxidant; it significantly decreases the adverse effect of reactive species such as reactive oxygen and nitrogen species that can cause oxidative damage to macromolecules such as lipids, DNA and proteins which are implicated in chronic diseases including cardiovascular disease, stroke and cancer [24]. Mature frond (3.4 mg g^{-1}) and young petiole (3.3 mg g^{-1}) of *Diplazium esculentum* contain higher amount of ascorbic acid than young frond (1.2 mg g^{-1}) and mature petiole (2.1 mg g^{-1}). Analysis showed that high amount of tocopherol was present in mature frond (4.2 mg g^{-1}) compared to other plant parts (Fig. 35.1d). Thiamine, riboflavin and niacin were present in sufficient amount in plant parts (Fig. 35.1d).

Phenolic compounds are secondary metabolites which possess biological properties such as antioxidant, anti-apoptosis, anti-ageing, anticarcinogen, anti-inflammation, anti-atherosclerosis and cardiovascular protection activity. Most of these biological actions have been attributed to their intrinsic reducing capabilities [25]. Analysis of *Diplazium esculentum* showed high phenol content in the mature frond (0.32 mg g^{-1}) of the plant (Fig. 35.2) compared to other parts. Trypsin and tannin are in higher concentrations in mature frond than in other parts (Fig. 35.2). The knowledge of phytate levels in food is necessary because high concentration can cause adverse effect on digestibility [26]. Phytic acid also influence oxidative stress by altering cell signalling pathways or influencing the activity and expression of antioxidant enzymes and also may be beneficial in preventing many types of cancers [27]. Petioles (0.12 mg g^{-1}) contain more phytic acid compared to fronds (0.03 mg g^{-1}) (Fig. 35.2). Trypsin, phytate and tannin in foods are anti-nutritional factors which affect the complete absorption of many minerals in the body [3]. Amino acids like

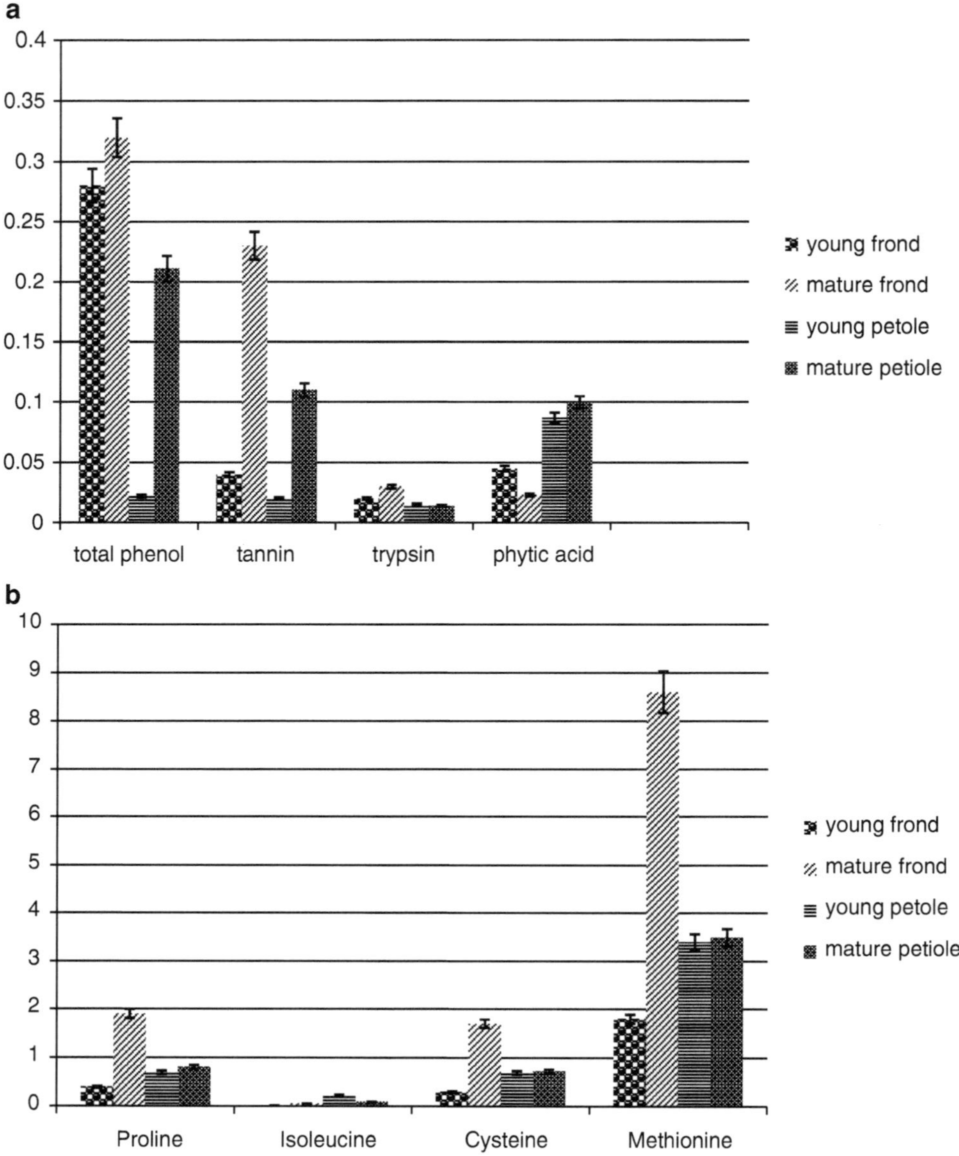

Fig. 35.3 (**a**) Amino acid content in *Diplazium esculentum* (in mg g^{-1}). (**b**) Amino acid content in *Diplazium esculentum* (in mg g^{-1})

ascorbic acid, tyrosine and methionine are present in higher quantities in mature frond compared to other parts (Fig. 35.3a and b).

The qualitative phytochemical analysis of young frond, mature frond, young petiole and mature petiole of *Diplazium esculentum* showed the presence of saponins, alkaloids, anthraquinones, flavonoids, tannins, phenols and glycosides at varied degrees (Table 35.1). Amount of flavanoids, glycosides and steroids present in different parts of the plant were higher than other phytochemicals. Results of mineral composition (Table 35.2) clearly indicate that *Diplazium esculentum* leaves constitute a rich source of mineral elements although the bioavailability of these elements has to be established. The elemental composition showed increased levels in iron and magnesium while zinc, manganese, copper and aluminium were relatively low.

Table 35.1 Phytochemical evaluation of young frond, mature frond, young petiole and mature petiole of *Diplazium esculentum*

	Young frond					Mature frond					Young petiole					Mature petiole				
Phytochemicals	M	A	P.E.	C	D.W.	M	A	P.E.	C	D.W.	M	A	P.E.	C	D.W.	M	A	P.E.	C	D.W
Reducing sugar	+	–	–	+	+	+	–	–	+	+	+	+	–	+	+	+	+	–	+	+
Glycosides	+	+	+	+	+	+	+	–	+	+	+	+	–	+	+	+	+	+	+	+
Flavonoids	+	+	+	+	+	+	+	+	+	+	+	+	+	+	+	+	+	+	+	+
Alkaloids	+	+	–	+	+	+	+	–	+	+	+	+	–	+	+	–	–	–	–	+
Tannins	–	+	–	+	–	+	+	–	+	+	–	+	–	+	–	+	–	–	–	+
Steroids	+	+	+	+	+	+	+	+	+	+	+	+	+	+	–	–	–	–	+	–
Terpenoids	+	–	+	–	+	+	–	–	–	+	+	+	–	–	–	+	+	–	+	–
Coumarins	–	–	+	+	+	+	–	–	–	+	+	–	–	+	+	+	+	+	+	+
Saponins	–	–	–	+	+	+	–	–	–	+	+	–	–	+	+	–	–	–	–	+
Anthraquinones	–	–	–	–	–	–	–	–	–	–	–	–	–	–	–	–	–	–	–	–
Phlobatannins	–	–	–	–	–	–	–	–	–	–	–	–	–	–	–	–	–	–	–	–
Iridoids	–	–	–	–	–	–	–	–	–	–	–	–	–	–	–	–	–	–	–	–

M methanol, *A* acetone, *P.E.* petroleum ether, *C* chloroform, *D.W.* distilled water

Table 35.2 Mineral element composition of young frond, mature frond, young petiole and mature petiole of *Diplazium esculentum*

Samples	Iron (mg g^{-1})	Manganese (mg g^{-1})	Magnesium (mg g^{-1})	Copper (mg g^{-1})	Zinc (mg g^{-1})	Aluminium (mg g^{-1})
Young frond	23.455	0.11	12.11	1.284	1.33	0.716
Mature frond	20.520	0.157	10.006	1.041	0.189	0.734
Young petiole	20.169	0.043	10.010	1.031	0.178	–
Mature petiole	21.321	0.384	10.016	1.221	0.365	0.1

Conclusions

Diplazium esculentum leaves contain high levels of protein, carbohydrate and reducing sugar; however, other proximate composition parameters such as crude fibre, lipids and fats were low. Phytic acid and trypsin content were very low although the total phenol and tannins were moderately high but not considered to be lethal to man. Phytochemical screening showed the presence of saponins, alkaloids, anthraquinones, flavonoids, tannins, phenols and glycosides at varied degrees. So studies on *Diplazium esculentum* which is rich in bioactive compounds and its nutritional value are of considerable significance since it may help to identify long-forgotten food resources.

Acknowledgements Authors are thankful to the Western Ghat Development Cell, Government of Kerala, for the financial support. Dr T. S. Swapna thank Dr N. Anilkumar, Community Agro Biodiversity Centre, MSSRF, Wayanad, for fruitful suggestions and technical support.

References

1. Ujowundu CO, Igwe CU, Eremor VA, Nwaogwu LA, Okafor OE. Nutritive and anti-nutritive properties of *Boerhavia diffusa* and *Commelina nucliflora* leaves. Pak J Nutr. 2008;7(1):90–2.
2. Oshodi AA. Comparison of protein, minerals and vitamin C content of some dried leafy vegetable. Pak J Sci Ind Res. 1992;35:267–9.
3. Ogontona T. Green leafy vegetable. In: Osagie UA, Eka OU, editors. Nutritional quality of plant food, post harvest research unit. Benin city: Department of Biochemistry, University of Benin; 1998.
4. Price KR, Johnson LI, Feriwick H. The chemical and Biological significance of saponins in food and feeding stuffs. CRC Crit Rev Food Sci Nutr. 1987;26:127–35.
5. Cho E, Seddom J, Ronser B, Willet W, Hunkison S. Prospective study of intake of fruits, vegetables, vitamins and carotenoids and risk of age-related maculopathy. Arch Opthalmol. 2004;122:883–92.
6. Vidyarthi LP. Role of forest in tribal life. In: Sinha SP, editor. Tribals and forest. Ranchi: Bihar Tribal Welfare Research Institute; 1987. p. 323.
7. Haynes JD. Botany. An introductory survey of the plant kingdom. New York: Wiley; 1975.
8. AOAC. Official methods of analysis. 11th ed. Washington, DC: Association of Analytical Chemist; 1975.
9. AOAC. Official methods of analysis. 15th ed. Arlington: Association of Analytical Chemist; 1990.
10. Miller GL. Use of dinitrosalicylic acid reagent for determination of reducing sugar. Anal Chem. 1959;31: 426–8.
11. Hedge JE, Hofreiter BT. In: Whistler RL, Miller JN, editors. Carbohydrate chemistry, 17. New York: Academic; 1962.
12. Lowry OH, Rosebrough NJ, Farr AL, Randall RJ. Protein measurement with the Folin Phenol reagents. J Biol Chem. 1951;193:265–75.
13. Moore S, Stein WH. In: Colowick SP, Kaplan ND, editors. Methods in enzymology. New York: Academic; 1948. 3.468.
14. Malik EP, Singh MP. Plant enzymology and histoenzymology. New Delhi: Kalyani Publishers; 1980. 286.
15. Kakade ML, Rackie JJ, McGhee JE, Puski G. Determination of trypsin inhibitor activity of soy products: a collaborative analysis of an improved procedure. Cereal Chem. 1974;51:376.
16. Wheeler EL, Ferrel RE. A method of phytic acid determination in Wheat and Wheat fractions. Cereal Chem. 1971;48:312–6.
17. Schanderl SH. Method in food analysis. New York: Academic; 1970. p. 709.
18. Harbone JB. Phenolic Glycosides and their natural distribution in the biochemistry of Phenolic compounds. New York/London: Academic; 1977. p. 152–62.
19. Uraih N, Izuagbe Y. Public health food and industrial microbiology. Benin: Uniben Press; 1990.
20. Maynard AJ. Method in food analysis. New York: Academic; 1970. p. 176.
21. Cheney SG. Principle of nutrition: macronutrient. In: Delvin TM, editor. Textbook of biochemistry with clinical correlation. New York: Wiley; 2006.
22. Witham FH, Blaydes DF, Devlin RM. Experiments in plant physiology. New York: Van Nostrand; 1971. p. 245.
23. Shankar A. Nutritional modulation of immunity and infection. In: Bowman BARR, editor. Present

knowledge in nutrition. 8th ed. Washington, DC: International Life Sciences Institute Press; 2001.

24. Halliwell B, Gutteridge JMC. JMC: free radicals in biology and medicine. Oxford: Oxford University Press; 1999.
25. Han X, Shen T, Lou H. Dietary polyphenols and their biological significance. Int J Mol Sci. 2007;8: 950–88.
26. Nwokolo EN, Boaggs BB. Influence of phytic acid and crude fibre on the availability of minerals from four protein supplements in growing chicks. J Anim Sci. 1977;57:475–7.
27. Vucenik I, Tomazic VJ, Fabian D, Shamsuddin AM. Antitumor activity of phytic acid (inositol hexaphosphate) in murine transplanted and metastatic fibrosarcoma, a pilot study. Cancer Lett. 1992;65(1):9–13.

36 *Agrobacterium*-Mediated Genetic Transformation and Somatic Embryogenesis from Leaf Callus of *Hevea brasiliensis*: Effect of Silver Nitrate

R.G. Kala, V. Abraham, S. Sobha, P.K. Jayasree, A.M. Suni, and A. Thulaseedharan

Abstract

Somatic embryogenesis and plant regeneration systems developed in *Hevea brasiliensis* are being used in *Hevea* crop improvement through transgenic approaches. In the present study, the effect of silver nitrate in improving the efficiency of the somatic embryogenesis pathway from leaf explants as well as *Agrobacterium*-mediated genetic transformation was studied. Experiments were done to find the effect of silver nitrate in callus proliferation and embryogenic callus initiation. Silver nitrate (0–30.0 mg/l) was added in the medium used for second and third subcultures for callus proliferation and in the embryo induction medium. Experiments were also carried out to find the effect of silver nitrate in controlling bacterial overgrowth and improve callus texture when freshly proliferated callus induced from leaf explants in the above medium was used as target tissue for *Agrobacterium* infection. *Agrobacterium tumefaciens* strain EHA 101 carrying a gene-encoding isopentenyltransferase (*ipt*) from the Ti plasmid of *Agrobacterium tumefaciens* was used for the study. Silver nitrate (0–30 mg/l) was added to the *Agrobacterium* infection as well as the cocultivation and selection medium. Addition of 20.0 mg/l silver nitrate in the callus proliferation medium used for the second and third subcultures and 10 mg/l in the embryo induction medium helped in the texture improvement of the callus which in turn increased the efficiency of the system by aiding embryogenic callus initiation. It was also observed that inclusion of 10.0 mg/l silver nitrate in the infection and cocultivation medium and 20 mg/l in the selection medium significantly controlled bacterial overgrowth, reduced tissue damage and improved the texture of callus in newly emerged putatively transgenic cell lines. Proliferation of the transgenic callus and somatic embryogenesis were also achieved. From the results, it has become evident

R.G. Kala • V. Abraham • S. Sobha • P.K. Jayasree • A.M. Suni • A. Thulaseedharan (✉)
Rubber Research Institute of India,
Kottayam, 686 009, Kerala, India
e-mail: thulaseedharan@rubberboard.org.in

A. Sabu and A. Augustine (eds.), *Prospects in Bioscience: Addressing the Issues*,
DOI 10.1007/978-81-322-0810-5_36, © Springer India 2013

that silver nitrate could improve somatic embryogenesis as well as *Agrobacterium*-mediated genetic transformation when included in the respective media used in the plant regeneration system developed from leaf explants.

Keywords
Agrobacterium • Genetic transformation • *Hevea brasiliensis* • Somatic embryogenesis

Abbreviations

BA	6-benzyl aminopurine
2,4-D	2,4-dihydrophenoxyacetic acid
GA_3	gibberellic acid
Kin	kinetin
MS	Murashige Skoog
NAA	naphthalene acetic acid

Introduction

Hevea brasiliensis, a cross-pollinated perennial tree species belonging to the family Euphorbiaceae, is the only commercially cultivated species as a source of natural rubber. In *H. brasiliensis,* plant regeneration through somatic embryogenesis has been obtained from immature anther, inner integument and immature inflorescence [1–5]. All the protocols developed earlier for somatic embryogenesis in *H. brasiliensis* were from floral-/fruit-derived explants. Since flowering in *H. brasiliensis* is seasonal and adverse environments during the season may hinder normal flowering, explant availability and viability is unpredictable. Recently, a system for somatic embryogenesis and plant regeneration was also developed from leaf explants, an easily available explant [6, 7]. Somatic embryogenesis is affected by numerous factors, notably ethylene. Ethylene is known to reduce somatic embryogenic competence in many plants. Silver nitrate ($AgNO_3$) has been reported to have influence in inducing embryogenic potential to fresh callus in several crops [8]. The use of silver nitrate, an ethylene action inhibitor, has been shown to increase *in vitro* embryogenesis and regeneration rates [9]. Presence of silver nitrate in culture medium has shown to have important effects in tissue culture of cassava, such as improving somatic embryogenesis and plant regeneration [10]. However, the influence of ethylene on embryogenic response is genotype specific [11].

The success in generation of transgenic plants depends on two major factors, transgene integration and regeneration capacity of explants. Due to the simplicity of the transformation system and precise integration of transgenes, *Agrobacterium* Ti plasmid-based vectors continue to offer the best system for plant transformation [12]. The major requirements for the development of transgenic plants include the availability of reliable and reproducible *in vitro* regeneration systems and efficient transformation protocols [13]. The main challenge with genetic transformation of tree species is achievement of high transformation efficiency and efficient plant regeneration for desired clones or cultivars [14]. It can be hypothesised that successful transgenesis occurs when regeneration capacity of a cell coincides with its ability to accept a transgene. *H. brasiliensis* being a tree crop, conventional agriculture relying on selection and traditional breeding programmes needs more time to deploy new varieties with modified traits into the field.

All the plant regeneration systems developed in *Hevea* were used for crop improvement through *Agrobacterium*-mediated genetic transformation. Transgenic plants incorporated with MnSOD gene were regenerated from anther calli of *H. brasiliensis* [15, 16] through transformation technology. Leaf-derived embryogenic callus has also given good response as a suitable target tissue for *Agrobacterium*-mediated infection [17]. Silver nitrate has been found to suppress overgrowth of *Agrobacterium* facilitating plant cell recovery that resulted in increased transformation efficiency [18]. In apple, it has been reported that during *Agrobacterium*-mediated transformation, ethylene production was increased which resulted in reduced efficiency of gene transfer mechanism [19]. Ozden et al. [20] have reported that involvement of ethylene production leads to tissue browning which can be reduced in the presence of silver nitrate. Opabode [21] reported that silver nitrate is an anti-necrotic compound which can reduce oxidative burst during the interaction between plant tissue and *Agrobacterium*. The objective of this study was to evaluate the effects of $AgNO_3$ on somatic embryogenesis from leaf explants and *Agrobacterium*-mediated genetic transformation of *H. brasiliensis*.

Materials and Methods

Somatic Embryogenesis from Leaf Explants

A system for somatic embryogenesis and plant regeneration has been developed from leaf explants collected from glass house-grown bud-grafted plants of *H. brasiliensis* (clone RRII 105) [6, 7]. The optimised medium for callus induction, embryo induction, embryo maturation and plant regeneration was used for the experiments (Table 36.1). Leaf explants collected from 6-month-old glass house-grown bud-grafted plants of *H. brasiliensis* were cultured for callus induction. Callus was proliferated by repeated subculture in medium where cytokinin/auxin ratio and sucrose concentration were increased (Table 36.1). The cultures were incubated at 25 ± 2 °C and kept in dark.

Effect of Silver Nitrate on Callus Friability and Embryogenic Callus Initiation

The effect of silver nitrate in improving callus friability and making the calli more embryogenic was examined. Silver nitrate stock solution (1.0 mg/ml) was prepared in water and the required quantity was filter sterilised and added to the autoclaved medium and was stored in the dark. Three different concentrations of silver nitrate (10, 20 and 30 mg/l) were added to the optimised proliferation medium during the second subculture. Embryogenic callus formation was tried in medium optimised earlier. In experiments on callus proliferation, silver nitrate has been found to have influence in changing the callus texture. Hence, different concentrations of silver nitrate (5, 10, 20 mg/l) were filter sterilised and added to the autoclaved medium, optimised for embryogenic callus formation.

Agrobacterium-Mediated Genetic Transformation

Bacterial Strain

Agrobacterium tumefaciens strain EHA 101 harbouring a binary plasmid vector carrying a gene-encoding isopentenyltransferase (*ipt*) from the Ti plasmid of *A. tumefaciens* was used. The binary vector also contains β-glucuronidase (GUS) gene as the reporter gene and *npt*II gene for kanamycin resistance as plant selection gene.

Agrobacterium Infection and Cocultivation: Influence of Silver Nitrate

50 μl of *Agrobacterium* suspension from glycerol stock was spread over semisolid AELB (Luria-Bertani) medium containing 50 mg/l kanamycin and 20 mg/l gentamicin and incubated at 28 °C for 2 days. *Agrobacterium* single colony was selected with sterile loop and suspended in 10 ml sterile liquid AELB medium containing the antibiotics 50 mg/l kanamycin and 20 mg/l gentamicin. The cultures were

Table 36.1 Media used for somatic embryogenesis and plant regeneration

Medium component	Callus induction	Callus proliferation	Embryogenic callus initiation	Embryo induction	Embryo maturation	Plant regeneration
Basal salts	MS + $Ca(NO_3)_2$, $4H_2O$ –1,200 mg/l	MS + $Ca(NO_3)_2$, $4H_2O$ –800 mg/l	MS + $Ca(NO_3)_2$, $4H_2O$ –800 mg/l	MS + $Ca(NO_3)_2$, $4H_2O$ –250 mg/l	MS modified ½x	MS modified ½x
Vitamins	B_5	B_5	B_5	B_5	B_5	MS
Sucrose	20 g/l	40 g/l	60 g/l	60 g/l	60 g/l	30 g/l
Organic supplements	Casein hydrolysate (1 g/l), coconut water (5%)	Casein hydrolysate (1 g/l), coconut water (5%)	Casein hydrolysate (300 mg/l), coconut water (5%)	Casein hydrolysate (300 mg/l), coconut water (5%)	Casein hydrolysate (400 mg/l), coconut water (10%), malt extract (150 mg/l)	–
L-cysteine HCl	50 mg/l	50 mg/l	50 mg/l	50 mg/l	–	–
Proline	–	–	200 mg/l	200 mg/l	–	–
Glutamine	–	–	300 mg/l	300 mg/l	–	–
Arginine	–	40 mg/l	40 mg/l	40 mg/l	40 mg/l	–
Hormones	2,4-D (5.4 μM), BA (4.4 μM) and NAA (1.08 μM)	2,4-D (1.8 μM), BA (2.2 μM) and NAA (1.08 μM)	BA (2.2 μM), GA_3 (2.9 μM), Kin (1.25 μM) and NAA (1.08 μM)	BA (2.2 μM), GA_3 (2.9 μM), Kin (1.25 μM), ABA (0.75 μM), and NAA (0.54 μM)	BA (4.4 μM), Kin (1.3 μM), GA_3 (4.4 μM) and IBA (0.49 μM)	
pH	5.6	5.6	5.7	5.7	5.6	5.6

incubated at 28 °C in a shaker with 200 rpm for 24 h to reach an optical density of 0.5 at 420 nm. Volume of bacterial suspension required for 10 ml infection medium was calculated, and bacteria were pelleted by centrifugation at 5,000 rpm for 10 min. The bacterial pellet was resuspended in the infection medium so as to get a bacterial density of 5×10^8 cells/ml. Murashige and Skoog (MS) [22] medium containing 100-μm acetosyringone and 1.0 mM each of proline and betaine hydrochloride (Table 36.2) was used as the infection medium. pH of the medium was adjusted to 5.2 with 1 N KOH and filter sterilised. Infection of the bacterial culture with the target tissue was done after 4 h incubation in an incubator shaker at 28 °C.

100 mg of proliferated friable *H. brasiliensis* calli was transferred to a small sterile petri dish and air-dried for 10 min in the laminar flow hood. Bacterial solution (5 ml) was poured over the callus, and infection was done for 15 min. The callus in the bacterial solution was injured with a sterile scalpel blade to facilitate entry of the bacteria. After infection, the bacterial solution was drained out and the infected tissues were blotted dry by transferring them to sterile filter papers. For cocultivation, the infected tissues were spread over filter paper kept on the surface of a semisolid cocultivation medium. The infected tissues were incubated for 72 h in dark at 28 °C [15–17]. The tissues were transferred to the preselection medium containing the antibiotic carbenicillin to control bacterial overgrowth. After 2 weeks, the tissues were subcultured to selection medium containing both selection antibiotics carbenicillin (400 mg/l) and kanamycin (300 mg/l) (Table 36.2). Transformation efficiency was scored based on the number of callus lines that emerged from the infected tissues. After the second subculture, carbenicillin was avoided from the selection medium used for subculture of infected tissues devoid of bacterial overgrowth. After 1 month, the newly emerged lines were proliferated in selection medium containing only kanamycin (200 mg/l). About 30 callus clumps were cultured in petri plates (90×15 mm) containing the selection medium.

Experiments were carried out to control overgrowth of bacteria by modifying the infection, cocultivation and selection media. The effect of silver nitrate in controlling bacterial overgrowth and improving transformation efficiency were studied. Different concentrations of silver nitrate were tried and the effects on percentage of proliferating lines with transient GUS expression were evaluated. Silver nitrate (10 mg/l) was supplemented in the infection and cocultivation medium. Selection medium was also supplemented with different concentrations of silver nitrate (0–30 mg/l) to find its effect on controlling bacterial growth and improvement in callus texture. pH of the infection medium was adjusted to 5.2 with 1 N KOH and the whole solution was filter sterilised before tissue infection. The pH of the cocultivation and selection media was adjusted to 5.7 with 1 N KOH before autoclaving at 121 °C for 10 min. Silver nitrate was added filter sterilised along with phytohormones and antibiotics to the autoclaved cocultivation and selection media. Each experiment was repeated thrice with five replications.

Somatic Embryogenesis of Transgenic Callus

The newly emerged lines were screened histochemically for GUS expression. Somatic embryogenesis was tried with all GUS positive lines. The proliferated transgenic calli were subcultured on embryo induction medium containing kanamycin (200 mg/l) for embryogenic callus initiation and subsequent embryogenesis. Embryogenic callus initiation and embryo induction were tried from proliferated callus in medium standardised earlier [7] which was modified MS basal medium ($Ca(NO_3)_2 \cdot 4H_2O$ – 500 mg/l and KH_2PO_4 – 270 mg/l) containing Gamborg's B_5 vitamins [23], amino acids, organic supplements such as coconut water (5%), malt extract (50 mg/l), casein hydrolysate (300 mg/l), sucrose 60 g/l and phytohormones.

Table 36.2 Media used for *Agrobacterium* infection and transformation

Medium component	Callus induction	Callus proliferation	*Agrobacterium* infection	Cocultivation	Preselection	Selection	Transgenic callus proliferation
MS basal	MS modified	MS modified	MS modified ½ x	MS modified ½ x	MS modified ½ x	MS modified ½ x	MS modified ½ x
Vitamins	B_5	B_5	B_5	B_5	B_5	B_5	B_5
Sucrose	30 g/l	40 g/l	50 g/l	50 g/l	50 g/l	50 g/l	50 g/l
Organic supplements	Casein hydrolysate (1 g/l), coconut water (5%)	Casein hydrolysate (1 g/l), coconut water (5%)	–	–	–	–	–
Hormones	5.4 μM 2,4-D, 4.4 μM BA and 1.08 μM NAA	1.8 μM 2,4-D, 2.2 μM BA and 1.08 μM NAA	1.8 μM 2,4-D, 2.2 μM BA and 1.14 μM IAA	1.8 μM 2,4-D, 2.2 μM BA and 1.14 μM IAA	1.8 μM 2,4-D, 2.2 μM BA and 1.08 μM NAA	1.8 μM 2,4-D, 2.2 μM BA and 1.08 μM NAA	1.8 μM 2,4-D,2.2 μM BA and 1.08 μM NAA
Acetosyringone	–	–	100 uM	100 uM	–	–	–
Proline	–	–	1.0 mM	1.0 mM	–	–	–
Betaine HCL	–	–	1.0 mM	1.0 mM	–	–	–
pH	5.7	5.7	5.2	5.6	5.7	5.7	5.7
Antibiotics	–	–	–	–	Carbenicillin (400 mg/l)	Carbenicillin (400 mg/l), kanamycin (300 mg/l)	Kanamycin (200 mg/l)

GUS Histochemical Assay

GUS staining was performed following the method reported by Jefferson [24]. Newly emerged lines from the infected tissues were immersed in X-Gluc solution (2 mM X-Gluc, 100 mM NaH2PO$_4$, 0.5 mM potassium ferricyanide and 50 mM ferrocyanide). The tissues were incubated at room temperature. GUS expression frequency was examined and scored as blue spots.

PCR Amplification

DNA was isolated from transgenic as well as control calli of the *Hevea* clone RRII 105 following CTAB procedure [25]. PCR amplification was carried out using 20 ng of genomic DNA with isopentenyltransferase gene-specific primers. Gene primers were designed from the coding region sequence of isopentenyltransferase gene of *Agrobacterium tumefaciens.* The forward primer 5′-CTTGCACAGGAAAGACGTCG-3′ and reverse primer 5′-CGTAAGCGGCTGCG-3′ were used for PCR amplification.

Amplifications were carried out in 20 μl reactions, which contained 1.0 μl template DNA (50 ng), 2.0 μl reaction buffer (Tris-HCl, pH 9–10, KCl 50 mM, $MgCl_2$ 15 mM), 2.0 μl dNTP mix(100 μM), 1.0 μl of each forward and reverse primer (250 nm each) and 0.5 U *Taq* DNA polymerase. Volume of the reaction mix was made up to 20 μl with sterile distilled water. The reaction mix was overlaid with a drop of mineral oil and amplification was carried out in MJ Research PTC 200 Peltier Thermal Cycler. The PCR conditions were 2-min initial denaturation at 94 °C followed by 36 cycles of denaturation (1 min, 94 °C), annealing (1 min, 55 °C) and extension (2 min, 724 °C). Final elongation was 10 min at 72 °C. The PCR products were analysed in 1.5% agarose gels.

Statistical Analysis

Data were subjected to square root transformation and analysed using one way ANOVA. All analyses were performed at 5% level.

Results and Discussion

Effect of Silver Nitrate on Callus Friability

Callus induction could be obtained within 40 days in the earlier standardised medium (Fig. 36.1a). Callus proliferation was obtained in the callus induction medium with reduced calcium nitrate (800 mg/l) and 2,4-D (1.8 μM) along with 40 g/l sucrose (Table 36.1). Proliferation of the callus was noticed after 3 weeks in this medium (Fig. 36.1b). The proliferated callus had a loose and friable texture. With 2–3 subcultures in proliferation medium, the texture of the callus could be improved further. When three different levels of silver nitrate (10, 20 and 30 mg/l) were tried in the proliferation medium during the second subculture, a concentration of 20 mg/l helped in improving the texture of the callus. Addition of 20 mg/l silver nitrate to the proliferation medium during the second and third subcultures made the calli yellow, more friable and nearly embryogenic compared with the control calli in which no silver nitrate was added (Fig. 36.1c).

Effect of Silver Nitrate on Embryogenic Callus Initiation

Embryogenic calli originated from more than one region of each primary callus clump as a small yellow lump which further proliferated within a month to form a mass of friable golden yellow callus (Fig. 36.1d). The frequency of embryogenic callus formation was 38% in the control. Embryogenic calli could be obtained from the primary callus clump when subcultured in fresh medium. It was observed that in the medium containing 10 mg/l silver nitrate, proliferating callus became more friable. At 5.0 and 10.0 mg/l concentrations of silver nitrate, significant improvement in the frequency of embryogenic callus formation was noted. Embryogenic callus formation could be obtained with improved frequency (43%) from the proliferated leaf callus in medium when 10 mg/l silver nitrate was added (Table 36.3). No significant difference was observed in the

Fig. 36.1 (**a–f**) Different stages of somatic embryogenesis. (**a**) Callus induction from leaf explants. (**b**) Callus proliferation in medium without $AgNO_3$ (**c**) Callus proliferation in medium supplemented with $AgNO_3$. (**d**) Embryogenic callus initiation. (**e**) Proliferated embryogenic callus. (**f**) Somatic embryo induction

frequency of embryogenic callus formation at higher levels of silver nitrate. Somatic embryogenesis could be obtained from the proliferated embryogenic callus in medium standardised earlier (Table 36.1) (Fig. 36.1e and f).

Culture medium compounds such as silver nitrate [26] have also been reported to modify callus texture in *Zea mays*. Addition of silver nitrate to the culture medium greatly improved the regeneration of both dicot and monocot plants in tissue culture. Silver ions in the form of nitrate, such as $AgNO_3$, play a major role in influencing somatic embryogenesis, shoot formation and efficient root formation which are the prerequisites for successful genetic transformation [27]. In the present work, good callus proliferation with improved texture was obtained in medium containing 20 mg/l silver nitrate. In *Coffea canephora*

Table 36.3 Effect of silver nitrate on embryogenic callus formation

Silver nitrate (mg/l)	*Embryogenic callus (mean %)
0	38.33
5	40.00
10	43.33
20	43.33
CD	**1.41**

*Each value is the mean % of 20 replicate samples repeated thrice

Pierre genotypes, it has been reported that $AgNO_3$ concentrations between 30 and 60 μM improved embryo yield during somatic embryogenesis [11]. In *Hevea* also, addition of 10 mg/l silver nitrate to the embryogenic callus induction medium favoured embryogenesis. In recent years, $AgNO_3$ has been employed in tissue culture studies for inhibiting ethylene action because of its water solubility and lack of phytotoxicity at effective concentrations. In carrot, $AgNO_3$ at concentrations of 10–20 μM caused a twofold increase in the number of somatic embryos without causing adverse effects on cell survival [28].

Influence of Silver Nitrate on Genetic Transformation

When transformation experiments were carried out using proliferated fresh leaf callus as the target tissue for *Agrobacterium* infection, following the earlier developed protocol [17], a transformation frequency of 9% was obtained with *Agrobacterium* strain EHA 101. Earlier bacterial overgrowth in the infected tissue was common causing hindrance to transgenic tissue regeneration after *Agrobacterium* infection. In the present study, when silver nitrate was supplemented in the infection, cocultivation and selection media, it was found that overgrowth of bacteria could be reduced without compromising the transformation frequency (Fig. 36.2a).

Infection medium containing silver nitrate (10 mg/l) was used for tissue infection. The infected tissues were transferred to cocultivation medium which was also supplemented with silver nitrate (10 mg/l). After 72 h of cocultivation, the tissues were again subcultured in selection medium containing the antibiotic carbenicillin (400 mg/l) for controlling bacterial overgrowth. When different concentrations of silver nitrate (0–30 mg/l) was supplemented in the selection medium, it was observed that with increasing concentration of silver nitrate up to 20 mg/l the bacterial overgrowth could be reduced significantly (Table 36.4). In the control medium, when no silver nitrate was provided, overgrowth of bacteria was persistent and sometimes 50% or more of the cultures had to be discarded. Bacterial overgrowth was reduced significantly at silver nitrate concentrations starting from 5.0 mg/l onwards. At concentrations above this, significant difference was noted in control of overgrowth. Addition of silver nitrate (10 mg/l) in the infection, cocultivation and selection media controlled bacterial overgrowth by 80% (Table 36.4). Maximum control was obtained at a silver nitrate concentration of 20 mg/l. Increasing the silver nitrate concentration above this did not give significant difference. The optimum concentration of silver nitrate to be used in the different steps of transformation was identified as 10.0 mg/l in infection and cocultivation media and 20 mg/l in the selection medium. By supplementing silver nitrate in the selection medium, the texture of the putatively transgenic callus lines was also found to be improved, with most of them friable, yellow and showing good proliferation (Fig. 36.2b). The tissue recovered free of bacterial overgrowth was subcultured every 3 weeks to fresh medium.

In apple, it has been reported that during *Agrobacterium*-mediated transformation, ethylene production was increased which resulted in reduced efficiency of gene transfer mechanism [18]. Ozden et al. [20] have reported that involvement of ethylene production leads to tissue browning which can be reduced in the presence of silver nitrate. Inclusion of silver nitrate in coculture medium has been proven for its anti-ethylene activity which is common with *in vitro* plant cultures. Opabode [21] reported that silver nitrate is an anti-necrotic compound which can reduce oxidative burst during the interaction between plant

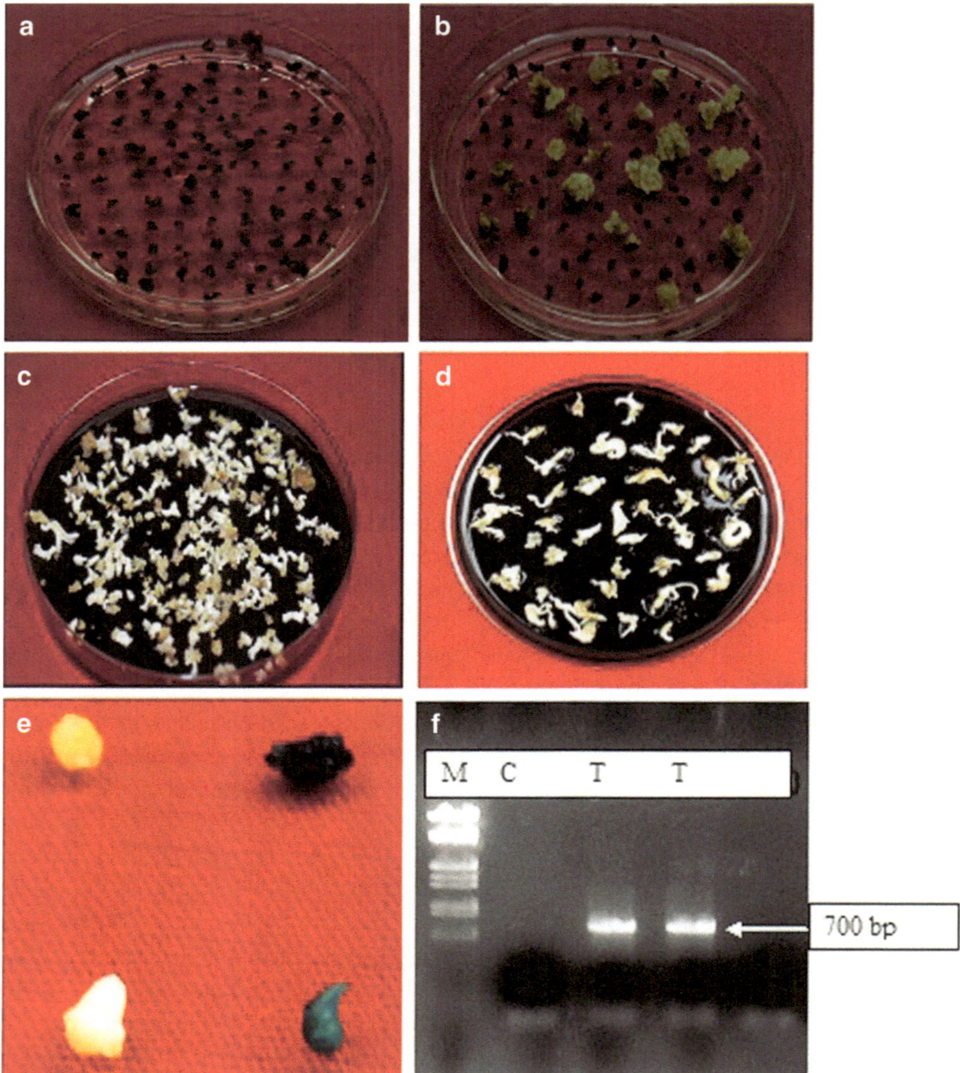

Fig. 36.2 (**a–f**) Different stages of *Agrobacterium* infection and somatic embryogenesis. (**a**) Infected callus in presence of $AgNO_3$ without bacterial overgrowth. (**b**) Infected tissues showing transgenic callus emergence in $AgNO_3$-containing medium. (**c**) Somatic embryogenesis from transgenic callus. (**d**) Transgenic somatic embryos. (**e**) GUS expression (control and transgenic calli and embryos). (**f**) PCR amplification (Lane, *M* Marker, *C* Control, *T1* Transgenic callus, *T2* Transgenic embryos)

tissue and *Agrobacterium*. Addition of silver nitrate in the coculture medium has shown to enhance stable gene transfer in maize [26, 32], carrot [28] and Fuji apple [29]. The suppressed *Agrobacterium* growth on target explants could facilitate plant cell recovery that resulted in increased efficiency of transformation. Orlikowska reported that addition of silver nitrate stimulated direct shoot regeneration in leaves taken from *in vitro* cultures and infected with *Agrobacterium* [30]. It also inhibited bacterial growth after cocultivation with the explants. In the present study, presence of silver nitrate in the selection medium (20 mg/l) prevented overgrowth of bacteria for about 3 weeks. Similar observation was also made by Orlikowska [30] where presence of silver

Table 36.4 Effect of silver nitrate in controlling bacterial overgrowth

Silver nitrate (mg/l) (selection medium)	Overgrowth (mean %)
0	50.0 (3.87)
5	30.0 (2.59)
10	22.0 (1.72)
15	22.5 (1.72)
20	20.2 (1.57)
30	20.4 (1.57)
CD	0.3

Data from 30 replicate samples, experiment repeated thrice
Values in parenthesis indicate transformed values

nitrate (100 mg/l) in the rose regeneration medium completely retarded bacterial overgrowth for about 1 month. Thus, by the addition of silver nitrate, they were able to replace or reduce the concentration of antibiotics such as cefotaxime or carbenicillin used for controlling bacterial growth, which are often phytotoxic to tissues. Tissue damage by A. *tumefaciens* infection has been reported earlier also and seems to be one of the major obstacles for *Agrobacterium*-mediated transformation. In control experiments, the texture of the callus in the putatively transgenic lines that emerged from the infected tissues also varied, with most of the calli being hard and showing no proliferation. In the present study when silver nitrate was supplemented in the medium, the texture of the putatively transgenic callus lines that emerged from the infected tissue was found to be improved, with most of them friable, yellow and showing good proliferation. The use of silver nitrate has shown to have other important effects in plant tissue culture of cassava, such as improving somatic embryogenesis. It is also reported to have significant influence on micropropagation in many other species [10].

GUS Histochemical Assay

Approximately 2-mg callus from each emerging line was used for GUS histochemical staining. β-Glucuronidase catalyses the hydrolysis of the substrate liberating indoxyl and indolyl groups which gets dimerised to form insoluble indigo causing blue colouration. Newly emerged putatively transgenic calli were antibiotic resistant and showed proliferation in selection medium and were found to be GUS positive (Fig. 36.2e).

Somatic Embryogenesis from Transgenic Callus

Proliferation of the putatively transgenic callus, embryogenic callus initiation and further embryo induction were obtained in medium standardised earlier for leaf explants [7]. Kanamycin (200 mg/l) was also added to all the media used for the subculture of transgenic callus. Callus proliferation could be obtained in modified MS medium with the addition of calcium nitrate (800 mg/l) and containing Gamborg's B_5 vitamins, sucrose (40 g/l) and growth regulators 2,4-D (1.8 μM), BA (4.4 μM) and NAA (1.08 μM). Embryogenic callus initiation and embryo induction were obtained from proliferated callus in modified MS basal medium solidified with 0.5% phytagel. Rate of embryo induction from the transformed callus was low (40%) compared to untransformed controls (60%). Embryos obtained had a different appearance, with most of them having a shooty nature (Fig. 36.2c and d). Maturation of the embryos could be obtained in medium standardised earlier containing basal salts of Woody Plant Medium [31]. Plant regeneration from these embryos is being attempted.

PCR Amplification of *ipt* Gene

The genomic DNA isolated from the control and transgenic calli was used for PCR amplification using *ipt* gene-specific primers. When PCR amplification was carried out with gene-specific primers for isopentenyltransferase gene, an amplified band of approximately 700 bp was obtained. Amplification of the gene was obtained in the transgenic callus showing gene integration. No amplification could be obtained in the primer controls and in untransformed callus (Fig. 36.2f).

Summary and Conclusion

Efficient plant regeneration and transformation systems using easily available explants are indispensible for genetic modification of important crop plants. The present study reports the effect of silver nitrate in improving efficiency of the somatic embryogenesis system from leaf explants. Supplementing silver nitrate in the medium during the second and third subcultures made the calli more friable. Presence of silver nitrate in the embryo induction medium helped in embryogenic callus initiation, thereby increasing the efficiency of the system. The influence of silver nitrate on transformation efficiency by improving callus texture in the newly emerging transgenic cell lines and control of bacterial overgrowth in infected tissues were also studied. It was observed that addition of silver nitrate in the infection (10.0 mg/l), cocultivation (10.0 mg/l) and selection (20.0 mg/l) medium significantly suppressed bacterial overgrowth and improved the texture of callus in newly emerged lines. The transgenic callus obtained could be proliferated and somatic embryo induction obtained. Hence, silver nitrate was found to be a suitable medium supplement in improving somatic embryogenesis from leaf explants and in *Agrobacterium*-mediated genetic transformation using proliferated leaf callus in *Hevea brasiliensis.*

Acknowledgement The authors wish to thank Dr. James Jacob, Director, RRII, for his encouragement to carry out this study. Assistance rendered for statistical analysis from Aneesh P. and technical assistance rendered from Leda P. are greatly acknowledged.

References

1. Wan AR, Ghandimathi WY, Rohanim H, Paranjothy K. Recent developments in tissue culture of *Hevea*. In: Rao AN, editor. Tissue culture of economically important plants. Singapore: Costed; 1981. p. 152–8.
2. Carron MP, Enjalric F. Studies on vegetative micropropagation of *Hevea brasiliensis* by somatic embryogenesis and *in vitro* microcutting. In: Fujiwara A, editor. Plant tissue culture. Tokyo: Maruzen; 1982. p. 751–2.
3. Kumari JP, Asokan MP, Sobha S, Sankari Ammal L, Rekha K, Kala RG, Jayasree R, Thulaseedharan A. Somatic embryogenesis and plant regeneration from immature anthers of *Hevea brasiliensis (*Muell.Arg). Curr Sci. 1999;76:1242–5.
4. Seneviratne P, Withange SP, Nugawela A, Wijesekera GAS, de Oysa GM. Somatic embryogenesis for *Hevea*. The technique for anther culture. J Rubber Res Inst Sri Lanka. 1996;78:79–99.
5. Sushamakumari S, Sobha S, Rehka K, Jayasree R, Asokan MP. Influence of growth regulators and sucrose on somatic embryogenesis and plant regeneration from immature inflorescence of *Hevea brasiliensis*. Indian J Nat Rubber Res. 2000;13:19–29.
6. Kala RG, Kumari JP, Sushamakumari S, Jayashree R, Rekha K, Sobha S, Thulaseedharan A. *In vitro* regeneration of *Hevea brasiliensis* from leaf explants. In: Keshavachandran R et al., editors. Recent trends in horticultural biotechnology. New Delhi: New India Publishing Agency; 2007. p. 223–8.
7. Kala RG, Gimisha GC, Kumari JP, Sushamakumari S, Sobha S, Jayashree R, Rekha K, Thulaseedharan A. Somatic embryogenesis in leaf cultures of *Hevea brasiliensis* – effect of source plant. Nat Rubber Res. 2009;22(1&2):117–26.
8. Giridhar P, Indu EP, Vijaya Ramu D, Andravishankar GA. Effect of silver nitrate on *in vitro* shoot growth of coffee. Trop Sci. 2003;43(3):144–6.
9. Beyer Jr EM. A potent inhibitor of ethylene action in plants. Plant Physiol. 1976;58:268–71.
10. Zhang P, Phansiri S, Puonti-Kaerlas J. Improvement of Cassava shoot organogenesis by the use of silver nitrate *in vitro*. Plant Cell Tissue Organ Cult. 2001;67:47–54.
11. Fuentes SL, Calheiros MBP, Manettifilho J, Vieira LGE. The effects of silver nitrate and different carbohydrate sources on somatic embryogenesis in *Coffea canephora*. Plant Cell Tissue Organ Cult. 2000; 60(1):5–13.
12. Veluthambi K, Aditya KG, Arun S. The current status of plant transformation technologies. Curr Sci. 2003;84(3):368–80.
13. Sharma KK, Bhatnagar-Mathur P, Thorpe TA. Genetic transformation technology: status and problems. In vitro cellular and developmental biology. Plant. 2005;41:102–12.
14. Arokiaraj P, Leelawathy R, Yeang HY. The supervirulence plasmid pToK47 from *Agrobacterium tumefaciens* A281 improves transformation efficiency of *Hevea brasiliensis*. Am J Biochem Biotechnol. 2009;5(3):137–41.
15. Sobha S, Sushamakumari S, Thanseem I, Kumari JP, Rekha K, Jayasree R, Kala RG, Asokan MP, Sethuraj MR, Dandekar M, Thulaseedharan A. Genetic transformation of *Hevea brasiliensis* with the gene coding for superoxide dismutase with FMV 34S promoter. Curr Sci. 2003;85(12):1767–73.
16. Jayasree R, Rekha K, Uratsu SL, Venketachalam P, Jayasree PK, Kala RG, Priya P, Sushamakumari S, Sobha S, Asokan MP, Sethuraj MR, Thulaseedharan A, Dandekar M. *Agrobacterium* mediated genetic transformation and regeneration of transgenic plants *Hevea brasiliensis* Muell Arg, via somatic embryogenesis. Plant Cell Rep. 2003;22:201–9.

17. Kala RG, Anu KS, Manesh K, Saleena A, Kumari JP, Narayanan PR, Thomas G, Thulaseedharan A. *Agrobacterium* mediated genetic transformation in *Hevea brasiliensis* for recombinant protein production. J Plant Crops. 2006;34(3):582–6.
18. Zhao ZY, Gu W, Cai T, Tagliani L, Hondered D, Bond D, Schroeder S, Ruder M, Pierce D. High throughput genetic transformation mediated by *Agrobacterium tumefaciens* in maize. Mol Breed. 2001;8:323–33.
19. Zeong P, Vadnais DA, Zhang Z, Polacco JC. Refined glufosinate selection in *Agrobacterium*-mediated transformation of soybean *Glycine max* (L.) Merrill. Plant Cell Rep. 2004;22(7):478–82.
20. Ozden-Tokatli Y, Ozudogru EA, Akrin A. *In vitro* regeneration of *Pistachio* through organogenesis: effect of silver nitrate, polyvinylpyrrolidone and citric acid. In: Proceedings of '2004 World Congress on In Vitro Biology', 40; 2004. pp. 46–50.
21. Opabode TJ. *Agrobacterium* mediated transformation of plants: emerging factors that influence efficiency. Biotechnol Mol Biol Rev. 2006;1(1):12–20.
22. Murashige T, Skoog F. A revised medium for rapid growth and bioassays with tobacco tissue culture. Physiol Plant. 1962;15:473–97.
23. Gamborg OL, Miller RA, Ojima K. Nutrient requirements of suspension cultures of soya bean root cells. Exp Cell Res. 1968;50:151–8.
24. Jefferson RA, Karanagh TA, Beven MW. GUS fusion: beta-Glucuronidase as a sensitive and versatile gene fusion marker in higher plants. EMBO J. 1987; 6(13):3901–7.
25. Doyle JJ, Doyle JL. A rapid total DNA preparation procedure for fresh plant tissue. Focus. 1990; 12:13–5.
26. Vain P, Yean H, Flament P. Enhancement of production and regeneration of embryogenic type 11 callus in *Zea mays* L. by $AgNO_3$. Plant Cell Tissue Organ Cult. 1989;18:143–51.
27. Bais HP, Sudha G, Suresh B, Andravishankar GA. $AgNO_3$ influences *in vitro* root formation in *Decalepis hamiltonii* Wight, Arn. Curr Sci. 2000;79:894–8.
28. Roustan J-P, Latche A, Fallot J. Control of carrot somatic embryogenesis by $AgNO_3$, an inhibitor of ethylene action: effect on arginine decarboxylase activity. Plant Sci. 1990;67(1):89–95.
29. Seong ES, Song KJ, Jegal S, Yu CY, Chung IM. Silver nitrate and amino ethoxyvinylglycine affect *Agrobacterium* mediated apple transformation. Plant Growth Regul. 2005;45:75–82.
30. Orlikowska T. Silver nitrate inhibits bacterial growth in plant tissue cultures. Agric Cell Rep. 1997;29(4):25.
31. Lloyd G, Mc Cown G. Commercially feasible micropropagation of Mountain Laurel, *Kalmia latifolia*, by use of shoot tip culture. Comb Proc Int Plant Propag Soc. 1980;30:421–7.
32. Armstrong CL, Rout JR. A novel *agrobacterium*-mediated plant transformation method. Int Patent Publ 2001.WOO1/09302 A2.

37 Immunoinformatics Prediction and Structure-Based Modeling of HLA-II Binding Epitopes of Iron Surface Determinant B (IsdB) Protein of *Staphylococcus aureus*

Francis Dileep and Surekha Kuyyalil

Abstract

Most of the attempts to develop a vaccine to prevent *Staphylococcus aureus* infections to date have not translated into clinical success. To date, all vaccine attempts have been based upon the development of opsonic antibodies. Research information suggests that cell-mediated immunity (CMI) may be critical for protection against *Staphylococcus aureus*. Iron surface determinant B (IsdB) protein is identified as a potent immunogen against *Staphylococcus aureus*. This protein, which scavenges iron from hemoglobin and myoglobin, is highly conserved among *Staphylococcus aureus* isolates from diverse clinical and taxonomical backgrounds including MRSA and VRSA and is expressed on the surface of all isolates tested so far. Adoptive transfer of CD4+ cells protected SCID (severe combined immunodeficiency) mice challenged with a lethal dose of *S. aureus* via tail vein while neither CD8+ T cells nor plasmacytes were protective. Immunization with IsdB was shown to be critical for activating the transferred cells as T cells from nonimmunized mice were not protective. To dissect the immune response further, it was shown that Th17 cells were necessary for IsdB-generated protective immunity. Th17 cells are a class of T helper cells that are activated upon encounter of an antigenic peptide coupled to a HLA (human leukocyte antigen) class II molecule. The HLA class II molecules are highly polymorphic. Immunoinformatics tools enable the prediction of T cell epitopes which can bind to the HLA-DR, HLA-DQ, and HLA-DP alleles using artificial neural networks. Here we describe a combined immunoinformatics and structure-based modeling approach for the prediction of T cell epitopes in the IsdB protein that can bind to 26 different alleles. The IsdB protein from *S. aureus* was analyzed computationally for the presence of HLA-II binding peptides. All possible overlap-

F. Dileep • S. Kuyyalil (✉)
Department of Biotechnology and Microbiology,
Kannur University, Thalassery Campus,
Palayad P.O. 670 661, Kannur, India
e-mail: skuyyalil@yahoo.com

A. Sabu and A. Augustine (eds.), *Prospects in Bioscience: Addressing the Issues*,
DOI 10.1007/978-81-322-0810-5_37, © Springer India 2013

ping 15-mer peptide sequences were generated *in silico* and analyzed for their ability to bind to 14 HLA-DR alleles covering 9 HLA super types, 6 HLA-DP alleles, and 6 HLA-DQ alleles. Approximately Twenty four, eight and six percents of the generated peptides binds to HLA-DR, HLA-DQ and HLA-DP alleles respectively with an IC-50 value less then 50 nM. The structural basis for recognition of these high-affinity peptides was studied using structural modeling of HLA class II peptide complexes, and there exists a good correlation between structural analysis and binding prediction.

Keywords
Staphylococcus aureus • Iron surface determinant protein B • T cell epitopes • HLA class II binding • Vaccine

Introduction

Staphylococcus aureus is a gram-positive human pathogen of serious concern due to the wide range of diseases caused and the emergence of multiple antibiotic resistance [1, 2]. The organism is capable of inflicting diseases ranging from minor skin infections to life-threatening infections such as bacteremia, endocarditis, and toxic shock syndrome and has evolved with time to produce antibiotic-resistant strains that are virtually resistant to almost all antibiotics available [3–5]. This necessitates the development of a vaccine against staphylococcal infections [6]. Earlier attempts to develop a vaccine against *Staphylococcus aureus* focused on humoral immunity mediated by B cells [7–9]. Unfortunately none of these strategies translated into a clinically viable vaccine. It is now postulated that a T-cell-mediated approach, particularly targeting $CD4^{+}$ T cells, may be an alternative strategy to the prevention of staphylococcal infection, as T helper cells were found to be involved in the natural immune response elicited against a staphylococcal infection [10–12]. Specifically, IsdB protein originally administered to boost a B-cell-mediated immunity was found to trigger Th17 activation, and this had a strong correlation with protective efficacy [10, 13, 14]. Based on the understanding that Th17 cells are activated via an HLA-class-II-mediated pathway – where an exogenous protein is taken up by an antigen presenting cell, cleaved into peptides, and transported to the cell surface coupled with receptive HLA-II molecule to be recognized by T cell receptors [15, 16] (Fig. 37.1) – the present study was designed to predict and validate peptides from IsdB that can bind HLA class II alleles. A bioinformatics-based strategy was employed for prediction and analysis. A set of potential peptides that can be used as peptide epitopes vaccines were obtained.

Material and Methods

Prediction of HLA Class II Binding Peptides

The amino acid sequence of IsdB protein was obtained from the TIGR database (BQF78003.1). All possible overlapping 15-mer peptides were generated and analyzed for the potential to bind to 24 different HLA-II alleles using the NetMHCII prediction server (http://www.cbs.dtu.dk/services/NetMHCII) [17, 18]. This algorithm identifies 15-mer peptides that contain allele-specific motifs for HLA class II alleles. The binding is measured in terms of IC50 expressed in nM.

Identification of Self Peptides

The peptides with sequence homology to human proteins annotated so far were removed from the analysis pool using the BLAST algorithm (70% homology cutoff).

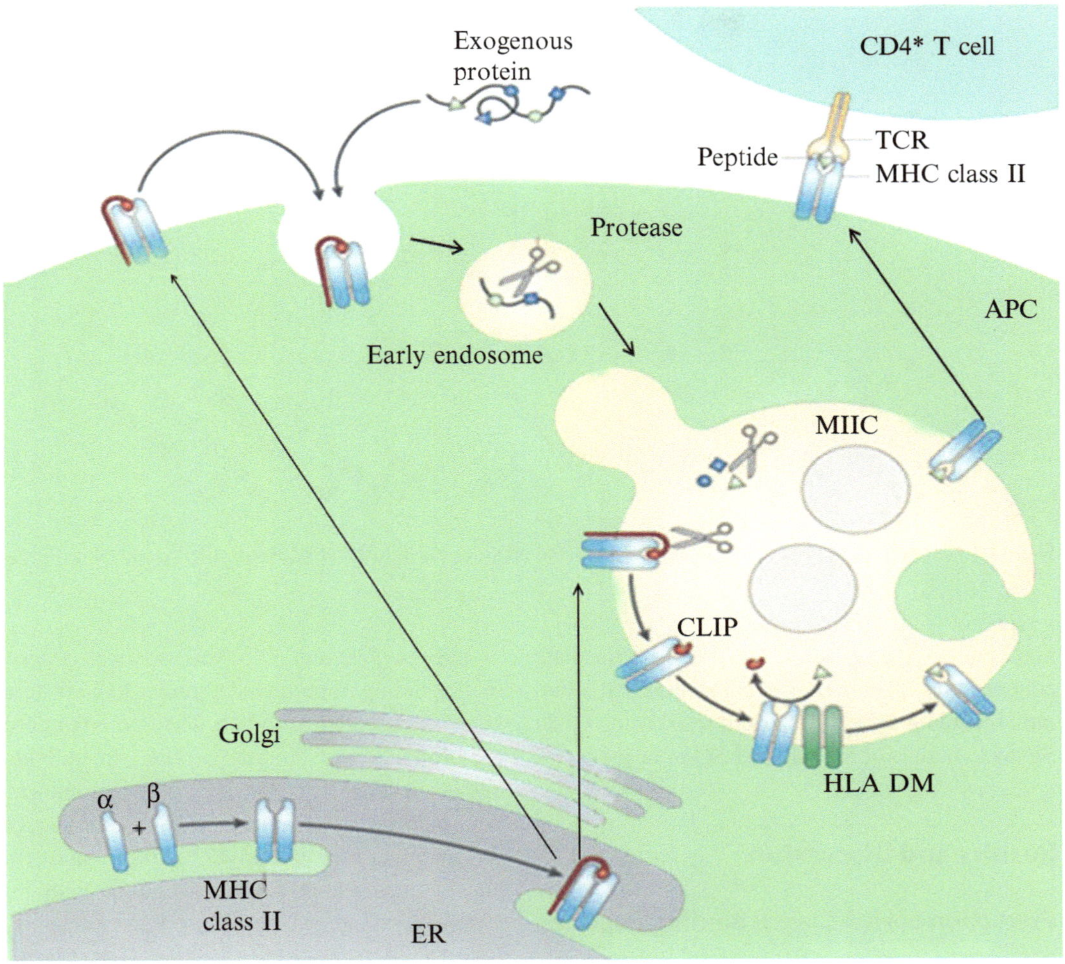

Fig. 37.1 MHC class II α- and β-chains assemble in the endoplasmic reticulum (ER) and form a complex with the invariant chain (Ii). The Ii–MHC class II heterotrimer is transported through the Golgi to the MHC class II compartment (MIIC), either directly and/or via the plasma membrane. Endocytosed proteins and Ii are degraded by resident proteases in the MIIC. The class II-associated Ii peptide (CLIP) fragment of Ii remains in the peptide-binding groove of the MHC class II dimer and is exchanged for an antigenic peptide with the help of the dedicated chaperone HLA-DM (known as H2-M in mice). MHC class II molecules are then transported to the plasma membrane to present antigenic peptides to CD4⁺ T cells (*APC* antigen-presenting cell, *TCR* T cell receptor)

Molecular Modeling of Peptide: HLA Complexes

The binding affinities predicted were verified using a molecular modeling approach. The structural feasibility of replacement of the resident peptide with predicted peptides was analyzed employing the 3D crystallographic structures of the HLA–peptide complex as templates. Crystal structures of peptide-bound alleles HLA-DRB1*0101 (1AQD), HLA-DPA1*0103-DPB1*0201 (3LQZ), and HLA DQA 1*0301-DQB1*0302 (1JK8) were obtained from the PDB. The peptides predicted to bind with high, moderate, and low energies were modeled onto their respective templates replacing the resident peptide. The raw amino acid sequences were fed to the Swiss PDB Viewer to generate peptide conformations that were docked with the binding site of the HLA alleles. The receptor grid box was set at 50 A* (x, y, and z axes) with the resident peptide as the centroid of the grid box. Moreover, peptides with maximum allele coverage were identified form the pool of predicted peptides and were modeled onto representative alleles from each locus. Such promiscuous peptides

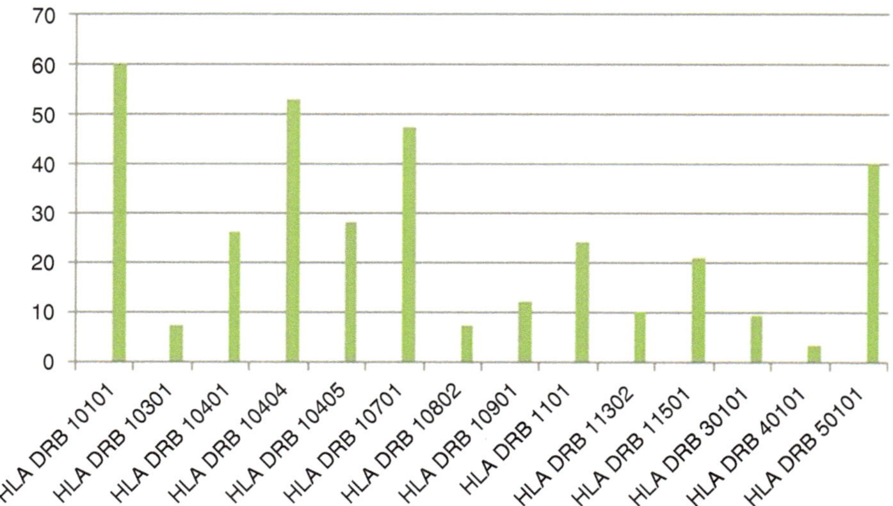

Fig. 37.2 Observed variation in the binding of predicted peptides to 14 different HLA-DR alleles studied

may enable identification of vaccine candidate epitopes that covers a larger population. All structural modeling studies were performed using Glide SP dock module of Maestro 9.0, Schrodinger.

Results and Discussion

Prediction of HLA Class II Binding Peptides

A total of 631 15-mer peptides were generated from the IsdB protein sequence. Each of these peptides was analyzed for binding to 26 alleles including 14 HLA-DR, 6 HLA-DQ, and 6 HLA-DP alleles. About 29% of the generated peptides bind to one or more HLA alleles with an IC50 value less than 50 nM. Twenty four, eight and six percentages of the generated peptides bind to atleast one of the HLA-DR, HLA-DQ and HLA-DP alleles respectively. About 54, 17, and 3% of the predicted strong binders bind to multiple alleles of DR, DQ, and DP loci, respectively. Most of the alleles were found to be monoallele specific. However, some of them exhibited multiple allele specificities. The highest number of alleles a peptide could bind was 8 in the case of HLA-DR and 2 each in cases of DP and DQ. This gives a clear idea of the specificity and promiscuity exhibited by individual peptides.

In Figs. 37.2 and 37.3, variation was observed in the binding of predicted peptides to 24 different HLA class II alleles studied. The figures represent the proportion of binding peptides to individual alleles. Alleles from the DR locus are shown to bind to large numbers of peptides as compared to DP and DQ loci. HLA-DRB 10101 shows the highest number of peptides binding with high affinity, around 60 peptides. Few alleles like *HLA-DQA1*0102-DQB1*0602*, *HLA-DQA1*0401-DQB1*0402*, and *HLA-DQA1*0501-DQB1*0301* do not bind any peptides.

Molecular Modeling of Peptide: HLA Complexes

The feasibility of the binding of peptides to their respective targets was investigated using molecular modeling and structural analysis. All of the strong binding peptides docked against their respective crystallographic templates revealed binding energies similar to that of the resident peptides (Fig. 37.4). Moreover, strong, weak, and non-binding peptides docked with high, medium, and low Glide scores, respectively, thus validating the prediction results. The modeled structures with high-affinity peptides show more interaction in terms of hydrogen bonds when compared to peptides with low affinity (Fig. 37.5). A comparison

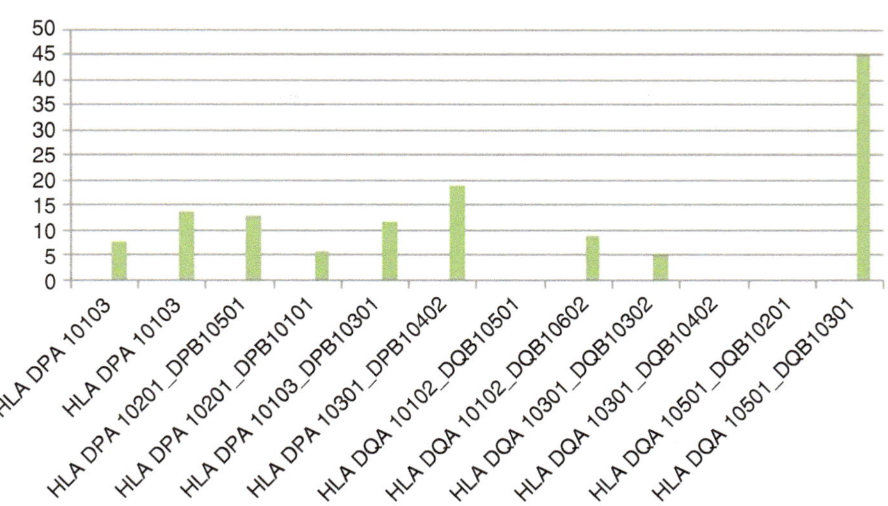

Fig. 37.3 Observed variation in the binding of predicted peptides to HLA-DP and DQ alleles studied

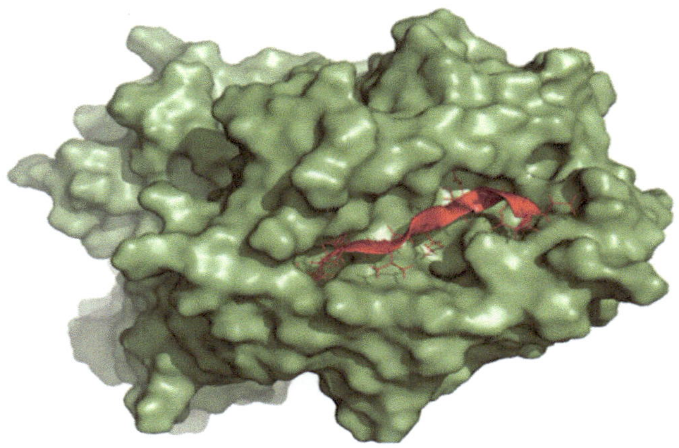

Fig. 37.4 Binding of the peptide core LLLMSNGEA to its receptor, HLA-DRB10101, as modeled by Glide SP docking module of Maestro 9.0, visualized in PyMOL. This peptide predicted to be a strong binder docked with a Glide score −14.4802644 compared to a less negative −13.382989 with the resident crystallographic peptide

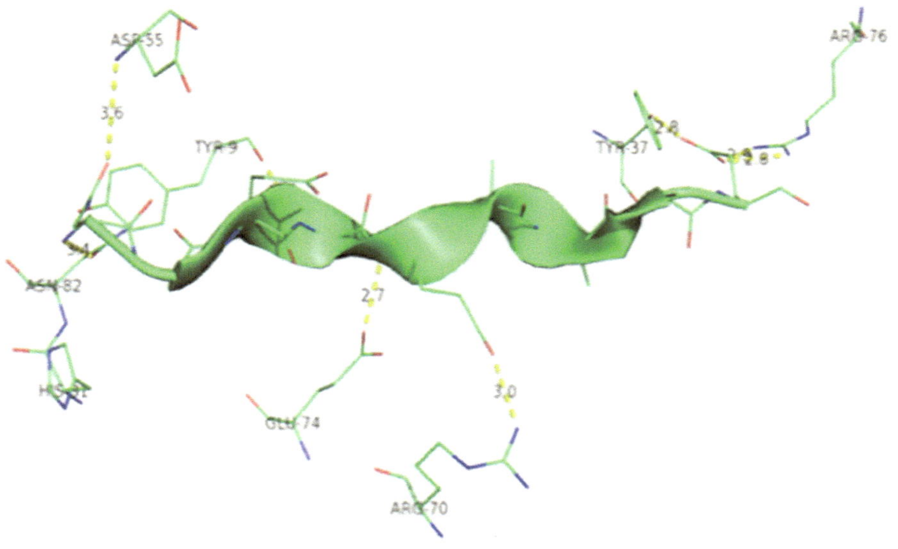

Fig. 37.5 The figure depicts the hydrogen bond interactions established by the strong binding peptide NGEAQAAAE, with the side chains of the residues in the binding pocket of *HLA- DQA 10301-HLA DQB 10302*.The hydrogen bonds are shown as *yellow dotted lines* with distances

Table 37.1 Comparison of prediction scores and intermolecular interaction energies of different peptides binding three different classes of HLA alleles

Sl. No	Allele	Type of predicted binding	Peptide core sequence	IC50 in nM	Glide score
1.	HLADRB1*0101	Strong	LLLMSNGEA	4.5	−14.4802644
		Weak	TLERQVYEL	480	−9.397090
2.	HLADPA1* 0103-DPB1*0201	Strong	YAYIRFSVS	9.5	−8.260415
		Weak	DYWKDFMVE	479.1	−10.826826
3.	HLADQA1* 0301-DQB1*0302	Strong	NGEAQAAAE	29.8	−11.479850
		Weak	DYTLMEFAQ	495.3	−9.012936

Table 37.2 Peptides with maximum allele coverage

Sl No.	Peptide core	No of alleles to which strong binding is predicted
1.	QFYHYASSV	9(8 DR + 1 DP)
2.	LMEFAQPIY	4(2 DR +1 DP +1 DQ)
3.	MALLALSSI	4(2 DR +1 DP +1 DQ)

of intermolecular energies and IC50 values shows that the peptides with high affinity, as judged by the IC50 values, also show high (more negative) interaction energies for most of the peptides (Table 37.1).

Humans carry only a limited number of codominant HLA alleles in their genome out of hundreds of polymorphic alleles present in the population [19]. It is important that the candidate vaccine must incorporate peptides that bind to a wide range of HLA alleles to provide a good population coverage [20]. This problem was addressed in the current study. The highest number of multiple allele binders was predicted for HLA-DR alleles suggesting that the DR loci may serve as the best target for an epitope vaccine design (Table 37.2).

The fact that most of the peptides are monoallele binders stresses the need of incorporation of peptides from multiple proteins that can cover maximum number of allelic variants. Studying each antigen experimentally for its potential as a vaccine candidate is an expensive and tedious process. Therefore, an initial bioinformatics filtering of probable candidate antigens is cost and time effective and is supported by the correlation between prediction results and structural modeling results. To conclude, a set of peptides were identified that may elicit T cell immunity if administered as part of a vaccine cocktail.

References

1. Easmon CSF, Adlam C. Staphylococci and staphylococcal infections. Vols 1 and 2. London: Academic Press; 1983.
2. Lyon BR, Skurray R. Antimicrobial resistance of *Staphylococcus aureus*: genetic basis. Microbiol Rev. 1987;51:88.
3. Robert SD. Skin and soft-tissue infections caused by Methicillin-Resistant *Staphylococcus aureus*. N Engl J Med. 2007;357:380–90.
4. Pettit CA, Fowler VG. *Staphylococcus aureus* bacteremia and endocarditis. Cardiol Clin. 2003;21(2):219–33.
5. Sanjai S, Charu G. Surmounting antimicrobial resistance in the Millennium Superbug: *Staphylococcus aureus*. Cent Eur J Med. 2010;5(1):12–29.
6. Foster TJ. Potential for vaccination against infections caused by *Staphylococcus aureus*. Vaccine. 1991;9(4):221–7.
7. Harro JM, Peters BM, O'May GA, Archer N, Kerns P, Prabhakara R, et al. Vaccine development in *Staphylococcus aureus*: taking the biofilm phenotype into consideration. FEMS Immunol Med Microbiol. 2010;59:306–23.
8. Middleton JR. *Staphylococcus aureus* antigens and challenges in vaccine development. Expert Rev Vaccines. 2008;7:805–15.
9. Projan SJ, Nesin M, Dunman PM. Staphylococcal vaccines and immunotherapy: to dream the impossible dream? Curr Opin Pharmacol. 2006;6:473–9.
10. Proctor R. Is there a future for a *Staphylococcus aureus* vaccine? Vaccine. 2011;30(19):2921–7.
11. Milner JD, Sandler NG, Douek DC. Th17 cells, Job's syndrome and HIV: opportunities for bacterial and fungal infections. Curr Opin HIV AIDS. 2010;5:179–83.
12. Popovich KJ, Weinstein RA, Aroutcheva A, Rice T, Hota B. Community-associated methicillin resistant *Staphylococcus aureus* and HIV: intersecting epidemics. Clin Infect Dis. 2010;50:979–87.
13. Cho JS, Pietras EM, Garcia NC, Ramos RI, Farzam DM, Monroe HR, et al. IL-17 is essential for host defence against cutaneous *Staphylococcus aureus* infection in mice. J Clin Invest. 2010;120:1762–73.
14. Joshi A, Pancari G, Smith S, Cope L, McNeely T. Exploring the role of T cells in IsdB mediated protection against *Staphylococcus aureus*, ISSSI Abst #168; 2010.
15. Thomas K, Estelle B, Mohamed O, Vijay KK. IL-17 and Th17 Cells. Annu Rev Immunol. 2009;27:485–517.

16. Jacques N, Marlieke L, Jongsma M, Petra P, Oddmund B. Towards a systems understanding of MHC class I and MHC class II antigen Presentation. Nat Rev Immunol. 2011;11:823–36.
17. Nielsen M, Lundegaard C, Lund O. Prediction of MHC class II binding affinity using SMM-align, a novel stabilization matrix alignment method. BMC Bioinformatics. 2007;8:238.
18. Nielsen M, Lund O. NN-align. An artificial neural network-based alignment algorithm for MHC class II peptide binding prediction. BMC Bioinformatics. 2009;10:296.
19. Borghans JAM, Joost BB, Rob JB. MHC polymorphism under host-pathogen coevolution. Immunogenetics. 2001;55(11):732–9.
20. Van Lierop MJ, Nilsson PR, Wagenaar JP, Van Noort JM, Campbell JD, Glass EJ, Joosten I, Hensen EJ. The influence of MHC polymorphism on the selection of T-cell determinants of FMDV in cattle. Immunology. 1995;84:79–85.

Damage Assessment of the Gall Mite *Aceria pongamiae* Keifer 1966 (Acari: Eriophyidae) on *Pongamia pinnata* (L.) Pierre

38

P.N.M. Nasareen, Y. Shibu Vardhanan, and N. Ramani

Abstract

Pongamia pinnata is well known for its use in traditional system of medicine, for the treatment of varied human diseases like bronchitis, whopping cough, rheumatism, diarrhoea, dyspepsia, flatulence, gonorrhoea, leprosy and even tumours. *Aceria pongamiae* is a highly host specific eriophyid mite, producing varying numbers of fingerlike leaf galls on *P. pinnata*. The number of galls on infested leaf varies, quite often individual galls fused to form complex, irregularly shaped, massive structures, covering entire laminar area including the midrib, vein and vein lets. Each gall carries hundreds of mites in different stages of development, namely, the egg, 1st nymph, 1st quiescent stage, 2nd nymph, 2nd quiescent stage and adult male and female. The galls induced by the mite adversely affect the quality of the leaves of *P. pinnata*, in turn reducing its economic utility in preparation of Ayurvedic medicines. Loss/reduction of leaf lamina would also drastically affect the photosynthetic machinery of the plant, thereby leading to biomass loss. The present study describes the morphological and anatomical changes induced by the feeding activity of various developmental stages of the mite *A. pongamiae* on the leaves of *P. pinnata*.

Keywords

Pongamia pinnata • *Aceria pongamiae* • Leaf structure damage • Polyphenol

Introduction

Eriophyid mites represent an exclusively phytophagous group with vermiform body [1, 2] and high host specificity [1, 3, 4], inducing diverse forms of abnormalities like galls , blisters, erinea, big buds, leaf rolling, stem and foliar deformation, silvering and bronzing of leaves on their respective host plants [5]. Galls are highly

P.N.M. Nasareen (✉) • N. Ramani
Division of Acarology, Department of Zoology, University of Calicut, Thenhipalam, 673 635, Kerala, India
e-mail: drramani@gmail.com

Y.S. Vardhanan
Department of Zoology, University of Calicut, Thenhipalam, 673 635, Kerala, India

A. Sabu and A. Augustine (eds.), *Prospects in Bioscience: Addressing the Issues*, DOI 10.1007/978-81-322-0810-5_38, © Springer India 2013

regulated growth manifestations developed by feeding stimuli of herbivores like insects, mites and nematodes on the host plant phenotype. A single female is able to cause a gall suitable for itself and its progeny.

Gross malformations of plant organs induced by eriophyid mites have been well described earlier [2, 4, 6–9]. Despite the morphological data on plant malformations induced by eriophyid mites [10, 11], information on the anatomical, i.e. cytological and/or histological, changes in plants infested by these mites is rather scanty [8, 12, 13]. Apart from damaging the epidermis, some eriophyid mites are known to damage the mesophyll tissue of the host plant as well [14].

P. pinnata is a medium-sized glabrous tree, mainly distributed in tidal forests of India [15]. Historically, *Pongamia* has been used as folk medicinal plant, particularly in Ayurvedic and Siddha systems of Indian medicine [16] for the treatment of varied human diseases like bronchitis, whooping cough, rheumatism, diarrhoea, dyspepsia, gonorrhoea, leprosy and even tumours. Recent recognition of the importance of the seeds of this plant as a source for biofuel production [17] extends the economic utility of the plant to industrial purposes as well. Considering the multiple economic utility of the plant, the present study was undertaken to analyse the cellular damages induced by *A. pongamiae* on the above plant.

Materials and Methods

Histological Studies

Studies on the anatomical features of the leaf tissue of normal, healthy vs. galled leaves of *P. pinnata* were made by making leaf sections and making comparative assessment of the cellular characteristics of both. This was achieved by making a collection of normal and galled leaves in different stages of development from plants of age groups 3–5, growing in and around the Calicut University Campus. The galled and healthy leaves were cut with the help of scissors, put in polythene bags and transported to the laboratory. In the laboratory, the control and experimental leaf samples were washed in water and cut into pieces of 5-mm size and fixed in FAA (formaldehide-acetic acid-ethyl alcohol) for subsequent dehydration and wax embedding. Cross-sectioning of paraffin-mounted tissues (5–7 μm thickness) was made with a microtome, and staining was done separately in toluidine blue/safranin/haematoxylin for getting better differentiation. Stained sections of normal and galled leaf tissues were examined under LEICA optical microscope for making comparative assessment of the anatomical changes induced by the mite. Microphotography of the stained sections was made with LEICA digital camera (DFC 295) attached to a LEICA fluorescent research microscope. Morphological details of the mite were studied by taking scanning electron micrograph (HITACHI SU6600-FESEM).

Estimation of Polyphenol

Comparative assessment of damage induced by *A. pongamiae* was also made quantitatively by estimating the polyphenol contents present in crude extracts of normal and galled leaf samples. For taking crude extracts, 10 leaves each representing the normal and galled leaves were taken and weighed out separately, for getting wet weight. The samples were then kept in an oven at 60°C for 5 days, and subsequently dry weight of the samples was determined. The dried samples were then crushed and used for the preparation of extracts using water as the solvent. Solvent extraction of the leaf samples was done in a Soxhlet apparatus. The weighed leaf samples were taken in a thimble made of thick filter paper and loaded in the main chamber of the Soxhlet extractor. The Soxhlet extractor was then placed in to a flask containing 250 ml solvent (water) and boiled using electric power supply. The solvent vapour travelled up through the distillation arm was flooded into the chamber housing the thimble of dried plant material. The solvent vapour was cooled by the condenser and dripped back down into the chamber housing the dried powder. When all the components in the plant materials were dissolved in the warm solvent and when the Soxhlet chamber was almost full, the chamber was automatically emptied by a siphon side arm, with the solvent running back down to the distillation flask. The cycle was repeated 40 times. Then the filtrate was taken out and weighed. The crude extract thus prepared in water was used for polyphenol estimation. Different concentrations of the crude extract samples like 50, 100, 250 and 500 μl were taken and made up to 5 ml with distilled water, with 0.1 ml Folin-Ciocalteu reagent and 1 ml of carbonate-tartrate solution added. Samples were incubated for 20 min. at room temperature, and the absorbance was read at 700 nm, using a UV spectrometer.

The polyphenol contents present in the experimental and control samples were estimated separately using the following equation:

Microgram polyphenol/milligram sample

$$\text{Concentration of sample} = \frac{\text{OD of sample}}{\text{OD of standard}} \times \text{concentration of the standard}$$

Fig. 38.1 Plate I Elongate fingerlike galls induced by *Aceria pongamiae* on the leaf of *Pongamia pinnata*

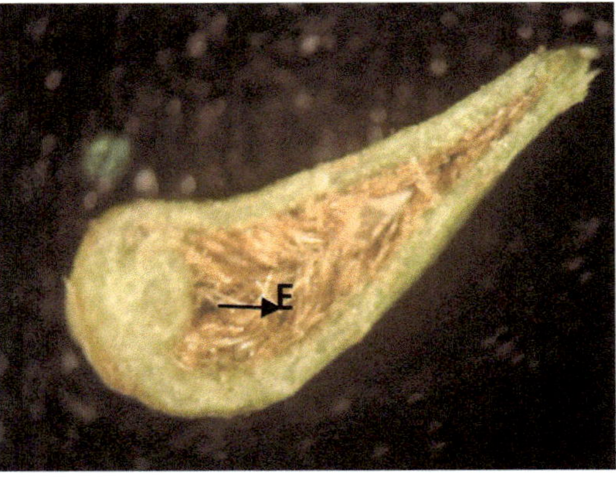

Fig. 38.2 Plate I Section of gall showing inner growth of erinea (Erinea)

Results and Discussion

Mite-infested leaves of *P. pinnata* appeared highly distorted, bearing fingerlike galls developed both on the abaxial and adaxial surfaces (Plate 1, Fig. 38.1). However, the abaxial surface revealed more number of galls. The number of galls often ranged from 4 to 25 or even more, and no regularity could be observed in the arrangement of galls on individual leaf. The galls were found mainly confined to tender leaves. The interior of the galls on observation in the laboratory showed the presence of the eriophyid mite *Aceria pongamiae* in varied stages of development, namely, the egg, I nymph, I quiescent phase, II nymph, II quiescent phase and the adult males and females, along with profuse growth of glandular hairs, the so-called erinea (Plate 1, Figs. 38.2 and 38.3).

Results of comparative studies on the stained sections of normal and galled leaves showed drastic anatomical differences. Normal leaves consisted of single layers of compactly arranged upper and lower epidermal cells. Lower epidermis was found interrupted by the presence of stomata at certain regions. The ground tissue of the leaf lying between the upper and lower epidermis, generally known as the mesophyll, was found comprised of parenchymatous cells containing chloroplast. The mesophyll tissue was found to comprise 2–3 layered palisade and loosely arranged spongy tissues, with conspicuous interspaces. The vascular bundles, which were visible

Fig. 38.3 Plate I Scanning electron micrograph of adults of *Aceria pongamiae* within the gall

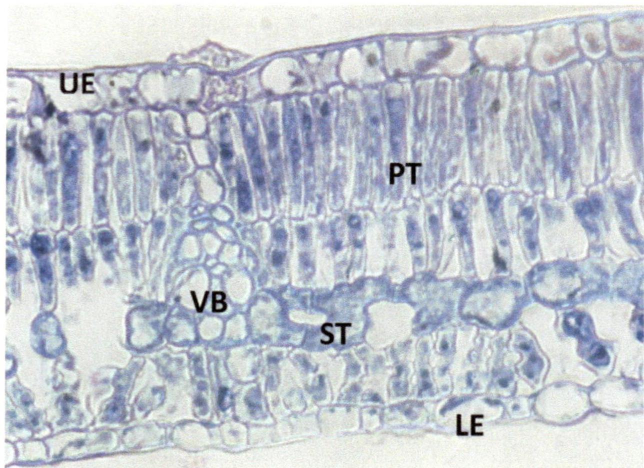

Fig. 38.4 Plate I Section of normal leaf of *P. pinnata* (*UE* upper epidermis, *LE* lower epidermis, *PT* palisade tissue, *ST* spongy tissue, *VB* vascular bundles)

externally as veins of the leaves, were observed in the mesophyll tissue and comprised of xylem and phloem – the conductive elements of the leaf (Plate 1, Fig. 38.4).

The leaf galls, developed on the adaxial surface, disclosed anatomical similarity with that of the normal, healthy leaf, in the initial stage of development. With the advancement of gall development, destruction of the mesophyll tissue was found initiated (Plate 1, Fig. 38.5). However, the xylem vessels remained similar to those of the normal leaves. The upper epidermis lining the interior of the gall cavity was found devoid of any glandular hairs/papillae in the initial stage of development. Ostiolar protrusion was found just initiated on the abaxial surface of leaf, and no tannin deposition could be observed in early galls.

An increase in the width and height of gall tissue was observed in medium-sized galls. High concentration of phenolic compounds could be located in the gall tissue. Mitotic divisions were evident, resulting in the modification of the mesophyll tissue into closely packed and hypertrophied meristematic cells, bearing granular cytoplasm and prominent nuclei (Plate 1, Fig. 38.6). Simultaneously, a large number of uniseriate, multicellular, elongated glandular hairs, the so-called erinea/papillae, were found protruded into the gall, from the lower epidermis. Tannin deposition could be located in the phloem vessels (Plate 1, Fig. 38.7). The epidermis of the adaxial surface of the galled leaf showed the presence of well-developed ostiolar protrusion also (Plate 1, Fig. 38.8).

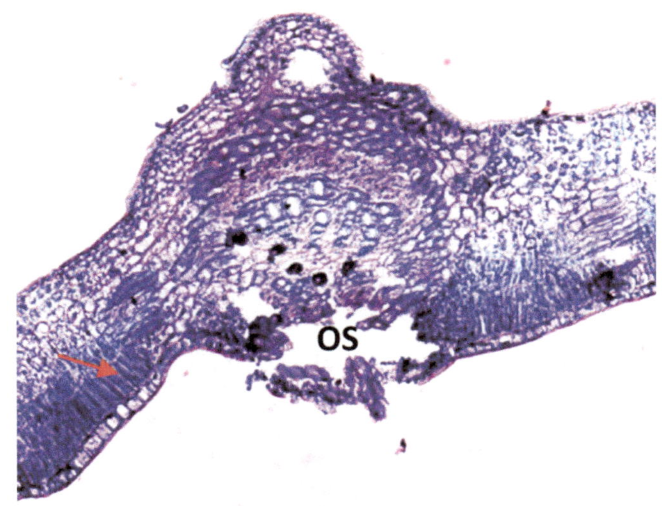

Fig. 38.5 Plate I Section of gall in the initial stage of development. *Arrow* indicates initiation of mesophyll tissue destruction, (*OS*) ostiolar protrusion

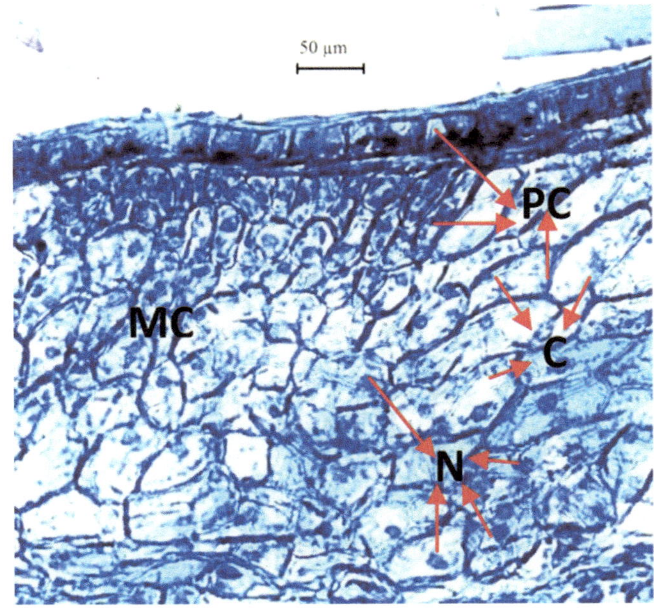

Fig. 38.6 Plate I Medium-sized gall showing accumulation of phenolic contents (*PC*) and hypertrophied meristematic cells (*MC*) with granular cytoplasm (*C*) and prominent nuclei (*N*)

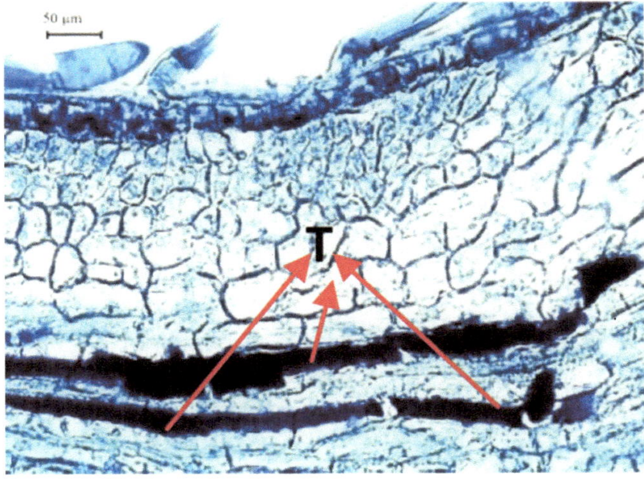

Fig. 38.7 Plate I Medium-sized gall showing tannin deposition (*T*) on phloem vessels

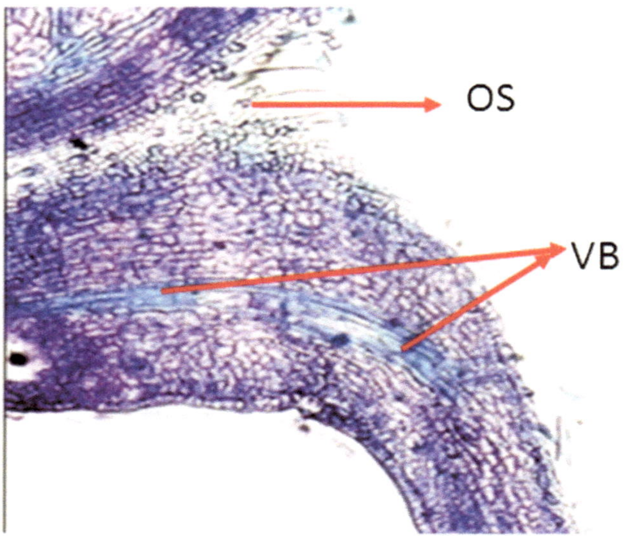

Fig. 38.8 Plate I Medium-sized gall showing ostiolar protrusion (*arrow*)

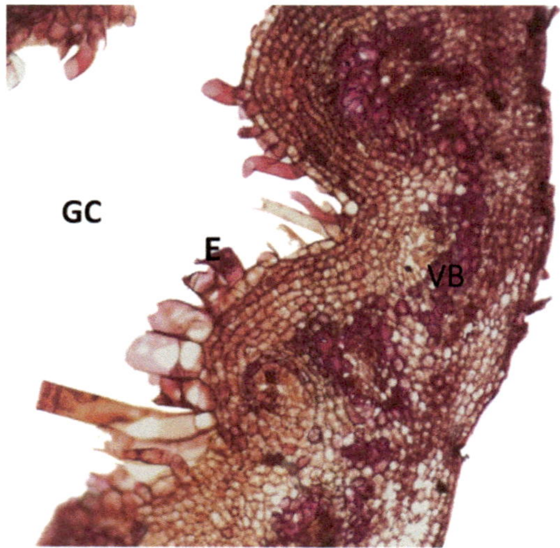

Fig. 38.9 Plate I Cross section of fully developed gall showing elongate erineal hairs/papillae (*E*) and enlarged gall cavity (*GC*)

The fully developed galls showed anatomical similarity to that of the medium-sized gall. However, the gall cavity appeared more enlarged with elongated papillae lining the gall cavity. This resulted in the invagination of the leaf into a cavity-like structure, filled with erineal hairs/papillae (Plate 1, Fig. 38.9). The gall cavity also got elongated with the progressive size of the gall.

Results of quantitative studies on of polyphenolic contents of the normal and galled leaves enabled to record an increase in concentration of polyphenols in mature gall tissue (Plate 2, Fig. 38.10).

A. pongamiae represents a highly host-specific eriophyid mite inducing varying numbers of fingerlike galls on the leaves of *P. pinnata*. The number of galls on a single leaf often ranged from 4 to 25, with a general preference to the abaxial surface, though adaxial galls also could be observed. The present study was focussed to assess the damage potential of *A. pongamiae* by comparing the anatomical features of the normal, healthy leaves vs. galled leaves, through microscopic observation of the stained sections and by estimating the polyphenolic contents.

Results of the study enabled to locate drastic anatomical alterations in the galled tissue, when compared to the normal tissue, owing to the feeding activity of *A. pongamiae.* Probably, the salivary phytotoxins secreted by the eriophyid species, while sucking the leaf sap, would be the root cause for making the anatomical and morphological deviations, leading to the production of galls [18]. The phenotype of the gall, however, is found determined by the complex interaction between the host plant, genotype of the mite and environment [3]. In the present study, *A. pongamiae* was found inducing elongated fingerlike galls, mostly on the abaxial surface, subtended by long, narrow opening with numerous hairs, lining the interior of gall cavity.

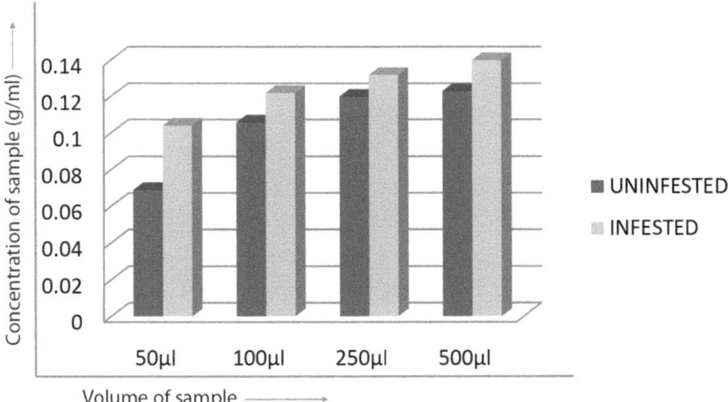

Fig. 38.10 Concentration of polyphenols in the healthy and galled leaf samples of *Pongamia pinnata*

Results of comparative anatomical studies made on the stained sections of normal and galled leaves enabled to locate marked variations in cell structure and form. Extensive injuries could be observed in the structure of mesophyll in the galled tissue, which was found modified into undifferentiated meristematic zone .This could be accounted due to the injection of salivary toxins [18] by the deutogyne of *A. pongamiae*, for digesting the cuticle and cellulose of the cell wall, as reported earlier [19–22]. Subsequently, the deutogyne would pierce the leaf tissue by inserting its cheliceral stylets, resulting in the formation of feeding punctures to suck out cell sap.

The undifferentiated meristematic tissue was found comprised of closely packed, hypertrophied cells, with elongate nuclei and dense cytoplasm. Hyperplasia and hypertrophy have been designated as the unique features of galls, irrespective of variations in the cecidogenic organisms [22]. Closely packed cells at the meristematic zone observed in the galled tissue of *P. pinnata* would be the result of mitotic activity, which would lead to the formation of multiple layers of nutritive tissue, to meet the nutritive requirements of the mite [23, 24]. In *P. pinnata*, the nutritive tissue was recognized to have 8–12 layers thickness. Further, it was also observed that, when the population density of the mite decreased with the advent of unfavourable season in the field, the nutritive tissue also got dried up. This would suggest that prolonged cheliceral puncturing is necessary for the maintenance of nutritive tissue [22].

In the galled tissue, a large number of uniseriate, multicellular and hairlike structures, the so-called erinea, were found developed, lining the interior of the gall cavity. Erineal formation is a characteristic feature of eriophyid galls, and the occurrence of such lignified hairs/papillae and proliferation of labiate tissue surrounding the ostiole was reported earlier [25, 26]. In mature gall tissue of *P. pinnata,* besides hyperplasia and hypertrophy, the meristematic cells showed prominent nuclei and granular cytoplasm, with tannin deposition along the wall of phloem elements as evidenced through toluidine blue staining. An increase in tannin deposition with the progressive development of gall was already reported in several eriophyid galls [26]. Tannins are produced by plants as defensive compounds against herbivory [27], and ingestion of even small amounts of tannin has been shown to make significant reduction in the growth rates of life stages of herbivores [28] as they bind with protein and serve as digestive inhibitors. This would be the reason for the absence of tannin in initial stages of development, as observed in the current study. Tannin deposition became evident in the galls of the medium stage of development [8, 22, 26].

Results of comparative studies pertaining to the estimation of polyphenol contents of normal and healthy leaves revealed an increase in concentration of polyphenols in galled tissues. An increase in concentration of phenols, especially o-dihydroxy phenols, is shown to be accompanied by an increase in indoleacetic acid within galled leaves of *P. glabra* and *Zizyphus jujuba* induced by *Eriophyes cheriani* and *E.cernuus* [29, 30]. In both cases, the formation of galls is reported to be the result of altered hormonal balance, as supported by the plant growth and development by regulating the rate of enzymatic oxidation of IAA and IAA polar transport [31, 32]. Condensed tannins and other polyphenolic compounds accumulated within the galled leaves infested by the mite might protect auxin from oxidation [22].

In conclusion, the result of the present study enabled us to establish the anatomical alterations and enhancement of phenolic compounds, induced by the mite *A. pongamiae* on its host, *P. pinnata*, by recording undifferentiated parenchymatous cells, with multilayered nutritive zone containing hypertrophied nuclei, granular cytoplasm, hyperplasia and hypertrophy and increased levels of tannin and polyphenol contents. Such morphological and anatomical deformations could be detrimental to the plant, adversely affecting the quality of leaves and thereby reducing the economic utility of the plant in the preparation of Ayurvedic medicines. Loss/reduction in the leaf area would also drastically affect the photosynthetic machinery of the plant, thereby leading to biomass loss.

Acknowledgements The first author is grateful to the Ministry of Minority Affairs, Government of India for providing Maulana Azad National Fellowship. The authors are most grateful to Dr A.B. Remasree, Scientist, CMPR, Kottakkal, for providing guidelines and technical assistance for doing the photography work. The help rendered by the Nanotechnology Division of NIT, Calicut, in taking scanning electron micrograph of the mite is also gratefully acknowledged. We are also grateful to the technical help rendered by the Mycology Division of the Department of Botany, University of Calicut.

References

1. Keifer H, Baker E, Kono T, Delfinado M, Styer W. An illustrated guide to plant abnormalities caused by Eriophyid mites in North America, USDA, Agriculture handbook number 573. Washington, DC: U.S. Dept. of Agriculture/Agricultural Research Service; 1982.
2. Westphal E. Cecidogenesis and resistance phenomena in mite-induced galls. In: Shorthouse J, Rohfritsch O, editors. Biology of insect-induced galls. New York: Oxford University Press; 1992. p. 141–56.
3. Weis A, Walton R, Crego C. Reactive plant tissue sites and the population biology of gall makers. Annu Rev Entomol. 1988;33:467–86.
4. Boczek J, Griffiths DA. Structure and systematics of eriophyid mites (Acari: Eriophyoidea) and their relationship to host plants. In: Williams MAJ, editor. Plant galls, systematics association special volume 49. Oxford: Calerdon Press; 1994. p. 119–29.
5. Castagnoli M. Ornamental coniferous and shade trees. In: Lindquist EE, Sabelis MW, Bruin J, editors. Eriophyoid mites – their biology, natural enemies and control. Amsterdam: Elsevier Science Publ; 1996. p. 661–71.
6. Mani MS. Ecology of plant galls. The Hauge: W. Junk Publishers; 1964.
7. Jeppson LR, Keifer HH, Baker BW. Mites injurious to economic plants. Berkeley: University of California Press; 1975.
8. Larew HG. A comparative anatomical study of galls caused by the major cecidogenic groups, with special emphasis on nutritive tissue. Dissertation, Oregon State University, Corvallis USA; 1982.
9. Westphal E, Manson DCM. Feeding effects on host plants: gall formation and other distortion. In: Lindquist EE, Sabelis MW, Bruin J, editors. Eriophyoid mites—their biology, natural enemies and control. Amsterdam: Elsevier; 1996. p. 231–41.
10. Boczek J, Petanovic R. Eriophyid mites as an agent for the biological control of weed. In: Moran VC, Hoffmann JH, editors. Proceedings of the 10th international symposium on Biological Control of Weeds. Stellenobosch South Africa, University of Cape Town; 1996. pp. 127–463, 19–26 Jan 1996.
11. Craemer C, Naser S, Smith Meyer MKP. Eriophyid mites (Acari:eriophyoideae; Eriophoidae) as control agents of weeds in South Africa. SA Tydskrif Vir Naturewetenskap en Technologie. 1996; 15:99–109.
12. Westphal E, et al. Adaptation of gall mites (Acari, eriophyoidea) to live in galls. In: Margaris NS, editor. Plant, animal, and microbial adaptation to terrestrial environment. New York: Plenum Publishing Corporation; 1983. p. 69–75.
13. Royalty RN, Perring TM. Morphological analysis of damage to tomato leaflets by tomato russet mite (Acari: Eriophyidae). J Econ Entomol. 1988;81(3): 816–20.
14. McCoy CW, Albrigo LG. Feeding injury to the orange caused by the citrus rust mite, *Phyllocoptruta oleivora* (Prostigmata: Eriophyoidea). Ann Entomol Soc Am. 1975;68:289–97.
15. Krishnamurthi A. The wealth of India, Vol. VIII. Publication and information Directorate CSIR, New Delhi, India; 1969.
16. Shoba GF, Thomas M. Study of anti-diarrheal activity of four medicinal plants in castor-oil induced diarrhea. J Ethnopharmacol. 2001;76(1):73–6.

17. Naik M, Meher LC, Naik SN, Dasa LM. Production of biodiesel from high free fatty acid Kranja (*Pongamia pinnata*) oil. Biomass Bioenerg. 2008;32:354–7.
18. Styer WE, Nault LR. Gall mites in Ohio. Ohio Rep. 1975;60(1):188–91.
19. Thomsen J. Feeding behavior of *Eriophyes tiliae* Pgst. and suction track in the nutritive cells of the galls caused by mites. Entl Medd. 1988;56(2):73–8.
20. Petanović R, Kielkiewicz M. Plant–eriophyoid mite interactions: specific and unspecific morphological alterations. Part II. Exp Appl Acarol. 2010; doi 10.1007/s 10493-009-9328-1.
21. Bronner R, Westphal E. Modifications des noyaux des cellules epidermiques foliaires de Solanacees soumises a l'action parasitaire d'un acerien cecidogene, *Eriophyes cladophthirus*. Premier Congres de la Societe Francaise de Phytopathologie, 1987; Rennes, Reumes; 39.
22. Bronner R, Westphal E, Dreger F. Chitosan, a component of the compatible interaction between *Solanum dulcamara* L. and the gall mite *Eriophyes cladophthirus* Nal. Physiol Mol Plant Pathol. 1989;34:117–30.
23. Westphal E. Morphogenese, ultrastructure et etiologie de quelque galles d'Eriophyes (Acariens). Marcellia. 1977;39:193–375.
24. Kane NA, Jones CS, Vuorisalo T. Development of galls of *Alnus glutinosa* and *Alnus incana* (Betulaceae) caused by the eriophyid mite *Eriophyes laevis* (Nalepa). Int J Plant Sci. 1997;158(1):13–23.
25. Thompson J. The coevolutionary process. Chicago: University of Chicago; 1994; p. 376.
26. Haukioja E, Niemela P, Siren S. Foliage phenols and nitrogen in relation to growth, insect damage and ability to recover after defoliation in the mountain birch *Betula pubescens* ssp. *tortosa*. Oecologia. 1985;65:214–22.
27. Greene E. A diet- induced developmental polymorphism in a caterpillar. Science. 1989;243:643–6.
28. Feeny P. Effect of oak leaf tannins on larval growth of the winter moth, *Operophtera brumata*. J Insect Physiol. 1968;14:805–17.
29. Balasubramanian M, Purushothaman D. Indole acetic acid in the eriophyid mite gall on *Pongamia glabra* Vent. caused by *Eriophyes cheriani* Massee (Eriophyidae: Acarina). Labdev J Sci Technol. 1972;10-B(3–4):172–3.
30. Tandon P, Arya HC. Presence of auxin protectors in *Eriophyes* induced *Zizyphus* stem galls. Experientia. 1980;36(8):958–9.
31. Lee TT, Starratt AN, Jevnikar JJ. Regulation of enzymatic oxidation of indol-3-accetic acid by phenols: structure activity relationships. Phytochemistry. 1982;21(3):517–23.
32. Hutangura P, Mathesius U, Jones MGK, Rolfe B. Auxin induction is a trigger for root gall caused by root-knot nematodes in white clover and is associated with the activation of the flavonoid pathway. Aust J Plant Physiol. 1999;26:221–31.

39 *Streptomyces* sp.VSMGT1014-Mediated Antifungal Activity Against Fungal Plant Pathogens

H. Harikrishnan and V. Shanmugaiah

Abstract

Presently actinomycetes from rice rhizosphere are screened for novel bioactive secondary metabolites and for biocontrol potential against fungal plant pathogens. A total of 102 actinomycetes were isolated from rice rhizosphere soils collected from Tamil Nadu, India. All the isolates were screened against sheath blight pathogen *Rhizoctonia solani* by dual culture method with inhibition zone range from 0.6 to 3.0 cm. Among 102, 56 have shown the antagonistic activity against *R. solani*. Due to superior biocontrol activity against fungal pathogens over the other actinomycetes, an efficient strain VSMGT1014 was selected and identified as *Streptomyces* sp. based on the biochemical and atomic force microscope observation. The culture filtrate of *Streptomyces* sp. VSMGT1014 remarkably inhibited the mycelial growth and conidial and sclerotial germination of many fungal pathogens such as *Macrophomina phaseolina*, *Fusarium oxysporum*, *Fusarium udum,* and *Alternaria alternata*. The strain *Streptomyces* sp. VSMGT1014 produced five different compounds with Rf value ranges: 0.46, 0.51, 0.61, 0.85, and 0.89. Ethyl acetate extract of crude metabolite from VSMGT1014 significantly inhibited the mycelia growth of most of the fungal pathogens as well as sclerotial germination of *R. solani*.

Keywords

Streptomyces • Antifungal • Metabolites • Biocontrol

H. Harikrishnan • V. Shanmugaiah (✉)
Department of Microbial Technology, School of Biological Sciences, Madurai Kamaraj University, Madurai, 625 021, Tamil Nadu, India
e-mail: vshanmugaiah@yahoo.co.in

A. Sabu and A. Augustine (eds.), *Prospects in Bioscience: Addressing the Issues*,
DOI 10.1007/978-81-322-0810-5_39, © Springer India 2013

Introduction

Actinomycetes represent a large portion of the rhizospheric microbial community and are productive producers of different bioactive secondary metabolites. Several strains of actinomycetes have been found to protect plants against plant diseases. A number of rhizobacteria that were shown to act as effective biocontrol agents by suppressing a variety of economically important phytopathogens often promote overall plant vigor and yield, either when applied to crop seed or incorporated into the soil [1–7]. The worldwide efforts in the search of natural products for the crop protection market have progressed significantly. The actinomycetes, especially those belonging to the genus *Streptomyces,* appear to be good candidates to find new approaches to control plant diseases [8].

Actinomycetes belong to the order *Actinomycetales;* a division of the Gram-positive bacteria characterized by a high genomic G+C content (means 74 mol %) [9–12]. Actinomycete species are well-known saprophytic bacteria that decompose organic matter, especially biopolymers such as lignocellulose, starch, and chitin in soil [13]. Several actinomycetes have characteristic biological features such as a mycelial growth that culminates in sporulation. They also possess the ability to biosynthesize a wide variety of antibiotics as secondary metabolites [14, 15]. The aerial spores of most actinomycete genera were found to resist desiccation and show a slightly higher resistance to wet or dry heat than the corresponding vegetative hyphae. *Streptomyces* and other actinomycetes are major contributors to biological buffering of soils and have roles in organic matter decomposition conducive to crop production.

Biological control is slow but can be long lasting, inexpensive, and harmless to living organisms and the ecosystem; it eliminates neither the pathogen nor the disease but brings them into natural balance. Intensive research on plant growth-promoting bacteria (PGPB) is underway worldwide for developing biofertilizers and biocontrol agents (BCAs) as better alternatives to agrochemicals, as the latter harm the environment and human health besides demanding high costs.

Materials and Methods

Rice Rhizosphere Soil Samples

A total of 18 rice rhizosphere soil samples were collected from paddy fields in southern parts of Tamil Nadu. Rice plants were uprooted gently, placed in sterile polythene bags, and brought to the laboratory. The soils adhered the root surface and were dislodged under aseptic condition and collected in sterile polypropylene bags.

Isolation of Actinomycetes

10 g of rice rhizosphere soil samples was dissolved in 90-ml sterile distilled water, and the flasks were placed on rotary shaker at 150 rpm for 30 min. Then the soil suspension were serially diluted up to 10^{-9} and plated (0.1 ml) on yeast-malt extract agar medium (ISP-2) [16] supplemented with cycloheximide and nystatin (50 μg/ml each) in order to minimize bacterial and fungal contamination. The plates were incubated for 3–14 days at 37 °C. Colonies of actinomycetes on the agar plates were picked on the basis of their morphological characteristics such as coloration of aerial mycelium (on the surface of agar), substrate mycelium (underside of agar), and diffusible pigments. The strain was purified on ISP-2 medium and maintained on the same medium by subculturing at monthly intervals.

Screening of Actinomycetes

All the 102 actinomycetes were subjected to screening against fungal pathogens such as *Rhizoctonia solani, Alternaria alternata, Fusarium oxysporum, Fusarium udum, and Macrophomina phaseolina* by dual culture method [17] on PDA and inoculated at the center of a fresh PDA Petri plate. Four actinomycetes were streaked in the periphery of each Petri plate and incubated at room temperature. Among the 102 actinomycetes, a strain designated as VSMGT1014 isolated from Madurai district was selected for further studies as it effectively inhibited most of the fungal pathogens than the other isolates.

Antifungal Activity of *Streptomyces* sp. VSMGT1014 Against Phytopathogens

Streptomyces sp. VSMGT1014 was grown in ISP-2 broth at 150 rpm in a rotary shaker at room temperature. After 6 days, the culture was harvested and centrifuged at 10,000 rpm for 15 min. After centrifugation the culture filtrate of VSMGT1014 was bioassayed against all the above fungi by agar well diffusion method at the concentrations of 25, 50, 75, and 100 μl, and the mycelial growth inhibition was recorded.

Biochemical Characterization of Streptomyces sp. VSMGT1014

The potent actinomycetes were characterized by morphological and biochemical methods. Morphological methods consist of macroscopic and microscopic observations. The microscopic characterization was done by cover slip culture method [18]. The mycelium structure, color, and arrangements of conidiospores and arthrospores on the mycelium were observed through atomic force microscope. Different physiological tests suggested in Bergey's Manual of Determinative Bacteriology [19] were carried out to identify the isolate. Various biochemical tests performed for the identification of potent isolate are as follows: indole test, methyl red test, Voges-Proskauer test, starch hydrolysis, and gelatin liquefaction. Lytic enzyme production tests were also performed for protease, amylase, and cellulase. Optimization of temperature, pH, and growth incubation time for the isolated strain was carried out.

Light Microscopic Study of VSMGT1014 and *R. solani* Interaction

The interaction of VSMGT1014 and *R. solani* was studied using light microscopy. Both the pathogen and antagonist were co-inoculated on PDA and PDB up to 2 and 5 days, respectively. Samples were taken from the pathogen and antagonist interaction zone and mycelia from PDB observed under the light microscope after staining with lactophenol cotton blue.

Extraction of Antifungal Metabolites

Crude metabolites were extracted from *Streptomyces* sp. VSMGT1014 by normal solvent extraction method [20]. The crude metabolites were tested against plant pathogens by well diffusion assay and sclerotial germination assay.

Thin-Layer Chromatography

Thin-layer chromatography of the crude compounds was performed using precoated TLC silica plates to detect the compounds produced by *Streptomyces* sp. VSMGT1014. The crude was eluted with different solvent system and visualized under UV and iodine and the Rf value of each spot was calculated.

Effect of Culture Filtrate and Crude Compound on Sclerotial Germination

Uniform-sized sclerotia collected from 7-day-old culture of *R. solani* plates were soaked with filter-sterilized 3-day-old culture filtrate, crude metabolite of VSMGT1014, positive control carbendazim, and universal control sterile distilled water for different time intervals: 4, 8, 12, 16, 20, and 24 h. Then sclerotia were washed in sterile distilled water, blot dried using sterile blotting papers in aseptic conditions, and inoculated on PDA plates. Plates were incubated at 28 °C; after incubation the plates were observed 1–3 days.

Results and Discussion

Isolation and Screening of Actinomycetes

A total of 102 actinomycetes were isolated from rice rhizosphere soils and screened for

Table 39.1 Antagonistic isolates from rice sample identification of VSMGT1014

District	Soil samples	Isolates
Madurai	10	70
Dindigal	1	4
Kanyakumari	3	14
Sivagangai	4	14
Total	18	102

Table 39.2 *In vitro* antagonism of actinobacterial isolates against plant pathogens

Plant pathogens	No. of antagonistic isolates	Zone of inhibition(cm)
Rhizoctonia solani	56	0.6–3.0
Alternaria alternate	57	0.4–2.4
Fusarium oxysporum	32	0.2–2.5
Fusarium udum	39	0.2–1.2
Macrophomina phaseolina	41	0.4–2.5

antifungal activity against *R. solani* (Table 39.1). Among 102, 56 isolates exhibited antagonistic activity and produced inhibition zones ranging from 0.6 to 3.0 cm. Among 56, the isolate-designated VSMGT1014 showed highest antifungal activity, a broad spectrum of phytopathogens such as *A. alternata, F. oxysporum, F. udum, and M. phaseolina* (Table 39.2, Fig 39.1).

Identification of *Streptomyces* sp. VSMGT1014

The highly potential antagonistic strain VSMGT1014 was identified as *Streptomyces* sp. based on the physiological, morphological, biochemical characteristic tests and atomic force microscopic studies (Table 39.3).

Effect of Culture Filtrate of *Streptomyces* sp. VSMGT1014 Against Fungal Pathogens

The culture filtrate of *Streptomyces* sp. VSMGT1014 effectively inhibited the conidial and sclerotial germination of fungal pathogens. The results indicated that VSMGT1014 grown on PDA plates released extracellular diffusible metabolites that inhibited the hyphal growth of *R. solani*. The antagonistic potential of actinomycetes pathogenic fungi involved the production of extracellular hydrolytic enzymes and moreover secondary metabolites or antifungal compounds. The decrease in the degree of fungal growth inhibition corresponds to the decrease in concentration of the culture filtrate (Table 39.4).

Effect of Crude Metabolite Against Sclerotial Germination of *R. solani*

The presence of five compounds with Rf value range such as 0.46, 0.51, 0.61, 0.85, and 0.89 was detected by thin-layer chromatography (TLC). This crude metabolite with five compounds remarkably inhibited the sclerotial germination of *R. solani*.

In recent years, the use of secondary metabolites of microbial origin is gaining great momentum in crop protection, and these metabolites may be a supplement or an alternative to chemical control [14, 21]. Thus, worldwide interest in them has been renewed, and presently about 40 different plant diseases are being managed with microbial metabolites [22]. Indian soils are rich in microbial diversity, especially actinomycetes [23], and the wealth of indigenous microflora of India has not been fully explored. In the present investigation, a total of 102 actinomycetes were isolated from the rice rhizosphere, of which 56 strains exhibited antifungal activity against plant pathogens. Among the 52 antagonists, a strain designated as VSMGT1014 was selected based on its broad-spectrum activity against various phytopathogens. The selected strain VSMGT1014 was identified as *Streptomyces* sp. based on the morphological, biochemical, and atomic force microscope observation. Out of 52, VSMGT1014 shown to be highly effective against *R. solani* as it remarkably inhibited their growth. Further, it has been observed that the culture filtrate of *Streptomyces* sp. VSMGT1014 completely inhibited the conidial and sclerotial germination of *R. solani*. The presence of five major compounds with the R_f value range such as 0.46, 0.51, 0.61, 0.85, and 0.89 was detected by analytical TLC.

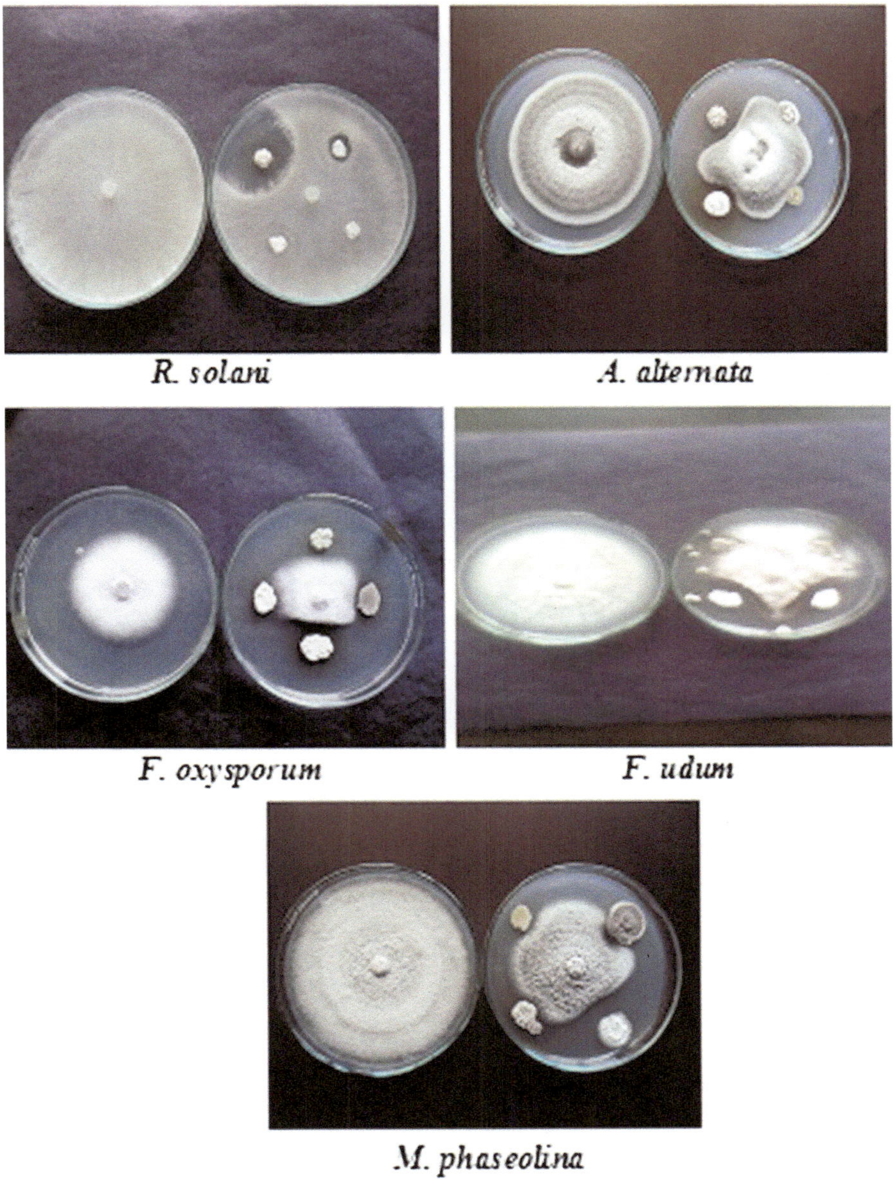

R. solani *A. alternata*

F. oxysporum *F. udum*

M. phaseolina

Fig. 39.1 Screening of actinobacteria against fungal pathogens

These partially purified compounds significantly inhibited the conidial and sclerotial germination of *R. solani*.

In spite of the use of many modern chemical fungicides, the management of these diseases in the farmers' fields is still a difficult task. Further, many of those fungicides although effective had narrow spectrum of activity, e.g., tricyclazole, a systemic fungicide, is effective for blast disease but is ineffective against sheath blight. Similar narrow-spectrum activity was reported with some *Streptomyces* metabolites such as blasticidin S and kasugamycin, which are effective only against blast of rice [24], and validamycin A is effective only against sheath blight of rice [25]. The present investigation demonstrates the role of rice rhizosphere actinomycetes in biocontrol of various fungal pathogens by *in vitro* antagonistic action

Table 39.3 Physiological and biochemical characteristics of VSMGT1014

Biochemical test	Characteristics
Gram staining	+
Motility	Nonmotile
Aerial mycelium	White
Production of diffusible pigment	±
Optimum temperature for growth	37 °C
Range of pH for growth	6–8
Optimum pH for growth	7.5
Indole test	–
Methyl red test	+
Voges-Proskauer test	+
Starch hydrolysis	+
Gelatin liquefaction	–
Lytic enzyme production	Characteristics
Amylase	+
Protease	+
Cellulase	+

+ positive response, – negative response, ± feeble response

Table 39.4 Antifungal activity of culture filtrate on plant pathogens

	Zone of inhibition in cm			
Plant pathogens	25 µl	50 µl	75 µl	100 µl
R. solani	2.2	2.3	2.4	2.5
A. alternata	2.3	2.4	2.5	2.6
F. oxysporum	2.3	2.5	2.6	2.7
F. udum	2.3	2.6	2.6	2.7
M. phaseolina	2.2	2.3	2.4	2.5

due to the production of secondary metabolites. Hence, the biological treatment not only reduces the diseases but also increases the plant growth, especially within the framework of integrated disease management system. Also the results strongly suggest that *Streptomyces* sp. VSMGT1014 can be employed not only as a biocontrol agent for the management of plant diseases but also promotes the plant growth.

Acknowledgement The authors greatly acknowledge the Head, Department of Microbial Technology; the Coordinator, NRCBS; and the Chairperson, School of Biological Sciences, Madurai Kamaraj University, Madurai, Tamil Nadu, India.

References

1. Bashan Y. *Azospirillum* plant growth promoting strains are non-pathogenic on tomato, pepper, cotton, and wheat. Can J Microbiol. 1998;44:168–74.
2. Brown AE, Surgeoner R. Enhancement of plant growth by *Zygorrhynchus moelleri*. Ann Appl Biol. 1991;118:39–46.
3. Burr TJ, Schroth MN, Suslow T. Increased potato yields by treatment of seed species with spécific strains of *Pseudomonas fluorescens* and *P. putida*. Phytopathology. 1978;68:1377–83.
4. Kloepper JW, Leong J, Teintze M, Schroth MN. Enhanced plant growth by siderophores produced by plant growth-promoting rhizobacteria. Nature. 1980;286:885–6.
5. Kloepper JW, Schroth MN, Miller TD. Effects of rhizosphere colonization by plant growth-promoting rhizobacteria on potato plant development and yield. Phytopathology. 1980;70:1078–82.
6. Kloepper JW, Lifshitz R, Zablotowicz RM. Free-living bacterial inocula for enhancing crop productivity. Trends Biotechnol. 1989;7:39–44.
7. Turner JT, Backman PA. Factors relating to peanut yield increases after seed treatment with *Bacillus subtilis*. Plant Dis. 1991;75:347–53.
8. Behal V. Bioactive products from *Streptomyces*. Adv Appl Microbiol. 2000;47:113–57.
9. Fox GE, Stackebrandt E. The application of 16S rRNA cataloguing and 5S rRNA sequencing in bacterial systematics. Methods Microbiol. 1987;19:405–58.
10. Goodfellow M, Cross T. Classification. The biology of actinomycetes. London: Academic; 1984. p. 7–164.
11. Goodfellow M, Ferguson EV, Sanglier JJ. Numerical classification and identification of *Streptomyces* species- a review. Gene. 1992;115:225–33.
12. Stackebrandt E, Woese CR. Towards a phylogeny of the actinomycetes and related organisms. Curr Microbiol. 1981;5:197–220.
13. Crawford DL, Lynch JM, Whipps JM, Ousley MA. Isolation and characterization of actinomycete antagonists of a fungal root pathogen. Appl Environ Microbiol. 1993;59:3899–905.
14. Franklin TJ, Snow GA, Barrett-Bee KJ, Nolan RD. Antifungal, antiprotozoal and antiviral agents. In: Franklin TJ, Snow GA, Barrett-Bee KJ, Nolan RD, editors. Biochemistry of antimicrobial action. 4th ed. New York: Chapman & Hall; 1989. p. 137–61.
15. Lechevalier HA, Waksman SA. The actinomycetes. III. Antibiotics of actinomycetes. Baltimore: Williams & Wilkins; 1962. 430.
16. Shirling EB, Gottlieb D. Methods for characterization of *Streptomyces* species. Int J Syst Bacteriol. 1966; 16:313–40.
17. Huang MC, Hoes JA. Penetration and infection of *Sclerotinia sclerotiorum* by *Coniothyrium minitans*. Can J Bot. 1976;54:406–10.
18. Carrera LM, Buyer JS, Vinyard B, Abdul-Baki AA, Sikora LJ, Teasdale JR. Effects of cover crops, compost and manure amendments on soil microbial

community structure in tomato production systems. Appl Soil Ecol. 2007;37:247–55.

19. Williams ST, et al., editors. Bergey's manual of determinative bacteriology. 9th ed. Baltimore: Williams and Wilkins Co; 1994.
20. Shanmugaiah V, Mathivanan N, Varghese B. Purification, crystal structure and ntimicrobial activity of phenazine-1-carboxamide produced by a growth-promoting biocontrol bacterium, *Pseudomonas aeruginosa* MML2212. J Appl Microbiol. 2010;108:703–11.
21. Colins JE, Chafik Z. Comparison of biological and chemical treatments for control of bacterial speck of tomato under field conditions in Morocco. Plant Dis. 1986;70:1048–50.
22. Shanmuganathan K, Yasin J, Jayaprakasam M. Antibiotics in agriculture. Agric Today. 2001;6:40–1.
23. Malarvizhi K, Mathivanan N. Biodiversity and antimicrobial activity of soil actinomycetes. Abstract of the National Conference on Emerging Trends in Mycology, Plant Pathology and Microbial Biotechnology. Osmania University, Hyderabad, India; 2004. 29–31 Dec 2004.
24. NHMRC. Antibiotics in agronomy and horticulture. National Health and Medical Council, Commonwealth of Australia. Page content last modified: 5 Feb 2001, URL: http://www.nhmrc.health.gov.au/publicat/fullhtml/dp15-p2.html
25. Yamaguchi I. Pesticides of microbial origin and applications of molecular biology. In: Copping LG, editor. Crop protection agents from nature: natural products and analogues. Cambridge: The Royal Society of Chemistry; 1996. p. 27–49.

40 Taxonomic Significance of Foliar Micromorphology and Their Systematic Relevance in the Genus Solanum (Solanaceae)

V.S. Anil Kumar and K. Murugan

Abstract

Micromorphology of vegetative and reproductive plant organs is the object of research to resolve the taxonomic problems of critical species and genera. Trichomes are distributed at the surface of aerial plant parts, having various functions and are extremely variable in their presence across plant, location on plant organs, density, form, etc., and, therefore, their morphology and structure can be used as taxonomic markers in the infrageneric classification of the genus. The purpose of this investigation was to analyse the degree of differentiation in micromorphological characters between the six species and to assess the expression of these characters in individuals with intermediate macromorphology. Of the several traits on leaf surface, the stomata are perhaps the most significant from the point of view of systematics and phylogeny. Stomata that are highly characteristic of the epidermis occur in widely divergent parts of the plants including common foliage leaves. Stomatal size is an ecologically important attribute. The size, distribution and frequency of stomata have been recognized as species specific and these characters were used as significant parameters in the angiosperm taxonomy as well as phylogeny; qualitative and quantitative micro morphological characters, distribution and systematic relevance of both glandular and eglandular foliar trichomes using scanning electron microscope (SEM) in species of genus Solanum (Solanaceae) including *S. melongena* var. *incanum*, *S. aculeatissimum*, *S. erianthum*, *S. macrocarpon*, *S. mammosum* and *S. virginianum* were characterized. Both the abaxial and adaxial surfaces were characterized by anisocytic stomata which were more prevalent on the abaxial surface than the adaxial surface. Glandular trichomes were capitate while nonglandular trichomes were stellate with many arms or linear fingerlike. Glandular trichomes play important roles in protecting plants from biotic attack by producing defensive compounds.

V.S. Anil Kumar • K. Murugan (✉)
Department of Botany, University College,
Thiruvananthapuram, Kerala, India
e-mail: harimurukan@gmail.com

A. Sabu and A. Augustine (eds.), *Prospects in Bioscience: Addressing the Issues*,
DOI 10.1007/978-81-322-0810-5_40, © Springer India 2013

The trichomes in the family Solanaceae vary from unicellular to multicellular, conical to elongated and smooth to ridges, with or without flattened disk at the base. Glandular capitate trichomes were the most unique on leaf surfaces of the examined taxa. In summary, the variation in morphology and distribution of foliar trichomes emerged as an important supportive taxonomic tool in delimiting species of genus Solanum.

Keywords
Leaf epidermal anatomy • Trichomes • Capitate glands • Solanum

Introduction

Epidermal micromorphology of the leaves is used in emphasizing the interrelationships, and segregation into major clades [1, 2] has supported the earlier grouping for various taxa from different families of plant systematics. Epidermal micro-characters are quite important to delineate the different taxa in terms of phylogenetic and taxonomic considerations. Angiosperms leaves are the most suited and studied for these purpose [3]. Several authors recognized the importance of these epidermal features in recognizing the different taxa belonging to different families [4–6]. More importantly this study useful to locate the markers present within the circumcision of the micromorphology [7] utilized epidermal characters for understanding the interrelationship of the various taxa. Indeed the foliar epidermal characters of the angiosperms depict the sufficient diversity of details due to its genetic and environmental make-up. Since the micromorphological characteristics of foliar trichomes have played an important role in plant taxonomy, especially of particular groups at generic and specific levels, more and more studies in this field have attracted the attention of plant morphologists and systematists to resolve the taxonomic conflicts. Similarly, of the several traits on leaf surface, the stomata are perhaps the most significant from the point of view of systematics and phylogeny. Stomata that are highly characteristic of the epidermis occur in widely divergent parts of the plants including common foliage leaves [8]. Stomatal size is an ecological important attributes. The size, distribution and frequency of stomata have been recognized to be specific to the taxa below the family, and these characters were used as significant parameters in the angiosperm taxonomy as well as phylogeny .The importance of epidermal cell characters is now well established in taxonomic considerations of angiosperms[5]. Therefore, in this chapter, the micromorphological characteristic of foliar trichomes in *Solanum* was studied by means of scanning electron microscopy (SEM). The specific objective of this chapter was to compare the micromorphological characteristics of foliar nature in different species of this genus.

Materials and Methods

Leaf materials used in this study were obtained from freshly collected plants during the expeditions to various parts of Kerala. The plants were identified at the Department of Botany, University of Calicut, and a voucher specimen was prepared and deposited in the herbarium of the Department of Botany, University College, Thiruvananthapuram.

Scanning electron microscopy: Fresh leaf pieces (10×10 mm^2) from all the species such as *S. melongena* var. *incanum, S. aculeatissimum, S. erianthum, S. macrocarpon, S. mammosum* and *S. virginianum were* immersed in a fixative solution of 3% glutaraldehyde in 0.1-M phosphate buffer for 24 h. Samples were washed for 15–30 min with the buffer and dehydrated in graded ethanol series. Samples were then critical point dried using CO_2, sputter coated with gold under vacuum and viewed with Hitachi (S-450) scanning electron microscope operating at 15 kV. Images were captured digitally with an Image Slave computer program for Windows.

Results and Discussion

SEM Study of the Leaf Reveals the Following Features

Surface of Leaf

S. aculeatissimum adaxial leaf surface shows irregular ridges and furrows appearing like chains of hillocks. Each ridge is shallowly undulated and striated. Flakes of wax particles are randomly distributed throughout the surface (Fig. 40.1a). In the case of *S. erianthum*, both surfaces were densely pubescent with short-stalked multiradiate stellate nonglandular hairs (Fig. 40.1b and c). In *S. melongena var. incanum*, both the surfaces were sparsely pubescent with sessile multiradiate stellate nonglandular hairs. Short-stalked multicellular glandular hairs were dense in the upper surface (Fig. 40.1d and e). *S. macrocarpon* leaf was ridged with short-stalked multicellular glandular hairs on both surfaces, whereas unicellular nonglandular pointed hairs are restricted along the margin of the leaf (Fig. 40.1f and g). Both the leaf surfaces of *S. mammosum* were studded with long-stalked multicellular glandular hairs intermingled with long multicellular nonglandular hairs (Fig. 40.1h and i). Interestingly, the lower leaf surface showed the presence of short-stalked multicellular glandular hairs along the veins. *S. virginianum* upper leaf surface showed the presence of long multicellular nonglandular hairs with bulbous tip, short-stalked multicellular glandular hairs and multicellular nonglandular stalked hairs with pointed tip (Fig. 40.1j). Lower leaf surface showed few stellate hairs, long multicellular nonglandular hairs with bulbous tip and multicellular nonglandular stalked hairs with pointed tip (Fig. 40.1k).

Generally the abaxial surface is uneven due to differential cuticularization. The cuticular striations are riblike with variable curvatures and apparent interlocking or articulations. *S. aculeatissimum* showed glandular trichomes all along the surface with sparse distribution of long multicellular nonglandular hairs with pointed tip (Fig. 40.2a). The ribs are in somewhat parallel orientation xent directions from stomatal rim. Each rib is differentially thickened and curved and has undulated surface. The leaf is amphistomatic with anomocytic stomata.

Stomata

Each stoma is lodged in a shallow depression. It is oval in shape, the aperture being elliptic with an outline of waxy deposition. The guard cells are slightly protuberant and the lateral walls of the guard cells are with uneven waxy coating (Fig. 40.3a to d).

The epidermal cells have sinuous anticlinal walls on the adaxial surface and waved on the abaxial surface. In cross section, the epidermis is one-layered with rounded cells, and the stomata are located on the same level as the other epidermal cells. The analysis of epidermal prints has shown that the epidermal tissue consisted of cells with rugose anticlinal walls, especially abaxially. Indumentum, made of nonglandular and glandular hairs, was sparse on both epidermal sides. Hairs were more numerous along veins and on the abaxial lamina side.

There were two types of glandular hairs. The hairs of one type consisted of a short, multicellular stalk and a multicellular secretory head, round to oval in shape (Fig. 40.1f). The hairs of the second type had a longer stalk, made of two, three or four cells, and a unicellular secretory head at the top (Fig. 40.1h). The stalk cell under the head was narrow. It has been noticed that this cell dried faster than the other stalk cells. Stalk cells closer to the hair base were widened.

Glandular trichomes are characterized by having 'heads' (glands) that release, on contact, sticky and/or toxic exudates that may entrap, irritate or potentially kill some pests [9]. These glands contain important secondary metabolites including terpenes, essential oils, flavonoids and lipophilic components [10, 11]. In most species, the source of these secondary metabolites has been attributed to the trichomes [12]. The possession of glandular trichomes is characteristic of the genus *Solanum* and of many other members

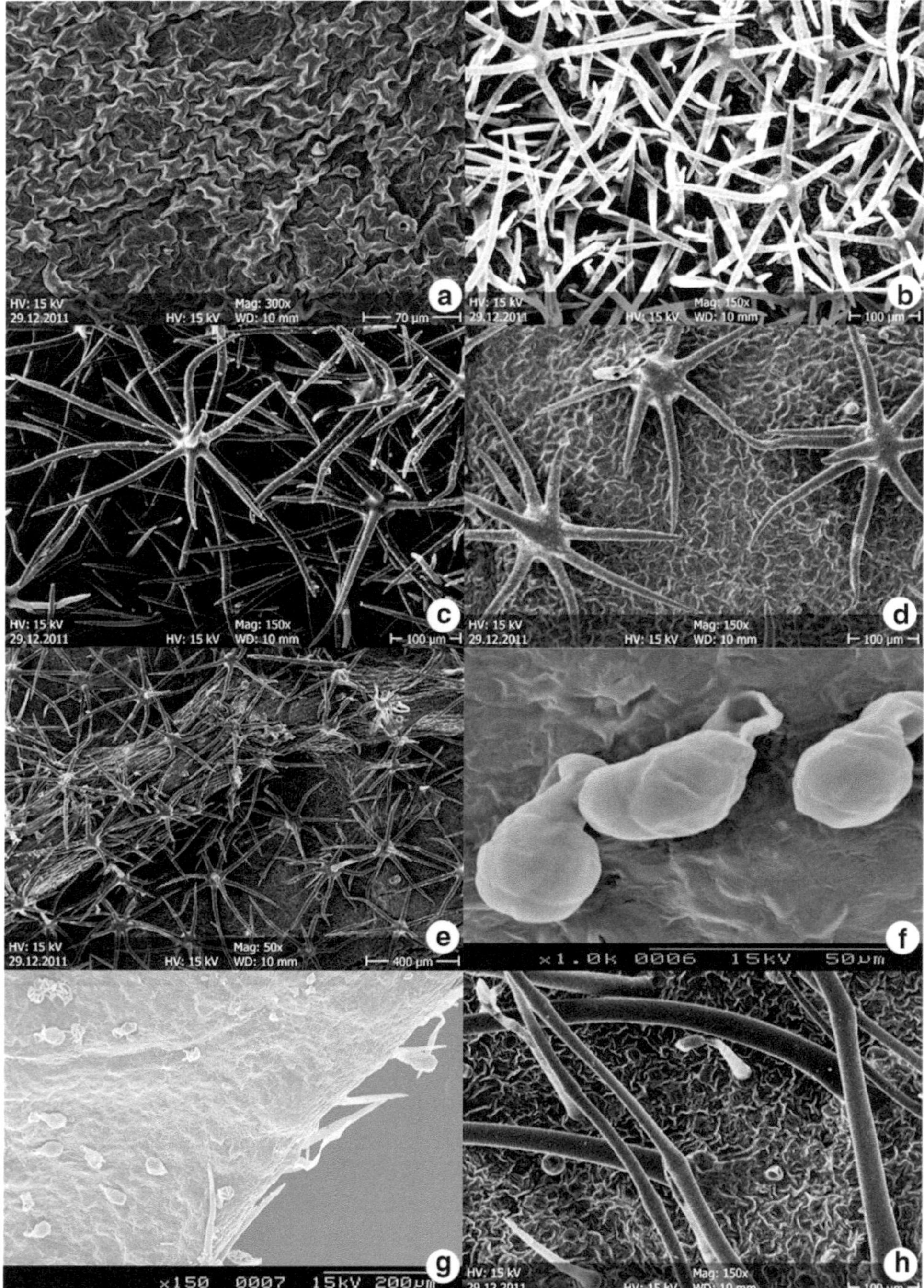

Fig. 40.1 Fig. 1a – h. 1a. Upper leaf surface of *Solanum aculeatissimum* showing ridges and furrows. Fig. 1b and c Upper and lower leaf surfaces of *Solanum erianthum* showing multiradiate nonglandular stellate hairs. Fig. 1d Upper leaf surface of *Solanum melongena var. incanum* showing multiradiate nonglandular stellate hairs and short-stalked glandular hairs. Fig. 1e Lower leaf surface of *Solanum melongena var. incanum* showing multiradiate stellate nonglandular hairs. Fig. 1f. Leaf surface *of Solanum macrocarpon* showing short-stalked multicellular glandular hairs. Fig.1g. Leaf margin of *Solanum macrocarpon* showing unicellular nonglandular pointed hairs along with glandular hairs. Fig. 1h Upper leaf surface of *Solanum mammosum* showing long-stalked multicellular glandular hairs along with long multicellular nonglandular hairs

Fig. 40.1 (continued)

of Solanaceae, with the exception of *Nicotiana glauca* and *Solandra nitida* [13]. The two types of glandular trichomes identified on the leaves of *Solanum* might be responsible for the production, accumulation and release of volatile and secondary metabolites such as the saponins and steroid alkaloids reported by Drewes and Van Staden [14]. Although micromorphological studies alone do not provide the information required to establish sites of synthesis in cells [10], it is plausible to assume that the therapeutic compounds in *Solanum* are produced by the glandular trichomes.

The leaf epidermis surfaces were investigated to evaluate their taxonomic significance to be used for separation and delimitation of the species of the section. Micro morphological characters of the leaves such as density, distribution and type of stellate trichomes, the anticlinal walls of epidermal cells and also the type and distribution of stomata proved to be the most useful and distinctive characters for the separation and delimitation of the species and also may contribute as an additional support to the interspecific taxonomy and systematics of *Solanum*.

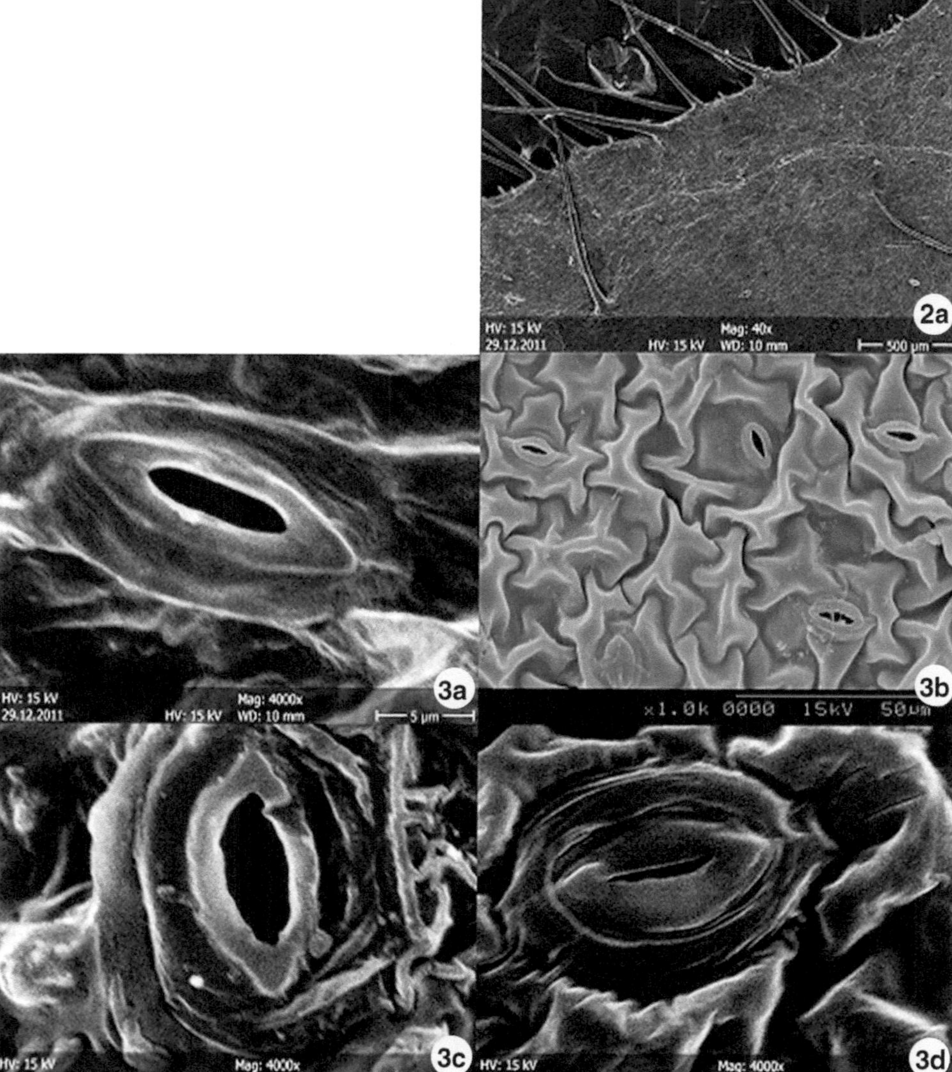

Fig. 40.2 1i. Lower leaf surface of *Solanum mammosum* showing long-stalked multicellular glandular hairs, long multicellular nonglandular hairs and short-stalked multicellular glandular hairs along the veins. Fig. 1j Upper leaf surface of *Solanum virginianum* showing multicellular nonglandular hairs with bulbous tip, short-stalked multicellular glandular hairs and multicellular nonglandular hairs with pointed tip. Fig. 1k Lower leaf surface of *Solanum virginianum* showing stellate hair, long multicellular nonglandular hairs with bulbous tip and multicellular nonglandular hairs with pointed tip. Fig. 2a. Leaf surface of *Solanum aculeatissimum* showing glandular hairs with sparse distribution of multicellular nonglandular hairs with pointed tip. Fig. 3a Stoma of *Solanum aculeatissimum.* Fig. 3b Stoma of *Solanum macrocarpon.* Fig. 3c Stoma of *Solanum melongena var. incanum.* Fig. 3d Stoma of *Solanum virginianum*

References

1. Metcalfe CR. Anatomy of the monocotyledons I-Gramineae. London: Oxford University Press; 1960. p. 1–711.
2. Stebbins GL, Khush GS. Variation in the organization of the stomatal complex in the leaf epidermis of monocotyledons and its bearing on their phylogeny. Am J Bot. 1961;48:51–9.
3. Watson L, Clifford HT, Dallwitz MJ. The classification of Poaceae, subfamilies supertribes. Aust J Bot. 1985;33:433–84.
4. Inamdar JA, Mohan JSS, Subramanium RB. Stomatal classification – a review. Feddes Rep. 1986;97(3 & 4):147–60.
5. Parveen NS, Murthy KSR, Pullaiah T. Leaf epidermal characters in *Crotalaria* sp. Papilionoideae from Eastern Ghats. Phytomorphology. 2000;50:205–12.
6. Edeoga HO, Ikem CI. Comparative morphology of leaf epidermis in three species of *Boerhhavia* L. (Nyctaginaceae). In: Maheshwari JK, Jain AP, editors. Recent researches in plant anatomy and morphology. Jodhpur: Scientific publishers; 2001.
7. Hilu KW. A century of progress in grass systematics. Kew Bull. 2007;62:355–73.
8. Badmus AA, Afolayan AJ. The foliar micromorphology of Arctotis arctotoides (L.f.) *O. Hoffm.* J Med Plant Res. 2010;4(16):1643–6.
9. Simmons AT, Gurr GM, McGrath Nicol HI, Martin PM. Trichomes of *Lycopersicon* spp. and their effect on *Myzus persicae* (Sulzer), Hemiptera: Aphidae. Aust J Entamol. 2003;42:373–8.
10. Afolayan AJ, Meyer JJM. Morphology and ultra structure of secreting and non secreting foliar trichomes of *Helichrysum aureonitens* (Asteraceae). Int J Plant Sci. 1995;156:481–7.
11. Ascensao L, Mota L, De Castro MM. Glandular trichomes on the leaves and flowers of *Plectranthus ornatus*. Morphology, distribution and histochemistry. Ann Bot. 1999;84:437–44.
12. Buta JG, Lusby WR, Neal JW, Waters RM, Pittarelli GW. Sucrose esters from *Nicotiana gossei* active against the green house whitefly *Trialeurodes vaporarium*. Phytochemistry. 1993;32:859–64.
13. Maiti RK, Villarreal LR, Trevino V, Vallades-Cerda MC. Some aspects on pharmacognosy of ten species of the family solanaceae utilized in traditional medicine. Caldasia. 2002;24:317–21.
14. Drewes FE, Van Staden J. Aspects of the extraction and purification of solasodine from *Solanum aculeastrum* tissues. Phytochem Anal. 1995;6:203–6.

41 Investigation of Functional Microbial Diversity and Related Abiotic Factors in Coastal and Desert Ecosystem of Gujarat

D. Biswas, A. Mandal, H. Akbari, and N. Munshi

Abstract

Soil microbial diversity is of utmost importance for functioning of any ecosystem. Samples from coastal region of Gujarat, that is, from Alang, Daman, Mandvi, Okha, Porbandar and Veraval, and from desert areas of Bhuj, Dhordo and Khavda were collected. The purpose of our study was enumeration and functional diversity analysis of 14 functional microbial groups of microorganisms present in both ecosystems and correlation of microbial data with abiotic parameters as well as with important soil enzyme activities. Functional diversity was analysed using CLPP. Finally, correlation between abiotic parameters and microbial data was done using SPSS, and various diversity indices values were calculated. We concluded that in comparison to coastal soil, desert soil exhibited high diversity, total activity and richness. PCA analysis displayed that coastal samples were clustered together, while the desert samples were separated from each other and from the coastal samples group. The coastal sites of Okha, Porbandar and Veraval seemed to be deteriorating with low microbial diversity.

Keywords

Functional microbial diversity • CLPP • PCA

D. Biswas
Department of Biotechnology and Environmental Sciences, Thapar University, Patiala, Punjab, India

A. Mandal • H. Akbari • N. Munshi (✉)
Institute of Science, Nirma University, Sarkhej-Gandhinagar Highway, Ahmedabad, Gujarat, India
e-mail: nasreenhaque@hotmail.com

A. Sabu and A. Augustine (eds.), *Prospects in Bioscience: Addressing the Issues*,
DOI 10.1007/978-81-322-0810-5_41, © Springer India 2013

Table 41.1 Different categories of soil microorganisms and their roles

Main division	Subdivision	Functional role	References
Bacteria	Prototrophic	Metabolise single carbon source like malate or acetate	[6]
	Nitrogen fixers	Nitrogen fixation, nitrification and denitrification	[7, 8]
	Nitrifiers		
	Denitrifiers		
	Ammonia oxidisers	Conversion of ammonium to nitrite	[8, 9]
	Element transformers	Ferric reducers	[7, 9]
		Sulphate reducers	
		Phosphate solubilisers	
	Gram-positive decomposers	Actinomycetes (mesophilic and thermotolerant)	[8–10]
		Involved in decomposition, mineralisation cycle and produce extracellular enzymes and antibiotics	
Fungi		Important degraders	[11–14]
		Yeast – breakdown of complex molecules	
Protozoa		Predation of bacteria	[15]

Introduction

Soil microorganisms play a major role in soil fertility as a result of their involvement in nutrient cycles like carbon, nitrogen and others which are required for plant growth. Microbial diversity is a general term used to include genetic diversity, that is, the amount and distribution of genetic information, within microbial species; diversity of bacterial and fungal species in microbial communities; and ecological diversity, that is, variation in community structure, complexity of interactions, number of trophic levels and number of guilds [1].

The microorganisms play a vital role in the soil ecosystem; some of the microbes reduce organic matter and utilise it for their growth and thus also eradicate organic pollutants from soil. Hence, microbial diversity is of utmost importance for functioning of any ecosystem. The diversity index is a measure of the relationship between the number of species collected and the evenness of their distribution. Diversity indices may be useful to further understand the status of soil microbial communities. Diversity is a function of two components: (1) the total numbers of species present, species richness, and (2) the distribution of individuals among those species, evenness [2]. Various diversity indices include diversity index given by Shannon and Weaver [3], richness (total number of species) given by Margalef (1958) and Menhinick (1964) and evenness (distribution of individual within the species) given by Pielou (1977) [2–5].

In any ecosystem, functional groups of microorganisms are important in contributing to the sustained functions of ecosystem. Hence, studying different functional groups of microorganisms may be significant representation of the total microbial communities contributing to functioning of that ecosystem. The major groups of microorganisms belong to bacteria, including actinomycetes, and fungi, including yeast. Table 41.1 displays different microbial groups along with their functional roles.

There are various methods to study microbial diversity which are as follows: (a) plate count method [16]; (b) molecular methods like DGGE or TGGE [17] and microarray [18]; (c) phospholipid fatty acid analysis (PLFA) [19] and (d) community-level physiological profiling (CLPP) [20]. Microbial diversity needs to be studied due to the following reasons: (a) to expand the frontiers of knowledge about the strategies and limits of life, especially those thriving at extreme conditions; (b) microorganisms are of critical importance to the sustainability of life on our planet; (c) the untapped diversity of microorganisms is a key resource for new genes and organisms of value to biotechnology; (d) diversity patterns of microorganisms can be used for monitoring and predicting

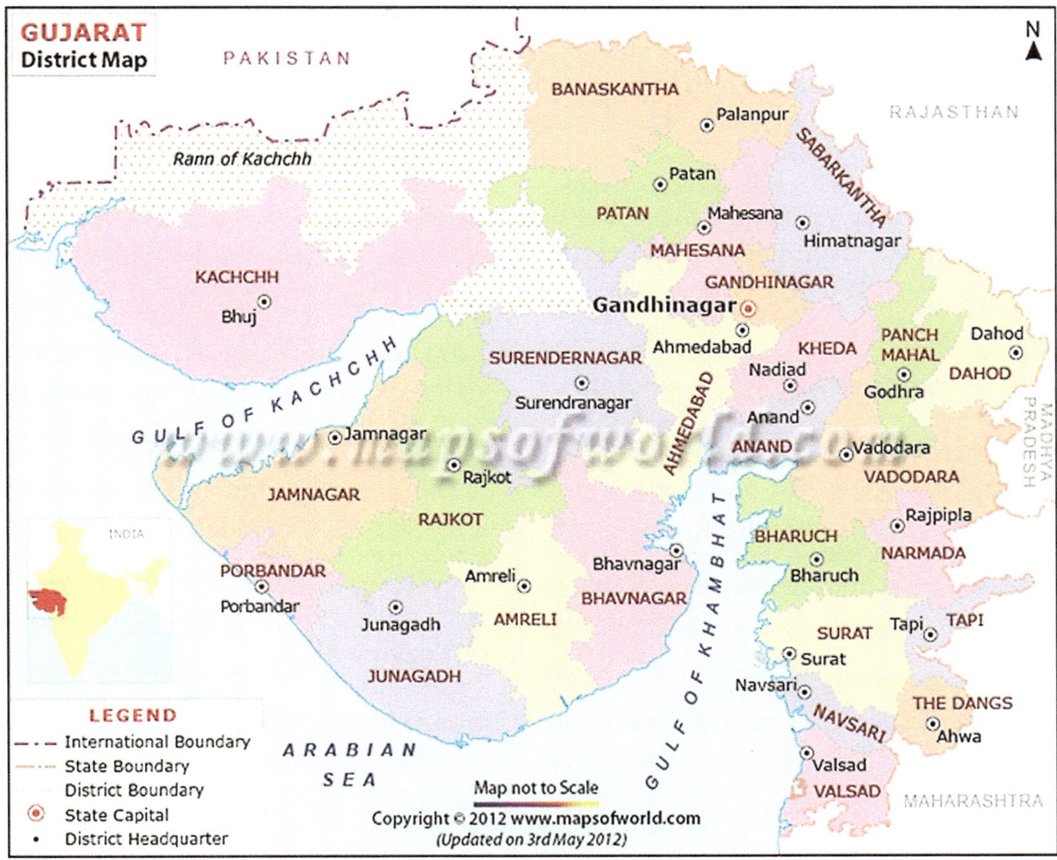

Fig. 41.1 Location of sites selected for study (desert and coastal areas are shown in *different colours*)

environmental change and (e) microbes play a role in conservation and restoration biology of higher organisms and degraded landscapes. The present work was undertaken to study the microbial diversity by different approaches in coastal and desert area and correlate them with the physico-chemical parameters.

The basis for selection of such area for the study was salinity as the common factor. Functional microbial diversity approaches adopted were 'enumeration of different functional groups of microorganisms', 'enzyme assay for whole soil microbial community' and 'functional diversity determination by CLPP analysis'.

Enzymes are among the most remarkable biomolecules because they show extraordinary specificity in catalysing biological reactions. In soil, the enzyme may be present as an extracellular soluble molecule, temporarily associated in enzyme-substrate complexes, adsorbed to clay minerals or associated with humic colloids. Enzyme measurements answer qualitative questions about specific metabolic process and, in combination with other approaches, may increase our understanding of the effect of agrochemicals, cultivation practices and environmental and climatic factors on the microbiological activity of soil [21].

Materials and Methods

Sample Collection

Soil samples were collected from coastal (Daman, Alang, Veraval, Porbandar, Okha and Mandvi) and desert regions (Bhuj, Khavda, Dhordo) of Gujarat, and virgin soil of Ahmedabad (Nirma University) was taken as control (Fig. 41.1). Sediment samples from coastal areas were also collected. From all locations, composite soil samples till depth of 6 cm were collected twice, during winter (December 2010) and at the end of winter (February 2011) [22]. Sea water samples were also collected for comparison of physico-chemical parameters of water.

From desert two types of soils were collected, one virgin and one under some plantation (desert bush). The samples were brought back to the laboratory within 6–8 h of collection and were preserved in the refrigerator (4 °C).

Determination of Abiotic Parameters

The physico-chemical parameters, namely, pH, temperature, moisture, TDS in soil and sediment suspension, salinity, conductivity, organic carbon and total nitrogen, were determined from soil and sediment samples, whereas from the sea water samples, pH, temperature, TDS, salinity, conductivity, chloride content, sulphate content, organic carbon and total nitrogen were estimated. The temperature was recorded at the surface and at the depth of 6 cm; moisture content, TDS, hardness, chloride content and sulphate content were determined by the standard procedures [23]. Salinity and conductivity were measured by PCSTestr™ 2122 (Eutech instruments) in the suspended solution. Total organic carbon and Kjeldahl nitrogen were analysed by M. K. Soil Testing Laboratory, Ahmedabad, Gujarat, India, using the standard procedure [23].

Microbial Characterisation

The microbial characterisation using 14 groups of microorganisms under the study can be functionally broadly categorised as follows: (1) general bacterial population (total viable count, gram-negative bacteria and prototrophic bacteria), (2) nitrogen fixers and nitrifiers (*Rhizobium*, *Azotobacter*, *Azospirillum* and ammonium oxidisers), (3) functional groups of bacteria of other elemental cycles (ferric reducers, sulphate reducers and phosphate solubilisers), (4) filamentous bacteria (mesophilic actinomycetes and thermotolerant actinomycetes) and (5) fungi (fungi and yeast). They were enumerated using one of the three basic methods: spread plate, pour plate [24] and most probable number method [25].

Prototrophic bacteria, ferric-reducing bacteria, sulphate-reducing bacteria and the nitrifiers were enumerated using MPN methods; *Azospirillum* was enumerated by pour plate method, while rest of the populations were determined by spread plate count methods. The ready-made media used were procured from Himedia, India. The media, incubation time, incubation temperature and detection method for all these microbial groups are listed in Table 41.2. The data obtained were used for statistical analysis by one of the multivariate analysis method of PCA (principal component analysis). The CLPP data were used for calculations of various diversity indices [3].

Correlation coefficient values were calculated for microbial population parameters and physico-chemical parameters by SPSS (Ver. 7.5).

Soil Enzyme Assay

The enzymes with important ecosystem functions were selected for the diversity study. The selected enzymes and the method used for analysis were protease, which catalyses the hydrolysis of proteins to polypeptides and oligopeptides to amino acids [26]; urease, which catalyses the hydrolysis of urea to CO_2 and NH_3 [27]; L-asparaginase, which has an important role in nitrogen mineralisation in soils and catalyses the hydrolysis of L-asparagine, producing L-aspartic acid and ammonia [28]; L-glutaminase, which supplies available nitrogen to plants by hydrolysis of L-glutamine yielding L-glutamic acid and NH_3^+ [29]; phosphomonoesterase, which is playing an important role in plant P nutrition [30]; and β-glucosidase, by which microbes degrade cellulose to glucose and is an important component for carbon cycling [31]. Correlation between enzyme activities and physico-chemical parameters of soil was statistically analysed.

Functional Diversity Determination

For functional diversity estimation, Biolog-based method is frequently used. But due to unavailability of Biolog, the method was simulated in tubes. The C sources selected belonged to GN/GP/EcoPlates of Biolog®.

Table 41.2 Media, incubation time, incubation temperature and method of detection of various microbial groups

Group of organism	Media used	Incubation time (days)	Temperature (°C)	Method for detection of microorganism
Total viable population	High plate count agar	3–4	30±2	Spread plate method
Gram-negative bacteria	MacConkey agar	2–3	30±2	Spread plate method
Prototrophic bacteria	Synthetic acetate- glutamate (AG) broth	14	30±2	MPN method – change of medium colour from green to blue
Rhizobium	Congo red yeast extract mannitol agar	3–4	30±2	Spread plate method
Azotobacter	*Azotobacter* agar	3–4	30±2	Spread plate method
Azospirillum	*Azospirillum* medium	4–6	30±2	Pour plate method – colonies showing blue halo in green medium were considered
Ammonium-oxidising bacteria	Mineral salts (MS) medium	4–6 weeks	30±2	MPN method – NO_3 formation was detected
Ferric-reducing bacteria	Glucose-asparagine Fe_2O_3 (GAF) broth	6–7	30±2	MPN method – red colour detected by adding (0.2%) 2,2-dipyridyl solution
Sulphate-reducing bacteria	Lactate-thioglycolate sulphate (LTS) broth	4–5	30±2	MPN method – blackening of the medium
Phosphate-solubilising bacteria	Pikovskaya agar	4–5	30±2	Spread plate method – colonies showing solubilisation zone were considered
Mesophilic actinomycetes	Actinomycetes isolation agar	3–4	30±2	Spread plate method – only actinomycete colonies were considered
Thermotolerant actinomycetes	Actinomycetes isolation agar	2–3	55	Spread plate method
Fungi	Potato dextrose agar	4–5	30±2	Spread plate method
Yeast	Glucose yeast extract agar	2–4	30±2	Spread plate method

Table 41.3 Different carbon sources used for CLPP assay

Tween 80	D-Xylose
Sodium pyruvate	Citric acid monohydrate
Ketoglutaric acid	D- and L-lactic acid
Adenosine 5′-monophosphoric acid Na salt	Dextrin
D-Glucose-6-phosphate disodium salt dihydrate	Sucrose
L-Arginine	DL-Malic acid
i-Erythritol	L-Alanine
L-Glutamic acid	Inositol
Dextrose	D-Fructose
Dulcitol	Acetic acid
D-Galactose	D-Mannitol
Maltose monohydrate Bacto	Formic acid
D-Sorbitol	L-Arabinose
D-Trehalose dihydrate	Tween 20
L-Asparagine monohydrate	Lactose
N-acetyl D-glucosamine	D-Melezitose
L-Proline	D-Melibiose
Xylitol	D-Ribose
L-Rhamnose monohydrate	

Substrate utilisation assay was done using 38 different carbon sources to check carbon source utilisation by soil microbial community upon incubation. For this assay, basic mineral medium was used with tetrazolium red dye, which is a redox indicator dye. In a set of 38 tubes, dilutions of soil or sediment suspension were inoculated. A tube without any C source was used as control, and upon incubation the OD_{600} response in this tube was subtracted from the responses of all other tubes.

After incubation period, this colourless media turned to red due to respiration of microbial communities present in the inoculum which indicated the utilisation of carbon sources. Potential substrates included sugars, amino acids, organic acids, sugar alcohols, polymers, etc. (Table 41.3).

The data obtained were used for calculation of various diversity indices [2–5] as well as for multivariate analysis by PCA to see the difference in microbial communities in various coastal and desert samples.

Results and Discussion

Abiotic Factors

During sampling, the pH of both desert and coastal soils was found to be lying in the range from 5.13 to 7.70, that is, acidic to neutral in nature. Moisture content was higher in coastal soil than in desert soils because of the xeric conditions. Temperature was recorded higher in the deserts (~30 °C) in comparison to coasts (~25 °C).

Total dissolved solids (TDS) were found to be negligible in the deserts. Salinity (g%) generally was found to range between 1 and 3% in desert soils, whereas the values were 3–5% in coastal soils and were highest in Daman polluted coast (9.36%) and lowest in deserts (0.31% in Khavda soil under plantation). Conductivity (μS) was found to be highest in coasts (1,887 μS in Daman polluted coast) and lowest in deserts (50.8 μS in Khavda S.P.). Organic carbon (%) and total nitrogen (%) were found to be highest in coasts in comparison to deserts. In desert soils, less than 1% organic carbon content was observed, whereas in the coastal soils, the values ranged between 1 and 2%.

Biotic Factors

Among the 14 groups of organisms, there were less significant differences in the CFU count of some of the groups of organisms. This includes total viable count, prototrophic bacteria, *Azotobacter* and sulphate reducers, but *Rhizobium,* phosphate solubilisers, thermotolerant actinomycetes and fungi showed more significant differences in the CFU counts.

Rhizobia are acid producers and alkali producers. Apart from fixing N_2 symbiotically, they can also solubilise sulphur and phosphorus because they produce exo-acid [8]. As seen in Fig. 41.2, the *Rhizobium* population was found to be higher in desert soils in comparison to coastal soils; as in all the desert soils, namely, Dhordo, Dhordo soil under plantation, Khavda, Khavda soil under plantation and Bhuj, count of almost 10^5 cfu g^{-1} soil was obtained.

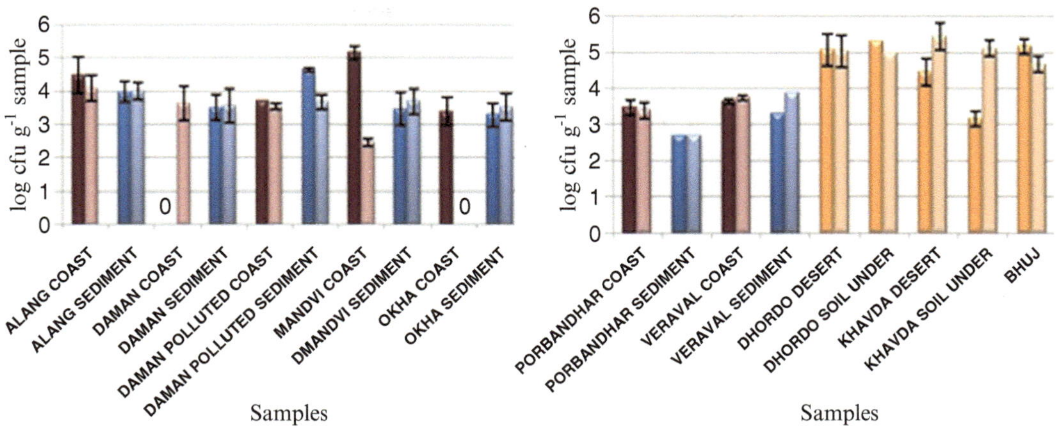

Fig. 41.2 Population of *Rhizobium* in coastal and desert samples

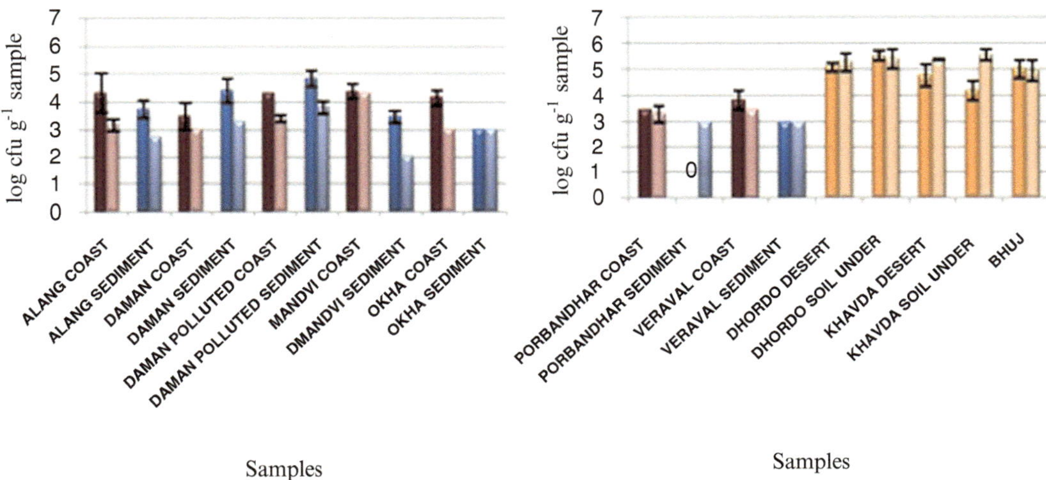

Fig. 41.3 Phosphate-solubiliser populations in coastal and desert samples

The two bars on each sample show values obtained during the two sampling times. Many *Rhizobium* species are known to be resistant to adverse environmental conditions, such as high temperature and desiccation [32, 33].

Phosphorus is not available in free form in soil because it is restricted by its tendency to precipitate in the presence of bivalent metals (calcium, magnesium, etc.). Thus, solubilisation of phosphorus by *phosphate solubilisers* makes it bioavailable to organism in soil by diverse groups of microorganisms [10]. Similarly, the population of phosphate solubilisers was found to be higher (~10^5 cfu g^{-1} soil) in desert soils in comparison to coastal soils (Fig. 41.3).

Actinomycetes able to grow at temperatures of 50 °C and above, classified as *thermotolerant actinomycetes,* have been reported to be major degrading organisms during composting. They produce exoenzyme and can degrade many polymorphic substances such as starch, cellulose, lignin, etc. [10]. They produce cellulolytic and hydrolytic enzymes, for example, *Actinosporangium, Kitasatoa,* etc. [12].

Their population was prominent and exclusively higher in all the desert soils although some low count was found in coastal soils of Alang, Mandvi and Daman (Fig. 41.4). This may be because of the higher temperature in the desert and the fact that only those organisms that can

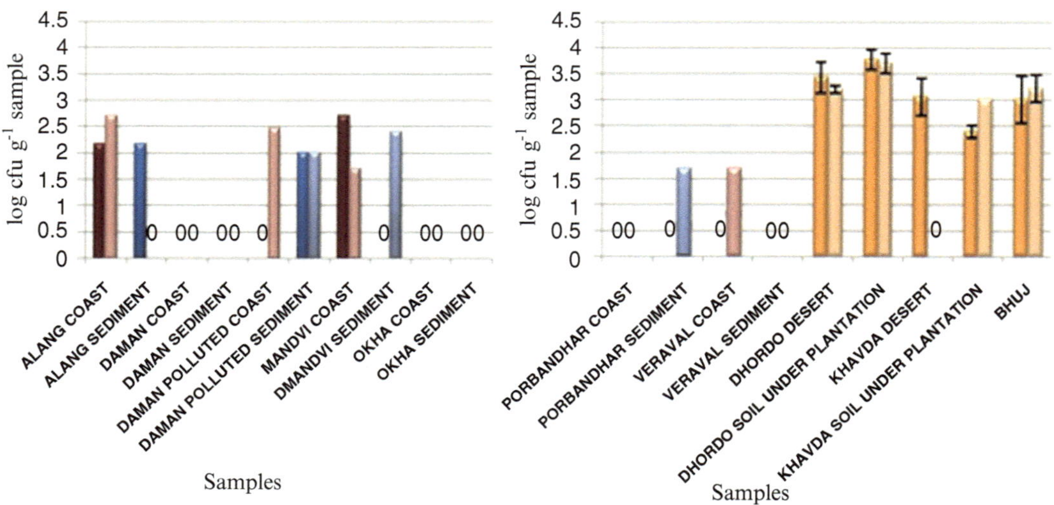

Fig. 41.4 Thermotolerant actinomycetes population in coastal and desert samples

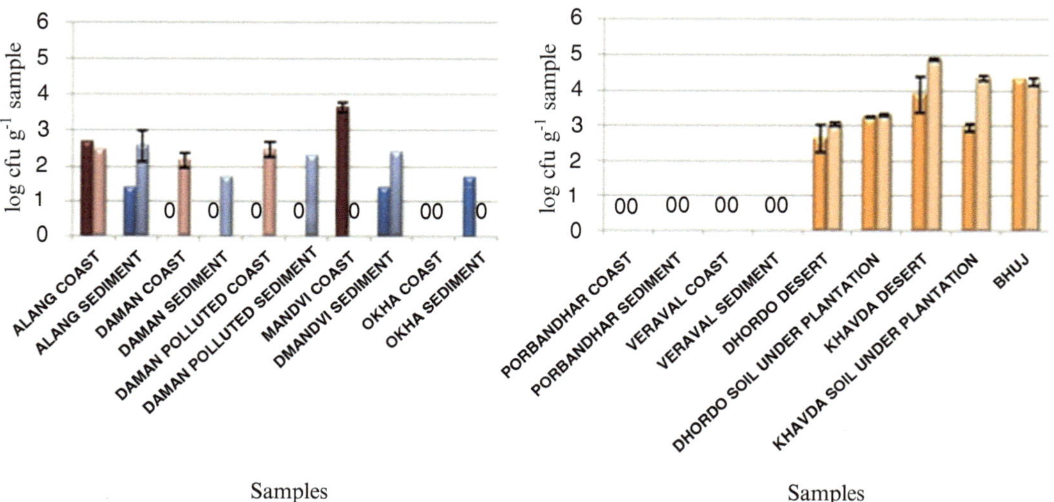

Fig. 41.5 Population of fungi in coastal and desert soil samples

adapt to this higher temperature can survive. According to another report, average cfu count of thermophilic actinomycetes in coastal sediment was found to be ranging from 10^3 to 10^5 cfu g^{-1} dry weight of soil, and with increase in temperature, they are found to take place [32].

Fungi, although, were not exclusively found in desert soils, but in comparison to coastal soils, in all the desert soils, fungal count was found to be higher (Fig. 41.5). Soil fungi are less numerous and are as diverse as bacteria, but may contribute up to 70% of the microbial biomass [33]. Higher fungal population in desert soil may be due to the fact that fungi are found free in soil, in association with plant roots or as saprophytes on detrital material [16], whereas coastal soils are flushed by sea water leaving little organic material.

Enzyme Assay

Highest activity of protease was found in coastal soil of Daman coast (1.33 μg tyrosine g^{-1}dwt 2 h^{-1}) in comparison to highest value in desert soil of Khavda (0.84 μg tyrosine g^{-1}dwt 2 h^{-1}). Highest activity of urease was found high in coastal soil of Daman (13.2% total nitrogen) in comparison to desert soil (5.71% total nitrogen). The urease activity in soil is very stable and rarely influenced by air-drying, irradiation or storage at temperatures between −22 and 60 °C [34]. Highest activity of L-asparaginase was found in Alang sediment (0.005% total nitrogen) in comparison to desert soil (0.0008% total nitrogen). Temperature sensitivity of the enzyme also affects the enzyme activity; when temperature goes beyond optimum temperature, the activity ceases [35].

Surprisingly, highest activity of L-glutaminase was found in desert soil of Dhordo desert, Dhordo soil under plantation and Bhuj. All were having the value of 0.0006% total nitrogen, while no enzyme activity was found in the coastal soils.

Phosphomonoesterase activity includes activity of acid phosphatase and alkaline phosphatase. Highest activity of acid phosphatase was found in the coastal soil of Daman coast (69.47 mg g^{-1} dwt 2 h^{-1}) in comparison to desert soil of Khavda soil under plantation (51.25 mg g^{-1} dwt 2 h^{-1}), while highest activity of alkaline phosphatase was found in Daman polluted sediment (99.7 mg g^{-1} dwt 2 h^{-1}) in comparison to maximum value among desert soils which was in Bhuj (65.42 mg g^{-1} dwt 2 h^{-1}). This enzyme activity is negatively dependent on salinity and positively depends on organic carbon present [36], which may be the reason for low level detected in coastal samples.

Highest activity of β-glucosidase was found in coastal soil of Daman polluted sediment (90.82 μg g^{-1} dwt h^{-1}) in comparison to highest among desert soil found in Dhordo (88.56 μg g^{-1} dwt h^{-1}). Coastal soil gets continuous input of autochthonous and allochthonous community, which are degraded by the microorganisms present in the soil, and utilise it as carbon source. Thus, the activity of enzymes depends on the amount of organic matter which enters into the system, while in case of desert ecosystem, the organism utilises the available source and respectively enzyme activity is observed [37]. More or less, the enzyme activities were found to be dependent on soil total organic carbon.

Functional Diversity Scenario

Functional diversity based on CLPP was analysed first by comparing AWCD (average well colour development) patterns of C substrate utilisation displayed by all samples along the incubation till 170 h. Average well colour development is represented in Fig. 41.6 for all coastal and desert samples.

Desert soil of Bhuj showed maximum utilisation of substrate (total activity, $n=6.893$), while least has been shown by the coastal soil of Okha (0.57) at 97 h of incubation, when the highest total activities were compared in all the samples. Another important observation was that the desert soils displayed higher utilisation in terms of AWCD as compared to coastal samples.

The graph of AWCD for all samples displayed that analysis of data of 97 h would be useful for examining the differences among the samples and sites. Hence, the data of 97 h of incubation was analysed by principal component analysis, and results are presented in Fig. 41.7.

Principal component analysis of the 97 h CLPP data showed the difference between coastal soil and desert soil. Here, all coastal soils were clustered together, while the desert soils were placed distantly. CLPP in 97 h incubation time resulted in extraction of total 6 PCs. Here, first 2 PCs extracted total 61.95% of variance present in the data.

Here, the soil of Porbandar, Okha, Alang and Mandvi is nearly similar to each other; Veraval coastal soil and Daman coastal soil are nearly similar in terms of functional microbial community. In case of desert soils, they were not found to be similar to each other and so were placed apart. Even Nirma soil showed quite different profile of C source utilisation and thus was shown

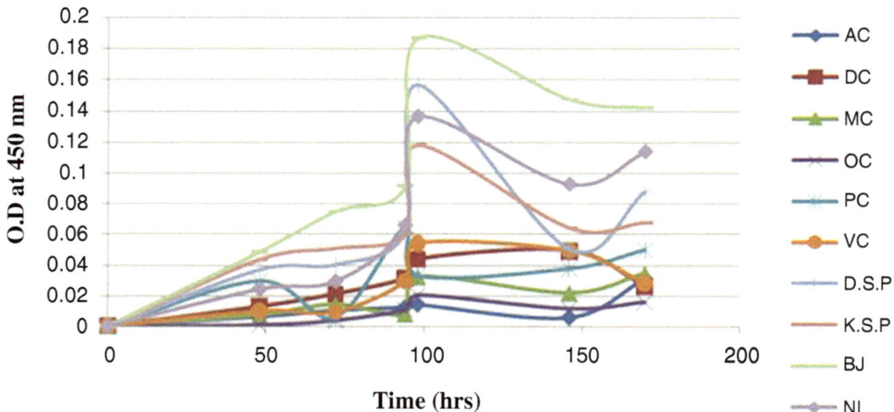

Fig. 41.6 Average well colour development (AWCD) from utilisation of 37 different sole C sources for coastal and desert soils (*AC* Alang coast, *DC* Daman coast, *MC* Mandvi coast, *OC* Okha coast, *PC* Porbandar coast, *VC* Veraval coast, *DSP* Dhordo soil under plantation, *KSP* Khavda soil under plantation, *BJ* Bhuj, *NI* Nirma)

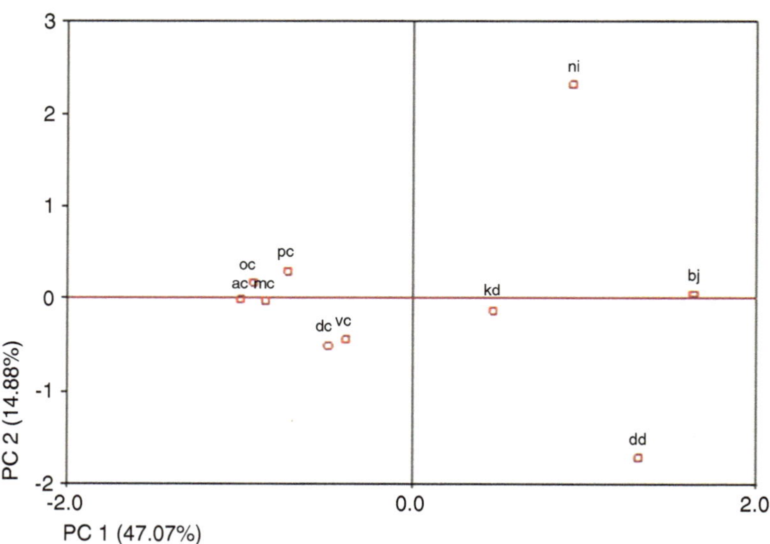

Fig. 41.7 Principal component analysis of CLPP data of 97 h of incubation time

to be different from all other samples. CLPP data of 97 h was used for calculation of diversity indices as well, and the results are presented in Table 41.4.

In general, desert soils exhibited high functional diversity (Shannon diversity H') as compared to coastal samples as highest value of H' was obtained in Bhuj (3.23) and lowest in Daman and Okha (2.44 in each). Microbial communities of the desert soils were found to be utilising maximum number of substrates, that is. 32 in Khavda soil under plantation. This signifies the presence of many different types of species present in the community of desert soil. But the total activity was found to be low in Khavda. In terms of evenness, all the microbial communities were found to be evenly distributed with high E_{pielou}, except for Daman coast.

On the other hand, Okha and Porbandar showed low richness and overall activity indicating highly disturbed conditions prevailing leading to low diversity, whereas Daman, Mandvi and

Table 41.4 Various diversity index values for coastal and desert samples

Sample	Diversity $D_{shannon}=H'$	Total activity (n)	Substrate (S)	Evenness E_{pielou}	Richness $R_{margalef}$	Richness $R_{menhinick}$
Alang coast	3.15	5.003	29	0.93	17.3	12.9
Daman coast	2.44	1.742	21	0.80	36.0	15.9
Mandvi coast	2.92	1.065	21	0.96	317.5	20.3
Okha coast	2.44	0.57	13	0.95	−21.3	17.2
Porbandar coast	2.71	1.086	18	0.94	206.0	17.2
Veraval coast	3.14	1.923	26	0.96	38.2	18.7
Dhordo S. P.	3.22	5.753	30	0.94	16.5	12.5
Khavda S. P.	3.17	4.366	32	0.91	21.0	15.3
Bhuj	3.23	6.893	31	0.94	15.5	11.8
Nirma	3.15	5.003	29	0.93	17.4	12.9

Veraval showed moderate levels of richness and diversity values.

On the basis of diversity indices and salinity values in desert and coastal samples, it can be said that the desert displayed high diversity when the salinity values were lower comparatively and vice versa. It proves that salinity values are negatively correlated with diversity, substrate richness and total activity. Moreover, diversity (H') correlated with substrate richness significantly ($r=0.9$, $p<0.01$) as well as with total activity ($r=0.783$, $p<0.01$). Hence, response obtained for sole carbon source utilisation proves to be an important analytical tool for assessment of microbial diversity in any ecological niche.

Conclusions

It was concluded that in comparison to coastal soil, desert soil exhibited high diversity, total activity and richness. One of the main reasons one can hypothesise is that the continuous sea water's flushing impact on coastal soil has lead to the loss of organic matter and nutrients from it, while in the desert, even with low soil organic matter, nutrients remain intact because of the absence of such effect. Also the high clay and silt content of the desert soil was helpful in making a meshwork which provided the suitable environment for the flourishing microbial diversity. Since microbial population depends more on organic matter and nutrients for their growth, they were found to be high in deserts in comparison to coastal soil. PCA analysis displayed that coastal sediments were clustered together, while the desert samples were separated from each other and from the coastal samples. The coastal soils of Okha, Porbandar and Veraval seemed to be deteriorating accompanied by low microbial diversity.

Acknowledgement The authors are grateful to Nirma Education and Research Foundation (NERF) for funding the research work.

References

1. Kennedy AC, Smith KL. Soil microbial diversity and the sustainability of agricultural soils. Plant Soil. 1998;170:75–86.
2. Margalef R. Information theory in ecology. Gen Syst. 1958;3:36–71.
3. Zak JC, Willig MR, Moorhead DL, Wildman HG. Functional diversity of microbial communities: a quantitative approach. Soil Biol Biochem. 1994;26(9):1101–8.
4. Menhinick EF. A comparison of some species – individual diversity indices applied to samples of field insects. Ecology. 1964;45:859–61.
5. Pielou EC. Mathematical ecology. New York: Wiley; 1977. 385 p.
6. Lorch HJ, Lorenz S, Ottow JCG. Populationsdichten verschiedener Bakteriengruppen in einer belüfteten Abwasserteichanlage. Forum Städte-Hyg. 1990;41:133–8.
7. Tiedje JM, Sextone AJ, Parkin TB, Revsbech NP, Shelton DR. Anaerobic process in soil. Plant Soil. 1984;76:197–212.
8. Joanne MW, Linda MS, Christopher JW. Prescott, Harley and Klein's microbiology. 7th ed. New York: McGraw Hill; 2008. p. 519–642.

9. Rodriguez H, Fraga R, Gonzalez T, Bashan Y. Genetics of phosphate solubilization and its potential applications for improving in plant growth promoting bacteria. Plant Soil. 2004;102:15–21.
10. Subba Rao NS. Soil microbiology. 4th ed. New Delhi: Oxford and IBH Publication; 2008. p. 292–300.
11. Atlas RM, Bartha R. Microbial ecology. Fundamentals and applications. 6th ed. New York: Benjamin Cummings Publishing; 2009.
12. Sollins P, Crowmack Jr K, Li CY, Fogel R. Role of low-molecular weight organic acids in the inorganic nutrition of fungi and higher plants. In: Carroll GC, Wicklow DT, editors. The fungal community: its organization and role in ecosystem. New York: Marcel Dekker; 1981. p. 607–19.
13. Lawrence JR, Gupta VVSR, Germida JJ. Impact of elemental sulphur fertilization on agricultural soils effects on sulphur oxidizing populations and oxidation rates. Can J Soil Sci. 1988;68:475–83.
14. Maria V, Mercedes S, Vicente B, Encarnación J, Lana N, Vanessa M, Nikolay V. Multifunctional properties of phosphate solubilizing microorganisms grown on agro-industrial wastes in fermentation and soil conditions. Appl Microbiol Biotechnol. 2009;85:1287–99.
15. Opperman MH, Wood M, Harris PJ. Changes in microbial populations following the application of cattle slurry to soil at two temperatures. Soil Biol Biochem. 1989;21:263–8.
16. Bakken LR. Culturable and nonculturable bacteria in soil. In: van Elsas JD, Trevors JT, Wellington EMH, editors. Modern soil microbiology. New York: Marcel Dekker; 1997. p. 47–61.
17. Heuer H, Wieland G, Schonfeld J, Schnwal der A, Gomes NCM, Smalla K. Bacterial community profiling using DGGE or TGGE analysis. In: Rochelle PA, editor. Environmental molecular microbiology: protocols and applications. Wymondham: Horizon Scientific Press; 2001. p. 177–90.
18. Rondon MR, August PR, Bettermann AD, Brady SF, Grossman TH, Liles MR, et al. Cloning the soil metagenome: a strategy for accessing the genetic and functional diversity of uncultured microorganisms. Appl Environ Microbiol. 2000;66:2541–7.
19. Tunlid A, White DC. Biochemical analysis of biomass, community structure, nutritional status, and metabolic activity of microbial communities in soil. In: Stotzky G, Bollag J-M, editors. Soil biochemistry, vol. 7. New York: Marcel Dekker; 1992. p. 229–62.
20. Garland JL. Analysis and interpretation of community-level physiological profiles in microbial ecology. FEMS Microbiol Ecol. 1997;24:289–300.
21. West NE, Skujins J. Nitrogen in desert ecosystems. Stroudsburg: Dowden, Hutchinson and Ross; 1978.
22. Alef K, Nannipieri P. Enzyme activities. In: Alef K, Nannipieri P, editors. Methods in applied soil microbiology and biochemistry. London: Academic Press Ltd; 1995. p. 311–66.
23. Eaton AR, Greenberg AE. Standard methods for the examination of water and wastewater. 21st ed. Washington, DC: American Public health association (APHA); 2005.
24. Hoben HJ, Somasegaran P. Comparison of the pour, spread, and drop plate methods for enumeration of *Rhizobium* spp. in inoculants made from pre-sterilized peat. Appl Environ Microbiol. 1982;44:51246–7.
25. Flemming V, Kjeld I. Improved most-probable-number method to detect sulfate reducing bacteria with natural media and a radiotracer. Appl Environ Microbiol. 1998;64(5):1700.
26. Burns RG. Enzyme activity in soil: location and a possible role in microbial ecology. Soil Biol Biochem. 1982;14:423–7.
27. Frankenberger Jr WT, Tabatabai MA. Amidase and urease activity in plants. Plant Soil. 1982;64:153–66.
28. Frankenberger Jr WT, Tabatabai MA. L-Asparaginase activity of soils. Biol Fertil Soils. 1991;11:6–12.
29. Frankenberger Jr WT, Tabatabai MA. L-glutaminase activity of soils. Biol Fertil Soils. 1991;23:869–74.
30. Tabatabai MA, Bremner JM. Use of p-nitrophenyl phosphate for assay of soil phosphatase activity. Soil Biol Biochem. 1972;1:301–7.
31. Andrea M. Diversity of bacillus like organism isolated from deep sea hyper saline anoxic sediments. Saline Syst. 2008;4:8.
32. Goodfellow M, Simpson KE. Ecology of streptomycetes. Front Appl Microbiol. 1987;2:97–125.
33. Whipps JM, Lynch JM. Substrate flow and utilisation in the rhizosphere of cereals. New Phytol. 1983;95:605–23.
34. Pancholy KS, Rice EL. Soil enzymes in relation to old field succession. Amylase, Cellulase, Invertase, Dehydrogenase and Urease. Soil Sci Soc Am J. 1973;37:47–50.
35. Privalov PL, Tsalkova TN. Micro-stabilities and macro-stabilities of globular-proteins. Nature. 1979; 280:693–6.
36. Cookson P. Variation in phosphatase activity in soil: a case study. Agric Sci. 2002;7:65–72.
37. Caruso G. Leucine amino peptidase, β-glucosidase and alkaline phosphatase activity rates and their significance in nutrient cycles in some coastal mediterranean sites. Mar Drugs. 2010;8:916–40.

Experimental and Theoretical Investigations on the Anti-oomycetous Activity of Paromomycin Targeted Against *Pythium* Tubulin

42

A.K. Resna, C. Geethu, S. Sumna, U.C. Jaleel, and R. Aswati Nair

Abstract

Pythium myriotylum is an oomycetous necrotrophic fungus that causes diseases in a wide variety of plants. Paromomycin, an aminoglycoside antibiotic, has been previously demonstrated to be effective against oomycetes. We undertook detailed studies to investigate the mode of inhibition of paromomycin against *P. myriotylum* since its mode of action against eukaryotes is not well defined. Mycelial radial growth and biomass yield were inhibited with increasing paromomycin concentration. The EC_{50} value for biomass production was found to be 8.07 μg/ml. Microscopic examination of paromomycin-treated *P. myriotylum* zoospores (1×10^7 spores/ml) also showed significant decrease in zoospores compared to control. Further evaluation of hyphae after treatment with inhibitory concentrations of paromomycin for 2 h revealed loss of membrane integrity and coagulation of cytoplasmic materials. As integrity of plasma membrane is a crucial prerequisite for maintaining cellular homeostasis, we further carried out molecular docking simulation to analyze the effect of paromomycin on β-tubulin that plays an important role in preserving membrane integrity. β-tubulin has been suggested as a potential target for developing antifungal agents, and since very little information is available on oomycete tubulin, a homology model of *P. ultimum* β-tubulin (gb id:AAF22515.1) was constructed using *Ovis aries* β-tubulin (PDB ID: 3RYC) as template. Validation of stereochemical quality of the final refined model revealed 90% identity to template with 95% residues falling in most favorable region, while no residues were in the disallowed region in the Ramachandran plot. Molecular docking revealed that the amino acids, Gln 11, Asp 67,

A.K. Resna • C. Geethu • S. Sumna • R. Aswati Nair (✉)
School of Biotechnology, National Institute of Technology Calicut (NITC), Calicut, Kerala, India
e-mail: aswati@nitc.ac.in

U.C. Jaleel
Centre for Cheminformatics, Malabar Christian College, Calicut, Kerala, India

A. Sabu and A. Augustine (eds.), *Prospects in Bioscience: Addressing the Issues*,
DOI 10.1007/978-81-322-0810-5_42, © Springer India 2013

Leu 68, Asn 99, Gly 142, and Asp 177, in the active site are consistently involved in strong hydrogen bonding with the target. The present study provides to our knowledge the first data on molecular interactions of paromomycin with tubulin that are involved in mediating its anti-oomycetous activity.

Keywords

Pythium myriotylum • Paromomycin • Membrane integrity • Tubulin • Homology modeling • Molecular docking

Introduction

Pythium myriotylum Drechsler is a ubiquitous soilborne pathogen [1] with a wide host range causing damping-off and root rot in many crop plants and resulting in substantial economic loss [2, 3]. *Pythium* is an oomycete and has little taxonomic affinity to fungi but is phylogenetically related to brown algae and diatoms within the Kingdom Stramenopiles [4–7]. Thus, many fungicides used to treat soilborne fungal diseases are not effective against *Pythium* due to the absence of fungicide targets in oomycetes [4].

Due to the detrimental effects associated with the use of chemical fungicides, previous studies have investigated on various biological control methods (BCA) [8, 9] and antibiosis has been the most widely studied and important biocontrol mechanisms [10] involved in antifungal activity and disease suppression. Previous studies have demonstrated that many of the antibiotics produced by BCAs have broad-spectrum activity against various phytopathogenic fungi [11–15]. Following evaluation of anti-oomycete activity of various aminoglycoside antibiotics, paromomycin was identified to have potent anti-*Pythium* activity. Paromomycin is an aminoglycoside antibiotic, first isolated from *Streptomyces rimosus* ssp. *paromomycinus*, an actinomycete species [16] and is known to inhibit prokaryotic translation by binding to 30S ribosomal subunits. However, its mode of action on eukaryotic oomycetes is not well defined. Previous studies have reported that the disease-suppressive and anti-oomycete action of antibiotics produced by BCAs is mediated by inducing malformations in fungi such as stunting and deformation of germ tubes of germinating cysts [13, 17], destroying germinating oospores, and damaging cell walls of fungal hyphae [18, 19]. Research is needed to understand the molecular basis of paromomycin action so that the fundamental knowledge gained can be applied to anticipate and prevent the breakdown of biocontrol.

Fungicides with antitubulin activity have been observed to be highly effective on a broad spectrum of oomycetes [20, 21] as interference to microtubule will consequently affect various cellular processes, such as mitosis and cytoskeleton formation. The present study provides detailed information for the first time on evaluation of inhibitory effect of paromomycin on *P. myriotylum* and discusses on the likely anti-*Pythium* mode of action of paromomycin. For the latter part of the study, we used β-tubulin sequence of a related species, *viz*., *P. ultimum,* and for which the genome sequence information is available [22] and is also responsible for soft rot disease [23].

Materials and Methods

Anti-*Pythium* Activity

Qualitative and quantitative analysis of anti-*Pythium* activity was done by radial diffusion assay. Briefly, a 5-mm-diameter agar plug containing actively growing *Pythium* was placed on a 5-mm filter paper disc (Whatman no. 4) impregnated with increasing concentrations of paromomycin (10, 25, 50, 75, 100 μg/ml) and placed in the center of Petri dish containing potato dextrose agar (PDA) medium. The plates were incubated at 25°C for 3 days and percentage of inhibition was measured by comparing mycelial growth in control plate containing water instead of paromomycin. Results obtained are presented as the mean ± SD of the inhibition zone (I % = [(C–T) C^{-1}]100: I % = relative inhibition, C = control colonial diameter, T = colonial diameter from the paromomycin-treated plate). Experiments were done in triplicate.

Effect of Paromomycin on Biomass

Mycelial discs were inoculated in Czapek Dox minimal media (30g sucrose, 2g $NaNO_3$, 1g KH_2PO_4, 0.5 KCl, 0.5g $MgSO_47H_2O$, 0.1g $FeSO_47H_2O$ per liter) with trace element solution (1g $ZnSO_4$, 0.5g $CuSO_4{\cdot}5H_2O$ per liter) containing different concentrations of paromomycin (5, 10, 25, 50, 75, and 100μg/ml) and incubated at 25°C for 5 days. Growth inhibitory effect of paromomycin on *P. myriotylum* was determined by measuring the mycelial biomass from triplicate samples for each treatment on pre-weighed filter papers. Dry weight was determined after 6h oven drying at 60°C. EC_{50} value was calculated by transforming the data for each experiment for regression analysis. The percentage of inhibition compared to control was transformed into probits [24], and paromomycin concentrations were log transformed. EC_{50} was then estimated from the regression line.

Zoospore Lytic Activity

Zoospores of *P. myriotylum* were produced from 7- to 10-day-old cultures grown on potato dextrose agar (PDA) plates at 25°C by the methods of Rahimian and Banihashemi [25] and Pacumbaba et al. [26] with modifications. Briefly, mycelial discs (5mm diameter) were transferred to Petri dish containing sterile water (pH 7.0) and placed under fluorescent light at room temperature for 24h to induce zoospore release. Zoospore concentrations were estimated with a hemocytometer and appropriate dilutions containing 1×10^7 zoospores were made with sterilized water. *P. myriotylum* zoospore viability at each paromomycin concentration (5, 10, 25, 50, 75, and 100 μg/ml) was evaluated in a hemocytometer after incubation for 16h at 25°C.The experiments were repeated three times.

Evaluation of Hyphal Integrity

Mycelia from 5-day-old *P. myriotylum* plates were harvested and incubated for 2h with optimal growth inhibitory concentration of paromomycin (25μg/ml) determined from radial diffusion experiments. Mycelia were harvested, washed in minimal medium, and observed under bright field microscope (Olympus BX51, Japan) at 40X.

Molecular Modeling

Gene sequence encoding 446 amino acid β-tubulin of *P. ultimum* was selected from GenBank (accession number: AAF22515.1). BLASTP search of NCBI Protein Data Bank [27] was done to identify homologous proteins. The three-dimensional (3-D) structure of β-tubulin was generated using Modeller v9.10 [28]. Reliability of the modeled structure of β-tubulin was assessed by calculating root-mean-square deviation (RMSD) by superimposing it on template structure using a three-dimensional structural superposition (3-DSS) tool [29]. Stereochemical quality of the 3-D structure was evaluated by inspecting the phi (Φ) and psi (ψ) torsion angles using PROCHECK [30], as determined by Ramachandran plot statistics. The ERRAT tool [31] was used to check the overall quality of modelled protein structure for potential errors and to assesses distribution of different types of atoms with respect to one another in the protein model.

Docking Simulations

Protein preparation wizard Epik v2.5 [32] was used for energy minimization of the modeled β-tubulin using the optimized potentials for liquid simulations all-atom (OPLS2005) force field [33] to discard the high-energy intramolecular

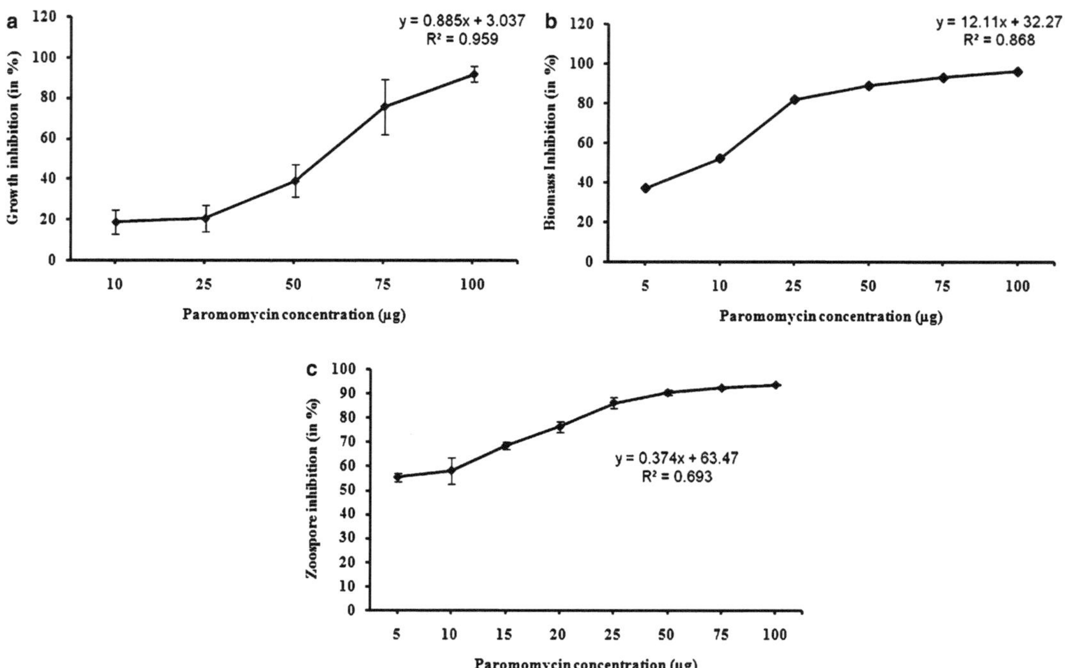

Fig. 42.1 Effect of different concentrations of paromomycin on (**a**) mycelial radial growth (**b**) hyphal biomass, and (**c**) zoospore viability of *P. myriotylum*

interactions. Chemical structure of paromomycin was drawn using ACD/ChemSketch Freeware v12.01. The active sites of modeled β-tubulin were predicted using Computed Atlas of Surface Topography of proteins (CASTp) program [34] and SiteMap v2.4 (Schrödinger) followed by docking studies. Docking procedure was performed using extra precision (XP) Glide v5.6 (Schrödinger), which produces the least number of inaccurate poses and calculates the accurate binding energy of the protein with the ligand [35, 36]. Good poses were selected by rejecting RMSD (root-mean-square deviation) greater than 0.5Å and metal score greater than 10. Binding affinity of protein and ligand was predicted using GlideScore [35, 36]. Other binding interactions like van der Waals interaction energy, Coulomb interaction energy, Emodel (model energy), and cumulative van der Waals interaction energy were also determined.

Results and Discussion

Pythium Growth Inhibition

Following disc diffusion assays, all concentrations significantly inhibited hyphal growth relative to control (ANOVA, $F=50.67$, $p<0.01$) with the percentage inhibition increasing with increasing paromomycin concentration (Fig. 42.1a). A concomitant reduction in hyphal biomass was also observed with increasing paromomycin concentration (Fig. 42.1b). The concentration inhibiting hyphal growth by 50% (EC_{50}) after 96 h of incubation was observed to be 8.07 µg/ml. Microscopic examination of paromomycin-treated *Pythium* zoospores (1×10^7 spores/ml) revealed significant reduction ($p<0.01$) in spore number with increasing paromomycin concentration (Fig. 42.1c). Fewer than 50% of spores survived at the lowest paromomycin concentration (5 µg/ml) compared

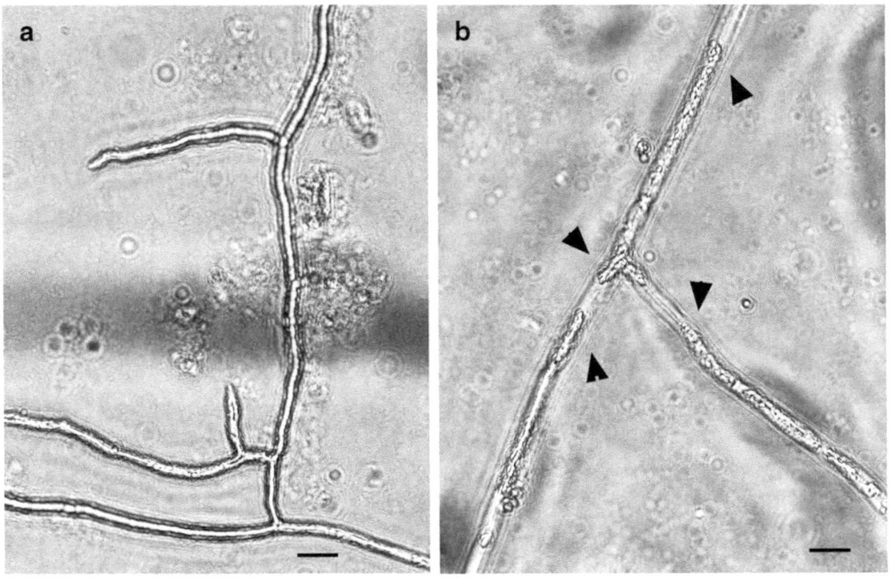

Fig. 42.2 Microscopic analysis of *P. myriotylum* hyphae (**a**) without paromomycin addition and (**b**) addition of paromomycin (5μg/ml) with *arrows* indicative of cytoplasmic coagulation. *Scale bars*: 20μm

to control. A regression analysis indicated significant correlation between paromomycin concentration and reduction in spore number. Morphological alterations to hyphae after exposure to paromomycin were detected compared to control. The hyphae of *P. myriotylum* control without paromomycin treatment were observed as aseptate and coenocytic with a continuous mass of cytoplasm (Fig. 42.2a), which is typical of oomycetes [37]. After exposure to paromomycin, hyphal morphology was damaged, and inside the shriveled hyphae, large vesicles were seen (Fig. 42.2b). These experiments provide strong evidence for changes to the hyphal integrity. Disruption of microtubules which are the key components of cytoskeletal elements leads to disruption of fungal growth and morphogenesis. To investigate the direct inhibitory effect of paromomycin on β-tubulin, molecular docking experiments were performed.

Comparative Homology Modeling

Since crystal structures of *Pythium* β-tubulin are not available, *P. ultimum* β-tubulin sequence retrieved from NCBI database (accession number: AAF22515.1) was used to identify homologous proteins from Protein Data Bank (PDB). *Ovis aries* tubulin (PDB ID: 3RYC) was selected as a promising template based on the low e-value and considerable similarity of 90% to *P. ultimum* β-tubulin with a resolution of 2.1 Å. After alignment of template and target using ClustalW [38], homology modeling was carried out using Modeller v9.10 [39]. After further optimization and validation of modeled *P. ultimum* β-tubulin (Fig. 42.3a), the RMSD value was calculated using 3-DSS tool as 0.163 Å (Fig. 42.3b). Geometric evaluations of the modeled 3-D β-tubulin structure were performed using PROCHECK by calculating the Ramachandran plot. The main chain conformations for 94.9% of the amino acid residues were within the favorable or allowed regions, whereas no residues were located in the disallowed regions of the Ramachandran plot (Fig. 42.3c). The overall observed G-factor (goodness factor) score determined after PROCHECK analysis for the present model was 0.14 (Table 42.1). Through ERRAT, overall quality factor of the modeled protein was calculated and it was 90.802. All these values are in the range suggestive of an

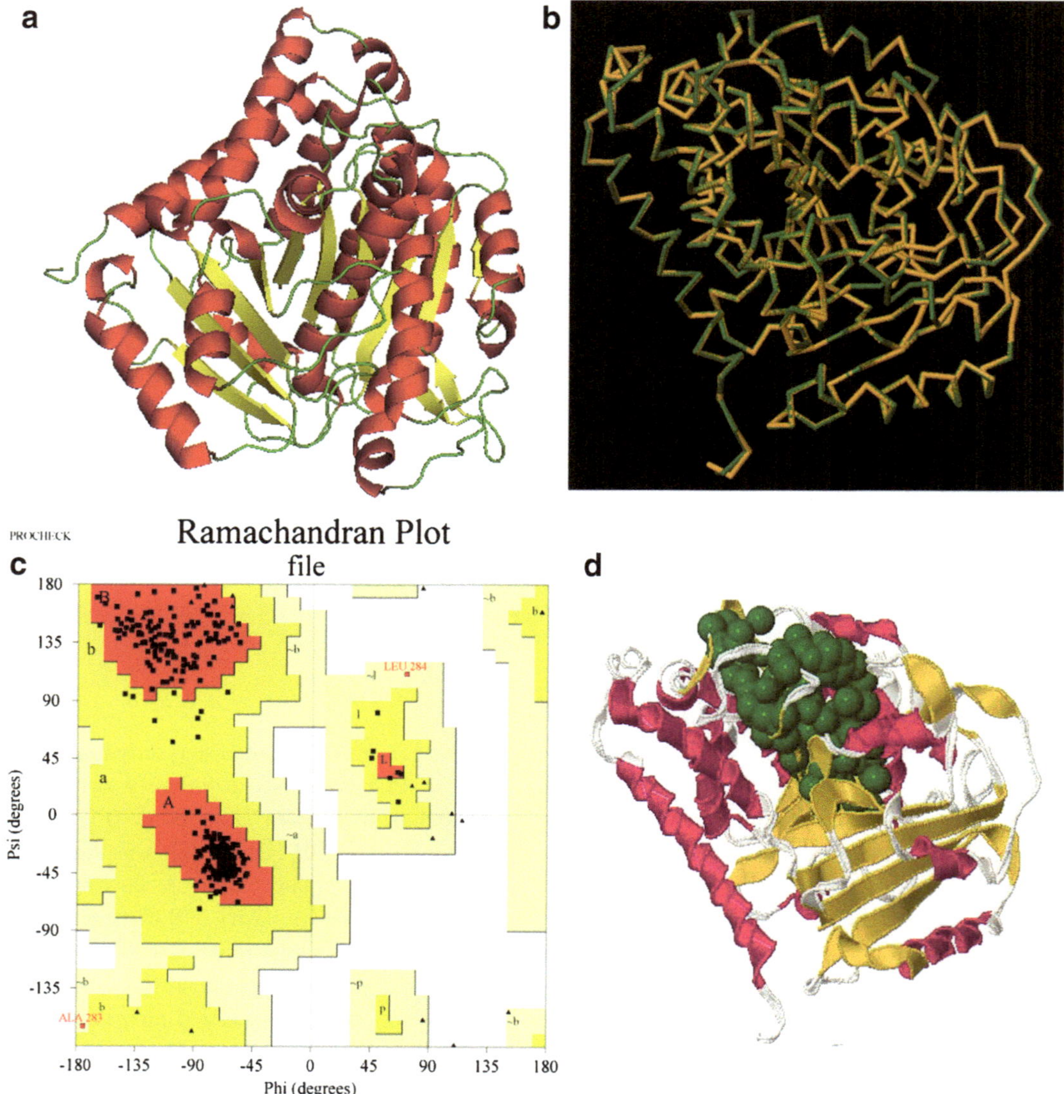

Fig. 42.3 Comparative homology modeling (**a**) ribbon representation of 3-D structure of *P. ultimum* β-tubulin; (**b**) superimposition of modeled *P. ultimum* β-tubulin (target; *yellow*) on 3RYC (template; *green*) drawn using 3-DSS tool; (**c**) *Ramachandran plot* showing residues in favored (*red*), allowed (*yellow*), and disallowed (*white*) regions; (**d**) catalytic active sites of β-tubulin predicted by CASTp calculation with *green*-colored spheres representing active site cavity

acceptable 3-D model. Analysis of secondary structure revealed the presence of 24 α-helices, 18 β-strands, and 40 turns in *P. ultimum* β-tubulin. Molecular modeling revealed that 3-D structure of *P. ultimum* β-tubulin possesses N-terminal Rossmann fold with parallel β-strands alternating with α-helices [40], and presence of this domain is typical of nucleotide-binding proteins.

Molecular Docking Studies

The validated and refined model was then used for docking paromomycin after predicting binding site with CASTp and SiteMap (Schrodinger) (Fig. 42.3d). Docking was performed within a radius of 12 Å. Analysis of RMSD values validates the docking protocol as the RMSD value for the best predicted binding pose gave a value of 0.045 Å, and RMSD values less than 2 Å are considered acceptable [41]. Molecular docking results revealed that six of the active site amino acid residues, *viz.*, Gln 11, Asp 67, Leu 68, Asn 99, Gly 142, and Asp 177, are involved in seven strong hydrogen bondings within 5 Å of paromomycin with a glide score of −11.02 kcal/mol. Other binding parameters like van der Waals interaction energy (−33.87), Coulombs interaction energy (−23.80), Emodel (−81.97), and cumulative van der Waals interaction energy (−57.67) showed good orientations at the binding sites with paromomycin occupying the whole cavity (Fig. 42.4).

Table 42.1 Ramachandran plot statistics calculated using PROCHECK for the 3-D model of *P. ultimum* β-tubulin

Parameter	Value
Core (%)	94.9%
Allowed (%)	4.5%
General (%)	0.5%
Disallowed (%)	0.0%
G-factor	0.14
Main chain bond lengths	−0.03
Main chain bond angles	−0.13

The ubiquitous, necrotrophic *P. myriotylum* poses a constant threat to a broad range of crop plants. Paromomycin, which has been identified to possess potent anti-oomycete activity [16], is known to interfere with translation in prokaryotic ribosomes, whereas its mode of inhibitory action against eukaryotes is not well defined. Understanding the interactions of this prokaryotic inhibitor at the molecular level is critical to sustain disease suppression with long-term use. This is essential to develop disease control strategies to minimize resistance and prevent its spread. In the present study, we observed that besides inhibition of growth of *P. myriotylum* hyphae, paromomycin was also observed to cause substantial reductions in zoospore number, and the inhibitory effect was evident even at low concentrations. These results are of significance as pathogen's infective potential as well as its spatial dissemination relies on copious production of asexual zoospores [42]. Inhibitory effects of paromomycin on zoospores as well as hyphal growth were concentration dependent. Detailed

Fig. 42.4 Molecular docking results showing the H-bonds (*yellow dotted lines*) between active site residues of β-tubulin and paromomycin

microscopic studies of *P. myriotylum* hyphae treated with inhibitory concentrations of paromomycin show distortions in hyphal morphology including cytoplasmic coagulation and vesiculation compared to the control. These observed hyphal abnormalities may be evoked by perturbations to the hyphal cytoskeleton. To understand the rapidly induced anti-*Pythium* effect, we carried out further investigation by studying the effect of paromomycin on the major proteins associated with the cytoskeleton. Hyphae of oomycetes have microtubules running longitudinally throughout the cytoplasm and are involved in cytoplasmic transport [43, 44]. Microtubules are highly dynamic structures made of tubulin proteins and play central role in hyphal growth and morphogenesis [45]. The β-tubulin sequence retrieved from *P. ultimum* genome sequence resource [22] was used for comparative homology modeling and subsequently for molecular docking to study the interactions of paromomycin with β-tubulin. Validation results of the 3-D β-tubulin structure suggest that the model performed well in terms of main chain stereochemistry and amino acid environment. This overall good quality of the modeled protein was supported by high ERRAT scores (90.802) compared to the template. Subsequent docking results indicate that the best conformation of paromomycin inside β-tubulin makes seven H-bonds with six amino acids. The amino acid residues constituting the β-tubulin active site and involved in H-bonds with paromomycin lie well within the consensus regions that include residues from 140 to 146 forming part of GTP binding site and responsible for GTPase activity of β-tubulin as well as residues 8–17, 61–70, 155–174, 178–184, and 295–298 which are implicated in base binding also [46]. Thus, interaction of paromomycin with functionally important amino acid residues in active site may block the catalytic activity of β-tubulin leading to destabilization of hyphal membrane resulting in cytoplasmic coagulation and altered morphogenesis. Hence, analysis of the results of homology modeling and molecular docking indicated that the theoretical predictions were consistent with the known set of experimental results. This first report detailing upon the inhibitory mechanism of paromomycin can serve as a tool for development of anti-*Pythium* agents with greater selectivity. In this direction treatment of rhizome with the corresponding BCA *viz.*, *S. rimosus* ssp. *paromomycinus*, prior to planting would be an economic method for disease control.

Acknowledgements RAK, GC and SS acknowledge the research fellowship received from Department of Biotechnology (DBT; No.: BT/PR13248/GBD/27/232/2009), Govt. of India, MHRD (Ministry of Human Resource and Development, Govt. of India) and Department of Science and Technology (DST; No. SR/FT/LS-052/2009), Govt. of India, respectively. AN acknowledges the financial support received under NITC FRG (Faculty Research Grant; No. NITC/Dean (C&SR)/FRG10/0112) scheme.

References

1. van der Plaats-Niterink AJ. Monograph of the genus *Pythium*. Stud Mycol. 1981;21:1–242.
2. Martin FN, Loper JE. Soilborne plant diseases caused by *Pythium* spp. ecology, epidemiology, and prospects for biological control. Crit Rev Plant Sci. 1999;18:111–81.
3. Dohroo NP. Diseases of ginger. In: Ravindran PN, Nirmal Babu K, editors. Ginger, the genus *Zingiber*: CRC Press, USA. 2004; 304–40.
4. Tyler BM. Genetics and genomics of the oomycete-host interface. Trends Genet. 1997;17:611–4.
5. Govers F. Misclassification of pest fungus puts vital research on wrong track. Nature. 2001;411:633.
6. Adl SM, Simpson AG, Farmer MA, Andersen RA, Anderson OR, Barta JR, Bowser SS, Brugerolle G, Fensome RA, Fredericq S, James TY, Karpov S, Kugrens P, Krug J, Lane CE, Lewis LA, Lodge J, Lynn DH, Mann DG, McCourt RM, Mendoza L, Moestrup O, Mozley-Standridge SE, Nerad TA, Shearer CA, Smirnov AV, Spiegel FW, Taylor MF. The new higher level classification of eukaryotes with emphasis on the taxonomy of protists. J Eukaryot Microbiol. 2005;52:399–451.
7. Birch PRJ, Rehmany AP, Pritchard L, Kamoun S, Beynon JL. Trafficking arms: oomycete effectors enter host plant cells. Trends Microbiol. 2005;14:8–11.
8. Cook RJ. Making greater use of introduced microorganisms for biological control of plant pathogens. Annu Rev Phytopathol. 1993;31:53–80.
9. Whipps JM. Developments in the biological control of soil-borne plant pathogens. Adv Bot Res. 1997;26: 1–134.
10. Zhang Y, Fernando WGD, de Kievit TR, Berry C, Daayf F, Paulitz TC. Detection of antibiotic-related

genes from bacterial biocontrol agents with polymerase chain reaction. Can J Microbiol. 2006;52: 476–81.

11. Nishida M, Matsubara T, Watanabe N. Pyrrolnitrin, a new antifungal antibiotic. Microbiological and toxicological observations. J Antibiot. 1965;18:211–9.
12. Keel C, Schnider U, Maurhofer M, Voisard C, Laville J, Burger P, Wirthner P, Haas D, Défago G. Suppression of root diseases by *Pseudomonas fluorescens* CHA0: importance of the secondary metabolite 2,4-diacetylphloroglucinol. Mol Plant Microbe Interact. 1992;5:4–13.
13. Silo-Suh LA, Stabb EV, Raffel SJ, Handelsman J. Target range of Zwittermicin A, an aminopolyol antibiotic from *Bacillus cereus*. Curr Micobiol. 1998;37: 6–11.
14. Silo-Suh LA, Lethbridge BJ, Raffel SJ, He H, Clardy J, Handelsman J. Biological activities of two fungistatic antibiotics produced by *Bacillus cereus* UW85. Appl Environ Microbiol. 1994;60:2023–30.
15. Ligon JM, Hill DS, Hammer PE, Torkewitz NR, Hofmann D, Kempf HJ, Pee KH. Natural products with antifungal activity from *Pseudomonas* biocontrol bacteria. Pest Manage Sci. 2000;56:688–95.
16. Coffey GL, Anderson LE, Fisher MW, Galbraith MM, Hillegas AB, Kohberger DL, Thompson PE, Weston PE, etal. Biological studies of paromomycin. Antibiot Chemother. 1959;9:730–8.
17. Milner JL, Silo-Suh LA, Lee JC, He H, Clardy J, Handelsman J. Production of kanosamine by *Bacillus cereus* UW85. Appl Environ Microbiol. 1996;62:-3061–5.
18. Yuan WM, Crawford DL. Characterization of *Streptomyces lydicus* WYEC108 as a potential biocontrol agent against fungal root and seed rots. Appl Environ Microbiol. 1995;61:3119–28.
19. Getha K, Vikineswary S. Antagonistic effects of *Streptomyces violaceusniger* strain G10 on *Fusarium oxysporum* sp. cubense race 4: indirect evidence for role of antibiosis in the antagonistic process. J Ind Microbiol Biotechnol. 2001;28:303–10.
20. Toquin V, Barja F, Sirven C, Gamet S, Latorse MP, Zundel JL, Schmitt F, Beffa R. A new mode of action for fluopicolide: modification of the cellular localization of a spectrin-like protein. Pflanzenschutz Nachr Bayer. 2006;59(2–3):171–84.
21. Jackson KL, Yin J, Csinos AS, Ji P. Fungicidal activity of fluopicolide for suppression of *Phytophthora capsici* on squash. Crop Prot. 2010;29:1421–7.
22. Levesque CA, Brouwer H, Cano L, Hamilton JP, Holt C, Huitema E, Raffaele S, Robideau GP, etal. Genome sequence of the necrotrophic plant pathogen *Pythium ultimum* reveals original pathogenicity mechanisms and effector repertoire. Genome Biol. 2010;11:R73.
23. Sarma YR. Rhizome rot disease of ginger and turmeric. Adv Hortic. 1994;10:1134–6.
24. Finney DJ. Probit analysis. A statistical treatment of the sigmoid response curve. Ed. University Press, Cambridge; 1952.
25. Rahimian MK, Banihashemi Z. A method for obtaining zoospores of *Pythium aphanidermatum* and their use in determining cucurbit seedling resistance to damping-off. Plant Dis Rep. 1979;63:658–61.
26. Pacumbaba RP, Wutoh JG, etal. Production of zoospores, mode of infection, and inoculum potential of *Pythium myriotylum* propagules on cocoyam. J Phytopathol. 1994;140(1):49–54.
27. Altschul SF, Madden TL, Schäffer AA, Zhang J, Zhang Z, Miller W, Lipman DJ. Gapped BLAST and PSI-BLAST: a new generation of protein database search programs. Nucleic Acids Res. 1997;25:3389–402.
28. Eswar N, Marti-Renom MA, Webb B, Madhusudhan MS, Eramian D, Shen M, Pieper U, Sali A. Comparative protein structure modeling with MODELLER. Current Protocols Bioinformatics Unit 5.6; 2006.
29. Sumathi K, Ananthalakshmi P, Roshan MN, Sekar K. 3dSS: 3D structural superposition. Nucleic Acids Res. 2006;34:W128–32.
30. Laskowski RA, MacArthur MW, Moss DS, Thornton JM. PROCHECK: a program to check the stereochemical quality of protein structures. J Appl Cryst. 1993;26:283–91.
31. Colovos C, Yeates TO. Verification of protein structures: patterns of nonbonded atomic interactions. Protein Sci. 1993;2(9):1511–9.
32. Schrodinger LLC. Glide: molecular docking tool, v.5.6. New York: Schrodinger LLC; 2010.
33. McDonald NA, Jorgensen WL. Development of an all-atom force field for heterocycles. Properties of liquid pyrrole, furan, diazoles, and oxazoles. J Phys Chem B. 1998;102:8049–59.
34. Dundas J, Ouyang Z, Tseng J, Binkowski A, Turpaz Y, Liang J. CASTp: computed atlas of surface topography of proteins with structural and topographical mapping of functionally annotated residues. Nucleic Acids Res. 2006;34:W116–8.
35. Kontoyianni M, McClellan LM, Sokol GS. Evaluation of docking performance: comparative data on docking algorithms. J Med Chem. 2003;47:558–65.
36. Friesner RA, Banks JL, Murphy RB, Halgren TA, Klicic JJ, Mainz DT, Repasky MP, Knoll EH, Shelley M, Perry JK, Shaw DE, Francis P, Shenkin PS. Glide: a new approach for rapid, accurate docking and scoring. 1. Method and assessment of docking accuracy. J Med Chem. 2004;47:1739–49.
37. Kortekamp A. Growth, occurrence and development of septa in *Plasmopara viticola* and other members of the Peronosporaceae using light and epifluorescence microscopy. Mycol Res. 2005;109:640–8.
38. Thompson JD, Gibson TJ, Higgins DG. Multiple sequence alignment using ClustalW and ClustalX. Curr. Protoc. Bioinformatics, Unit 2.3; 2002.
39. Sali A, Blundell TL. Comparative protein modelling by satisfaction of spatial restraints. J Mol Biol. 1993;234:779–815.
40. Rossmann M, Moras D, Olsen K. Chemical and biological evolution of a nucleotide-binding protein. Nature. 1974;250:194–9.

41. Warren GL, Webster AC, Capelli AM, Clarke B, LaLonde J, Lambert MH, Lindvall M, Nevins N, Semus SF, Senger S, Tedesco G, Wall ID, Woolven JM, Peishoff CE, Head MS. A critical assessment of docking programs and scoring functions. J Med Chem. 2006;49:5912–31.
42. Stanghellini ME, Rasmussen SL, Kim DH, Rorabaugh PA. Efficacy of nonionic surfactants in the control of zoospore spread of *Pythium aphanidermatum* in a recirculating hydroponic system. Plant Dis. 1996;80:-422–8.
43. Heath IB. Preservation of a labile cortical array of actin filaments in growing hyphal tips of the fungus *Saprolegnia ferax*. Eur J Cell Biol. 1987;44:10–6.
44. Gow NAR. Control of the extension of the hyphal apex. In: McGinnis MR, Borgers M, editors. Current topics in medical mycology 3. New York: Springer; 1989. p. 109–52.
45. Heath IB. Integration and regulation of hyphal tip growth. Can J Bot. 1995;73(1):S131–9.
46. Burns RG, Farrell KW. Getting to the heart of β-tubulin. Trends Cell Biol. 1996;6:297–303.

Prenatal Genetic Studies on Intrauterine Growth Retardation

43

P. Jisha, R. Manikkumar, D. Dinesh Roy, and K. Ittyavirah Alex

Abstract

Intrauterine growth retardation/restriction (IUGR) implies that fetal growth is being inhibited and that fetus does not attain its growth potential. Prenatal diagnosis enables early diagnosis of congenital anomalies and genetic disorders in utero. Thirty-eight high-risk pregnancies whose gestational age ranged from 14 to 30 weeks with a mean gestational age of 20.05 were selected. Chromosome analysis and cytokinesis-block micronuclei assay were performed. The mean CBMN frequency of study subjects was 13 and the control subjects showed 10.14 ($p<0.05$). Chromosome analysis revealed 65.78% of normal karyotype and 34.21% abnormal karyotypes. The abnormal karyotypes include 92.30% of numerical abnormalities and 7.69% structural abnormalities. The incidence of abnormal karyotype and CBMN frequency increased with increased paternal and maternal age, duration of married life, previous history of abortion and IUGR, socioeconomic status, etc. Genetic counseling in association with modern prenatal diagnostic procedures constitutes a basic element of prevention of congenital anomalies and genetic disorders. The magnitude of the risk of occurrence, the impact of the disease on the patient and family, and the anticipated future developments are the main informations conveyed during genetic counseling. Hence from the present study, it can

P. Jisha (✉)
School of Health Sciences, Kannur University, 670661, Thalassery Campus Kannur, Kerala, India
e-mail: jishaprasanth.2007@gmail.com

R. Manikkumar
Vinayaka Missions University, Salem 636 308, Tamil Nadu, India

D. Dinesh Roy • K.I. Alex
Ittyavirah Scan & Genetic Research, Thiruvananthapuram 695 011, Kerala, India

A. Sabu and A. Augustine (eds.), *Prospects in Bioscience: Addressing the Issues*,
DOI 10.1007/978-81-322-0810-5_43, © Springer India 2013

be concluded that proper diagnosis of chromosomal abnormalities can lead to the prevention of future birth of similarly affected children. This can be achieved by instituting genetic counseling. Parents with an IUGR fetus should be counseled that their neonates may have some immediate complications at birth but also some long-term complications including impaired cognitive function such as learning disabilities and spastic cerebral palsy.

Keywords
Intrauterine growth retardation • TORCH infection • Prenatal diagnosis • Karyotype analysis

Introduction

Intrauterine growth restriction (IUGR) is defined as a failure of the fetus to reach its growth potential. It is a major contributor to perinatal mortality, being etiologically responsible for 50% of perinatal deaths occurring preterm and 20% at term [1]. The term intrauterine growth restriction has largely replaced the term intrauterine growth retardation (IUGR). The definition of IUGR is a problematic one because we do not know the inherent growth potential of the fetus. The most common definition used is fetal weight below the 10th percentile for gestational age. Fetal growth restriction (FGR) is defined as a fetal growth rate that prevents an infant from obtaining the complete genetic growth potential [2]. It is surprisingly common with placental dysfunction occurring in about 3% of pregnancies. Impaired growth of the fetus is a serious complication in pregnancy, being a major determinant of perinatal morbidity and mortality. Early recognition of fetuses at increased risk of being growth-restricted enables more appropriate surveillance, thereby optimizing management, which has been shown to reduce the risk of adverse fetal outcome [3]. Defining fetal growth restriction has been a longstanding challenge. Currently, most clinicians and researchers use small for gestational age (SGA) (i.e., the smallest 10% of fetuses or newborns at a given gestational week) as a surrogate for fetal growth restriction [4]. One of the primary deficiencies of the current definition of fetal growth restriction is that it is based on an absolute fetal size, irrespective of maternal and fetal genetic and physiologic factors.

Intrauterine growth restriction is one of the most common and complex problems in modern obstetrics. In fact, there is some confusion in terminology for the lack of uniform diagnostic criteria. Furthermore, almost all authors use the terms small for gestational age (SGA) and intrauterine growth restriction (IUGR) as synonymous. Others think that the term SGA is more appropriate referring to infant while IUGR referring to fetus. By definition, 10% of people in any population have weights, as well as heights, below the 10th percentile. This is the cutoff value mainly used for defining the IUGR. A minority of authors define the cutoff value at the 5th or at 3rd percentile. There are many evidences demonstrating that the adverse perinatal outcome is mainly confined to infants below the 5th or 3rd percentile [5].

The fetal growth is determined by a plethora of maternal, fetal, and placental factors [6, 7]. Available literature recognized many risk factors like hypertension, renal disease, diabetes, restrictive lung disease, cyanotic heart disease, multiple gestation, hemoglobinopathies, smoking, substance abuse, malnutrition, low socioeconomic status, low prepregnancy maternal weight, and extremes of reproductive age are strictly related with IUGR. Sometimes, it is difficult to identify a specific cause for IUGR. The main cause for the onset of IUGR can be divided into three categories: maternal, fetal, and placental. Among the maternal ones, it is very important to remind all medical conditions affecting the microcirculation that cause fetal hypoxemia, vasoconstriction, or a reduction in fetal perfusion [8].

Hypertension, typically in preeclampsia, is a relatively common example as well as severe chronic diseases like renal insufficiency, systemic lupus erythematosus, chronic anemia, or pregestational diabetes [9, 10]. Other principal reasons are the behavioral conditions including substance abuse (alcohol or heroin) or smoking. Among the placental ones, we remember the most common causes of SGA in nonanomalous fetus: impaired placental perfusion and placenta. Placenta previa is one of the well-recognized pathology methods determining IUGR, as consequence of the abnormal placental implantation that is very important for a good oxygenation and nutrition of fetus. Finally, we underline some relatively rare primary pathological condition, such as mosaicism, chorioangioma [11] infarcts, or partial abruptions. The last category of pathological condition comprises the fetal diseases. The major cause of IUGR is the chromosome anomalies. Infective diseases or exposure to teratogens are other very well-known causes.

It is largely accepted that once an IUGR is suspected sonographically, a specific examination should be performed. A prenatal confirmative test should be done in case of early or severe IUGR or if there are associated anomalies. Assessment of chromosomal defects should be done if there are anomalies; however, once a nonanomalous IUGR is identified, many authors recommend some antenatal surveillance, even if there is a lack of randomized controlled trials regarding which is the best method to perform this supervision.

IUGR is the term used to describe a condition in which the fetus is smaller than expected for the number of weeks of pregnancy. IUGR is a major public health problem worldwide. It is the second leading cause of prenatal morbidity and mortality. IUGR is caused by maternal factors, placental insufficiency, infections, teratogens, and chromosomal aberrations. In India according to recent UNICEF surveys, the incidence of IUGR is 20–30%. Incidence varies depending on the population. Approximately 4–8% of all infants born in developed countries and 6–30% in developing countries are classified as growth restricted. Prenatal diagnosis is a most widely used and helpful technique to detect chromosomal abnormalities of unborn child. Hence, the present study was undertaken to evaluate the chromosomal abnormalities, if any, in fetuses with IUGR by prenatal diagnostic techniques to evaluate various socioeconomic and demographic characteristics associated with IUGR. Cytokinesis-block micronuclei assay was also performed to evaluate the genotoxic/teratogenic effects if any in developing fetus.

Materials and Methods

Thirty-eight high-risk pregnancies with proven IUGR were taken for the study. These subjects were referred to Ittyavirah Scan and Genetic Research, Trivandrum, for genetic studies from various maternity centers of Kerala. Detailed and relevant informations were recorded. Fourteen children without any chronic illness were selected as control for this study. 1–2 ml of umbilical/cord blood samples were collected from all the subjects under maximum sterile condition into sodium-heparinized Vacutainers by cordocentesis after obtaining written consent. Lymphocyte microcultures were set up as described by Moorhead et al. (1960), and GTG-banded karyotypes were prepared according to ISCN pattern 1995.

The lymphocyte CBMN assay using whole-blood cultures was also performed as described by Fenech [12]. Cytochalasin B was added after 44 h of culture, and cells were harvested 28 h later and then fixed and stained. The CBMN assay slides were labeled with the patient's code number and scored by a single scorer who was not aware of the risk group and the eventual pregnancy outcome of the participants when the slides were scored. Frequency of BN cells with micronuclei (MN-BN) was determined by scoring 1,000 BN cells from each subject.

Results and Discussion

Thirty-eight high-risk pregnancies whose gestational age ranged from 14 to 30 with a mean gestational age of 20.05 were selected for the study. Cordocentesis was performed in all the cases. Chromosome analysis revealed 65.78% of normal karyotypes and 34.21% abnormal karyotypes. The abnormal karyotype includes 92.30%

Table 43.1 Distribution of karyotypes and mean CBMN frequencies according to various demographic characters

Demographic characters		Number	Normal karyotype	Abnormal karyotype	Mean CBMN frequency
Age of husband	<35	14 (36.84 %)	12 (85.71 %)	2 (14.28 %)	12.43
	≥35	24 (63.16 %)	13 (54.16 %)	11 (45.83 %)	13.33
Age of wife	<35	29 (76.31 %)	21 (72.41 %)	8 (27.58 %)	12.76
	≥35	9 (23.68 %)	4 (44.44 %)	5 (55.55 %)	13.78
Duration of married life	<3	17 (44.74 %)	11 (64.70 %)	6 (35.29 %)	13
	≥3	21 (55.26 %)	14 (66.66 %)	7 (33.33 %)	13
Residence after marriage	Rural	26 (68.42 %)	17 (65.38 %)	9 (34.61 %)	12.96
	Urban	11 (28.94 %)	7 (63.63 %)	4 (36.36 %)	12.9
	Coastal	1 (2.63 %)	1 (100 %)	0 (0 %)	15
Socioeconomic status	High	11 (28.93 %)	9 (81.81 %)	2 (18.18 %)	12.5
	Average	23 (60.52 %)	13 (56.52 %)	10 (43.47 %)	13.09
	Low	4 (10.52 %)	3 (75 %)	1 (25 %)	14
Religion	Hindu	32 (84.21 %)	20 (62.5 %)	12 (37.5 %)	13.06
	Muslim	4 (10.52 %)	3 (75 %)	1 (25 %)	12.75
	Christian	2 (5.26 %)	2 (100 %)	0 (0 %)	12.5

(n=12) of numerical abnormalities and 7.69% (n=1) structural abnormalities.

Cytokinesis-block micronuclei assay was also performed to evaluate the extent of teratogenic/genotoxic effects, if any, in these developing fetuses. The mean CBMN frequency of study subjects was 13 and the control subjects showed a mean CBMN frequency of 10.14. This difference has a statistical significance (t=5.763, p=<0.05).

The distribution of karyotypes and mean CBMN frequencies according to various demographic characters was analyzed and represented in Table 43.1. The age of the husband ranged from 29 to 50 with a mean age of 35.81. The study revealed that the incidences of abnormal karyotypes of fetuses (45.83%) were increased with increasing age of the husband. This is also true in the case of CBMN assay. The micronuclei frequencies of these fetuses were increased with increasing paternal age. The age of wife ranged from 22 to 39 with a mean age of 30.86. Women above 35 years showed increased incidences of abnormal karyotypes (55.55%) and the CBMN frequency (13.78) also increased with increasing maternal age. The study also observed that the incidences of abnormal karyotypes of fetuses (33.33%) were increased with increased duration of married life of the couple. However, no significant differences were observed between the CBMN frequency and duration of married life.

The subjects were grouped on the basis of area of residence as rural, urban, and coastal, and the distribution of karyotypes was analyzed. The incidences of abnormal karyotype (36.36%) were higher among subjects who belong to urban centers. However, the CBMN frequency (15) is higher among subjects who belong to the coastal area. On the basis of socioeconomic status, the study subjects were classified in to high, average, and low. Majority of the subjects belong to average socioeconomic status. The incidences of abnormal karyotype (43.47%) were higher among subjects who belong to average socioeconomic status followed by low and high socioeconomic status. However, the CBMN frequency is high (14) among subjects who belonged to poor socioeconomic status. On the basis of religion, the study subjects were grouped into Hindu, Muslim, and Christian. Majority of subjects belonged to Hindu community and showed increased incidence of abnormal karyotype (37.5%) and increased CBMN frequency (13.06).

Distribution of karyotypes and mean CBMN frequencies according to various biological risk factors was also analyzed and recorded in Table 43.2. 34.21% study subjects reported previous history of abortion, and the rest (65.78%)

Table 43.2 Distribution of karyotypes and mean CBMN frequencies according to various biological risk factors

Risk factors		Number	Normal karyotype	Abnormal karyotype	Mean CBMN frequency
Previous abortion	Yes	13 (34.21 %)	10 (76.92 %)	3 (23.07 %)	13.46
	No	25 (65.78 %)	15 (60 %)	10 (40 %)	12.76
Previous IUGR	Yes	9 (23.68 %)	6 (66.66 %)	3 (33.33 %)	13.78
	No	29 (76.31 %)	19 (65.51 %)	10 (34.48 %)	12.76
Live children	Yes	12 (31.57 %)	8 (66.66 %)	4 (33.33 %)	12.5
	No	26 (68.42 %)	17 (65.38 %)	9 (34.61 %)	13.23
Family history of BOH	Yes	4 (10.25 %)	1 (25 %)	3 (75 %)	14.75
	No	34 (89.47 %)	24 (70.58 %)	10 (29.41 %)	12.79
Consanguinity	Yes	4 (10.52 %)	1 (25 %)	3 (75 %)	13.75
	No	34 (89.47 %)	24 (70.58 %)	10 (29.41 %)	12.91
TORCH investigation	Negative	10 (26.31 %)	4 (40 %)	6 (60 %)	12.9
	Positive	7 (18.42 %)	6 (85.71 %)	1 (14.28 %)	14.14
	Not performed	21 (55.26 %)	15 (71.42 %)	6 (28.57 %)	12.67

reported no incidence of previous abortion. The incidences of abnormal karyotype (40%) were higher among mothers without previous history of abortion. However, mothers with previous abortions showed increased CBMN frequency of fetuses (13.46) than the rest.

Previous history of IUGR was noted among nine study subjects, and these fetuses showed increased CBMN frequency (13.78). However, fetuses without previous history of IUGR showed increased incidence of abnormal karyotype. The distribution of karyotype of study subjects was analyzed on the basis of incidence of live children among their parents. Twelve subjects belong to the parents with live children, and parents of 26 fetuses did not have any live children. The incidences of abnormal karyotype (34.61%) were higher among mothers without live children. The CBMN frequency is also higher (13.25) among these subjects.

The distribution of karyotype of the study subjects was analyzed according to history of bad obstetric history (BOH) in their family. Four out of 38 subjects reported family history of BOH and showed increased CBMN frequency (14.75) and increased incidence of abnormal karyotype (75%). Based on consanguinity, the subjects were grouped and their distribution of karyotypes was analyzed. The incidences of abnormal karyotypes (75%) were higher among consanguineous couples, and the CBMN frequency is also higher (13.75) among these subjects. Majority of abnormal karyotypes were detected within 20 weeks of gestational age.

The subjects were grouped on the basis of screening test for TORCH infection in their mothers. Among the 38 study subjects, 17 had undergone TORCH screen test. TORCH screen test was not performed in the remaining 21 subjects. 10 out of 17 subjects who had undergone the screen test gave negative result, and 7 were positive for the infection. TORCH-positive subjects showed increased CBMN frequency (14.14) than the rest.

Micronuclei (MN) in mammalian cells originate from lagging chromosome fragments or whole chromosomes during anaphase [13] in mitosis or meiosis. It is important to note that during meiosis, there are two anaphase stages during which chromosomal material is segregated. During anaphase I, one homologous chromosome, as two joined chromatids, moves to one pole of the cell, while the other homologous chromosome, also as two joined chromatids, segregates to the opposite pole of the cell. In anaphase II of meiosis, the two chromatids of each chromosome separate and move to the opposite poles of the cell. At each of these two meiotic anaphases, it is possible that

DNA damage and/or spindle/kinetochore defects may lead to lagging whole chromosomes and/or acentric chromosome fragments and result in MN formation in the germ line [14, 15]. These chromosomal damage events are mainly due to (1) inherited or acquired genetic defects in genome maintenance and/or (2) inadequacy of micronutrients required as cofactors for proteins involved in DNA synthesis or repair and/or (3) exposure to endogenous or environmental genotoxicants [13]. Formation of MN can occur in rapidly proliferating tissues such as the germ line in males, in the placenta, in the embryo, and in the various organs of the fetus during gestation. Induction of MN results in significant altered gene dosages (gains or losses) depending on the distribution of MN in daughter cells and is often associated with reduced proliferative potential and aberrant gene expression [16]. Such significant genetic events are likely to have important consequences on reproductive capacity both with respect to fertility as well as pregnancy outcomes. It is for this reason that there is a growing interest in the relationship between MN expression and reproductive health in humans.

Intrauterine growth restriction (IUGR) is an important problem in perinatal medicine. It is the second cause of perinatal mortality after prematurity. The two conditions are often associated in case of iatrogenic prematurity [17]. Over the last three decades, prenatal diagnosis – the ability to detect abnormalities in an unborn child – has been widely used. It is never easy for a couple to decide to pursue prenatal diagnosis, but it is only an option which is chosen by many couples at high risk of having a child with serious hereditary disorders. Birth defects have been present in babies from all over the world. IUGR has attracted lot of attention of the scientific community and extensive research is going on in many centers. With better understanding of the pathophysiology of IUGR, better fetal monitoring and therapeutic modalities will be available in the future to tackle this serious problem more effectively. Prevention and better management of IUGR will not only stop the reproductive wastage and reduce infant morbidity/mortality but reduce the incidence of serious cardiac and metabolic disorders thus preventing premature deaths. In the present study, cordocentesis was conducted in 38 high-risk pregnancies suspected with IUGR and performed fetal karyotype analysis and cytokinesis-block micronuclei assay to quantify the extent of somatic DNA damages. The study revealed that 13 out of 38 showed abnormal karyotype. The abnormal karyotype includes 92.30% ($n=12$) of numerical abnormalities and 7.69% ($n=1$) structural abnormalities. Cytokinesis-block micronuclei assay was also performed to evaluate the extent of teratogenic/genotoxic effects, if any, in these developing fetuses. The mean CBMN frequency of study subjects was 13 and the control subjects showed a mean CBMN frequency of 10.14. This difference has a statistical significance also ($t=5.763$, $p=<0.05$).

Genome damage can influence virtually any cellular process due to its ability to affect gene dosage, level of gene expression, and subsequent gene products. During fetal development, rapid cell proliferation, active gene transcription, and a high rate of DNA metabolism can increase genotoxic insult susceptibility [18]. Genome damage can occur via a variety of mechanisms, including point mutations, base modifications, chromosome breakage, chromosome rearrangement, and chromosome loss or gain [19]. Increased DNA damage at the molecular and chromosomal level has been associated with infertility [20, 21], recurrent miscarriage [22], and developmental defects [23].

According to Mandruzzato et al. [17], when IUGR is suspected or diagnosed, it is necessary to distinguish between fetuses that are small but otherwise healthy and those that are a consequence of an abnormal condition such as a maternal condition (chronic hypertension, pregestational diabetes, cardiovascular disease, substance abuse, autoimmune conditions, etc.), a fetal condition (infection, malformation, chromosomal aberration, etc.), or a placental condition (chorioangioma, infarction, circumvallate placenta, confined placental mosaicism, obliterative vasculopathy of the placental bed, etc.). Placental conditions are the most frequent etiology of IUGR. In the present study, it was found that the incidence of abnormal fetal karyotype increases with increasing maternal age. When the maternal age is more than 35 years, 57% of abnormal fetal karyotype was reported. The highest frequency of chromosome abnormalities was observed in couples with more than 5 years of married life. The couple with history of previous abortions had higher incidence of fetal chromosomal abnormalities than the couple without any previous abortions. This present study also showed that loss of a previous pregnancy increased the risk of having an IUGR fetus in the subsequent pregnancy. The incidence of chromosomes anomalies was found to be lower in subjects with live children, whereas in subjects without live children, the incidence of fetal chromosome anomalies was higher. This present study also observed that the mean CBMN frequency of children increased with increase in paternal and maternal age. The incidences of abnormal karyotypes increased with increased duration of married life of the couple, but no significant difference was observed between the CBMN frequency and duration of married life. The incidences of abnormal karyotype were higher among subjects without previous history of abortions, but subjects with previous abortions showed increased CBMN frequency than the rest. The subjects with previous IUGR showed higher CBMN frequency than those subjects without previous IUGR.

Intrauterine growth restriction (IUGR) is an important problem in perinatal medicine. It is the second cause of perinatal mortality after prematurity. The two conditions are often associated in case of iatrogenic prematurity. Fifty-two percent of stillbirths are associated with IUGR and 10% of perinatal mortality cases in Europe are the consequence of unrecognized severe growth restriction. Moreover, IUGR is also associated with increased neonatal mortality and morbidity, and it has been suggested that some diseases (metabolic and vascular) evident in adult life are the consequence of fetal growth restriction.

The increase in prevalence of intrauterine growth restriction observed between 1993 and 2004, especially among lower-income women, is also a point of concern, given the potential short- and long-term consequences associated with this condition. As demonstrated by previous studies of the 1982 cohort [24], children born small for their gestational age show greater morbidity and mortality in early life and may have a higher risk of chronic degenerative disease in adulthood [25]. The causes for such an increased need also be investigated so that preventive action can be taken. In the present study, majority of the subjects belong to average socioeconomic status. The incidences of abnormal karyotype (43.47%) were higher among subjects who belong to average socioeconomic status followed by low and high socioeconomic status. However, the CBMN frequency is high (14) among subjects who belonged to lower socioeconomic status.

Congenital CMV infection is almost invariably accompanied by positive histopathologic findings in the placenta [26], indicating that the placenta is the main entrance for CMV to the fetuses and that placental infection following maternal viremia is the initiating step of fetal infection. The placenta could have a more important role as a protective factor than maternal immunologic reactivity. One of the factors in limiting viral spread could be the ability of the placenta to produce interferon when challenged [27]. Interferons are constitutively expressed by several tissues, notably in the placenta of several species including human [28, 29]. Maternal infections play a critical role in pregnancy wastage and their occurrence in patients with BOH is a significant factor. In the current study, subjects were grouped on the basis of screening test for TORCH infection in their mothers. Among the 38 study subjects, 17 had undergone TORCH screen test. TORCH screen test was not performed in the remaining 21 subjects. 10 out of 17 subjects who had undergone the screen test gave negative result, and 7 were positive for the infection. TORCH-positive subjects showed increased CBMN frequency (14.14) than the rest.

Both CMV and HSV are known to have an intrauterine route of transmission with significant mortality and morbidity [30]. The study shows seropositivity rate of 5.33% for CMV-specific IgM in women with BOH. In other studies, seropositivity ranges from 3 to 12.9% [1, 31, 32]. It was suggested that pregnancy may reactivate the latent virus leading to further reproductive wastages. In the present study, 10 out of 17 subjects who had undergone the screen test were negative, and 7 were positive for the infection. TORCH-positive subjects showed increased CBMN frequency (14.14) than the rest.

Genetic counseling in association with modern prenatal diagnostic procedures constitutes a basic element of prevention of congenital anomalies and genetic disorders. The magnitude of the risk of occurrence, the impact of the disease on the patient and family, and the anticipated future developments are the main informations conveyed during genetic counseling. Hence, the present study can be concluded that proper diagnosis of chromosomal abnormalities can lead to the prevention of future birth of similarly affected children. This can be achieved by instituting genetic counseling. Though a new technique, cordocentesis is playing a major role in modern perinatology and plays a crucial role in antenatal decision making. Through the invasive diagnostic procedures, we can comment on the integration of genetics-based prenatal invasive diagnostic procedures with the accuracy of fetal anomalies diagnosed by ultrasound.

References

1. Kady M, Gardosi J. Perinatal mortality and fetal growth restriction. Best Pract Res Clin Obstet Gynaecol. 2004;18:397–410.
2. Odegard RA, Vatten LJ, Nilsen ST, Salvesen KA, Austgulen R. Preeclampsia and fetal growth. Obstet Gynecol. 2009;96:950–5.
3. Lindqvist PG, Molin J. Does antenatal identification of small-for-gestational age fetuses significantly improve their outcome? Ultrasound Obstet Gynecol. 2005;25:258–64.
4. Zhang J, Merialdi M, Platt LD, et al. Defining normal and abnormal fetal growth: promises and challenges. Am J Obstet Gynecol. 2010;202(6):522–8.
5. McIntire DD, Bloom SL, Casey BM, Leveno KJ. Birth weight in relation to morbidity and mortality among newborn infants. N Engl J Med. 1999;340(16):1234–8.
6. Maulik D. Fetal growth restriction: the etiology. Clin Obstet Gynecol. 2006;49(2):228–35.
7. Kinzler WL, Vintzileos M. Fetal growth restriction: a modern approach. Curr Opin Obstet Gynecol. 2008; 20:125–31.
8. Pedersen NG, Figures F. Early fetal size and growth as predictors of adverse outcome. Obstet Gynecol. 2008;112:765–71.
9. ACOG Practice Bulletin. Clinical management guidelines for obstetrician- gynecologists. Number 12. Jan; 2000.
10. Dunger DB, Petry CJ, Ong KK. Genetic variations and normal fetal growth. Horm Res. 2006; 65(suppl):34–40.
11. Zhang J. Defining normal and abnormal fetal growth: promises and challenges. Am J Obstet Gynecol. 2010, Jun 202(6):522–8.
12. Fenech M. Cytokinesis-block micronucleus cytome assay. Nat Protoc. 2007;2:1084–104.
13. Fenech M, Kirsch-Volders M, Natarajan AT, et al. Molecular mechanisms of micronucleus, nucleoplasmic bridge and nuclear bud formation in mammalian and human cells. Mutagenesis. 2010;26:125–32.
14. Handel MA, Schimenti JC. Genetics of mammalian meiosis: regulation, dynamics and impact on fertility. Nat Rev Genet. 2010;11:124–36.
15. Obradors A, Rius M, Cuzzi J, et al. Errors at mitotic segregation early in oogenesis and at first meiotic division in oocytes from donor females: comparative genomic hybridization analyses in metaphase II oocytes and their first polar body. Fertil Steril. 2010;93:675–9.
16. Van Leeuwen DM, Pedersen M, Knudsen LE, Bonassi S, Fenech M, Kleinjans JCS, Jennen DGJ. Transcriptomic network analysis of micronuclei-related genes: a case study. Mutagenesis. 2010;26:27–32.
17. Mandruzzato GP, et al. Intrauterine growth restriction. Recommendations and guidelines for perinatal practice. J Perinat Med. 2008;36:277–81.
18. Vinson RK, Hales BF. DNA repair during organogenesis. Mutat Res. 2002;509:79–91.
19. Fenech M. The Genome Health Clinic and Genome Health Nutrigenomics concepts: diagnosis and nutritional treatment of genome and epigenome damage on an individual basis. Mutagenesis. 2005;20:255–69.
20. Trkova M, Kapras J, Bobkova K, Stankova J, Mejsnarova B. Increased micronuclei frequencies in couples with reproductive failure. Reprod Toxicol. 2000;14:331–5.
21. Sankoff D, Deneault M, Turbis P, Allen C. Chromosomal distributions of breakpoints in cancer, infertility, and evolution. Theor Popul Biol. 2002;61:497–501.
22. Warren JE, Silver RM. Genetics of pregnancy loss. Clin Obstet Gynecol. 2008;51:84–95.
23. Morawiec Z, Janik K, Kowalski M, Stetkiewicz T, Szaflik J, Morawiec-Bajda A, Sobczuk A, Blasiak J. DNA damage and repair in children with Down's syndrome. Mutat Res. 2008;637:118–23.
24. Barros FC, Huttly SRA, Victora CG, Kirkwood BR, Vaughan JP. Comparison of the causes and consequences of prematurity and intrauterine growth retardation: a longitudinal study in southern Brazil. Pediatrics. 1992;90(2 Pt 1):238–44.
25. Barker DJ. The origins of the developmental origins theory. J Intern Med. 2007;261:412–7.
26. Altshuler G. Diseases of placenta and their effect on transport. Mead Johnson Symp Perinat Dev Med. 1981;18:35–43.

27. Banatvala JE, Potter JE, Best JM. Interferon response to Sendai and rubella virus in human foetal cultures. J Gen Virol. 1971;13:193–201.
28. Roberts RM, Farin CE, Cross JC. Trophoblast proteins and maternal recognition of pregnancy. Oxf Rev Reprod Biol. 1990;12:147–80.
29. Roberts RM, Cross JC, Leaman DW. Interferons as hormones of pregnancy. Endocr Rev. 1992;13:432–52.
30. Stagno S, Pass RF, Coud G. Primary Cytomegalovirus infection in pregnancy: incidence, transmission to fetus and clinical syndrome. JAMA. 1986; 256(14):1904–8.
31. Hossain A, Bakir TM, Ramia HS. Immune status to congenital infection by TORCH agents in pregnant Saudi women. J Trop Pediatr. 1986; 32:84–6.
32. Seth P, Balaya S, Mahapatra LN. Seroepidemiological study of rubella infection in female subjects of Delhi and its surrounding villages. Indian J Med Res. 1971;59:190–4.

Performance of West African Dwarf Sheep Fed with Selected Tropical Roughage Sources: Rumen Microbial Analysis and Nutrient Utilization

44

O.A. Isah and R.M. Akinbode

Abstract

Influence of selected tropical roughages on rumen microbial profile and nutrient utilization in West African Dwarf sheep was investigated. Twenty-four female sheep (mean body weight of 12.2 kg) were divided into four treatment groups in a completely randomized design. Treatments were Panicum hay (PH), urea-treated (5%) corncob (UCC), urea-treated (5%) sugar cane top (UST) and cassava leaf hay (CH). All animals were supplemented with the concentrate at 2% of body weight on a dry matter basis and fed their respective experimental diets *ad libitum* for 84 days. Performance characteristics, nitrogen utilization and rumen microbial count were studied, and polymerase chain reaction (PCR) was done. Data collected were analysed using one-way analysis of variance. Results showed that animals on UST and PH (89.73 and 87.82 g/kgB$W^{0.75}$respectively) performed best in terms of dry matter (DM) intake, while UCC (62.70 g/kgB$W^{0.75}$) group had least value. Highest weight gain was recorded for animals fed with UST (38.14 g/day). Nitrogen utilization revealed that animals on CH (10.61 g/day) had best nitrogen retention compared to animals on other treatments. Highest but similar count of cellulolytic bacteria was recorded for animals fed with PH and UST (7.25 and 7.06 × 10^8 CFU/g rumen content respectively), while animals on CH had least value (2.50×10^8 CFU/g rumen content). Similar trend was observed in the result of predominant cellulolytic bacterial presence analysed using PCR technique. From the gel electrophoresis bands, it was inferred the highest value in UST treatment with *Fibrobacter succinogenes* having highest intensity. All tropical roughages investigated could be used as feed for ruminant animals especially during dry season.

Keywords

Sheep • Rumen microbes • Nutrient utilization

O.A. Isah (✉) • R.M. Akinbode
Department of Animal Nutrition,
Federal University of Agriculture, PMB 2240,
Abeokuta, Ogun State, Nigeria
e-mail: bukkyisah@yahoo.com

A. Sabu and A. Augustine (eds.), *Prospects in Bioscience: Addressing the Issues*,
DOI 10.1007/978-81-322-0810-5_44, © Springer India 2013

Introduction

The rumen ecosystem is essentially a fermentative system that houses a vast array of different microbes which consist primarily of bacteria, protozoa and fungi [1]. The ability of ruminants to convert low-quality feeds into high-quality protein and to utilize feed from land not suitable to grow crops for human consumption is made possible by these microorganisms that synthesize and secrete the β-1,4 cellulase enzyme complex thereby allowing hydrolysis of plant cell walls. Without ruminal microorganisms or when the ruminal ecosystem is disturbed, there can be a cascade of detrimental effects on animal health and productivity. Local feed resources, particularly low-quality roughages and agricultural crop residues, are of prime importance for ruminant production in the tropics. These feeds exhibit close relationships with rumen ecology, microbes and rumen fermentation patterns. A number of dietary factors influence rumen fermentation especially the basal roughage sources, their physical form and fermentation end products. Chemical methods for identifying microorganisms based primarily on pure-culture isolation have numerous limitations. Thus, accurate methods for the identification and quantification of fibrolytic microorganisms are needed to study microbial interactions within the rumen or any other complex community. The study is therefore designed to test the effect of different roughage sources (Panicum hay, urea-treated corncobs, urea-treated sugar cane tops and cassava hay) on the rumen ecology and digestibility in sheep using new molecular biology technique for the identification of the microbes so as to supply fundamental understanding of the microbial ecology in the rumen which is essential for manipulation of ruminal microflora and subsequent improvement in animal production.

Materials and Methods

The study was carried out at small ruminant unit of Teaching and Research Farm Directorate (TREFAD), University of Agriculture, Abeokuta (7°10′N and 3°2′E), Nigeria. Twenty-four female sheep (mean body weight of 12.2 kg) were divided into four treatment groups in a completely randomized design. Treatments were Panicum hay (PH), urea-treated (5%) corncob (UCC), Urea-treated (5%) sugar cane top (UST) and cassava leaf hay (CH). All animals were supplemented with concentrate at 2% of body weight on a dry matter basis and fed their respective experimental diets *ad libitum* for 84 days. Proximate composition, performance characteristics, nitrogen utilization [2], rumen microbial count and PCR were investigated. Data collected were analysed using one-way analysis of variance. Table 44.1 shows the oligonucleotide primers used in the study.

Results and Discussion

As per the data presented in Table 44.2, Panicum hay contained 824.20 g/Kg dry matter, while 801.50, 805.00 and 857.00 g/Kg DM were obtained for urea-treated corncob, urea-treated sugar cane top and cassava hay respectively. Crude protein values of 61.70, 98.00, 107.30 and 201.00 g/Kg DM were obtained for Panicum hay, urea-treated corncob, urea-treated sugar cane top and cassava hay respectively. Crude protein values of 9.8 and

Table 44.1 Oligonucleotide primers used

Cellulolytic species	*rrs* gene forward primer	*rrs* gene reverse primer	Product *tm*/size
Fibrobacter succinogenes	5′-GGTATGGGATGAG CTTGC-3′	5′-GCCTGCCCCTGAACTATC-3′	89.4° C/446 bp
Ruminococcus albus	5′-CCCTAAAAGCAGT CTTAGTTCG-3′	5′-CCTCCTTGCGGTTAG AACA-3′	88.5° C/175 bp
Ruminococcus flavefaciens	5′-TCTGGAAACGGAT GGTA-3′	5′-CCTTTAAGACAGGAG TTTACAA-3′	87.8° C/295 bp

Table 44.2 Chemical composition of Panicum hay, urea-treated corncob, urea-treated sugar cane top, cassava hay and concentrate

Parameters (g/Kg Dm)	Panicum hay	Urea-treated corncobs	Urea-treated sugar cane top	Cassava hay	Concentrate	SEM
Dry matter	824.20^{c}	801.50^{d}	805.00^{d}	857.00^{b}	893.20^{a}	9.34
Organic matter	878.40^{d}	953.60^{b}	847.50^{e}	917.00^{c}	978.30^{a}	12.84
Crude protein	61.70^{e}	98.00^{d}	107.30^{c}	201.00^{a}	147.00^{b}	12.71
Ether extract	15.60^{c}	24.60^{b}	18.20^{c}	53.00^{a}	28.60^{b}	3.59
NDF	687.60^{c}	800.00^{a}	705.20^{b}	594.00^{d}	293.00^{e}	46.57
ADF	398.60^{a}	390.00^{b}	372.00^{c}	356.00^{d}	220.00^{e}	17.47
Ash	121.60^{b}	46.40^{d}	152.50^{a}	83.00^{c}	21.70^{e}	12.76
Hemicellulose	289.00^{c}	410.00^{a}	332.20^{a}	238.00^{d}	73.00^{e}	30.1
Condensed tannin	–	–	–	34.00	–	

SEM standard error of means

^{a,b,c,d,e}Means on the same row having different superscripts are significantly different ($p<0.05$)

10.73% obtained for urea-treated corncob and urea-treated sugar cane tops respectively were lower than 11.35 and 13.73% as reported by [3].

Panicum hay used in this study contained 15.60 g/Kg DM ether extract, while 24.60, 18.20 and 53.00 g/Kg DM were obtained for urea-treated corncob, urea-treated sugar cane top and cassava hay respectively. Urea-treated corncob had the highest neutral detergent fibre (800.00 g/kg DM) followed by urea-treated sugar cane top (705.20 g/kg DM) and Panicum hay with 687.60 g/kg DM NDF, while concentrate had the least value of 293.0 g/kg DM of NDF. Panicum hay contained 398.60 g/Kg DM of ADF, while urea-treated corncob, urea-treated sugar cane top and cassava hay contained 390.00, 372.00 and 356.00 g/Kg DM ADF respectively. In addition, cassava hay contained 34g/Kg DM condensed tannin. There were significant differences across treatments ($P<0.05$) in total dry matter intake, weight gain and feed conversion ratio. Animals on urea-treated sugar cane top recorded the highest dry matter intake of 632.09 gd^{-1} followed by those on Panicum hay (606.07 gd^{-1}) and cassava hay (561.31 gd^{-1}), while those on urea-treated corncob had the lowest intake value of 417.12 gd^{-1}. The higher total dry matter intake obtained for urea-treated sugar cane top treatment (632.09 gd^{-1}) could be due to the fact that sugar cane top is highly palatable forage with good voluntary consumption. Cassava hay treatment with lower DMI of 561.30 gd^{-1} could be attributed to residual bitter flavour contained in cassava hay since the cassava used in this experiment was the bitter variety and some researchers [4, 5] demonstrated that the bitter taste of cassava foliage could negatively influence its intake by animals. Highest weight gain was recorded for sheep fed with urea-treated sugar cane top treatment (38 g/day), while urea-treated corncob had the least weight gain of 11g/day. Highest feed conversion ratio value was recorded for sheep fed with urea-treated corncob, while urea-treated sugar cane treatment had the least value of 0.25 which is an indication of better utilization (Table 44.3).

Table 44.4 shows that nitrogen intake was significantly different across the treatments ($P<0.05$) with sheep on cassava hay treatment having the highest value of 16.28 gd^{-1} while sheep on urea-treated corncob treatment had the lowest value of 8.08 gd^{-1}. Also, the nitrogen retention was significantly different ($P<0.05$) among the treatments and values ranged from 3.4 to 10.61 gd^{-1}. The high nitrogen retention in cassava hay treatment could be as a result of condensed tannin contained in it, and this is in line with the earlier reports [4–6] that lower concentration of condensed tannin could lead to improvement in nitrogen utilization by protecting the protein in the rumen and making it available at the hindgut.

Cellulolytic bacterial count (Table 44.5) was highest and similar in sheep fed with Panicum hay treatment and urea-treated sugar cane top (7.25 and 7.06 × 10^{8} CFU/g rumen content respec-

Table 44.3 Feed intake and weight gain of WAD sheep fed with different roughage sources

Parameters (g/Kg Dm)	Panicum hay	Urea-treated corncobs	Urea-treated s ugar cane top	Cassava hay	SEM
Concentrate (gd^{-1})	214.37^{a}	196.50^{b}	205.44^{ab}	205.44^{ab}	2.75
Roughage ($gd^{-1)}$	391.70^{b}	220.61^{d}	426.65^{a}	355.87^{c}	23.88
Total dry matter intake($gd^{-1)}$	606.07^{b}	417.12^{d}	632.09^{a}	561.31^{c}	25.13
Weight gain(g/day)	24b	11c	38a	27b	0.34
Feed conversion ratio	0.37^{b}	0.56^{a}	0.25^{b}	0.31^{b}	0.39

SEM standard error of means
^{a,b,c,d}Means on the same row having different superscripts are significantly different ($p<0.05$)

Table 44.4 Effect of different roughage sources on nitrogen utilization in WAD sheep

Parameters (g/Kg Dm)	Panicum hay	Urea-treated corncobs	Urea-treated sugar cane top	Cassava hay	SEM
N intake	8.91^{c}	8.08^{d}	12.16^{b}	16.28^{a}	0.97
Faecal N	2.62^{b}	2.61^{b}	2.56^{b}	3.36^{a}	0.13
Urinary N	2.15	2.07	1.91	2.31	0.08
N absorption	6.28^{c}	5.47^{c}	9.59^{b}	12.92^{a}	0.89
N retention	4.14^{c}	3.40^{c}	7.69^{b}	10.61^{a}	0.88

SEM standard error of means
^{a,b,c,d}Means on the same row having different superscripts are significantly different ($p<0.05$)

Table 44.5 Effect of different roughage sources on cellulolytic bacteria count in WAD sheep

Parameters (g/Kg Dm)	Panicum hay	Urea-treated corncobs	Urea-treated sugar cane top	Cassava hay	SEM
Cellulolytic bacteria ($\times 10^8$ CFU/g rumen content)	7.25^{a}	6.28^{b}	7.06^{a}	2.50^{c}	0.21

SEM standard error of means
^{a,b,c}Means on the same row having different superscripts are significantly different ($p<0.05$)

Table 44.6 Comparative quantity of predominant cellulolytic bacteria analysed by PCR technique

	Fibrobacter succinogenes (copies/ml)		*Ruminococcus flavefaciens* (copies/ml)		*Ruminococcus albus* (copies/ml)	
Treatments	0 h	4 h	0 h	4 h	0 h	4 h
Panicum hay	1.48×10^{9}	1.76×10^{11}	7.24×10^{2}	4.79×10^{5}	–	–
Urea-treated corncob	7.74×10^{2}	4.04×10^{5}	6.13×10^{2}	1.59×10^{3}	–	–
Urea-treated sugar cane top	4.84×10^{9}	2.82×10^{11}	1.14×10^{3}	1.25×10^{6}	1.92×10^{3}	1.86×10^{4}
Cassava hay	–	–	–	–	–	–

SEM standard error of means
0 h samples taken before morning feeding
4 h samples taken 4 h after feeding
^{a,b,c}Means on the same row having different superscripts are significantly different ($p<0.05$)

tively). *Ruminococcus flavefaciens* was present in Panicum hay, urea-treated corncob and urea-treated sugar cane top treatments but absent in cassava hay treatment (Table 44.6).

References

1. Varga GA, Kolver ES. Microbial and animal limitations to fiber digestion and utilization. J Nutr. 1997; 127:819S.
2. AOAC. Official methods of analyses. 15th ed. Arlington: Association of Official Analytical Chemists; 1990.
3. Osakwe II. Effects of *Leucaena leucocephala* supplementation to basal hay on energy and protein metabolism in West African Dwarf Sheep. Nig J Anim Prod. 2006;33(1):94–101.
4. Megahed GA, Etman AHM. Effects of agricultural by-products in ration on productive and reproductive performance of Saidi Rams. Acta Vet Brno. 2006; 75:227–34.
5. Ravindran V, Korneguyand ET, Rajaguru ASB. Influence of processing methods and storage time on thyocyanide potential of cassava leaf meal. Anim Feed Sci Technol. 1987;17:227–34.
6. Wanapat M. Rumen manipulation to increase the efficient use of local feed resources and productivity of ruminants in the tropics. Asian Australas J Anim Sci. 2000;13:59–67.

Reproductive Toxicity of *Anamirta cocculus* (L.) Wight and Arn and *Sphagneticola trilobata* (L.) Pruski Extracts on *Culex pipiens* Linn. (Diptera: Culicidae)

45

E. Pushpalatha, D. Aiswarya, E.K. Sharihan, and B.K.P. Priya

Abstract

Acute toxicity of the acetone extracts of *Anamirta cocculus* (L.) Wight and Arn (fruits) and *Sphagneticola trilobata* (L.) Pruski (leaf) was tested against *Culex pipiens* (L.). Bioassay tests were conducted using different larval instars of *Culex pipiens* and determined the LC_{50} values. Fecundity of the *Culex pipiens* that emerged from the treatments and hatchability of the eggs were also estimated. I instar larvae of *Culex pipiens* were exposed in various sublethal concentrations and allowed them to moult and pupate. The pupae from each of the treatment set and control set were collected separately and allowed them to emerge in the emergence cages. A bowl with water was kept inside the emergence cage to enhance oviposition. The eggs laid by the females were collected, counted and compared, and fecundity rates were calculated. The data shows that the effect of the selected plant extracts is dose dependent and reduced the fecundity level very significantly when compared with the control. However, *A. cocculus* fruit extract in acetone showed a significant reduction in fecundity and hatchability of the *C. pipiens*. The findings of the present study evidenced that both the selected plant extracts have effective role in the reduction in fecundity and can be utilized further for developing formulations. The insecticidal property of these plant extracts may serve as a suitable alternative to synthetic insecticides in the future as they are relatively safe and easily degradable.

Keywords

Plant extracts • *Anamirta cocculus* • *Sphagneticola trilobata* • *Culex pipiens* • Reproductive toxicity • Mosquito control

E. Pushpalatha (✉) • D. Aiswarya • E.K. Sharihan • B.K.P. Priya
Department of Zoology, University of Calicut, Malappuram Dist., Kerala, 673 635, India
e-mail: drepushpalatha@yahoo.co.in

A. Sabu and A. Augustine (eds.), *Prospects in Bioscience: Addressing the Issues*, DOI 10.1007/978-81-322-0810-5_45, © Springer India 2013

Introduction

In view of the fact that mosquitoes are blessed with the ability to exploit adverse environmental conditions to reproduce and disperse [1] and also with the development of resistance in the insect vectors due to the indiscriminate use of synthetic insecticides, many of the mosquito-control programmes always face difficulties to achieve the total control of the target mosquito species. Moreover, mosquitoes directly affect man by spreading several dreadful diseases such as dengue fever, dengue haemorrhagic fever, filariasis, chikungunya and malaria. Phytochemicals obtained from plants with proven mosquito-control potential can be used as an alternative to synthetic insecticides or along with other insecticides under the integrated vector control programmes. Insecticidal properties of various plants have been reported by several workers such as mimics of insect ecdysone and juvenile hormone activity [2] and larvicidal and insect growth regulatory activity [3, 4] reduction in fecundity and hatchability of eggs [5]. Extensive research on mosquitocidal compounds from plants has been evidenced from the reports by several workers [6–8] on their larvicidal, adulticidal, repellent, feeding deterrent and oviposition deterrent properties. In general, plant extracts have been recognized as the important natural resources of insecticides [9]. The present study aims to determine the larvicidal activity of *Anamirta cocculus* (L.) Wight and Arn and *Sphagneticola trilobata* (L.) Pruski and their effect on the fertility (fecundity of females and hatchability of the eggs) of the *Culex pipiens* (L.).

Materials and Methods

Laboratory Culture of *Culex pipiens* (L)

The larvae/pupae of *C. pipiens* from the field were brought to the laboratory and transferred to plastic trays containing ¾ part of 0.08% saline water. The pupae were collected in the glass bowls and kept in the mosquito emergence cage. The freshly emerged adults were maintained at room temperature. The adult females were provided with blood meal from an immobilized hen and the males were fed with sucrose solution. Bowls with 0.08% saline water were provided in the cage for enhancing oviposition. The freshly laid eggs were taken out and the eggs were transferred into a trough and allowed to hatch. Freshly hatched larvae were fed by fine powder of dog biscuits and yeast in the ratio 3:1. Water was changed every alternate day and the feeding continued till the larvae moulted into pupae. The freshly hatched/moulted larvae were used for bioassay tests.

Preparation of Plant Extracts

Fresh leaves of *Sphagneticola trilobata* and fruits of *Anamirta cocculus* were thoroughly washed with water and dried under shade at room temperature. The dried materials were powdered in a kitchen machine and Soxhlet extracted with analytical grade acetone. The extracts obtained were then dried and weighed, and the yields of the plant materials were calculated. One percent stock solution was prepared out of each of the extracts and was stored in refrigerator.

Bioassay

The appropriate volume of 1% stock solution of the acetone extract was diluted to 100 ml of 0.08% saline water in a disposable tumbler to obtain desired concentration of the test medium. 20 larvae of each instar were released into each tumbler containing test medium. A control with 100 ml of 0.08% saline water and a control with highest volume of acetone were also set. Triplicates were maintained in the experiment and the control sets. Observations were taken after 24 h, and the mortality of the treated larvae was recorded at the end of 24 h of exposure, and the LC_{50}, LC_{90} and associated statistics were estimated using the software program developed by Reddy et al. [10].

Freshly hatched I instar larvae of *X. pipiens* reared separately in 100 ml of 0.08% saline containing four sublethal concentrations (2, 4, 6, 8 ppm of *A. cocculus* and 20, 40, 60 and 80 ppm of *S. trilobata*) of the acetone extracts. Normal control (0.08% saline alone) and acetone control (0.08 saline with maximum volume of acetone in the sample) were maintained. Three replicates for the controls and the different treatments were maintained. The larvae were provided with dry yeast powder and dog biscuit in the ratio of 3:1. After the larvae metamorphosed into pupae, they were transferred to emergence cages (30×30×30 cm) and allowed to emerge, and the number of males and females was counted. With the help of an aspirator, 10 males and 10 females were introduced into another cage, and the males were provided with 10% sucrose solution and the females with blood meal from an immobilized chicken. One bowl containing water was kept inside the cage to facilitate oviposition. The eggs laid were removed, counted and allowed to hatch in separate tray. Fecundity was monitored till the females were dead. The number of eggs that successfully hatched into the first instar larvae in each concentration was counted. Relating fecundity and hatchability of the treated females to those of the control females, reduction in fecundity and hatchability due to treatment was calculated.

Results and Discussion

The data on the 24-h LC_{50} (ppm) of the acetone extracts of the selected plants against the different instars of *C. pipiens* were provided in Table 45.1. The results show that fruit extract of *A. cocculus* has more activity than the leaf extract of *S. trilobata*. The 24-h LC_{50} values ranged from 23.25 to 30.71 ppm for *A. cocculus* and from 99.76 to 189.24 ppm for *S. trilobata*. Adult mosquitoes emerged from the larvae reared in the sublethal concentrations showed significant decrease in the fecundity.

Data on fecundity of the females of the selected mosquito and the hatchability of the eggs were provided in Table 45.2. Average realized fecundity of a female *X. pipiens* in the control series pertaining to the tests of different extracts ranged from 130.4 to 140.9. Rearing freshly moulted larvae at 2-ppm concentration of the acetone extract of *A. cocculus* decreased the fecundity to 99.8 eggs/female. At the highest concentration (8 ppm) of the acetone extract of *A. cocculus,* the fecundity decreased to 39.8 eggs/female. The lower and higher concentrations of the acetone extract of *S. trilobata* decreased to 98.9 and 26.8 respectively. Like fecundity, hatchability also displayed a linear inverse relationship with treatment concentrations (Table 45.2). Perusal of the data on fecundity reported by various workers [11–14] is by no means comparable with the data reported in the present study. However, Mohsen et al. [15] reported 86.3% decrease in fecundity of *C. quinquefasciatus* treated with 500 ppm of crude ethanol extract of *Imperata cylindrica*. The acetone extracts of the selected plant materials tested in the present study also considerably decreased the hatchability of the egg of *X. pipiens*. The observations made in the present study convincingly confirm that the acetone extracts of the selected plant materials exhibit the highest negative effect on fecundity and hatchability of *X. pipiens*.

Table 45.1 24-h Lc50(ppm) and Lc 90 (ppm) of the acetone extract of the selected plant materials against the different instars of *Culex pipiens*

Sl. no.	Name of the plant	Instar	LC50 (fiducial limits)	LC90 (fiducial limits)
1	*Anamirta cocculus*	I	27.61(16.5–58.04)	56.23(39.3–200.35)
		II	23.25(20.4–26.2)	53.96(47.48–63.89)
		III	29.52 (17.99–44.4)	81.19(61.6–138.7)
		IV	30.71(24.56–39.04)	69.55(55.32–104.8)
2	*Sphagneticola trilobata*	I	109.24(103.4–116.0)	168.61(155.8–186.8)
		II	99.76(94.1–106.4)	161.6(147.8–182.7)
		III	189.24(179.6–198.7)	282.31(267.7–301.1)
		IV	113.34(108.3–125.6)	226.48 (197.0–278.8)

Table 45.2 Fecundity and hatchability of *C. pipiens* as a function of sublethal concentrations of the acetone extract of the selected plant materials. Each value represents the mean (X ± SD) of three observations

Conc. (ppm).	Total no. of eggs	Fecundity (egg/female)	Hatchability		Decrease over control (%)	
			Eggs hatched (no.)	%	Fecundity	Hatchability
Anamirta cocculus						
Control	1,304 ± 4.0	130.4 ± 0.4	1,289 ± 4.0	98.84	0	0
2	998.8 ± 10	99.8 ± 1.0	960.6 ± 4.0	96.17	24.46	2.7
4	799.2 ± 7.0	79.9 ± 0.7	673.0 ± 12.0	84.21	38.73	14.8
6	576.0 ± 8.0	57.6 ± 0.8	398.0 ± 10.0	60.09	55.82	39.2
8	398.4 ± 12.0	39.8 ± 1.2	168.0 ± 6.0	42.21	69.47	57.3
Sphagneticola trilobata						
Control	1,409 ± 8.0	140.9 ± 0.8	1299.8 ± 12.0	92.24	0	0
20	1,203 ± 2.0	120.3 ± 0.2	989.0 ± 10.0	81.96	14.62	11.1
40	999.7 ± 10.0	99.9 ± 1.0	800.2 ± 8.0	80.02	30.45	13.2
60	690.4 ± 14.0	69.0 ± 1.4	530.4 ± 6.0	76.82	51.08	16.7
80	400.0 ± 10.0	40.0 ± 1.0	268.0 ± 8.0	67.01	58.84	27.4

References

1. Berlin G, Goswami D, Rubha B. Preliminary evaluation of mosquito larvicidal efficacy of plant extracts. J Vector Borne Dis. 1972;44:145–8.
2. Grainge M, Ahmed S. Handbook of plants with pest control properties. New York: John and Sons; 1988. p. 470.
3. Muthukrishnan J, Pushpalatha E. Larvicidal activity of few plant extract against Cx. quinquefasciatus and An. Stephensi. Indian J Malariol. 1995;32:14–23.
4. Thangam TS, Kathiresan K. Larvicidal activity of some mangrove plants against Ae. *aegypti*. Indian J Med Res. 1988;89:20–1.
5. Muthukrishnan J, Pushpalatha E. Effects plant extracts on fecundity and fertility of mosquitoes. J Appl Entomol. 2001;125:31–5.
6. Elttal IH, Guelbeogo WM, Neble R, Costantini C, Sagonon N, Kabore Z. Ovicidal and larvicidal activity against *Aedes aegypti* and *Anopheles gambiae*, complex mosquitoes of essential oils extracted from three spontaneous plants of Burkina Faso. Parassitologia. 1999;45(1):23–6.
7. Amer A, Mehlhorn H. Persistency of larvicidal effects of plant oil extracts under different storage conditions. Parasitol Res. 2006;99:473–7.
8. Senthikumar N, Varama P, Gurusubramanyian G. Larvicidal and adulticidal activities of some medicinal plants against the malarial vector *Anopheles stephensi*. Parasitol Res. 2009;104:237–44.
9. Rahuman AA, Venketesan P. Larvicidal efficacy of cucurbitaceous plant leaf extracts against mosquito species. Parasitol Res. 2008;103:133–9.
10. Reddy PJ, Krishna D, Murthy US, Jamil K. A microcomputer FORTRAN program for rapid determination of lethal concentration of biocides in mosquito control. CABIOS. 1992;8:209–13.
11. Saxena RC, Dixit OP, Sukumaran P. Laboratory assessment of indigenous plant extracts for anti-juvenile hormone activity in *C. quinquefasciatus*. Indian J Med Res. 1992;95:204–6.
12. Saxena RC, Harshan V, Saxena A, Sukumaran P. Larvicidal and chemosterilant activity of *Annona squamosa* alkaloids against Anopheles stephensi. J Am Mosq Control Assoc. 1993;9:84–7.
13. Dash AP, Ranjit MR. Comparative efficacy of aphid extracts and some juvenoids against the development of mosquitoes. J Am Mosq Control Assoc. 1992;8:247–51.
14. Pawar PV, Pisale SP, Sharma RN. Effect of some new insect growth regulators on metamorphosis and reproduction of *Aedes aegypti*. Indian J Med Res. 1995;101:13–8.
15. Mohsen ZH, Jawad AM, Al-Saadi M, Al-Naib A. Anti-oviposition and insecticidal activity of *Imperata cylindrical*. (Gramineae). Med Vet Entomol. 1995;9:441–2.

Epidemiological Typing of Methicillin-Resistant *Staphylococcus aureus* (MRSA) Isolated from Kerala Using Phage Typing

46

J. Jeshina and Kuyyalil Surekha

Abstract

Staphylococcus aureus is a pyogenic coccus responsible for various suppurative diseases with a wide distribution and carriage rate among hospital personnel as well as patients. To control the spread of MRSA, which has high virulence and seems to defy the usual infection control measures, it is important to understand the epidemiological relatedness between them. The present study was undertaken to study the epidemiology and prevalence of MRSA isolates in Kerala with respect to their currently existing phage types. A total of 104 isolates of MRSA were tested from various hospitals in Kerala between January 2006 and January 2009. The isolates were characterised using staphylococcal cassette chromosome typing (SCC*mec* typing). Phage typing was performed at the National Staphylococcal Phage Typing Centre, MAMC, New Delhi, using standard Blair and William method (1961) at routine test dilution (RTD) and 100XRTD. The results were also corroborated using pulsed field gel electrophoresis (PFGE). The strains carried type III cassette chromosome and were classified as group I – 29, 52, 52A, 79, 80; group II – 3A, 3C, 55, 71; group III – 6, 42E, 47, 53, 54, 75, 77, 83A, 84, 85; and group IV – 94, 96, and non-allocated 81, 95.53% of the isolates were nontypable at both RTD and 100XRTD. The remaining 47% strains were classified into 13 phage types by using RTD. Majority of the isolates were lysed by phage 75 of the lytic group III (87.2%). Seventeen isolates were lysed by phage 79 and 8 isolates by phage 29 of the lytic group I. The phage group III strains represented the largest proportion of the isolates in our study which has been found to be the predominant phage group among HA-MRSA in

J. Jeshina
Infectious Diseases Training and Research Centre,
Christian Medical College, Vellore, Tamil Nadu, India

K. Surekha (✉)
Department of Biotechnology and Microbiology,
Kannur University, Thalassery Campus, Kannur,
670661, Kerala, India
e-mail: skuyyalil@yahoo.com

A. Sabu and A. Augustine (eds.), *Prospects in Bioscience: Addressing the Issues*,
DOI 10.1007/978-81-322-0810-5_46, © Springer India 2013

several studies. Their PFGE patterns also confirmed the close relatedness among the strains. These results prove that phage typing can still be used as an inexpensive screening method especially for mass typing since the gold standard PFGE is time-consuming, cumbersome and expensive.

Keywords
MRSA • Phage-typing • Staphylococcal cassette chromosome • *mec*A gene • PFGE

Introduction

Staphylococcus aureus is a pyogenic coccus responsible for various suppurative diseases with a wide distribution and carriage rate among hospital personnel as well as patients. Methicillin-resistant *Staphylococcus aureus* (MRSA), a major pathogen and now endemic in many hospitals around the world, is reported to be increasing with incidence ranging between 40 and 50% in different institutions. Some of the serious clinical manifestations of MRSA infections, such as bloodstream infections (BSI) and ventilator-associated pneumonia (VAP), are frequently reported in the ICUs [1]. It has been increasingly reported from India as a coloniser in healthy individuals in community-acquired infections and is the predominant pathogen isolated in pyodermas [2]. The incidence of MRSA infections in India was reported to be between 18 and 33% in 1997 which increased to 51% in 2001 [3]. A high mortality rate still exists in the case of MRSA infections owing to the resistance to many potent antibiotics.

Strain typing is an integral part of surveillance networks. There are several typing techniques for MRSA, but a simple inexpensive technique with a good discriminatory power and wide availability is lacking. PFGE is considered the gold standard, but it is expensive and labour intensive, and it uses genetic markers which undergo rapid evolutionary changes overtime which makes it difficult to standardise at different times and between laboratories [4]. Epidemiological studies have shown high virulence among epidemic MRSA (EMRSA) strains of certain phage types which spread rapidly among hospitals and vary greatly in their antibiotic resistance patterns [5]. So typing of these isolates is important to understand the epidemiology and spread of such highly resistant organisms. The present study was undertaken to study the epidemiology and prevalence of MRSA isolates in Kerala with respect to their currently existing phage types.

Materials and Methods

A total of 104 isolates of methicillin-resistant *Staphylococcus aureus* were tested from various hospitals in Kerala between January 2006 and January 2009. Molecular confirmations of the strains were done using 16SrRNA typing and *mec*A gene amplification and sequencing. Phage typing was performed at the National Staphylococcal Phage Typing Centre, MAMC, New Delhi, using standard Blair and William method [6] at routine test dilution (RTD) and 100XRTD using the international basic set of 23 phages obtained from CHPL, Colindale, UK. The strains were classified as group I – 29, 52, 52A, 79, 80; group II – 3A, 3C, 55, 71; group III – 6, 42E, 47, 53, 54, 75, 77, 83A, 84, 85; group IV – 94, 96; and non-allocated 81, 95.53. The lytic reactions on the inoculated plates were read visually after overnight incubation at 30°C and were graded as negative, inhibition, weak lytic reaction or strong lytic reaction.

Pulsed Field Gel Electrophoresis

The isolates were also typed using PFGE with the *Sma*I restriction enzyme, as previously described [7]. The fragments were resolved on a 1% gel on a Chef Mapper (Bio-Rad, Richmond, Calif.) PFGE apparatus at 6 V/cm for 22 h, with switching times ramped from 5 to 35 s at 14°C. A lambda DNA PFGE molecular size standard (Amersham Pharmacia Biotech, Little Chalfont, United Kingdom) and an ATCC control strain were included in each gel. PFGE patterns were interpreted according to the criteria of Tenover et al. [8].

Results and Discussion

Majority of the isolates were nontypable (53%) at both RTD and 100XRTD. These isolates were concentrated in three districts. The remaining 47% strains were classified into 13 phages by using RTD. Majority of the isolates (41) were lysed by phage 75 of the lytic group III (87.2%). The isolates lysed only by phage 75 were concentrated in two districts. 17 isolates were lysed by phage 79, and 8 isolates were lysed by phage 29 of the lytic group I. Four isolates were lysed by phage 81 of lytic group IV, and 4 isolates were lysed by phage 3C of lytic group II and 2 isolates by phage 3A of group II. The major phage types obtained in the study are summarised in Table 46.1.

Table 46.1 Major phage types obtained in the study

Phage types	Group	No. of strains (%)
52A/75	I	1 (2.1)
3A/3C	II	2 (4.2)
79/75	I, III	1 (2.1)
75/81	III, NA	1 (2.1)
79/75/81	I, II, NA	2 (4.2)
54/75	III	10 (21.2)
29/79/54/75	I, III	7 (14.89)
29/79/75	I, III	1 (2.1)
79/54/75	I, III	1 (2.1)
75	III	17 (36.1)
3C	II	2 (4.2)
80	I	1 (2.1)
79	I	1 (2.1)

Pulse field gel electrophoresis revealed that the isolates were closely related, probably belonging to a single endemic clone, and they all carried type III cassette chromosome, thus confirming that they are HA-MRSA. Figure 46.1 shows the *Sma*I restriction pattern of the isolates.

Active surveillance and an understanding of the epidemiological relatedness between the isolates are essential components for control strategies especially in organisms such as *S. aureus* which has high virulence and seems to be unaffected by the usual infection control practices in hospitals [9]. Phage typing has been the classical typing method for epidemiological typing of *S. aureus* strains. Our isolates were confirmed to be hospital-acquired MRSA with type III cassette chromosome. Only 47% of the isolates were typable, and these strains were classified into 13 phage types with the majority of the isolates being lysed by phage 75 of lytic group III. This is higher than the percentage of typable isolates reported earlier in India where majority of the isolates

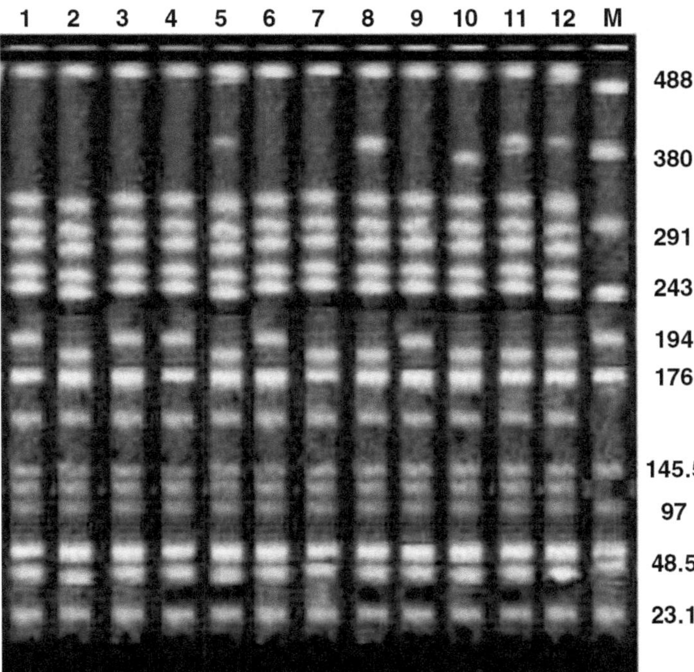

Fig. 46.1 PFGE patterns of representative MRSA isolates from Kerala, India, after digestion of chromosomal DNA with *SmaI*. Positions of marker bands are indicated on the *right*

belonged to group III followed by mixed phages [3]. Similarly, only 39.1% strains of MRSA isolated in Maharashtra area were typable with the majority again belonging to phage group III [10].

In a study by Lemaitre et al. MRSA isolates from France were also found to belong to phage group III, and they found that 60% strains were typable which is higher compared to the present study [11]. Similar studies from Spain had also reported phage group III as the predominant group [12, 13]. Trilla et al. [13] reported that majority of the isolates were lysed by phage 77 whereas most of our isolates were lysed by phage 75. Nontypable and phage type 77 were the most predominant in MRSA isolated from Thailand [14]. Recently, up to 52% of the isolates obtained from all over India were reported to be typable and 48% nontypable with distinct restriction fragment length polymorphism (RFLP) patterns. Majority of these isolates belonged to mixed phage groups [15]. Similar to several other reports from around the world, we found that phage group III strains represented the largest proportion of the isolates in our study which has been reported to be the predominant phage group among hospital-acquired MRSA (HA-MRSA) along with cyclic fluctuations in predominance of group I and III strains [16].

While phage group II strains have been associated with exfoliatin-related diseases and community-associated MRSA (CA-MRSA), group III has been predominant in HA-MRSA. Similar results were reported in a recent study from Assam where all the MRSA isolates in the study belonged to group III phage type and resembled the predominant strain in Asia which is the Brazilian/Hungarian strain CC8-MRSA-III. Their PFGE patterns also confirmed the close relatedness among the strains [17]. Isolates carrying type III SCC*mec* and predominantly lysed by group III phages with closely related PFGE patterns have also been reported from Hong Kong. These isolates were thought to be derived from ST8 MSSA isolates through the acquisition of type III SCC*mec* [4].

Phage typing is still preferred as the first-line approach in epidemiological typing of MRSA [18], and these results prove that it can still be used as an inexpensive screening method. Nevertheless, PFGE is essential for the examination of non-typable strains, identification and characterisation of epidemic strains or confirmation of ambiguous results, even though it is more time-consuming and cumbersome and requires expensive equipment [19].

Conclusions

We conclude that bacteriophage typing, even though less discriminatory compared to the currently available molecular techniques, can still be used as an inexpensive screening method but cannot be depended on solely for typing isolates. This is especially useful in an epidemiological context particularly for mass typing since the gold standard PFGE is time-consuming and requires expensive equipments and expertise. In the meantime, the introduction of new MRSA phages has further reduced the number of circulating nontypable strains which warrants further studies to perfect this technique.

Acknowledgements We thank the National Staphylococcal Phage Typing Centre, Maulana Azad Medical College, New Delhi, for phage typing the isolates and Kerala State Council for Science Technology and Environment (KSCSTE) for funding the study.

References

1. Humphreys H. Can we do better in controlling and preventing methicillin-resistant *Staphylococcus aureus* (MRSA) in the intensive care unit (ICU)? Eur J Clin Microbiol Infect Dis. 2008;27:409–13.
2. Patil R, Baveja S, Natraj G, Khopkar U. Prevalence of methicillin resistant *Staphylococcus aureus* (MRSA) in community acquired primary pyoderma. Indian J Dermatol Venereol Leprol. 2006;72:126–8.
3. Vidhani S, Mehndiratta PL, Mathur MD. Study of methicillin resistant *Staphylococcus aureus* (MRSA) isolates from high risk patients. Indian J Med Microbiol. 2001;19:13–6.
4. Ip M, Lyon DJ, Chio F, Enright MC, Cheng AF. Characterization of isolates of methicillin-resistant *Staphylococcus aureus* from Hong Kong by phage typing, pulsed-field gel electrophoresis, and fluorescent amplified-fragment length polymorphism analysis. J Clin Microbiol. 2003;41:4980.

5. Kareiviene V, Pavilonis A, Sinkute G, Liegiute S, Gailiene G. *Staphylococcus aureus* resistance to antibiotics and spread of phage types. Medicina (Kaunas). 2006;42(4):332–9.
6. Blair JE, Williams REO. Phage typing of staphylococci. Bull WHO. 1961;24:771–84.
7. Swaminathan B, Barrett TJ, Hunter SB, Tauxe RV, et al. PulseNet: the molecular subtyping network for food borne bacterial disease surveillance, United States. Emerg Infect Dis. 2001;7:382–9.
8. Tenover FC, Arbeit RD, Goering RV, Micklesen PA, Murray BE, Persing DH, Swaminathan B. Interpreting chromosomal DNA resolution patterns produced by pulsed field gel electrophoresis: criteria for bacterial strain typing. J Clin Microbiol. 1995;33:2333–9.
9. Struelens MJ, Hawkey PM, French GL, Witte W, Tacconelli E. Laboratory tools and strategies for methicillin-resistant *Staphylococcus aureus* screening, surveillance and typing: state of the art and unmet needs. Clin Microbiol Infect. 2009;15:112–9.
10. Kandle SK, Ghatole MP, Takpere AY, Hittinhalli VB, Yemul VL. Bacteriophage typing and antibiotic sensitivity pattern of *Staphylococcus aureus* from clinical specimen in and around Solapur (South Maharashtra). J Commun Dis. 2003;35:17–23.
11. Lemaitre N, Sougakoff W, Masmoudi A, Fievet MH, Bismuth R, Jarlier V. Characterization of gentamicin susceptible strains of methicillin resistant *Staphylococcus aureus* involved in nosocomial spread. J Clin Microbiol. 1998;36:81–5.
12. Cuevas O, Cercenado E, Bouza E, Castellares C, Trincado P, Cabrera R, Vindel A. Molecular epidemiology of methicillin-resistant *Staphylococcus aureus* in Spain: a multicentre prevalence study (2002). Clin Microbiol Infect. 2007;13:250–6.
13. Trilla A, Marco F, Moreno A, Prat A, Soriano E, Jiménez de Anta MT. Clinical epidemiology of an outbreak of nosocomial infection caused by *Staphylococcus aureus* resistant to methicillin and aminoglycosides: efficacy of control measures. Comité de Control de Infecciones. Med Clin (Barc). 1993;100:205–9.
14. Chomvarin C, Tishyadhigama P, Siripornmongcolchai T, Wongwanich S, Limpaiboon T, Chaicumpar K. Analysis of three phenotyping methods and pulsed-field gel electrophoresis for differentiation of methicillin resistant *Staphylococcus aureus* isolated from two hospitals in Thailand, Southeast Asian. J Trop Med Public Health. 2005;36:1221–8.
15. Mehndiratta PL, Bhalla P, Ahmed A, Sharma YD. Molecular typing of methicillin resistant *Staphylococcus aureus* strains by PCR RFLP of spa gene: a reference laboratory perspective. Indian J Med Microbiol. 2009;27:116–22.
16. Mehndiratta PL, Gur R, Saini S, Bhalla P. *Staphylococcus aureus* phage types and their correlation to antibiotic resistance. Indian J Pathol Microbiol. 2010;-53:738–41.
17. Sarma JB, Ahmed GU. Characterization of methicillin resistant *S. aureus* strains and risk factors for acquisition in a teaching hospital in northeast India. Indian J Med Microbiol. 2010;28:127–9.
18. Mehndiratta PL, Bhalla P. Typing of Methicillin resistant *Staphylococcus aureus*: a technical review. Indian J Med Microbiol. 2012;30:16–23.
19. Lee HJ, Suh JT, Kim YS, Lenz W, Bierbaum G, Schaal KP. Typing and antimicrobial susceptibilities of methicillin resistant *Staphylococcus aureus* (MRSA) strains isolated in a hospital in Korea. J Korean Med Sci. 2001;16:381–5.

47 Hypolipidemic Effect of Methanol Fraction of *Acorus calamus* Linn. in Diet-Induced Obese Rats

K.S. Arun and Anu Augustine

Abstract

Acorus calamus Linn. is a traditional medicinal plant included in "lekhaneyagana" (which means "reduce excess fat"), a pharmacological classification mentioned by Charaka in Charakasamhita. Traditionally, this plant has been prescribed as a first-line treatment for many ailments such as digestive problems, diabetes, obesity, and related problems. The purpose of the present study was to check the effect of methanolic fraction of *A. calamus* on serum lipids, Lecithin-cholesterol acyltransferase (LCAT) enzyme, and apolipoproteins. We found that treatment of *A. calamus* at concentrations of 200 and 400 mg/kg body weight was able to reduce total cholesterol (T-c), triglycerides (TG), and low-density lipoprotein cholesterol (LDL-c) levels and increase high-density lipoprotein cholesterol (HDL-c) levels compared to model control group. *A. calamus* treatment results in activation of LCAT enzyme, which helps in the maturation of HDL-c. ApoA1 level was found to be increased considerably by the administration of *A. calamus* extract, and the Apo B level was decreased. GCMS analysis of the *A. calamus* extract showed the presence of compound, Friedelan-3-one. The result suggests the use of methanol fraction of *A .calamus*, as a potential drug for controlling the cholesterol level.

Keywords

A. calamus • LCAT • GCMS • Apolipoproteins

K.S. Arun
Department of Biotechnology and Microbiology,
Kannur University, Thalassery Campus, Kannur,
670661, Kerala, India

A. Augustine (✉)
Inter University Centre for Bioscience,
Kannur University, Kannur 670661, India
e-mail: anuaugus@rediffmail.com

A. Sabu and A. Augustine (eds.), *Prospects in Bioscience: Addressing the Issues*,
DOI 10.1007/978-81-322-0810-5_47, © Springer India 2013

Abbreviations

Apo A1	Apolipoprotein A1
Apo B	Apolipoprotein B
GCMS	Gas chromatography mass spectrometry
HDL-c	High-density lipoprotein cholesterol
LCAT	Lecithin-cholesterol acyltransferase
LDL-c	Low-density lipoprotein cholesterol
MSMS	Tandem mass spectrometry
RT	Retention time
T-c	Total cholesterol
TG	Triglycerides

Introduction

Acorus calamus Linn. (Family: Araceae), also known as sweet flag, is a semiaquatic perennial herb with creeping and branched aromatic rhizome. The roots and rhizomes of this plant have been used in the Indian and Chinese systems of medicine for at least 2,000 years. The essential oils isolated from *A. calamus* showed anti-hypercholesterolemia, anti-inflammatory, and antioxidative activity, respectively [1–3]. Furthermore, *A. calamus* extract showed antifungal activity and insecticidal properties [4].

Obesity is an excess proportion of total body fat and is reaching epidemic proportions. Direct health costs attributable to obesity have been estimated at $52 billion in 1995 and $75 billion in 2003 and increasing in alarming rate [5]. *A. calamus* is a plant included in lekhaneyagana, a pharmacological classification mentioned in "Charakasamhita" which means "reduce excess fat" [6]. In the present study, we screened the methanolic fraction of *A. calamus* for hypolipidemic effects in diet-induced rat models. The fraction which shows hypolipidemic effect was used to perform GCMS analysis.

Materials and Methods

Preparation of the Root Extract of *A. calamus*

Root samples of *A. calamus* were purchased from local market and authenticated by the Centre for Medicinal Plants Research (CMPR), Kottakkal, India. Voucher specimen was processed and deposited (No: CMPR 1486). 2.3 kg of the fine powdered sample was used for Soxhlet extraction using methanol. The homogenate was then filtered using Millipore filtration system 2 (Millipore, USA) and dried using a rotary evaporator at 40°C (273 g). The methanol fraction was loaded to a silica gel column eluted with MeOH: water (18×500 mm column), increasing water gradually to supply five fractions, since the 30% MeOH-soluble fraction exhibited considerable hypolipidemic activity [1, 4]. The fourth fraction was used to feed the diet-induced obese rats, and the same was used for GCMS analysis.

Establishment of Experimental Model and Drug Treatment

Sprague Dawley (SD) strain rats of body weight 160±10 g were used for experiments. The animals were housed in polypropylene cages at controlled temperature (22±3°C) and humidity (50±10%) and were kept in 12-h light cycle. The experiment was approved by the Institutional Animal Ethics Committee (IAEC No-KULS/IAEC 2011-04) and performed based on CPCSEA accepted guidelines for care and use of laboratory animals.

Rats were randomly divided into two groups: normal control group ($n=10$) and high-fat group ($n=40$). The normal control rats were fed with standard diet, and hyperlipidemic rats were fed with high-fat diet [7, 8]. The rats in high-fat model group were randomly divided into four groups with ten rats per group: high-fat model control group, *A. calamus I* group (200 mg/kg body weight (BW)), *A. calamus II* group (400 mg/kg BW), and mevinolin (3.0 mg/kg body weight)-treated positive control group [7]. The acute toxicity of the extract was performed, and the doses were chosen according to the acute toxicological study [9, 10]. At the end of the experiment, the overnight fasted rats were sacrificed and blood samples were taken for the estimation of serum T-c, TG, HDL-c, LDL-c [11–14], Lecithin-cholesterol acyltransferase [15], and apolipoproteins [16].

GCMS Analysis of Methanolic Extract of *A. calamus*

GCMS and MSMS analysis of the methanolic extract of *A. calamus* was performed using Perkin Elmer Clarus 600 GCMS and Turbomatrix 40 trap GC-HS controlled by Turbo mass software version 5.4.2. One microliter of the sample was injected by direct liquid injection mode into Perkin Elmer Clarus 600 Gas Chromatogram equipped with Elite-5MS, 30 m × 0.25-mm ID × 0.25-um film thickness. The GC conditions were, injection temperature 275°C, helium as carrier gas with flow of 15 psi, and a split ratio of 10. GC effluent was introduced to the mass spectrometer for EI mass spectrometry analysis where data were recorded in full scan or linear mode in the mass range of 10–600 *m/z*.

Statistical Analysis

All values are presented as mean ± s.d. Statistical comparisons of the groups were made by ANOVA, and each group was compared with the others by Posthoc Fisher's PLSD test (SPSS Inc-IBM, USA). Statistical significance was defined as $P < 0.05$.

Results and Discussion

Effect of *A. calamus* on Serum Lipid Levels in Diet-Induced Obese Rats

The estimated lipid levels in normal control, high-fat model control, extract-treated groups, and positive control are given in Table 47.1. The T-c and TG levels of *A. calamus* extract-treated groups were lower than the high-fat group; the dose response was also obvious. Moreover, *A. calamus* facilitates enhancement of cardio-protective lipid HDL-c, and this effect may be due to the increase in activity of lecithin: cholesterol acyl transferase (LCAT) enzyme, which may contribute to the regulation of blood lipids. The effect of *A. calamus* on HDL-c may be largely attributed to its central function in the reverse cholesterol transport, a process whereby excess cell cholesterol is taken up and processed by HDL particles for further delivery to the liver for metabolism [17, 18].

Treatment of hyperlipidemic rats with *A. calamus* extract at a dose of 200 and 400 mg/kg body weight lowered the serum LDL-c levels. The mevinolin-treated group had also showed a significant hypocholesterolemic effect in serum as compared to model control group [19, 20].

Table 47.1 Effect of *A. calamus* on serum lipids, apolipoproteins, and LCAT activity in diet-induced obese rats

	Normal control	High-fat model	*A. calamus* *I* 200 mg/kg	*A. calamus* *II* 400 mg/kg	Positive control mevinolin
Serum lipids (mgdL⁻¹)					
T-c	72.34 ± 7.64[a]	111.81 ± 6.67	97.58 ± 2.32[b]	87.03 ± 2.07[b]	83.7 ± 1.93
TG	78.10 ± 4.14[a]	107.1 ± 6.62	95.03 ± 2.18[b]	85.11 ± 1.81[b]	81.81 ± 1.77
HDL-c	35.1 ± 3.49	22.43 ± 4.08[a]	24.1 ± 1.01	28.21 ± 0.94[b]	31.6 ± 1.65[b]
LDL-c	21.62 ± 8.02	67.96 ± 9.88[a]	54.47 ± 1.79[b]	41.79 ± 2.71[b]	35.75 ± 1.29[b]
Apolipoproteins (mgdL⁻¹)					
ApoA1	6.28 ± 0.37	4.26 ± 0.12[a]	4.83 ± 0.17[b]	5.62 ± 0.21[b]	5.31 ± 0.29
ApoB	2.82 ± 0.29	8.11 ± 0.41[a]	7.48 ± 0.27[b]	6.61 ± 0.5[b]	4.5 ± 0.32
LCAT	30.98 ± 2.53	37.01 ± 2.67	39.25 ± 1.44	41.53 ± 0.94[b]	42.66 ± 1.60

LCAT Lecithin-cholesterol acyltransferase, normal control, rat fed with normal diet; high-fat model, rat fed with high-fat diet; *A calamus I*-200 mg/kg and *A calamus II*-400 mg/kg, high-fat diet rats treated with 200 and 400 mg/kg body weight of *A calamus* extract, respectively; positive control mevinolin, high-fat diet rats treated with mevinolin (3.0 mg/kg). The values are mean ± s.d for ten rats

[a]$p < 0.05$ compared with control group

[b]$p < 0.05$ compared with untreated model group

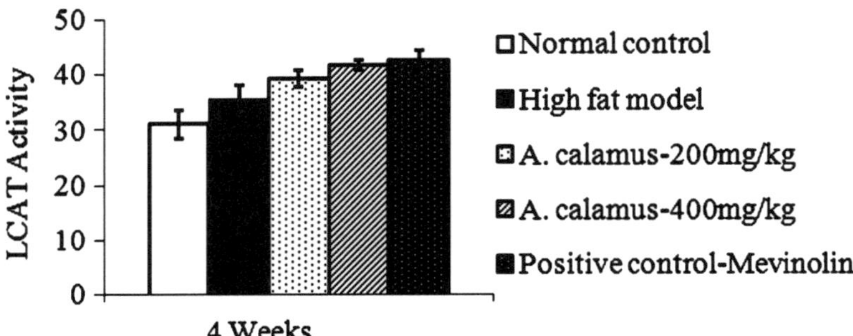

Fig. 47.1 *Graph* showing the effect on *A. calamus* on LCAT activity in diet-induced obese rats

Effect of *A. calamus* on LCAT Activity in Diet-Induced Obese Rats

Oral administration of *A. calamus* extracts that result in significant increase in LCAT activity supports the above result (Fig. 47.1). LCAT is an enzyme that catalyzes the formation of cholesteryl esters on HDL and by that promotes maturation of HDL particles in plasma and facilitates reverse cholesterol transport by maintaining a concentration gradient for the diffusion of cellular unesterified cholesterol to HDL. It helps to improve cholesterol exportation in biliary acids [18, 19]. Similar results were observed for lipid profiles treating *Curcuma kwangsiensis* polysaccharides [7] and LCAT activity in case of *Gymnema sylvestre* and Monascus-fermented soybean extracts [21, 22] (Fig. 47.1).

Effect of *A. calamus* on Serum Apolipoproteins in Diet-Induced Obese Rats

The administration of *A. calamus* extract significantly increases the apoA1 and drops down Apo B levels. Both lipid-poor apoAI and HDL act as acceptors for cellular lipids, and lipid efflux from cells can be mediated via the cell surface proteins. Apo A-I and Apo A-II are the main protein components of HDL cholesterol, which helps in removing excess cholesterol from the tissues for removal by the liver [23]. On the other hand, Apo B, which is a component of LDL-c, enables tissue cells to take up cholesterol [24].

GCMS Analysis of Methanolic Extract of *A. calamus*

The GCMS and MSMS analysis done using *A.calamus* extract shows the presence of Friedelan-3-one (Fig. 47.2). The spectrum and library search of 34.41 min RT peak represents the compound, Friedelan-3-one.

Conclusions

The administration of *A.calamus* extract reduced serum T-c, TG, and LDL-c levels. Furthermore, *A.calamus* helps to improve cardio-protective lipid HDL-c. The mechanism of hypolipidemic activity of *A. calamus* was not studied till date. Our study addresses the shortfall; it is presumed that a higher hypolipidemic effect of *A. calamus* might be due to the activation of LCAT enzyme. GCMS analysis data of same fraction of *A. calamus* extract shows that major peak present is Friedelan-3-one. The decrease of serum TG level is also a promising result, since triglycerides are independently related to cardiovascular disease and most of hypercholesterolemic drugs were not able to reduce TG levels. Based on this observation, it can be comprehended that the *A.calamus* fraction containing compound Friedelan-3-one is able to activate LCAT and exhibit potential hypolipidemic activity. The results constitute a valid scientific groundwork for its medicinal application and valid support for "lekhaneya" action of *A.calamus.*

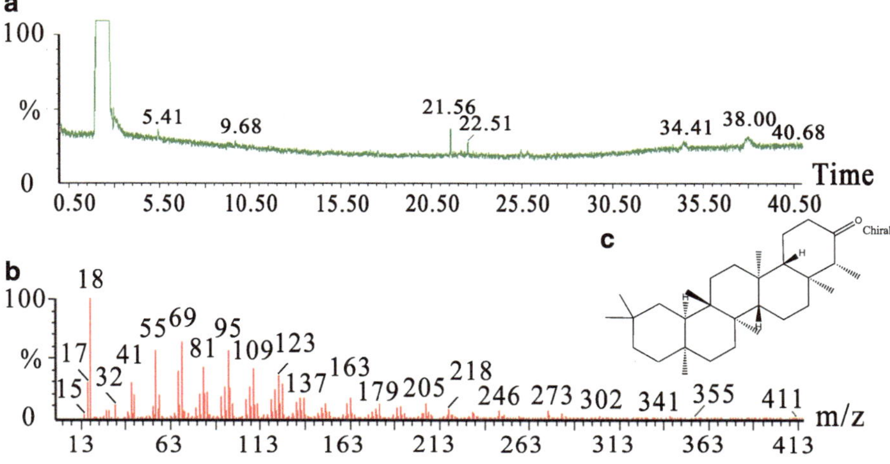

Fig. 47.2 (**a**) The total ion chromatogram of *A. calamus* extract. (**b**) The spectrum and library search of 34.41 min RT peak of Friedelan-3-one. (**c**) Structure of Friedelan-3-one

Acknowledgements DBT-BIF, Govt. of India is gratefully acknowledged for the support for computational and other allied facilities. Arun KS is thankful to UGC-BSR, New Delhi, for providing the fellowship to carry out the research work.

References

1. Parab RS, Mengi SA. Hypolipidemic activity of *Acorus calamus* L. in rats. Fitoterapia. 2002;73:451–5.
2. Kim H, Han TH, Lee SG. Anti-inflammatory activity of a water extract of *Acorus calamus* L. leaves on keratinocyte HaCaT cells. J Ethnopharmacol. 2009;122:149–56.
3. Manikandan S, Srikumar R, Jeya Parthasarathy N, Sheela Devi R. Protective effect of *Acorus calamus* Linn on free radical scavengers and lipid peroxidation in discrete regions of brain against noise stress exposed rat. Biol Pharm Bull. 2005;28:2327–30.
4. Sharma PR, Sharma OP, Saxena BP. Effect of sweet flag rhizome oil (*Acorus calamus*) on hemogram and ultra structure of hemocytes of the tobacco armyworm, *Spodoptera litura* (Lepidoptera: Noctuidae). Micron. 2008;39:544–51.
5. Wolf AM, Colditz GA. Current estimates of the economic cost of obesity in the United States. Obes Res. 1998;6:97–106.
6. Sharma PV. Caraka samhita. Varanasi: Chaukhambha Orientalia; 2000.
7. Xu C, Haiyan Z, Hua Z, Jianhong Z, Pin D. Effect of *Curcuma kwangsiensis* polysaccharides on blood lipid profiles and oxidative stress in high-fat rats. Int J Biol Macromol. 2009;44:138–42.
8. Lien EL, Boyle FG, Wrenn JM, Perry RW, Thompson CA, Borzelleca JF. Comparison of AIN-76A and AIN-93G diets: a 13-week study in rats. Food Chem Toxicol. 2001;39:385–92.
9. Tajuddin AS, Latif A, Qasmi IA, Amin KMY. An experimental study of sexual function improving effect of *Myristica fragrans* Houtt. (nutmeg). BMC Complement Altern Med. 2005;5:16.
10. Pahua-Ramos ME, Ortiz-Moreno A, Chamorro-Cevallos G, Hernández-Navarro MD, Garduño-Siciliano L, Necoechea-Mondragón H, Hernández-Ortega M. Hypolipidemic effect of Avocado (*Persea americana* Mill) seed in a Hypercholesterolemic mouse model. Plant Food Hum Nutr. 2012;67:10–6.
11. Rubinstein A. National cholesterol education program, second report of the expert panel on detection, evaluation, and treatment of high blood cholesterol in adults. Circulation. 1995;91:908–9.
12. Stein EA, Myers GL. National cholesterol education program recommendations for triglyceride measurement: executive summary. The national cholesterol education program working group on lipoprotein measurement. Clin Chem. 1995;41:1421–6.
13. Rifai N, Warnick GR, Dominiczak MH. Handbook of lipoprotein testing. 2nd ed. Washington, DC: AACC Press; 2001.
14. Friedewald WT, Levy RI, Fredrickson DS. Estimation of the concentration of low-density lipoprotein cholesterol in plasma, without use of the preparative ultracentrifuge. Clin Chem. 1972;18:499–502.
15. Nagasaki T, Akanuma Y. A new colorimetric method for the determination of plasma lecithin-cholesterol acyltransferase activity. Clin Chim Acta. 1977;75:1–5.
16. Marcovina SM, Albers JJ, Dati F, Ledue TB, Ritchie RF. International federation of clinical chemistry standardization project for measurements of apolipoproteins A-1 and B. Clin Chem. 1991;37:1676–82.
17. Rader DJ, Alexander ET, Weibel GL, Billheimer J, Rothblat GH. The role of reverse cholesterol transport

in animals and humans and relationship to atherosclerosis. J Lipid Res. 2009;50:189–94.
18. Bainton D, Miller NE, Bolton CH, Yarnell JW, Sweetnam PM, Baker IA, Lewis B, Elwood PC. Plasma triglyceride and high density lipoprotein cholesterol as predictors of ischaemic heart disease in British men. The caerphilly and speedwell collaborative heart disease studies. Br Heart J. 1992; 68:60–6.
19. Wilson PW. High-density lipoprotein, low-density lipoprotein and coronary artery disease. Am J Cardiol. 1990;66:7–10.
20. Oh PS, Lee SJ, Lim KT. Hypolipidemic and antioxidative effects of the plant glycoprotein (36 kDa) from *Rhus verniciflua* stokes fruit in Triton WR-1339-induced hyperlipidemic mice. Biosci Biotechnol Biochem. 2006;70:447–56.
21. Pyo YH, Seong KS. Hypolipidemic effects of Monascus-fermented soybean extracts in rats fed a high-fat and -cholesterol diet. J Agric Food Chem. 2009;57:8617–22.
22. Shigematsu N, Asano R, Shimosaka M, Okazaki M. Effect of administration with the extract of *Gymnema sylvestre* R. Br leaves on lipid metabolism in rats. Biol Pharm Bull. 2001;24:713–7.
23. Grundy SM, Cleeman JI, Merz CN, Brewer Jr HB, Clark LT, Hunninghake DB, Pasternak RC, Smith Jr SC, Stone NJ. Implications of recent clinical trials for the national cholesterol education program adult treatment panel III guidelines. J Am Coll Cardiol. 2004;44:720–32.
24. Smith JD. Apolipoprotein A-I and its mimetics for the treatment of atherosclerosis. Curr Opin Investig Drugs. 2010;11:989–96.

Biotyping and Phage Typing of *Salmonella enterica* Serotype Typhi Isolates from Kerala, South India

48

N. Ayana and Kuyyalil Surekha

Abstract

Typhoid fever, caused by *Salmonella enterica* serotype Typhi (*S. typhi*), is a major public health problem, particularly in developing countries. *Typhoid fever* causes 16 million cases with 600,000 deaths worldwide annually. Multidrug-resistant and fluoroquinolone-resistant strains of *S. typhi* have been reported from various parts of the country and created a major therapeutic problem. The present study was undertaken to document the serotype, biotype and phage types of S. *typhi* isolated from different regions in Kerala. A total of 65 strains of *S. enterica* serovar Typhi were isolated from blood cultures of febrile patients from various hospitals in Kerala, India, during the period from 2006 to 2008. Strains were tested biochemically, and confirmation was done using agglutinating O antisera. Strains were characterized by sequence analysis of quinolone resistance determining regions of *gyrA, gyrB, parA* and *parB*. Phage typing and biotyping were done at the National Salmonella Phage Typing Centre, Department of Microbiology, Lady Hardinge Medical College, New Delhi, using Kauffmann-White serotyping scheme of Popof and Le Minor. Phage typing was performed according to the recommendations of the Health Protection Agency Service (Colindale, UK) in the Pasteur Institute, Brussels, Belgium, using ten phages. All the isolates agglutinated with O antiserum. Four phage types were observed among the isolates (E1, Vi. neg, Deg Vi and UVS). The most prevalent phage type was E1 (70.8%) followed by Deg Vi (13.8%). Our report contradicts the findings of studies from Mumbai but similar to reports from Kolkata. Biotyping revealed all

N. Ayana
Department of Microbiology, Sree Narayana College, Kannur, Kerala, India

K. Surekha (✉)
Department of Biotechnology and Microbiology, School of Life Sciences, Kannur University, Thalassery Campus, Kannur 670661, Kerala, India
e-mail: skuyyalil@yahoo.com

A. Sabu and A. Augustine (eds.), *Prospects in Bioscience: Addressing the Issues*, DOI 10.1007/978-81-322-0810-5_48, © Springer India 2013

isolates belong to group I and group II isolates were not recovered. This finding was similar to earlier reports from different parts of the country. It could be concluded that majority of *Salmonella enterica* serotype Typhi isolates belong to E1 phage type and single biotype was prevalent among the strains.

Keywords
Salmonella typhi • Phage typing • Typhoid fever

Introduction

Salmonella enterica serovar Typhi is the causative agent of typhoid fever which causes 16 million cases with 600,000 deaths worldwide annually [1]. It is a major health problem in developing countries as a result of many interrelated factors including increased urbanization, inadequate supplies of clean water, antibiotic resistance, variable efficacy of vaccine preparation and increased regional movement of large numbers of migrant workers [2]. Multidrug-resistant and fluoroquinolone-resistant strains of *S. typhi* have been reported from various parts of the country and created a major therapeutic problem [3]. Effective epidemiological typing of serovar Typhi strains is needed to monitor the presence and spread of the disease. The phenotypic methods are used for epidemiological typing of serovar Typhi strains such as serotyping and phage typing. The present study was undertaken to document the serotype, biotype and phage types of S. *typhi* isolated from different regions in Kerala.

Materials and Methods

Bacterial Strains

A total of 64 strains of *S. enterica* serovar Typhi were isolated from blood cultures of febrile patients from various hospitals in Kerala, India, during the period from 2006 to 2008. Strains were tested biochemically, and confirmation was done using agglutinating O antisera.

Antimicrobial Susceptibility Testing

S. enterica serovar Typhi isolates were tested for susceptibility to antimicrobials by disc diffusion technique according to guidelines provided by the Clinical and Laboratory Standards Institute, CLSI (formerly National Committee for Clinical Laboratory Standards) [4]. The following antimicrobials were tested: chloramphenicol (30 μg), ampicillin (10 μg), amoxycillin (30 μg), gentamycin (10 μg), co-trimoxazole (25 μg), tetracycline (30 μg), nalidixic acid (10 μg), ciprofloxacin (5 μg), ofloxacin (5 μg) and ceftriaxone (25 μg). *Escherichia coli* ATCC 25922 was used as control strain for antimicrobial susceptibility testing.

Determination of Minimum Inhibitory Concentration (MIC)

Minimum inhibitory concentrations (MIC) of ampicillin, amoxycillin, tetracycline, gentamycin, chloramphenicol, ciprofloxacin, ofloxacin, nalidixic acid and ceftriaxone were determined for each strain by broth macro dilution method according to guidelines provided by CLSI [5].

Plasmid DNA Analysis

All isolates of *S. enterica* serovar Typhi isolates were subjected to plasmid profiling. Plasmid DNA was extracted by rapid procedure using alkaline lysis method [6, 7]. Plasmid DNA was analysed on its electrophoretic movement on

0.8% agarose gel at 100 V for 2 h. Molecular mass of plasmid DNA was estimated by Hind III-digested lambda DNA marker.

Amplification of QRDR Region of gyrA, gyrB, parC and parc Genes

Polymerase chain reaction (PCR) primers were designed using Primer programme. (http://frodo.wi.mit.edu/cgi-bin/primer3/primer3_results.cgi). PCR was performed in a total volume of 25 μl which contained 5 μl of supernatant, 25pmol of each primer, 200 μM dNTPs, 1.5 mM $MgCl_2$ and 0.5U of Taq polymerase (Bangalore Genei, India). After an initial denaturation step of 3 min at 94 °C, amplification was performed over 30 cycles, each consisting of 1 min at 94 °C, 1 min at hybridization temperature (56 °C for *gyrA*, 58 °C for *gyrB* and 53 °C for *parC* and *parE*) and 1 min at 72 °C,with a final extension step of 10 min at 72 °C. PCR products were purified according to standard procedures [7].

Sequencing and Analysis of QRDR Region of gyrA, gyrB, parC and parE Genes

Sequencing of amplified QRDR region of *gyrA, gyrB, parC* and *parE* genes was carried out by the method of Sanger et al. in an automated DNA sequencer (ABI 3100 Genetic Analyzer) with selected primers. DNA sequences of amplified regions were analysed using commercial software (Chromas software). Sequences were compared using BLAST analysis with nucleotide sequence database of *gyrA, gyrB, parC* and *parE* under the following accession numbers: *S. enterica* serovar Typhi *gyrA* accession no: AB071870, *S. enterica* serovar Typhi *gyrB* accession no: AB072396, *S. enterica* serovar Typhi *parC* accession no: AB072701 and *S. enterica* serovar Typhi *parE* accession no: AB072701.

Phage Typing and Biotyping

Phage typing and biotyping were done at the National Salmonella Phage Typing Centre, Department of Microbiology, Lady Hardinge Medical College, New Delhi, using Kauffmann-White serotyping scheme of Popof and Le Minor. Phage typing was performed according to the recommendations of the Health Protection Agency Service (Colindale, UK) in the Pasteur Institute, Brussels, Belgium, using ten phages.

Results and Discussion

Antibiotyping of *S. typhi* strains yielded 18 biotypes. Antimicrobial susceptibility pattern of *S. enterica* serovar Typhi revealed that 71.9% of isolates were resistant to nalidixic acid. 45.3% isolates showed susceptibility to ciprofloxacin, 54.7% were intermediately susceptible and none were resistant. 21.9% of isolates showed resistance to ofloxacin. None of the isolates exhibited multiple drug resistance. All strains were sensitive to chloramphenicol. 42.2% strains were resistant to tetracycline. 7.8% isolates showed resistance to ceftriaxone, and 20.3% showed intermediate resistance to ceftriaxone. Antimicrobial susceptibility pattern was given in Fig. 48.1.

MIC values of ciprofloxacin ranged between 0.25 and 16 mg/L. According to NCCLS breakpoint, the criteria for ciprofloxacin resistance was an MIC value of ≥4 mg/L; for decreased susceptibility, MIC value was in range between 1 and <4 mg/L, and for susceptible strains, MIC value was <1 mg/L. So in our study, out of 48 isolates, 23 (47.9%) isolates showed ciprofloxacin resistance and 10 (20.8%) isolates showed decreased susceptibility to ciprofloxacin. 36 (75%) isolates showed MIC value of >4 mg/L for ofloxacin. One strain had MIC value of 16 mg/L for ciprofloxacin and 64 mg/L for ofloxacin.

Plasmid profile analysis revealed that none of the isolates carried high molecular weight plasmids. Eighty-five percent of isolates carried plasmid with 23 Kb molecular weight, and all the isolates had smaller bands of plasmids.

QRDR of *gyrA, gyrB, parC* and *parE* genes of 30 ciprofloxacin-resistant isolates (23 strains that showed resistance to ciprofloxacin and 7 strains that showed decreased susceptibility to

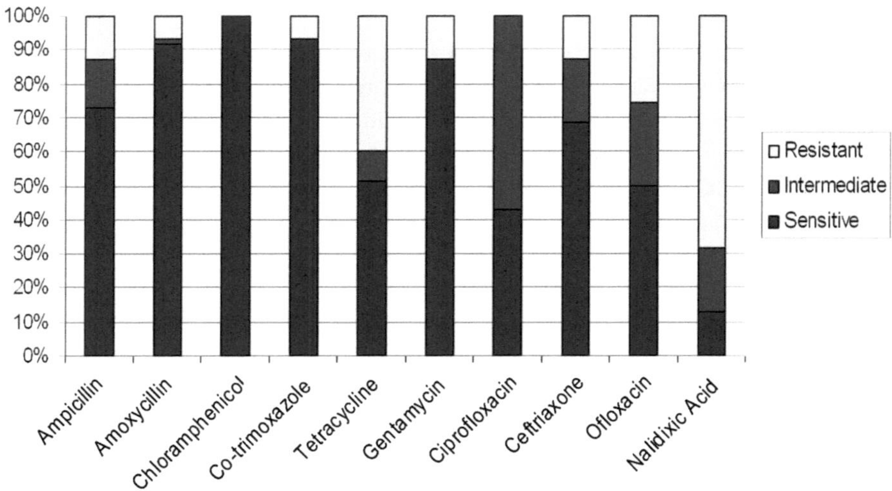

Fig. 48.1 Antimicrobial susceptibility pattern of *S. typhi* isolates

Table 48.1 Distribution of different phage types of *Salmonella enterica* serotype Typhi ($n=64$) during the period from 2006 to 2009

Phage type									Biotype	
Typable						Untypable			I	II
O	E1	C9	D1	E9	40	Vi neg	Degraded Vi	UVS		
–	70.8%	–	–	–	–	10%	13.8%	5.4%	100%	–

UVS untypable Vi strains

ciprofloxacin) were amplified and sequenced, and DNA sequences were compared using BLAST analysis with that of nucleotide sequence database of 4 genes. Ninety percent of isolates showed single mutation in QRDR of *gyrA*. 54.5% isolates showed double mutation both in *gyrA* and *parC*. The strains carried some nonspecific mutations. The isolate exhibited mutations that resulted in changes of amino acid sequences Ser83→Phe and Asp 87→Asn, Tyr in *gyrA*, or Glu84→Lys in *parC*.

A total of 64 isolates was subjected to phage typing. All the isolates agglutinated with O antiserum. Four phage types were observed among the isolates (E1, Vi.neg, Deg Vi and UVS). The most prevalent phage type was E1 (70.8%) followed by Deg Vi (13.8%). Biotyping revealed that all isolates belong to group I and group II isolates were not recovered. Table 48.1 shows the distribution of different phage types and biotypes of *Salmonella enterica* serotype Typhi isolated from Kerala during the period from 2006 to 2009.

Recently, there is an increase in the occurrence of antibiotic-resistant bacteria of all genera. So, scientific community should perform routine surveillance of microbial population to determine the extent of antibiotic resistance to provide suitable guidelines for treatment. MDR *S. enterica* serovar Typhi were not isolated in this study. Studies conducted at various regions in India also show that there is a decrease in the occurrence of MDR *S. typhi* [8–11]. Reoccurrence of resistant strains has also been reported [12]. Studies in the United States reported an increase in the number of MDR nalidixic acid-resistant *S. typhi* [13, 14] whereas studies in Bangladesh reported a decrease in MDR isolates with no corresponding increase in sensitive strains [15].

Majority of the isolates were sensitive to all antibiotics except nalidixic acid. 50% of isolates were identified as intermediately susceptible to ciprofloxacin by disc diffusion technique. There was an increased value of MICs of ciprofloxacin. MIC values of ciprofloxacin of isolates ranged between 0.25 and 16 mg/L. This finding was similar to earlier studies in the UK and India [16–19]. MIC values showed that 48% isolates were ciprofloxacin resistant and 21% showed decreased susceptibility to ciprofloxacin. 8.3% strains showed resistance and 23% strains showed intermediate resistance to ceftriaxone which was in contrast to earlier study in which all strains were reported as ceftrioxone sensitive [20].

Our phage typing report contradicts the findings of studies from Mumbai [21] but similar to reports from Kolkata [22]. This finding was similar to earlier reports from different parts of the country. It could be concluded that majority of *Salmonella enterica* serotype Typhi isolates belong to E1 phage type, and single biotype was prevalent among the strains.

Conclusions

Majority of the clinical isolates of *S. enterica* serovar typhi isolates showed resistance and decreased susceptibility to fluoroquinolones. The most prevalent phage type was E1 (70.8%) followed by Deg Vi (13.8%). Our report contradicts the findings of studies from Mumbai but is similar to reports from Kolkata. Biotyping revealed all isolates belong to group I and group II isolates were not recovered. This finding was similar to earlier reports from different parts of the country. It could be concluded that majority of *Salmonella enterica* serotype Typhi isolates belong to E1 phage type and single biotype was prevalent among the strains.

References

1. Crump JA, Luby SP, Mnitz ED. The global burden of typhoid fever. Bull WHO. 2004;82:346–53.
2. Pang T, Levine M, Ivanoff B, Wain J, Finlay BB. Typhoid fever important issues still remain. Trends Microbiol. 1998;6:131–3.
3. Rowe B, Ward LR, Threfall EJ. Multidrug resistant *Salmonella typhi*; a worldwide epidemic. Clin Infect Dis. 1997;24 Suppl 1:S106–9.
4. National Committee for Clinical Laboratory Standards. Performance standard for antimicrobial susceptibility testing; approved standard, NCC 2006; 16th ed. Document M100-S16, ol.26 No.3 villanova PA; 2006.
5. National Committee for Clinical Laboratory standards. Methods for dilution of antimicrobial susceptibility tests for bacteria that grow aerobically: approved standard M7-A5. In NCCLS 5th edition. Wayane, PA, USA; 2000.
6. Kado CI, Len ST. Rapid procedure for the detection of large and small plasmids. J Bacteriol. 1981;145:1365–73.
7. Sambrook J, Sritsch EF, Manaiatis T. Molecular cloning: a laboratory manual. New York: Cold Spring harbor Laboratory Press/Old Spring Harbor; 1989.
8. Madhulika U, Harish BN, Parija SC. Current pattern in antimicrobial susceptibility of *Salmonella typhi* isolates in Pondicherry. Indian J Med Res. 2004; 120(2):111-4.
9. Sanghavi SK, Mane MP, Nipadhkar KB. Multidrug resistance in Salmonella serotypes. Indian J Med Microbiol. 1999;17(2):88–90.
10. Chande C, Shrikande S, Kapale S, Agarwal S, Fule RP. Change in antimicrobial sensitivity pattern of *Salmonella typhi* in Central India. Indian J Med Res. 2002;115:248–50.
11. Saha MR, Dutta P, Niyogi SK, Dutta S, Mitra U, Ramamurthy T, et al. Decreasing trend in the occurrence of *Salmonella enterica* serotype Typhi amongst hospitalized children in Kolkata, India during 1990–2000. Indian J Med Res. 2002;115:46–8.
12. Kumar R, Aneja KR, Roy P, Sharma M, Gupta R, Ram S. Evaluation of minimum inhibitory concentration of quinolones and third generation cephalosporins to *Salmonella Typhi* isolates. Indian J Med Sci. 2002;56:1–8.
13. Aekers ML, Puhr ND, Tauxe RV, Mintz ED. Laboratory based surveillance of *Salmonella* serovar Typhi infections in United States: antimicrobial resistance on the rise. JAMA. 2000;283:2668–73.
14. Stevenson JE, Gay K, Barrett TJ, Medalla F, Chiller TM, Angulo FJ. Increase in nalidixic acid resistance among non-Typhi *Salmonella enterica* isolates in the United States from 1996 to 2003. Antimicrob Agents Chemother. 2007;51(1):195–7.
15. Rahman M, Ahmad A, Shoma S. Decline in epidemic of multi drug resistant *Salmonella Typhi* is not associated with increased incidence of antibiotic susceptible strain in Bangladesh. Epidemiol Infect. 2002;129:29–34.
16. Threlfall EJ, Ward LR. Decreased susceptibility to ciprofloxacin in salmonella enterica serotype typhi, United Kingdom. Emerg Infect Dis. 2001;7:448–50.
17. Baliga S, Shenoy S, Vidyalaxmi K, Pereira P. Ciprofloxacin resistant *Salmonella typhi*. Natl Med J India. 1999;12:138.
18. Bhat KG, Surek K. Ciprofloxacin resistant *Salmonella typhi*. Natl Med J India. 1999;12:88.
19. Jesudason MV, Malathy B, John TJ. Trend of increasing levels of minimum inhibitory concentration of

ciprofloxacin to *Salmonella typhi*. Indian J Med Res. 1996;103:247–9.

20. Pallab R, Jyothi S, Rugmei SK, Garg KK. Predictive efficacy of nalidixic acid resistance as a marker of fluoroquinolone resistance in *Salmonella enterica* var Typhi. Indian J Med Res. 2006;124:105–8.
21. Kumar R, Aneja KR, Punia AK, Roy PK, Sharma M, Gupta R, et al. Changing pattern of biotype, phage types & drug resistance of *Salmonella typhi* in Ludhiana. Indian J Med Res. 2001; 113:175–80.
22. Bhaswati S, Shanta D, Dipika S, Byomkesh M, Deb AK, Bhattacharya SK, Niyogi SK. Phage typing, biotyping & antimicrobial resistance profile of *Salmonella enterica* serotype Typhi from Kolkata. Indian J Med Res. 2007;125:685–8.

Lymphocyte Teamwork in Tumor Rejection

49

Anil Shanker

Abstract

Immune responsiveness depends on a complex interplay between the elements of innate and adaptive immunity. Several recent studies in various models of immune rejection, including ours in a mouse tumor model, yield new insights into the adaptive control of innate effector functions. This chapter provides a discussion of key observations indicating an indispensable T cell help for innate antitumor immunity.

Keywords

T lymphocytes • Natural killer cells • Functional cooperativity • Tumor rejection

Introduction

The presence of enlarged lymph nodes due to leucoreticular infiltrates in the vicinity of tumors, observed in the nineteenth century by Rudolf Virchow, suggested the first indication of a possible link between immune inflammation and cancer [1]. Since then, the idea that host immunity can control cancer has been the subject of intense debate for over a century. In 1909, Paul Ehrlich was perhaps the first to suggest that the immune system could control cancer development. He predicted that cancer would occur at incredible frequency if host defense did not prevent the outgrowth of continuously arising cancer cells [2]. This concept that the immune system can recognize and destroy nascent transformed cells was embodied in the cancer immuno-surveillance hypothesis formalized by Burnet and Thomas [3, 4]. This triggered the investigations to identify T cell-reactive antigens on various tumors [5, 6]. Indeed, many human tumor antigens were identified that could be recognized by T cells [7]. These observations validated the existence of tumor immunity in the host. Recently, the cancer immuno-surveillance hypothesis has been extended to cancer immuno-editing, which includes the notion that the immune system plays a role not only in cancer prevention but also in editing the phenotype of tumor cells toward antigen-loss immune escape variants [8].

A. Shanker (✉)
Laboratory of Lymphocyte Function, Department of Biochemistry and Cancer Biology, School of Medicine, Meharry Medical College,
Nashville, TN 37208, USA

Department of Cancer Biology, Vanderbilt-Ingram Comprehensive Cancer Center, Vanderbilt University, Nashville, TN, 37232, USA
e-mail: ashanker@mmc.edu

A. Sabu and A. Augustine (eds.), *Prospects in Bioscience: Addressing the Issues*,
DOI 10.1007/978-81-322-0810-5_49, © Springer India 2013

Among various tumor antigens, two major categories of unmutated self-protein antigens have been identified on tumor cells that are capable of eliciting an immune response. These antigens are derived from (1) tissue-specific differentiation antigens or (2) a class of "cancer-germline" gene products that are encoded on the X chromosome and have restricted expression in gametogenic tissue and tumor cells [9]. In the mouse, the P1A antigen characteristic of such cancer-germline genes has been shown to be the major rejection antigen of DBA/2-derived mastocytoma P815 and is known to encode a nonapeptide presented by H-2L^d to CD8$^+$ CTL that can kill P815 cells [10]. Using mice expressing as a transgene, the TCR (TCRP1A) from P1A$_{35-43}$-specific CTL clone P1.5 of DBA/2 origin, we established a model where CD8 T cell reactivity was restricted to a single epitope (P815AB) of the natural tumor Ag P1A presented by H-2L^d on the P815 mastocytoma [11].

We investigated the potential role of the tumor-infiltrating natural killer (NK) cells observed to accumulate at the site of ongoing tumor-reactive T cell response and evaluated whether the activation of Ag-specific T cells at the site of tumor might lead to a bystander effect with elimination of P1A$^-$ as well as P1A$^+$ tumors. Comparison of activation markers on NK cells present in P1A$^+$ and P1A$^-$ tumor-infiltrating lymphocytes (TILs) revealed that only the former expressed an increased level of surface 4.1BB, a marker previously associated with NK cell activation. To obtain a comprehensive analysis of the changes induced in tumor-infiltrating NK cells upon recruitment of activated CD8$^+$ T cells, NK cells were purified from P1A$^+$ and P1A$^-$ tumor infiltrate, as well as from LNs draining the P1A$^+$ and P1A$^-$ tumors 4 days after TCRP1A CD8$^+$ T cell transfer in tumor-bearing RAG$^{-/-}$ mice (endogenous T cell-deficient but NK cell competent). Gene profiling of the purified NK cells identified clusters of stronger expression of effector transcripts (*gzma*, *gzmb*, *prf1*, *and 4.1BB*), tissue migratory transcripts (*gpr33* and *ccr5*), and immune signaling transcripts (*Ifngr*, *Klhdc2*, *Eif3s6*, *Map3k6*, *Tnfrsf1b*, *Icos*, and *Nmi*) in P1A$^+$ TILs than P1A$^-$ TILs. Upregulated expression of granzyme B in NK cells found in the P1A$^+$ TILs versus their P1A$^-$ counterparts was further confirmed by intracellular staining. Moreover, the majority of the NK cells present in the P1A$^+$ tumor had their IFN-γR1 chain occupied by IFN-γ. No increase in NK cell IFN-γ mRNA was observed, which correlated with the absence of IFN-γ production as detected by intracellular immunofluorescence staining. Our results also revealed the coordinated expression of the *CCL3* and *CCL4* chemokine genes by CD8$^+$ TCRP1A cells and of the corresponding chemokine receptor *CCR5* gene on the NK cells. Unlike the critical role played by CCL3 in promoting NK cell recruitment and inflammation, we found no evidence, using CCL3$^{-/-}$ mice, for such an exclusive role for CCL3 in either NK cell recruitment or activation at the tumor site in the present study [12].

The requirement for an Ag-dependent reactivation of the CD8$^+$ T cells at the tumor site for their promotion of NK cell activation was further confirmed by experiments comparing NK cells present in P1A$^-$ and in P1A$^+$ TILs from the same mouse. TCRP1A CD8$^+$ T cells activated in the P1A$^+$ tumor-draining LN migrated to both P1A$^+$ and P1A$^-$ tumors in such mice, albeit more CD8$^+$ T cells accumulated in the P1A$^+$ tumor, and CD25 surface expression indicates their sustained activation state only in the P1A$^+$ tumor. Furthermore, only NK cells present in the P1A$^+$ tumor showed an activated (GzmB$^+$) phenotype. Evidence for *in situ* exocytosis by NK cells, as shown by detection of Lamp-1 *ex vivo* or after reactivation in the presence of YAC tumor cells, was also selectively increased in P1A$^+$ tumors in mice bearing both P1A$^+$ and P1A$^-$ tumors. In aggregate, this study showed that tumor Ag-specific CD8$^+$ T cells modulated the activation status of tumor-infiltrating NK cells toward a killer effector phenotype [12].

Treatment of transgenic TCRP1A RAG-1$^{-/-}$ mice or non-transgenic RAG-1$^{-/-}$ mice reconstituted with TCRP1A CD8$^+$ T cells with the NK cell-depleting NK1.1 mAb PK136 or NK cell-activating receptor NKG2D-blocking mAb MI-6 consistently abrogated the resistance of these mice to growth of the P1A$^-$ tumor in a mix-P1A$^+$/

P1A⁻ tumor condition. Hence, the NK cells along with tumor antigen-specific T cells are important contributors to tumor resistance through an NKG2D-dependent mechanism.

In the absence of a polyclonal T cell repertoire, precursor frequencies of natural killer cells and tumor-specific CTL affect tumor resistance by acting in a cooperative manner [13]. The activated CD8⁺ T cells provide a necessary "help" to dormant NK cells in eliciting their antitumor effector function. This cooperativity of T cell and NK cell effector mechanisms leads to complete tumor regression, including antigen-deficient tumor escape variants. These novel observations highlight a complex interplay between the elements of innate and adaptive immunity, wherein NK cell is an important player [14].

While most of the focus has been on the innate instruction of the adaptive immune responses, an equally important adaptive activation of the innate immunity appears important in the control of tumors as evident in this murine P815 tumor model. This bidirectional cooperativity in innate and adaptive immune functions has broad implications for immune responses in general as discussed recently [15]. Several recent studies in various models of immune rejection yield new insights into how the adaptive immunity by initiating an antigen-specific response can compensate, suppress, and activate innate responses at the site of tissue antigen [16]. Further advances require detailed mechanistic understanding of the adaptive control of innate immune effector functions in tumor and other pathophysiological conditions.

Acknowledgements AS is supported by funds from the NIH grants U54 CA091408 (NCI), 5 U54 RR026140-03 (NCRR), 8 U54 MD007593-03 (NIMHD), and 5P50 CA 090949 (NCI).

References

1. Virchow R. Handbuch der speciellen pathologie und therapie. Erlangen, 1; 1854.
2. Ehrlich P. Uber den jetztigen stand der Karzinomforschung. Ned Tijdschr Geneeskd. 1909;5:273–90.
3. Burnet F. Cancer – a biological approach. 1. The process of control. Br Med J. 1957;1:779–82.
4. Thomas L. Reactions to homologous tissue antigens in relation to hypersensitivity [discussion]. In: Lawrence HS, editor. Cellular and humoral aspects of the hypersensitive states. New York: Hoeber-Harper; 1959.
5. Van den Eynde B, Hainaut P, Herin M, Knuth A, Lemoine C, Weynants P, van der Bruggen P, Fauchet R, Boon T. Presence on a human melanoma of multiple antigens recognized by autologous CTL. Int J Cancer. 1989;44:634–40.
6. van der Bruggen P, Traversari C, Chomez P, Lurquin C, De Plaen E, Van den Eynde B, Knuth A, Boon T. A gene encoding an antigen recognized by cytolytic T lymphocytes on a human melanoma. Science. 1991;254:1643–7.
7. Boon T, Cerottini JC, Van den Eynde B, van der Bruggen P, Van Pel A. Tumor antigens recognized by T lymphocytes. Annu Rev Immunol. 1994;12:337–65.
8. Matsushita H, Vesely MD, Koboldt DC, Rickert CG, Uppaluri R, Magrini VJ, Arthur CD, White JM, Chen YS, Shea LK, Hundal J, Wendl MC, Demeter R, Wylie T, Allison JP, Smyth MJ, Old LJ, Mardis ER, Schreiber RD. Cancer exome analysis reveals a T-cell-dependent mechanism of cancer immunoediting. Nature. 2012;482:400–4.
9. Scanlan MJ, Gure AO, Jungbluth AA, Old LJ, Chen YT. Cancer/testis antigens: an expanding family of targets for cancer immunotherapy. Immunol Rev. 2002;188:22–32.
10. Van den Eynde B, Lethe B, Van Pel A, De Plaen E, Boon T. The gene coding for a major tumor rejection antigen of tumor P815 is identical to the normal gene of syngeneic DBA/2 mice. J Exp Med. 1991;173: 1373–84.
11. Shanker A, Auphan-Anezin N, Chomez P, Giraudo L, Van den Eynde B, Schmitt-Verhulst AM. Thymocyte-intrinsic genetic factors influence CD8 T cell lineage commitment and affect selection of a tumor-reactive TCR. J Immunol. 2004;172:5069–77.
12. Shanker A, Verdeil G, Buferne M, Inderberg-Suso EM, Puthier D, Joly F, Nguyen C, Leserman L, Auphan-Anezin N, Schmitt-Verhulst AM. CD8 T cell help for innate antitumor immunity. J Immunol. 2007;179:6651–62.
13. Shanker A, Buferne M, Schmitt-Verhulst AM. Cooperative action of CD8 T lymphocytes and natural killer cells controls tumour growth under conditions of restricted T-cell receptor diversity. Immunology. 2010;129:41–54.
14. Malhotra A, Shanker A. NK cells: immune cross-talk and therapeutic implications. Immunotherapy. 2011;3:1143–66.
15. Shanker A, Marincola FM. Cooperativity of adaptive and innate immunity: implications for cancer therapy. Cancer Immunol Immunother. 2011;60: 1061–74.
16. Shanker A. Adaptive control of innate immunity. Immunol Lett. 2010;131:107–12.

50 Biosorption of Nickel (II) from Aqueous Solution Using *Bacillus laterosporus*

Rajeswari M. Kulkarni, G. Srinikethan, and K. Vidya Shetty

Abstract

The presence of heavy metals in the environment is of major concern because of their extreme toxicity and tendency for bioaccumulation in the food chain even in relatively low concentrations. Nickel is toxic and relatively widespread in the environment. It is used in a wide variety of industries such as plating and cadmium–nickel battery, phosphate fertilizers, mining, pigments, stabilizers, and alloys and finds its way to the aquatic environment through wastewater discharge. Several researches in environmental biotechnology have shown that many biosorbents present in our environment have the capacity to remove heavy metals from solutions. This chapter presents the result of studies carried out on absorption of nickel ions from aqueous solutions by *Bacillus laterosporus.* The effect of various process parameters on the removal of Ni (II) has been investigated. The pseudo second-order model was used to describe the kinetic data. The results of the study showed that the organism can be efficiently used as biosorbent for the removal of divalent nickel from aqueous solution.

Keywords

Biosorption • *Bacillus laterosporus* • Ni (II) • Process parameters • Adsorption kinetic

Introduction

Water is considered to be a vital and limited resource. Population growth, industrial developments, and other pressures faced by developing countries have led to structured measures to ensure sustainable management of this important source. Many heavy metals and their compounds have been found toxic, while some are also subjected to biomagnifications.

R.M. Kulkarni (✉)
Department of Chemical Engineering, M.S.R.I.T,Bangalore-54, India
e-mail: rajeswari.kulkarni@yahoo.in

G. Srinikethan • K.V. Shetty
Department of Chemical Engineering, NITK,Surathkal, Karnataka, India

A. Sabu and A. Augustine (eds.), *Prospects in Bioscience: Addressing the Issues*,
DOI 10.1007/978-81-322-0810-5_50, © Springer India 2013

The presence of these heavy metals in wastewaters will pose an important problem for the environment. The discharge of heavy metal into aquatic ecosystem has become a matter of concern in India over the last few decades. The pollutants of serious concern include lead, chromium, mercury, uranium, selenium, zinc, arsenic, cadmium, gold, silver, copper, and nickel [1–5]. Nickel is toxic and relatively widespread in the environment. It is used in a wide variety of industries such as plating and cadmium–nickel battery, phosphate fertilizers, mining, pigments, stabilizers, and alloys and finds its way to the aquatic environment through wastewater discharge [6, 7].

Conventional methods that are used to treat the wastewater containing heavy metals are chemical precipitation, ion exchange, solvent extraction, adsorption, membrane processes, etc. The search for new technology led to biosorption which is a physicochemical process dealing with the interaction of material of biological origin and a contaminant [8]. This chapter reports the effect of contact time and adsorbent dose on Ni (II) sorption by using bacteria *Bacillus laterosporus*.

Materials and Methods

Preparation of Biosorbent

The biosorbent used for the study was *Bacillus laterosporus*. The strain was maintained on nutrient agar slants and stored at 4°C with routine transfers in a cycle of 1month. The bacteria were cultured in agitated and aerated liquid media by adopting the shake flask method. Dried biomass was used for the biosorption studies.

Preparation of Stock Solution

All reagents were of analytical reagent grade. A stock solution of 1,000 mg/l of Ni (II) was prepared by using nickel sulfate in double distilled water.

Biosorption Experiments

Biosorption experiments were performed in a rotary shaker using 250-ml Erlenmeyer flask containing 0.2 g of biomass and 50 ppm Ni (II) solution. The concentration of metal ion was determined by using atomic absorption spectrophotometer. The amount of Ni (II) adsorbed by the biomass was calculated from the differences between metal quantity added to the biomass and metal content of the supernatant using the following equation:

$$q = v(C_i - C_f) / m$$

where q is the metal uptake (mg metal/g of the biosorbent), v is the liquid sample volume (l), C_i is the initial concentration of the metal in the solution (mg/l), C_f, the final concentration of metal in the solution(mg/l), and m is the amount of the added biosorbent on the dry basis (g).

Results and Discussion

Effect of Contact Time

Performance of biosorbent on the biosorption of Ni (II) at different contact time was carried out with initial Ni (II) concentration of 50 ppm. Results (Fig. 50.1) show that capacity of biosorption increased with contact time.

Effect of Adsorbent Dose

Experiments were conducted to study the effect of biosorbent dosage on metal removal at room temperature at initial concentration of 50 ppm Ni (II) solution with contact time 1h. It can be observed from (Fig. 50.2) that the metal uptake decreases with increase in adsorbent dosage.

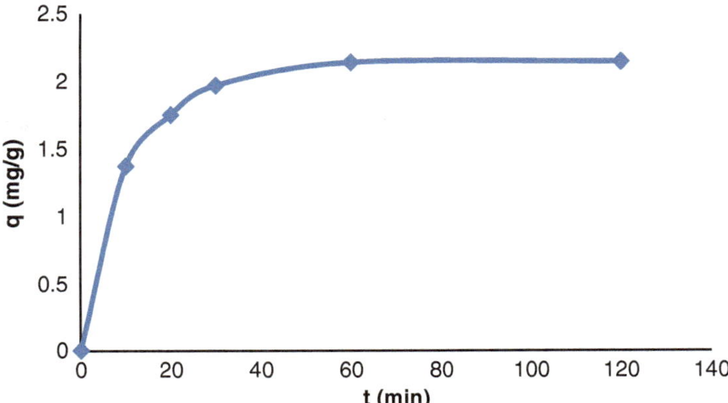

Fig. 50.1 Effect of contact time on Ni (II) sorption

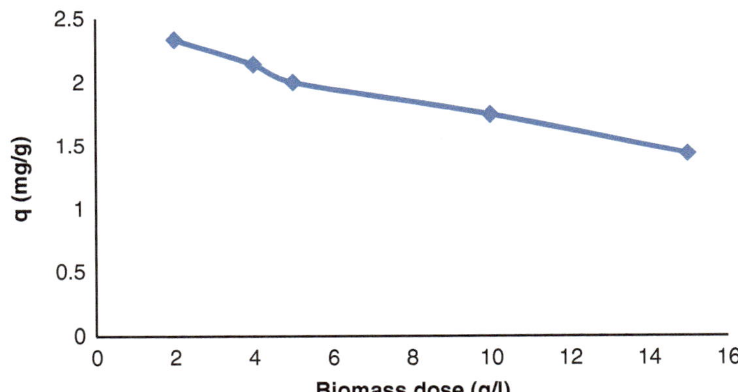

Fig. 50.2 Effect of biomass dose on Ni (II) sorption

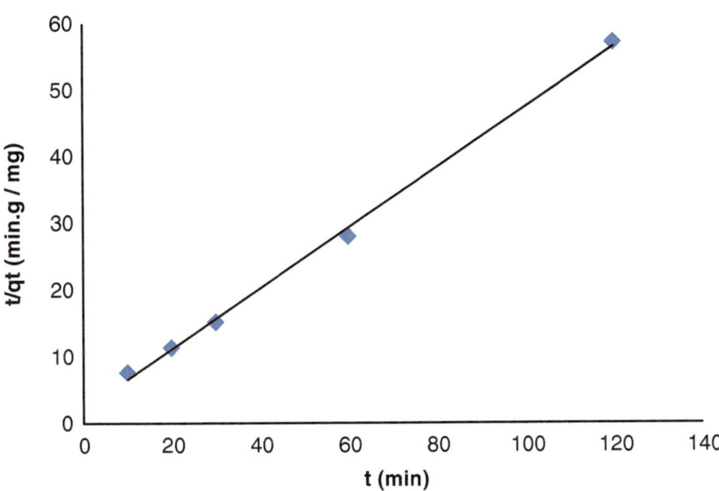

Fig. 50.3 The adsorption kinetic data of Ni (II) on *Bacillus laterosporus*

Kinetic Modeling of Ni (II) Adsorption

Kinetics describes the solute uptake rate which in turn controls the resident time of sorbate uptake at the solid–solution interface.

The pseudo second-order rate equation is expressed as

$$\mathrm{d}q_t/\mathrm{d}t = k_2\left(q_{\mathrm{eq}} - q_t\right)^2$$

where k_2 is pseudo second-order rate constant.

The experimental data fit well with second-order kinetic model (Fig. 50.3).

Conclusions

The present study shows that *Bacillus laterosporus* was an effective biosorbent for biosorption of nickel (II) ions from aqueous solutions. The effects of process parameters like contact time and adsorbent dosage were studied. The biosorption could be best described by pseudo second-order kinetic equation.

References

1. Ahalya N, Ramachandra TV, Kanamadi RD. Biosorption of heavy metals. Res J Chem Environ. 2003;7(4):71–9.
2. Ajaykumar AV, Naif A, Hilal N. Study of various parameters in the biosorption of heavy metals on activated sludge. World Appl Sci J. 2009;5(Special Issue for Environment):32–40.
3. Volesky B. Biosorption and me. Water Res. 2007;41: 4017–29.
4. Volesky B. Detoxification of metal-bearing effluents: biosorption for the next century. Hydrometallurgy. 2001;59:203–16.
5. Volesky B, Holan ZR. Biosorption of heavy metals. Biotechnol Prog. 1995;11:235–50.
6. Padmavathy V, Vasudevan P, Dhingra SC. Biosorption of nickel (II) ions on baker's yeast. Process Biochem. 2003;38:1389–1395.
7. Aksu Z, Donmez G. Binary biosorption of cadmium (II) & Nickel (II) onto dried chlorella vulgaris: coion effect on mano component isotherm parameters. Process Biochem. 41:860–868.
8. Wang J, Chen C. Biosorbents for heavy metals removal and their future. Biotechnol Adv. 2009;27:195–226.

Subject Index

A

Alternaria alternata, 285
Abiotic parameters, 354
Abiotic stress, 138
Abortion, 376, 377, 379
Aceria pongamiae, 325–332
Acetaldehyde, 213, 219
Acetocarmine, 274, 275
Acetone, 94, 95, 99, 133, 235, 259, 265, 286, 294, 299, 390–392
Acetonitrile, 202, 252
Achlorophyllous, 121
Achyranthus aspera, 118, 122
Acidity, 174–180
Acid-ninhydrin, 134
Acinetobacter, 14
Acorus calamus, 399–403
Actinomycetales, 336
Actinomycetes, 160–163, 165, 223, 336–339, 352, 354–358
Actinoplanes, 160
Acute toxicity, 400
Adaptive immunity, 413
Adoptive transfer, 317
Adriamycin, 160
Aerobic bacterium, 211–212
Agarose gel, 51, 162, 309, 407
Agglutinin, 170, 171
Agricultural waste, 222–224
Agriculture, 20, 142, 230, 231, 278, 304, 384
Agrobacterium, 303–314
Agrobacterium tumefaciens, 305, 309, 313
Agrochemicals, 63, 219, 336, 353
Alcaligenes sp.*d2*, 209–219
Aldehydes, 118, 218, 219
Alkaline protease, 64, 66–69, 71, 72, 74, 75
Alkaloids, 20, 92, 95, 98, 99, 118–120, 122, 294, 298–300, 347
Allostery prediction, 38
Alternanthera brasiliana, 118, 122
Alternanthera maritima, 118, 122
Alternanthera sessilis, 118, 122
Alternanthera tenella, 118, 122
Alternaria alternata, 338
Alternaria sesami, 243–249
Alternaria sp., 119, 121
Aluminium, 95, 98, 100, 294, 298, 300
Amaranthus spinosus, 117–122
Amaranthus tricolor, 118, 122
Amaranthus viridis, 118, 122
Amino acid, 6, 7, 31, 33–39, 41, 42, 92, 94, 95, 98–100, 153, 170, 194, 195, 197, 253, 254, 256, 258, 294, 297, 298, 307, 318, 319, 354, 356, 365, 367, 369, 370, 408
Amino acid metabolism, 41
Aminoglycoside, 364
Ammonium nitrate, 70, 200, 201, 205
Amoxicillin, 258, 260
Amphora lineolata, 266, 269
Amplification, 50–52, 162, 193, 195, 235, 240, 241, 246, 309, 312, 313, 394, 407
AMPs. *See* Antimicrobial peptides (AMPs)
Amycolatopsis, 160
Amylase, 279–281, 337, 340
Analysis of variance (ANOVA), 21, 134, 195, 223, 275, 309, 366, 384, 401
Anamirta cocculus, 389–392
Anatomical, 50, 326–328, 330–332
Angiosperms, 20, 31, 344
Angiosperm taxonomy, 344
A. niger, 122, 285, 289
Animal cell, 50
Animal waste, 222
ANOVA. *See* Analysis of variance (ANOVA)
Antagonistic, 162–163, 278–281, 338, 339
Anthraquinones, 95, 98, 99, 119, 294, 298–300
Anthrone method, 95, 294
Antibacterial, 2, 90, 91, 118–122, 160–162, 285, 286, 289
Antibacterial drugs, 160
Antibiotic resistance, 121, 318, 394, 406, 408
Antibiotics, 90, 91, 121, 152, 160–162, 252, 258–260, 278, 305, 307, 308, 311, 313, 318, 336, 352, 364, 394, 406, 408
Antibodies, 112, 114
Anticancer, 2, 160
Anticarcinogenic, 120
Antidote, 118
Antifungal, 119–122, 160, 161, 249, 280, 285–287, 335–340, 364, 400
Antihelmintic avermectin, 160

A. Sabu and A. Augustine (eds.), *Prospects in Bioscience: Addressing the Issues*, DOI 10.1007/978-81-322-0810-5, © Springer India 2013

Anti hypersensitive, 2
Anti-inflammatory, 89, 120, 400
Antimicrobial, 89–92, 117–122, 159–166, 245, 247, 251–261, 283–291, 406–408
Antimicrobial peptides (AMPs), 252
Anti nutritional factors, 94, 95, 97, 98, 100, 294, 297
Anti-oomycetous activity, 363–370
Antioxidants, 120, 132, 138, 219, 244, 246, 247, 249, 297
Antioxidative, 2, 400
Antioxidative enzymes, 243, 245
Anti rheumatic, 89
Antiseptic, 120
Antispasmodic, 89
Apolipoprotein, 400–402
Apolipoprotein A-I precursor, 171
APX. *See* Ascorbate peroxidase (APX)
Aquatic herb, 93–100
Aqueous two-phase system, 1–8
Arabidopsis halleri, 131
Arable land, 104
Arachis hypogaea, 231, 235
Arbuscular mycorrhizal (AM) fungi, 142, 145–148
Arid, 104, 231
Arnon's formula, 95, 294
Aroma, 174–179
Arsenic, 16, 132, 416
Arthrospores, 337
Artocarpus heterophyllus, 170
Ascorbate peroxidase (APX), 245, 247
Ascorbic acid, 94, 297, 298
Asian Gray Weevil, 191–198
A. flavus, 285
Aspergillus sp., 119, 121
Asterionella japonica, 266, 269
Asthma, 120
Astringent, 118
Aulacodiscus sp., 266, 269
AutoDock, 56, 57, 59, 82
AutoDock Vina, 82
AutoDock Vina 4.2, 82
Autoimmune, 112, 378
Avian influenza, 126
Ayurvedic medicines, 332
Azoreductase, 200, 201, 204, 206
Azospirillum sp., 227
Azotobacter, 14, 223, 225, 227, 354–356

B

Bacillus laterosporus, 415–418
Bacillus licheniformis, 64, 70
Bacillus mojavensis A21, 67
Bacillus SDR-10, 66–73, 75
Bacillus sp., 11–17, 64, 66–72, 74
Bacillus subtilis SDR-10, 66, 69
Bacillus thuringiensis, 191–198
Bacteremia, 318
Bambara groundnut, 229–241
BamHI, 195
6-Benzyl aminopurine, 306, 308, 313
Bialaphos, 160
Binary vector, 305
Bioaccumulation, 12–17
Bioactive molecules, 94
Bioagents, 192
Biocompatibility, 2
Biocontrol, 122, 278, 281, 336, 339, 340, 364
Biodegradation, 200, 202, 209–219
BioEdit, 193, 194
Bio fertilizers, 336
Bioinformatics, 38, 45, 80, 153, 194, 318, 322
Biomass, 13, 14, 16, 17, 29, 132, 136, 184, 224, 231, 233, 236–238, 270, 332, 358, 365, 366, 416, 417
Biomolecules, 2, 3, 6–8, 70, 162, 353
Bio-networks, 33
Bioparts, 118–121
Biopesticides, 192
Biophysical parameters, 103–110
Biopolymer production, 219
Bioremediation, 11–17
Biosorbents, 17, 416, 418
Biotechnology, 126, 276, 352
Biotic stress, 235
Biotransformation, 182–184, 186, 188, 189, 200
Biotreatment, 200
Biotyping, 405–409
Bisphenol-A, 210
Black gram husk, 71, 72
BLAST, 36, 64, 66, 162, 164, 193–195, 318, 407, 408
Blood pressure, 79
Blutaparon portulacoides, 118, 122
Boll opening, 107
Bovine serum albumin (BSA), 3, 4, 65
Brachymenium nepalense, 21
Brachythecium buchananii, 21
Bradford, 3, 4, 22, 170
Bradford reagent, 3
Brassica, 274
Brassicaceae, 132
Bronchitis, 89, 326
Brucella, 182
Bryophytes, 19–31
Bryum argenteum, 21, 23, 26, 30
BSA. *See* Bovine serum albumin (BSA)
Bt technology, 192
Buchnera aphidicola 5A, 42
Buffalo meat, 50, 52, 53
Buffers, 21, 51, 64, 65, 133, 162, 182–184, 194, 201, 212, 245, 252, 253, 255, 284, 309, 336, 344

C

Cadmium, 13, 16, 131–138, 416
Cairina moschata, 126, 127, 129
Cajanus cajan, 143, 145, 146
Callus, 122, 303–314
Campylopus pilifer, 21

Cancer, 56, 120, 170, 172, 297, 411
Cancer-germline, 412
Cancer immunoediting, 411
Candicidin, 160
Candida albicans, 161, 285, 286, 288
Carbapenems, 258
Carbendazim, 278, 279, 281, 337
Carbenicillin, 307, 308, 311, 313
Carbohydrates, 91, 92, 94–96, 98, 100, 294–296, 300
Carcinogenic, 200
Cardiovascular disease diabetes, 378
Carotene, 94
β-Carotene, 94, 95, 297
Carriage rate, 394
Casein, 64, 65, 67, 70, 72, 182, 279, 285, 306–308
Caseinomacropeptide, 2
Cassava leaf hay (CH), 384
CAT. *See* Catalase (CAT)
Catabolites, 183, 186, 205
Catalase (CAT), 13, 245, 247
Catechol, 210–215, 219, 249
CBD, 174
CBMN, 375–380
CD4+ cells, 318, 319
CDD search. *See* Conserved domain database search (CDD search)
Cell
  extract, 201, 206
  mass, 183, 184, 186, 189
  mediated immunity, 114, 297, 318
Cell-agglutinating proteins, 171
Cellulase, 279–281, 337, 340
Cellulolytic, 357, 384–386
Centrifugation, 2, 4, 64, 170, 194, 201, 307, 337
Cephalosporins, 258
Ceratium furca, 265, 266, 269, 270
CFS. *See* Correlation-Based Feature Selection (CFS)
CGR. *See* Chaos Game representation (CGR)
Chaos Game representation (CGR), 41–47
Charakasamhita, 400
Chemical pesticides, 192
Chemical stimulation, 284
Chicken, 2, 51, 52, 126–128, 391
Chicken egg white, 2
Chikungunya, 390
Chitinase, 279–281
Chlamydia, 182
Chloramphenicol, 90, 91, 406, 407
Chloramphenicol C30, 90
Chloroform, 57, 94, 95, 99, 118, 121, 122, 259, 286, 294, 299
Chlorophyll, 95–97, 108–110, 132, 133, 135, 136, 138, 234, 235, 238, 265, 266, 269, 294, 296, 297
Chlorophyll-a, 96, 97, 133, 135, 265, 269, 297
Chlorophyll-b, 96, 97, 135, 266, 297
Chlorosis, 132
Cholesterol biosynthesis, 56
Cholesteryl esters, 170, 402
Chromate, 16
Chromophore group, 205
Chromosome, 126, 192, 375, 377–379, 395, 412
Chronic anemia, 375
Chronic bronchitis, 89
Chrysaora quinquecirrha, 283–291
Chryseobacterium sp., 181–190
Chylomicrons, 170
Chymotrypsin, 29
Cicer arietinum, 143, 145, 148
Clavulanic acid, 258, 260
Clear zone, 66, 260, 279
Clitoria ternatea, 251–256
CLPP. *See* Community-level physiological profiling (CLPP)
Cnidarians, 284
Cobalt, 16, 194
Co cultivation, 305, 307, 308, 311, 312, 314
Coffea arabica, 173–180
Coffea canephora, 177–179
Coffee, 174–177
  leaf rust, 174
  quality, 174
Coinfection, 152
Coleoptera, 191–198
Colon, 80
Colorado potato beetles, 29
Community-level physiological profiling (CLPP), 352–354, 356, 359, 360
Community structure, 265, 269, 352
Compactin, 57, 60
Computational algorithms, 34
Computational biology, 38, 45
Computational docking, 153
Computational drug design, 80
Congenital anomalies, 380
Conidiospores, 337
Conserved domain database search (CDD search), 193
Contamination, 2, 12, 13, 295, 336
Coomassie brilliant blue, 4, 172
Copper, 16, 94, 95, 98, 100, 294, 298, 300, 416
Corn Cob, 384–387
Corn earworm, 20, 22–31
Cornish, 126–129
Correlation, 35–37, 114, 134, 143, 146–148, 162, 175, 177–180, 238, 263–271, 318, 322, 354, 367
Correlation-based feature selection (CFS), 34, 36–38
Cotton, 103–110, 143, 274, 337
Coturnix japonica, 126, 127, 129
Coumarins, 95, 99, 118–120, 294, 299
Cow dung, 222–225
Crassocephalum crepidioides, 98
Crassulaceae, 132
Crop growth, 231, 278
Crop plants, 97, 104, 141–148, 222, 231, 236, 278, 296, 314, 364, 369
Crude fibre, 95, 96, 98, 100, 294, 295, 300
cry3A gene, 191–198
cry genes, 192, 193
Cryo-preservation, 273–276
Cry proteins, 192
Crystal structure, 57, 82, 319, 367

CTAB
method, 64
procedure, 309
Culex pipiens, 389–392
Cultivars, 104, 105, 174, 175, 177–179, 231, 239, 244, 278, 304
Cumulative water transpired (CWT), 236
Cup quality, 173–180
Curated dataset, 34
Cutaneous macules, 112
CWT. *See* Cumulative water transpired (CWT)
Cyanotic heart disease, 374
Cycloheximide, 161, 162, 336
Cytochrome b gene, 50, 52
Cytokinesis, 375, 376, 378
Cytolytic, 284, 291
Cytoplasmic male sterile, 274
Cytoskeleton, 364, 370

D
Data mining, 36, 43, 45
DDBJ. *See* DNA database of Japan (DDBJ)
Decolorization, 199–206, 259
Deionised water, 234
Denaturation, 8, 50, 51, 203, 309, 407
Dengue fever, 390
Dengue haemorrhagic fever, 390
Denitrification, 121, 352
de novo sequencing, 252, 253, 256
Depigmenting, 112
Dermatophytes, 118, 122
Design Expert software 7.1.4, 211, 212
Detergent formulations, 63, 69
DGGE, 352
Diabetes, 80, 374, 375, 378
Diagnosis, 375, 378, 380
Diagnostics, 63, 112, 172, 374, 375, 380
Dialysis, 21, 284
Dialysis tubing, 21
Diaphoretic, 118
Diarrhoea, 326
Diel vertical migration (DVM), 264
Digestive problems, 399
2,4-Dihydrophenoxyacetic acid (2,4-D), 306, 308, 309, 313
Dimensionality reduction (DR), 36, 43, 320
2-[(3, 4-Dimethyl-2-oxo-2H-chromen-7-yl) oxy] propanoate, 59
Dimethyl sulfoxide (DMSO), 91, 235
Dinitrosalicylic acid, 95, 294
Dinoflagellate, 265, 266, 269–271
Di-peptide composition, 34–38
Diplazium esculentum, 293–300
Disc-diffusion, 90, 91, 118, 285, 366, 406, 409
Diuretic, 94, 118
Diversity index, 352, 361
DMSO. *See* Dimethyl sulfoxide (DMSO)
DNA
hybridization, 50
sequencing, 50
size marker, 51
DNA database of Japan (DDBJ), 126
Domestication, 125
DR. *See* Dimensionality reduction (DR)
Dropsical, 89
Drought
stress, 103–110
susceptibility, 105, 106, 110
tolerant, 104, 231
Drug
resistance, 152, 153, 407
targets, 34, 84, 153
Dryopteridaceae, 293–300
Dual culture method, 336
Duckweed, 132
DVM. *See* Diel vertical migration (DVM)
Dwarf sheep, 383–387
Dwarf x Tall crosses, 175–178
Dyes, 2, 171, 199–206, 285, 356
Dyspepsia, 326

E
Earthworm, 222–224
Ecdysones, 20, 390
Economy, 2
Ecosystem, 20, 147, 148, 263, 336, 351–361, 384, 416
Edaphic factors, 141–148
Edible fern, 294
Effluent, 1–8, 12, 15, 200, 210, 219, 410
Eichhornia crassipes, 132
Eisenia foetida, 222
Electrical conductivity, 143, 223
Electrophoresis, 51, 162, 170–172, 285, 289, 394, 395
Electroplating, 12, 14
Electrostatic interaction, 7, 17
ELISA, 50
EMBL. *See* European molecular biology laboratory (EMBL)
Embryogenic callus, 305–307, 309–311, 313, 314
Emmenagogue, 118
Emollient, 118, 120
Empirical orthogonal function (EOF), 43
The empirical orthogonal function (EOF), 43
Endocarditis, 318
Endospores, 13, 66
Enlarged lymph nodes, 411
Ensemble classifier, 34, 37
Enterobacter, 14, 42
Enterobacter 638, 42
Environment, 2, 12, 14, 16, 69, 71, 104, 118, 119, 131, 132, 148, 152, 160, 162, 174, 192, 200, 210, 219, 231, 235, 304, 330, 336, 361, 370, 416
Environmental biotechnology, 415
Environmental pollution, 12, 200
Environmental stresses, 104
Enzymatic, 2, 22, 57, 132, 138, 206, 211, 245, 332
EOF. *See* Empirical orthogonal function (EOF)
Epidemiological relatedness, 395

Epidemiological typing, 393–396, 406
Epidermal, 112, 290, 327, 344, 345, 347
Epitopes, 317–322, 412
Eriophyid mite, 325–327, 330
Erythromycin, 160
Escherichia coli, 42, 90, 91, 118, 120, 122, 161, 193–198, 240, 258, 259, 285, 406
Essential oils, 89, 118, 345, 400
Etiology, 112, 378
Eukaryotes, 369
Euphorbiaceae, 132, 304
European molecular biology laboratory (EMBL), 126
Evapotranspiration, 232, 233
Evolution, 16, 125, 126, 233, 238, 239
Exopeptidases, 23
Expression vector, 193–194
Extensively drug-resistant (XDR), 152
Exudates, 122, 278, 345

F

Fabaceae, 132, 252
Factor XIIa, 171
Faecal lipids, 57–58
Fall armyworm (FAW), 21–23
FASTA, 126
Fast correlation based feature selection (FCBFS), 34, 36–38
Fatty acid translocase LDL cholesterol, 80
FAW. *See* Fall armyworm (FAW)
FCBFS. *See* Fast correlation based feature selection (FCBFS)
Feasibility, 2, 34, 38, 319, 320
Febrifuge, 118
Fecundity, 390–392
Feed, 2, 22, 96, 223, 227, 295, 384–386, 400
Fermentation broth, 2
Fern, 20, 132–136, 138, 294
Ferulic acid, 20, 249
Fetal growth restriction (FGR), 374, 379
Fetuin-A, 171
Fetus, 374–379
Fibrobacter succinogenes, 384, 386
Filariasis, 390
Filtration, 2, 12, 200, 400
Finger printing, 50
Fish
  processing, 1–8
  proteins, 2, 3, 5–8
Fissidens asperifolius, 21, 23, 26, 30
Fissidens crispulus, 21
Fixed oils, 91, 92
Flame photometer, 4, 143
Flatulence, 325
Flavonoids, 20, 91, 92, 95, 99, 118–120, 122, 294, 298–300, 345
Flocculation, 2, 200
Fluorescent pseudomonads, 278–281
Fluoroquinolone resistant strains, 406
Foliar absorption, 132, 136, 138
Folic acid, 94
Folin–Ciocalteau, 64
Food
  chain, 132, 263
  poisoning, 120
Formaline-acetic acid, 326
Formulations, 2, 63, 69
Free fatty acids, 95, 96, 294, 295
Frequency chaos game representation (FCGR) matrix, 42, 47
Friedelan-3-one, 402, 403
Fruit length, 175, 177–179
FT-IR analysis, 211, 213, 216, 218
Funaria hygrometrica, 21
Functional annotation, 33
Functional diversity, 353, 354, 356, 359–361
Fungi, 118, 120–122, 142, 145–148, 160, 161, 163, 165, 166, 223, 225, 244, 249, 280, 337, 338, 352, 354–356, 358, 364, 384
Fusarium oxysporum, 278, 280, 281, 336, 338
Fusarium sp., 119, 121
Fusarium udum, 278, 280, 281, 336, 338

G

a-Galactose-specific lectin, 169–172
Gallus gallus, 126, 127
Galvanization, 12
Gas analyzer, 105
Gasteiger charges, 82
Gelatin, 51, 65, 70, 279, 337, 340
Gene ontology, 34
Genetic counseling, 380
Genetic disorders, 380
Genetic distance, 127–129
Genetic diversity, 352
Genetic engineering, 19
Genetic transformation, 303–314
Genetic variability, 229–241
GENETOOL software, 51
Genome mapping, 231
Genomic DNA, 50–53, 64, 161, 162, 193, 194, 235, 240, 309, 313
Genomics, 20, 38, 42, 47, 50–53, 64, 125–129, 153, 161, 162, 193, 194, 235, 240, 309, 313, 336
Genotypes, 105–110, 174, 176, 177, 231–233, 236, 238, 239, 304, 311, 330
Gestational age, 374, 375, 377, 379
Gibberellic acid (GA_3), 306
Glandular capitate trichomes, 344
Glandular trichomes, 345, 347
Glucose, 13, 65, 72–74, 161, 162, 183, 185–187, 284, 354–356
β-Glucuronidase (GUS), 305, 307, 309, 312, 313
Glycerol, 21, 65, 72, 161, 194, 278, 305
Glycine max, 21, 143, 145, 146, 148
Glycoconjugates, 170, 171
Glycosides, 91, 92, 95, 98, 99, 118, 119, 294, 298–300
Goat meat, 50
Gonorrhoea, 326

Gossypium hirsutum, 104
Gossypium sativum, 143–146, 148
Granulometric fractions, 13
GraphPad Prism5, 195
Green gram, 222–227
GROMACS, 153
Growth, 13, 23, 36, 65, 91, 94, 103, 120, 132, 142, 161, 178, 184, 192, 201, 230, 244, 264, 278, 305, 326, 336
Gs. *See* Stomatal conductance (Gs)
Guaiacol peroxidase, 243
Guilleminea densa, 118, 122
Guinea fowl, 52, 126–128

H

Haematoxylin, 285, 326
Halal food, 50
Hatchability, 390–392
Hazardous materials, 12
Health care, 118
Haemophilus, 182
Heavy metals, 12, 13, 16, 17, 20, 131, 132, 134, 136, 138, 182, 415, 416
Helicoverpa zea, 20, 24–28, 30
Hemocytometer, 365
Hemoglobin, 317
Hemoglobinopathies, 374
Herbivores, 132, 326, 331
Herbivory, 22, 331
Heritable, 112
Hevea brasiliensis, 303–314
Hexane, 118–122, 259
HindIII, 195
Histocompatibility complex, 112, 125, 126
Histological, 50, 326
Histopathological, 112
HIV. *See* Human immuno deficiency virus (HIV)
HLA. *See* Human leukocyte antigen (HLA)
HMG-CoA reductase (HMGR), 55–60
HMGR. *See* HMG-CoA reductase (HMGR)
Holoptelea integrifolia, 55–60, 259, 260
Homeostasis, 41, 80, 86, 132
Homology modeling, 367–368, 370
Homo sapiens, 34, 35
Hormones, 20, 80, 294, 306, 308, 390
Hot pepper, 273–276
Hub proteins, 33–39
Human immuno deficiency virus (HIV), 152, 153, 182
Human leukocyte antigen (HLA), 319–322
Humicola sp., 182, 184, 186
Humoral, 112, 297, 318
Humoral immunity, 112, 318
Hybrid progenies, 178
Hydrogen bonds, 2, 59, 60, 84–86, 153, 156, 218, 320, 321
Hydrolysates, 2
Hydrolysis, 2, 313, 337, 340, 354, 384
Hydrolyzable tannins, 20
Hydrophobic, 2, 3, 7, 8
Hyper cholesterolemic drugs, 402
Hyperlipidemia, 57, 58
Hypertension, 80, 374, 378
Hypertriglyceridemia, 80
Hyphae, 145, 336, 364, 367, 369, 370
Hypopterygium tamarisci, 21

I

IBP cryo-container, 274
Immune
  inflammation, 411
  rejection, 413
  response, 126, 290, 318, 412, 413
  responsiveness, 411
Immunity, 41, 112, 114, 125, 297, 318, 322, 411, 413
Immuno-diffusion, 114
Immuno-electrophoresis, 50
Immunogen, 317
Immunomodulatory, 283–291
Immunosuppressive, 288
Immunosurveillance, 411
Indian Elm, 56
Indian red jungle fowl, 126–129
Industrial effluent, 1–8, 15, 210, 219
Infrageneric, 343
Innate effector functions, 413
Insect(s), 19–31, 105, 174, 192–195, 198, 326, 390
  resistance, 20, 31
  vectors, 390
Insemination, 274
Instar larvae, 22, 30, 194, 390, 391
Integrated pest management (IPM), 192
Inter-alpha-trypsin, 171
Interfacial tension, 2
Intrauterine growth restriction (IUGR), 374, 375, 377–379
In utero, 373
Invasive weed, 117–122
Ionic interactions, 2
IPM. *See* Integrated pest management (IPM)
Iridoids, 95, 98, 99, 118, 119, 294, 299
Iron, 94, 95, 98, 100, 223, 294, 298, 300, 317–322
Iron surface determinant B (IsdB), 317–322
IR protein, 20, 22–30
Iso electric point, 5
Isolation and identification of bacteria, 13
Isoniazid, 151–157
Isopentenyltransferase (ipt), 305, 309, 313
Isoprenoids, 56
Isopterygium albescens, 21
Isozyme, 231
IUGR. *See* Intrauterine growth restriction (IUGR)

J

Jacalin, 170–172
Jacalin–Sepharose, 169–172
Jackfruit, 170, 171
Japanese quail, 126–128
Jelly fish, 283–291

K

Kadaknath, 126–129
Kanamycin, 160, 193, 194, 307, 308, 313
Kanamycin resistance, 305
Karhunen–Loève Transform (KLT), 43
Karim Koovalam. *See Monochoria vaginalis*
Karyotype, 375–379
Keratinocytes, 80
β-Ketoadipate pathway, 210
Kinetin (Kin), 304, 306
Kirby–Bauer disc diffusion. *See* β-Lactamase inhibitor assay
Klebsiella enteric a 342, 42
Klebsiella pneumonia, 90, 91, 121, 259, 285
Klebsiella sp., 118, 121
KLT. *See* Karhunen–Loève Transform (KLT)
Knop's medium, 133
Krebs cycle, 210

L

Lactagogue, 118
β-Lactam antibiotics, 258, 260
β-Lactamase, 257–261
β-Lactamase inhibitor assay, 257–261
β-Lactam ring, 258
Lactophenol cotton blue, 143, 337
Lamarckian genetic algorithm, 57
LC-ESI MS, 252, 253
LCMS. *See* Liquid chromatography mass spectrometry (LCMS)
Leaf area duration (LAD), 232, 233, 236, 237
Leaf epidermal anatomy, 347
Leaf galls, 328
Lecithin-cholesterol acyltransferase (LCAT), 170, 400–402
Lectins, 170–172
Legionella, 182
Lekhaneyagana, 56, 60, 400
Lemna minor, 132
Lepidopteran insect pests, 20
Leprous affections, 89
Leucoreticular infiltrates, 411
Leucubryum bowringii, 21
Liberica coffee, 174
Ligand, 57, 60, 80–86, 152–157, 366
Lineweaver burk plot, 184, 186, 188
Lipid homeostasis, 80, 86
Lipid peroxidation (LPX), 132, 245–247
Lipids, 60, 80, 86, 95, 96, 170, 247, 295, 352, 401–402
Lipoprotein, 170
Liquid chromatography mass spectrometry (LCMS), 56, 57, 59, 256
Liquid nitrogen, 21, 274, 275
Listeria, 14
Liver disorders, 93
LL-diaminopimelic acid (LL-DAP), 163, 164
Lowry's method, 65, 95, 134, 294
Lyophilization, 284
Lysozyme, 2, 194, 290

M

Machine learning algorithms, 37
Macromorphology, 343
Macrophomina phaseolina, 278, 280, 281, 336, 338
Magnesium, 95, 98, 100, 294, 298, 300, 357
Malaria, 390
Malnutrition, 94, 374
Manganese, 95, 98, 100, 132, 294, 298, 300
Marchantia linearis, 21, 23, 26, 30
Marine
  invertebrates, 138
  phytoplankton, 264
  toxins, 284, 290
Marker-trait analysis, 231
Mass spectrometry (MS), 57, 171, 172, 252–253, 256, 401
Maternal age, 376, 378, 379
Matlab, 45
Maximum likelihood, 127, 128
Maximum parsimony, 127–129
MDR. *See* Multidrug-resistant (MDR)
Mean transpiration rate (MTR), 232, 233, 236, 237
Melanocytes, 112, 114
Meleagris gallopavo, 126, 127, 129
Membrane integrity, 363 (Found only in abstract)
Meso-diaminopimelic (Meso-DAP) acid, 163, 164
Metabolic syndrome, 79–87
Metabolic syndrome-insulin resistance, 80
Metal ions, 2, 13, 17, 64, 132, 416
Meta pathway, 210–213, 218, 219
Methanol, 54, 57, 94, 95, 118–121, 182, 183, 245, 252, 259, 286, 294, 399–403
Methicillin-resistant *Staphylococcus aureus* (MRSA) strains, 259, 394
Methionine, 98, 231, 298
Mevalonate pathway, 56
Mevinolin, 57, 58, 400, 401
Michaelis–Menten kinetics, 186
Microarray, 352
Microbes, 12, 118, 120, 122, 160, 223, 224, 227, 352–354, 384
Microbial diversity, 338, 351–361
Microbial infection, 245, 247, 249
Microbial resistance, 251
Microfuge, 133
Micromonospora, 160
Micromorphology, 161, 343–348
Micronuclei (MN), 193, 194, 375–378
Microorganism, 12, 64, 66, 210, 355
Microsatellite DNA, 235, 240
Mitochondria, 50–53, 132
MN. *See* Micronuclei (MN)
Molecular biology, 33, 126, 284, 384
Molecular docking, 56, 57, 81, 367–370
Molecular docking crystallographic structure, 319
Molecular techniques, 50, 396
Monobactams, 258
Monochoria vaginalis, 93–100
Morphological traits, 175, 177–180, 231
Mosquitocidal compounds, 390

Mosquito control, 390
Mosses, 20
Mouse tumor model, 412
MSMS. *See* Tandem mass spectrometry (MSMS)
MTR. *See* Mean transpiration rate (MTR)
Multidrug-resistant (MDR), 152, 153, 252, 406, 408
Multifactorial, 112
Multigene, 126
Multiple gestation, 374
Municipal waste, 221
Muscovy duck, 126–129
Mustard oil, 184, 190
Mutagenic, 200
Mutations, 152–156, 170, 258, 378, 408
Mycelial growth, 336, 337, 365
Mycobacterium, 151
Mycobacterium tuberculosis, 152
Mycolic acid, 153
Myllocerus undecimpustulatus undatus, 191–198
Myoglobin, 317

N
Naïve Bayes classification, 41–47
Naïve Bayes classifier, 42, 43, 45, 47
Nanoparticles, 2
National Center for Biotechnology Information (NCBI), 126, 162, 164, 193, 197, 365, 367
National Staphylococcal phage typing centre, 394
Natural toxins, 283
NCBI. *See* National Center for Biotechnology Information (NCBI)
Necrotrophic, 244, 369
Neighbour-joining (N-J), 66, 67, 126–129, 162, 164, 165
Neisseria, 182
Nematocysts, 284
Nematodes, 174, 223, 326
Nephrolepis exaltata, 20
Neural Network classifier, 37
New-castle disease, 126
Nickel, 16, 95, 98, 131, 415–418
Nicoloson's formula, 143
Nitrite, 265, 352
Nitrogen sources, 64–66, 69–72, 182, 184, 201, 212
Nocardia mediterrani, 182
Nonimmune, 170
*npt*II gene, 305
Nuclear receptor. *See* Peroxisome proliferator activated receptor delta (PPAR δ)
Nucleotide, 42, 51, 126, 127, 129, 162, 165, 194, 195, 197, 368, 407, 408
Numida meleagris, 126, 127
Nutrient(s), 13, 29, 64, 71, 75, 91, 98, 100, 142, 148, 161, 200, 222, 224, 227, 244, 263–271, 278, 279, 293–300, 352, 361, 383–387
  agar, 91, 118, 161, 182, 223, 259, 260, 416
  agar endospore, 13
  utilization, 383–387
Nutritional factors, 96, 98, 295
Nystatin, 160, 336

O
Obesity, 56, 80, 400
Octoblepharum albidum, 21, 23, 26, 30
Ohmic heating, 2
Oleoresin, 89
Oligomers, 235
Oligonucleotides, 42, 384
Oomycetes, 364, 367, 370
Oomycetous, 363
Opsonic antibodies, 317
Organelles, 2
Organic load, 2
Organic pollutants, 352
Organic waste, 222, 223
Organoleptic, 50
Oryza sativa L., 277
Oviposition, 390, 391
Ovis aries β-tubulin, 367
Oxaloacetate, 213, 219
Oxidation/reduction, 12
Oxidative burst, 243–249, 305, 311

P
Pallavicinia lyelii, 21
Panicum hay (PH), 384–387
Paracoccus sp., 118, 121
Paromomycin, 363–370
Partition coefficient, 2–8
Partitioning, 1–8, 236
Pathogen, 126, 222, 244–247, 249, 277, 278, 280, 318, 336, 337, 364, 369, 394
Pathology, 112, 375
Paullinia cupana, 148
PCA. *See* Principal component analysis (PCA)
PCR. *See* Polymerase Chain Reaction (PCR)
PCR-RFLP, 49–53
PDA. *See* Potato dextrose agar (PDA)
PDB. *See* Protein Data Bank (PDB)
Pectinase, 280, 281
Penems, 258
Penicillinase, 258
Penicillins, 258
Peptides, 35, 37, 82, 118, 126, 132, 170, 194, 251–256, 318–322
Peptidoglycan, 163, 164
Peptone, 13, 64, 65, 70, 72, 118, 259
Peridinium sp., 266, 269, 270
Perinatal deaths, 374
Perinatal mortality, 374–379
Peroxisome proliferator activated receptor delta (PPAR δ), 79–87
Petroleum ether, 94, 95, 294, 299
Phage typing, 393–396, 405–409
Pharmaceutical industries, 63
Pharmaceuticals, 2, 63, 219
Pharmacological, 120, 291
Pharmacological classification, 56, 400
Pharmacophore, 80

Phenol, 91, 92, 95, 97, 209–219, 245, 247, 248, 289, 294, 297, 300
Phenol hydroxylase, 210–214
Phenolic acids, 20, 245, 247–249
Phenolic resins, 210
Phenolics, 20, 120, 210, 247–249, 297, 328, 329, 332
Phenological observations, 105–107
Phenylalanine dehydrogenase, 2
Philonotis fontana, 21
Phlobatannins, 95, 98, 99, 118, 119, 294, 299
Phosphate
  fertilizers, 131, 416
  solubilising bacteria, 221–227, 355
Phospholipid fatty acid analysis, 352
Phospholipid fatty acid analysis (PLFA), 352
Phosphorous, 94, 227, 265, 269
Phosphorus, 97, 132, 142, 143, 148, 222–224, 356, 357
Photic zone, 264
Photosynthate, 148
Photosynthesis, 95, 105, 108–110, 200, 233–234, 236, 238, 239, 246, 271
Phototactic movements, 264, 265, 271
pH shifting, 2
Phylogenetic analyses, 125–129, 159–166, 195
Phylogenetic trees, 66, 126, 162, 164
Phylogeny, 126–129, 344
Physiology, 20, 132, 264, 276, 378
Phytic acid, 95, 97, 98, 294, 297, 300
Phytoalexins, 244
Phytochemical, 20, 89–92, 94, 95, 98, 99, 118–120, 122, 294, 298–300, 390
Phytohormones, 278, 307
Phytopathogenic, 280, 364
Phytopathogens, 245, 281, 336–338
Phytophagous, 19–31, 325
Phytophagous insects, 19–31
Phytoremediation, 131–138
Pig meat, 50, 52
Pigments, 13, 94, 95, 135, 164, 186, 265, 269, 294, 336, 340, 416
Pigweed, 118
Piperacillin, 258
Pistia, 132
Pittosporaceae, 89
Pittosporum viridulum, 89–92
Plackett–Burman, 211–213
Plagiochila beddomei, 21
Plant defense, 29
Plant extracts, 91, 259–261, 390
Plant pathogens, 162, 244, 278, 335–340
Plasmacytes, 317
PLFA. *See* Phospholipid fatty acid analysis (PLFA)
Pollen
  germination, 275, 276
  viability, 274–275
Polyethylene glycol, 2, 3
Polygenic, 112
Polymerase chain reaction (PCR), 50–53, 64, 162, 163, 193–195, 239, 240, 309, 312, 313, 384, 386, 407
Polymer/polymer systems, 2
Polymer/salt systems, 2
Polymorphisms, 50, 126, 129, 152, 153
Polyphasic, 159
Polyphenol, 326–327, 330–332
Polytricum commune, 21
Pongamia pinnata, 325–332
Pontederiaceae. *See Monochoria vaginalis*
Porcine, 49–53
Porella acutifolia, 21
Postinoculation, 244
Potassium, 2–4, 94, 223, 224, 227, 259, 261, 309
Potato dextrose agar (PDA), 223, 278, 336–338, 355, 364, 365
Poultry droppings, 222, 224, 227
Poultry litter, 224, 225
PPIN. *See* Protein-protein interaction network (PPIN)
pQE-80L, 193, 194
Precipitation, 2, 5, 21, 50, 416
Pregestational diabetes, 375, 378
Pregnancy, 374, 375, 378, 379
Prenatal diagnosis, 375, 378
Principal component analysis (PCA), 42, 43, 45, 47, 354, 356, 359–361
Proanthocyanidins, 20
Progenies, 175–179
Proline, 132, 134, 136–138, 244, 306–308, 356
Protease, 23, 29, 54, 63–75, 279–281, 319, 337, 340, 359
Protein data bank (PDB), 34, 82, 153, 319, 337, 365, 367
Protein-protein interaction network (PPIN), 34, 35, 39
Proteins, 2, 20, 33, 41, 50, 56, 65, 80, 94, 132, 153, 170, 192, 231, 244, 258, 273, 285, 294, 331, 354, 365, 378, 384, 402, 412
Proteolysis, 29, 74
Proteomics, 20, 38, 252
Proteus mirabilis, 90, 91
Protozoa, 138, 223, 352, 384
Pseudomonas, 14, 91, 118, 120, 225, 227, 277–281
Pseudomonas aeruginosa, 90, 91, 122, 285
Pseudomonas fluorescens MTCC 103, 161
Pteridaceae, 132
Pteris sp., 131–138
Pteris vitatta, 132
Pterosins, 20
Public health, 12, 375
Pupae, 22, 29, 390, 391
Purification, 2, 5, 75, 122, 194–198, 200, 206, 252, 284
Pyodermas, 394
Pyogenic coccus, 394
Pyridoxine, 94
Pyruvate, 213, 356
Pythium myriotylum, 364
Python Molecular Viewer, 57

Q
Quantitative trait locus (QTL), 231
Quinines, 120

R

Racopilum cuspidigerum, 21
Rainfed conditions, 105–110
Ramachandran plot, 365, 367–369
Random amplified polymorphic DNA (RAPD), 50
Random forest, 34, 37, 38
Randomized block design, 134
Randomized design, 384
Rapamycin, 160
RBF network, 34, 37, 38
RCSB. *See* Research Collaboratory for Structural Bioinformatics (RCSB)
Reactive oxygen species (ROS), 132, 232, 244–247, 249
Receptor, 36, 56, 60, 80–87, 126, 319, 321, 412
Reducing sugar, 94–96, 98–100, 118, 119, 294, 295, 299, 300
Refractometer, 4
Religious taboos, 50
Remediation. *See* Phytoremediation
Renal disease, 374
Reproductive toxicity, 389–392
Research Collaboratory for Structural Bioinformatics (RCSB), 82
Restriction enzyme, 51, 52, 235, 394
Restriction fragment length polymorphism (RFLP), 50, 396
Restriction maps, 51
Restrictive lung disease, 374
Retention time (RT), 59, 202, 206, 213, 216, 402, 403
Reverse diel vertical migrations, 264
Reverse-phase high-performance liquid chromatography (RP-HPLC), 245, 248, 249
RFLP. *See* Restriction Fragment Length Polymorphism (RFLP)
Rheumatic swellings, 89
Rheumatism, 89, 326
Rhizobacteria, 278, 336
Rhizobium, 223, 225, 227, 354–357
Rhizoctonia solani, 278, 280, 336, 338
Rhizosphere, 142, 148, 278, 280, 281, 336–339
Riboflavin, 94, 297
Riccia frostii, 21
Rice, 72, 104, 277–281, 336–339
Rice bran, 65, 71, 72
Rifampicin, 182
Rifamycin B, 182–184, 186
Rifamycin oxidase (RO), 182–190
Rifamycin S, 182, 183
RIF resistance, 152
River Periyar, 12, 270
16S RNA fragment, 51–53
ROS. *See* Reactive oxygen species (ROS)
Roughages, 383–387
RP-HPLC. *See* Reverse-Phase High-Performance Liquid Chromatography (RP-HPLC)
12S rRNA gene, 50, 52
16S rRNA gene, 49–53, 64, 66, 67, 164
RSCB database, 79
RT. *See* Retention time (RT)
Rumen microbes, 384
Rumen microbial count, 384
Rumen microbial profile, 383–387
Ruminants, 20, 97, 384

S

Sabouraud's agar, 119, 161, 285
Saccharomyces cerevisae, 122, 161
Saccharopolyspora, 160
Saccharum officinarum, 143, 145, 148
Safranin, 326
Salmonella enteric, 405–409
Salmonella paratyphi, 285, 286, 289
Salmonella typhi, 405–409
Salmonellosis, 126
Saponins, 89, 91, 92, 95, 99, 118–120, 294, 298–300, 347
Sarcoplasmic proteins, 5
Sclerotial germination, 337–340
Sclerotium, 280, 281
SDS-PAGE. *See* Sodium dodecyl sulphate-polyacrylamide gel electrophoresis (SDS-PAGE)
Seafood, 2
Secondary metabolites, 20, 70, 95, 119, 122, 160, 278, 280, 294, 297, 336, 338, 340, 345, 347
Sediment, 12–15, 353, 354, 356, 358, 359, 361
Sedum alfredii, 132
Seed cotton, 104–106, 110
Selective hybridization, 235
Self-consistency test, 38, 46
Semi-arid, 104, 231
Sepharose, 169–172
Sequence likeness, 33–39
Sequences, 34–39, 42, 43, 45, 47, 51, 64, 66, 67, 126, 127, 129, 132, 162–165, 193–195, 197, 232, 234, 235, 240, 253, 254, 256, 258, 309, 318–320, 322, 364, 365, 367, 370, 407, 408
Serial dilution, 13, 223
Serotypes, 120, 405–409
Serovar Typhi, 406–409
Serum C3, 112–114
Serum IgG, 113, 114
Serum lipids, 57, 58, 401–402
Sesame, 184, 190, 243–249
Sesamum indicum, 243–249
Severe combined immunodeficiency (SCID), 317
Sheath blight disease, 278, 281
Sheep, 51, 52, 383–387
Sheep meat, 50
Shikimic acid, 120, 246, 247
Short Genes, 42, 47
Siderophore, 278–281
Signal transduction, 33, 41
Silicate, 265
Silver nitrate, 303–314
Simulation, 365–366
Single nucleotide polymorphisms (SNPs), 152, 153

Singular value decomposition (SVD), 43
Snake-bite, 93
SNPs. *See* Single nucleotide polymorphisms (SNPs)
Socio-economic, 284, 374–376, 379
SOD. *See* Superoxide dismutase (SOD)
Sodium dodecyl sulphate-polyacrylamide gel electrophoresis (SDS-PAGE), 22, 171, 194, 195, 285, 289, 290
Soil fertility, 142, 352
Solanaceae, 343–348
Solanum, 343–348
Solanum aculeatissimum, 344–346, 348
Solanum erianthum, 344–346
Solanum macrocarpon, 344–346, 348
Solanum melongena var. incanum, 344–346, 348
Solanum paratyphi, 285, 286, 289
Somatic embryogenesis, 303–314
Sorghum vulgare, 143–146, 148
Soxhlet apparatus, 56, 90, 94, 118, 259, 294, 326
Soya oil, 190
Soybean leaf disks, 22
Specialty coffee, 173
Speciation, 49–53, 125
Species richness, 352
Sphagneticola trilobata, 389–392
Spodoptera exigua, 29
Spodoptera litura, 20, 23–31
Spore development, 41
Sporocarps, 143
Sprague Dawley (SD), 56, 400
Sprains, 89
SSR markers, 240
Stand-alone BLAST, 36
Staphylococcal cassette chromosome (SCCmec) typing, 396
Staphylococcus aureus, 90, 91, 122, 161, 259, 317–322, 393–396
Starch, 65, 72, 259, 279, 336, 337, 357
Steroids, 99, 119, 299
Sterols, 56
Stigmasterol, 89
Stomach, 93
Stomatal conductance (Gs), 104, 233, 238
Strain typing, 394
Stramenopiles, 364
Streptomyces sp., 162, 163, 335–340
Streptomycin, 152, 160, 253, 256
Streptosporangium nondiastaticum, 164
Streptosporangium sp., 163
Structural biology, 80
Student's t-test, 112
Subcritical water hydrolysis, 2
Submerged fermentation, 64
Substance abuse, 374, 375, 378
Subtilisin Carlsberg, 64
Sucrose, 65, 72, 164, 183, 185–187, 279, 305–309, 313, 356, 365, 390, 391
Sudorific, 118
Sugar cane top, 384–387
Sulbactam, 258–260
Sulfated cinnamic acids, 20
Supercritical fluid extraction, 2
Superoxide dismutase (SOD), 245, 247
Suppurative diseases, 394
Surveillance networks, 394
SVD. *See* Singular value decomposition (SVD)
Symbionts, 12, 148
Synthetic insecticides, 390
Systematics, 235, 343–348
Systemic lupus erythematosus, 375

T

Tacrolimus, 160
Tamarindus indica, 259, 260
Tandem mass spectrometry (MSMS), 56, 57, 59, 401, 402
Tannery, 3, 6
Tannins, 20, 91, 95, 98, 99, 118–120, 294, 297–300, 328, 329, 331, 332, 385
Taxiphyllum taxirameum, 21
Tazobactam, 258–260
T-c. *See* Total cholesterol (T-c)
T cell epitopes, 318, 319
Terpenes, 20, 345
Terpenoids, 89, 91, 95, 99, 118–120, 294, 299
Tetracycline, 160, 406, 407
Textile dyes, 199–206
Textile effluent, 200
TG. *See* Triglycerides (TG)
TGGE, 352
Theobroma grandiflorum, 148
Thermal stability, 65
Thermodynamic parameters, 2
Thiaminase, 20
Thin layer chromatography (TLC), 337, 338
Thlaspi caerulescens, 132
Thymus praecox, 132
Ticarcillin, 258
Tie-line length (TLL), 2–4, 6, 8
Toluidine blue, 326, 331
Toothache, 93
Topologies, 34, 162
Total carbohydrate, 94–96, 98, 294, 296
Total cholesterol (T-c), 57, 58, 60, 400–402
Total nitrogen, 265, 354, 356, 359
Toxicity, 12, 17, 29, 31, 131, 132, 191–198, 389–392, 400
Transferring, 16, 64, 171, 307
Transgenes, 304, 412
Transpiration, 232, 233, 236, 237
Transpiration rate, 233, 236
Transposons, 16
TREECON®, 162, 164, 165
Trachurus japonicus, 50
Trachurus lepturus, 50
Trachurus sp., 50
Trichomes, 344, 345, 347

Triglycerides (TG), 57, 58, 60, 80, 400–402
Triterpenoid saponins, 89
Triticum aestivum, 143, 145, 146, 148
Trophic levels, 352
Trypsin inhibitor, 22, 95, 97, 98, 171, 294
Trypsins, 23, 24, 29, 252, 254, 297, 300
Tryptone, 200, 201, 205, 206
Tryptone-yeast extract broth, 161
Tuberculosis (TB), 151–157
β-Tubulin, 364–370
Tumor antigens, 413
Turkey, 52, 126–128
Typhoid fever, 406

U
Ubiquitin, 36, 38
Ubiquitous, 142, 278, 364, 369
Uniprot, 34
Unmutated self-protein antigens, 412
UPGMA, 126, 127, 129
Uterine. *See* Intrauterine growth restriction (IUGR)

V
Vaccine, 120, 318, 320, 322, 406
van der Waals, 2, 17, 366, 369
Vegetable waste, 222, 224
Venom, 283–291
Vermicompost, 221–227
Vermiform body, 325
Vibrio cholera, 285
Vigna subterranea, 229–241
Virtual screening, 81
Vitamin-A, 95, 96, 297
Vitamin-C, 95
Vitamin-E, 95, 96
Vitamins, 64, 94, 98, 100, 294, 306–308, 313
Vitiligo, 111–114

W
Waste management, 63
Wastewaters, 2, 3, 17, 200, 206, 416
Water
  deficit, 104
  hyacinth, 132, 222–227
  imbalance, 132
  potential, 108–110
Water use efficiency (WUE), 231–233, 236–239
Weathering, 12, 131
Western Ghats, 94, 98, 100, 159–166, 265, 293–300
Wheat bran, 65, 71, 72
White leghorn, 126–129
White stem borer, 174
Whopping cough, 326
Wild leafy vegetables, 94, 98, 294
Withania somnifera, 148
WUE. *See* Water use efficiency (WUE)

X
XDR. *See* Extensively drug-resistant (XDR)
Xylose, 65, 72, 200, 201, 205

Y
Yeast, 34, 64, 118, 122, 161, 162, 200–202, 205, 206, 223, 259, 279, 336, 352, 354, 355, 390, 391
Yeast extract, 161, 162, 200–202, 205, 206, 223, 259, 279, 355
Yersinia pestis KIM 10, 42

Z
Zea mays, 143–146, 148, 310
Zinc, 11–17, 36, 37, 95, 98, 100, 294, 298, 300, 416
Zinc analysis, 13, 14
Zooplankton, 265
Zoospores, 365, 366, 369